D0848498

Chemical Engineering

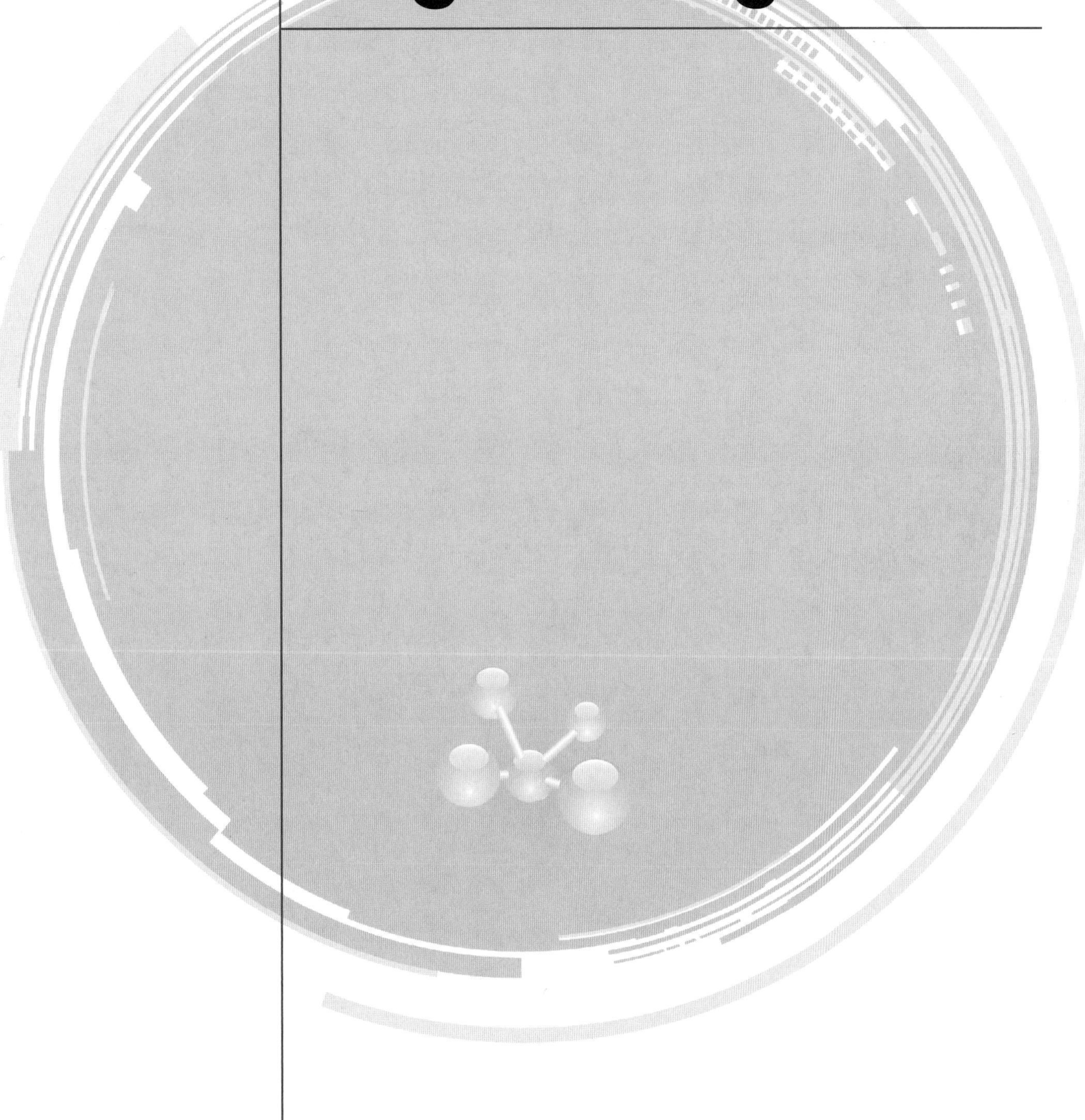

About the Author

Louis Theodore, Eng.Sc.D., is a former professor and Graduate Program Director in the Department of Chemical Engineering at Manhattan College. He is a consultant for Theodore Tutorials, which provides training solutions to industry, academia, and government. Dr. Theodore is the recipient of awards from the International Air & Waste Management Association and the American Society for Engineering Education. He is the author or coauthor of 96 books and is a section editor for *Perry's Chemical Engineers' Handbook*. Dr. Theodore is a member of IAABO (International Association of Approved Basketball Officials) and was recognized in 2008 at Madison Square Garden for his contributions to basketball and the youth of America.

Chemical Engineering

The Essential Reference

Louis Theodore

New York Chicago San Francisco
Athens London Madrid
Mexico City Milan New Delhi
Singapore Sydney Toronto

Library of Congress Cataloging-in-Publication Data

Theodore, Louis.
Chemical engineering : the essential reference / Louis Theodore.
pages cm
Includes index.
ISBN 978-0-07-183131-4 (alk. paper)
1. Chemical engineering. I. Title.
TP155.T52 2014
660–dc23

2013037457

McGraw-Hill Education books are available at special quantity discounts to use as premiums and sales promotions, or for use in corporate training programs. To contact a representative please visit the Contact Us page at www.mhprofessional.com

Chemical Engineering: The Essential Reference

Copyright ©2014 by McGraw-Hill Education. All rights reserved. Printed in China. Except as permitted under the United States Copyright Act of 1976, no part of this publication may be reproduced or distributed in any form or by any means, or stored in a database or retrieval system, without the prior written permission of the publisher.

1 2 3 4 5 6 7 8 9 0 DOW/DOW 1 0 9 8 7 6 5 4 3

ISBN 978-0-07-183131-4
MHID 0-07-183131-2

The pages within this book were printed on acid-free paper.

Sponsoring Editor
Michael Penn

Acquisitions Coordinator
Amy Stonebraker

Editorial Supervisor
David E. Fogarty

Project Manager
Harleen Chopra, Cenveo® Publisher Services

Copy Editor
Cathy Hertz

Indexer
Rita D'Aquino

Proofreader
Cenveo Publisher Services

Production Supervisor
Richard C. Ruzycka

Composition
Cenveo Publisher Services

Art Director, Cover
Jeff Weeks

Information contained in this work has been obtained by McGraw-Hill Education from sources believed to be reliable. However, neither McGraw-Hill Education nor its authors guarantee the accuracy or completeness of any information published herein, and neither McGraw-Hill Education nor its authors shall be responsible for any errors, omissions, or damages arising out of use of this information. This work is published with the understanding that McGraw-Hill Education and its authors are supplying information but are not attempting to render engineering or other professional services. If such services are required, the assistance of an appropriate professional should be sought.

To
P.S. 94
Joan of Arc Junior High School
Stuyvesant High School
The Cooper Union

It could have never happened without each of you

Brief Contents

Contents

Preface

Chemical engineering is one of the basic tenets of engineering, and contains many practical concepts that are utilized in countless real-world industrial applications. Therefore, the author considered writing a text that highlighted pragmatic rather than theoretical material. This book is intended to serve as a training tool for those individuals in academia and industry involved with chemical engineering applications. Although the literature is inundated with texts emphasizing theory and theoretical derivations, the goal of this book is to present the subject of chemical engineering from a strictly pragmatic perspective.

The book contains 30 chapters. It provides material required for the solution of most engineering problems particular to the chemical engineering profession. It also contains an overview of Accreditation Board for Engineering and Technology (ABET)-related topics as they apply to practicing engineers. The topics covered are listed in the Contents table.

This book is the result of several years of effort by the author. The first rough draft was prepared during the years 2010 through 2012. However, it underwent peripheral classroom testing during the author's 50-years tenure as a professor of chemical engineering. The manuscript underwent significant revisions during this past year, some of it based on the experiences gained from earlier class testing. However, it should be noted that the author cannot claim sole authorship to all of the essay material in this text. The present book has evolved from a host of sources, including notes, homework problems, and exam problems prepared by several faculty for undergraduate courses offered at Manhattan College, several other books authored by L. Theodore, and a host of Theodore tutorials published by Theodore Tutorials of East Williston, New York. Although the bulk of the material is original or taken from sources that the author has been directly involved with, every effort has been made to acknowledge material drawn from other sources.

In the final analysis, the problem of what to include and what to omit was particularly difficult. However, every attempt was made to offer most, if not all, chemical engineering course material to individuals at a level that should enable them to better cope with some of the problems encountered later in practice. As such, the book was not written for the student planning to pursue advanced degrees; rather, it was written primarily for those individuals who are currently working as practicing chemical engineers or plan to work as chemical engineers in the future solving real-world problems. It contains topics of importance to the chemical engineer of tomorrow—subject matter not previously addressed in other texts and handbooks such as Perry's—including operation, maintenance, and inspection (OMI) procedures; nanotechnology; how to purchase equipment; legal considerations; the need for a second language (e.g., Mandarin, Spanish, Hindi); and oral and written communication.

The author hopes that this book will place in the hands of academic, industrial, and government personnel a book that covers the principles and applications of chemical engineering in a thorough and clear manner. The author further hopes that, on completion of the text, readers will have acquired not only a working knowledge of the principles of chemical engineering but also experience in their application; and that they will find themselves approaching advanced texts, engineering literature, and industrial applications (even unique ones) with more confidence. The author strongly believes that, while understanding the basic concepts is of paramount importance, this knowledge may be rendered virtually useless to a practicing engineer who cannot apply these concepts to real-world situations. This is the essence of engineering.

The author is particularly indebted to Rita D'Aquino for effectively serving as the author's personal technical and editorial consultant on the project, and for preparing the index. Thanks are also due to Ronnie Zaglin for doing a superb job in "beautifying" the manuscript, for the extra set of eyes when it came time for proofreading.

Last, but not least, the author believes that this modest work will help the majority of individuals working and/or studying in the field of engineering to obtain a more complete understanding of chemical engineering. If you have come this far and read through the Preface, you have more than just a passing interest in this subject. I strongly recommend that you take advantage of the material available in this book. I think that it will be a worthwhile experience.

Louis Theodore
East Williston, New York

Introduction

A discussion centered on the field of chemical engineering is warranted before proceeding to some specific details regarding this discipline. A reasonable question to ask is: What is chemical engineering? An outdated, but once official definition provided by the American Institute of Chemical Engineers (AIChE) is:

> Chemical engineering is that branch of engineering concerned with the development and application of manufacturing processes in which chemical or certain physical changes are involved. These processes may usually be resolved into a coordinated series of unit physical "operations." The work of the chemical engineer is primarily concerned with the design, construction, and operation of equipment and plants in which these unit operations and processes are applied. Chemistry, physics, and mathematics are the underlying sciences of chemical engineering, and economics is its guide in practice.

This definition was appropriate until a few decades ago when the profession branched out from the chemical industry. Today, that definition has changed. Although it is still based on chemical fundamentals and physical principles, these have been de-emphasized in order to allow for the expansion of the profession to other areas. These areas include environmental management, health and safety, computer applications, project management, probability and statistics, ethics, and economics and finance, plus several other "new" topics referred to in the Preface. This has led to many new definitions of chemical engineering, several of which are either too specific or too vague. A definition proposed by the author is simply that "chemical engineers solve problems."

A classical approach to chemical engineering education, which is still used today, has been to develop problem-solving skills through the study of several traditional topics. Many of the topics have withstood the test of time. All of these topics receive treatment in this book. As noted in the Preface, this book contains 30 chapters, each serving a unique purpose in an attempt to treat a particular chemical engineering topic. From a practical perspective, systems and plants routinely apply chemical engineering principles. Hence, the student or practicing engineer is concerned with the subject matter that is presented in the book. These receive some measure of treatment in the chapters contained in this book.

In terms of history, the chemical engineering profession is usually considered to have originated shortly before 1900. However, many of the processes associated with this discipline were developed in antiquity. For example, filtration operations were carried out 5000 years ago by the Egyptians. *Mass transfer operations* (MTOs) such as crystallization, precipitation, and distillation soon followed. Other MTOs evolved from a mixture of craft, mysticism, incorrect theories, and empirical guesses during this period.

In a very real sense, the chemical industry dates back to prehistoric times when people first attempted to control and modify their environment. The chemical industry developed as did any other trade or craft. With little knowledge of chemical science and no means of chemical analysis, the earliest "chemical engineers" had to rely on previous art and superstition. As one would imagine, progress was slow. This changed with time.

The chemical industry in the world today is a sprawling complex of raw-material sources, manufacturing plants, and distribution facilities which supply society with thousands of chemical products, most of which were unknown over a century ago. In the latter half of the nineteenth century, an increased demand arose for engineers trained in the fundamentals of these chemical processes. This demand was ultimately met by chemical engineers.

The first attempt to organize the principles of chemical processing and to clarify the professional area of chemical engineering was made in England by George E. Davis, who organized a Society of Chemical Engineers in 1880 and presented a series of lectures in 1887 which were later expanded and published in 1901 as *A Handbook of Chemical Engineering*. In 1888, the first course in chemical engineering in the United States was organized at the Massachusetts Institute of Technology (MIT) by Lewis M. Norton, a professor of industrial chemistry. The course applied aspects of chemistry and mechanical engineering to chemical processes.

Chemical engineering began to gain professional acceptance in the early years of the twentieth century. The American Chemical Society (ACS) was founded in 1876 and, in 1908, organized a Division of Industrial Chemists and Chemical Engineers while authorizing the publication of the *Journal of Industrial and Engineering Chemistry*. A group of prominent chemical engineers, also in 1908, met in Philadelphia and founded the American Institute of Chemical Engineers (AIChE).

The mold for what is now called *chemical engineering* was fashioned at the 1922 meeting of the AIChE when A. D. Little's committee presented its report on chemical engineering education. The 1922 meeting marked the official endorsement of the unit operations concept and saw the approval of a "declaration of independence" for the profession.

A key component of the Little report included the following (see also Chap. 6):

> Any chemical process, on whatever scale conducted, may be resolved into a coordinated series of what may be termed "unit operations," as pulverizing, mixing, heating, roasting, absorbing, precipitation, crystallizing, filtering, dissolving, and so on. The number of these basic unit operations is not very large and relatively few of them are involved in any particular process . . . . An ability to cope broadly and adequately with the demands of this (the chemical engineer's) profession can be attained only through the analysis of processes into the unit actions as they are carried out on the commercial scale under the conditions imposed by practice.

The report also stated that "Chemical engineering, as distinguished from the aggregate number of subjects comprised in courses of that name, is not a composite of chemistry and mechanical and civil engineering, but is itself a branch of engineering."

A timeline diagram of the history of chemical engineering between the profession's founding and the present day is illustrated in the figure shown next. As can be seen from the timeline, the profession has reached a crossroads regarding the future education and curriculum for chemical engineers. This is highlighted by the differences between transport phenomena and unit operations (see also Chap. 6).

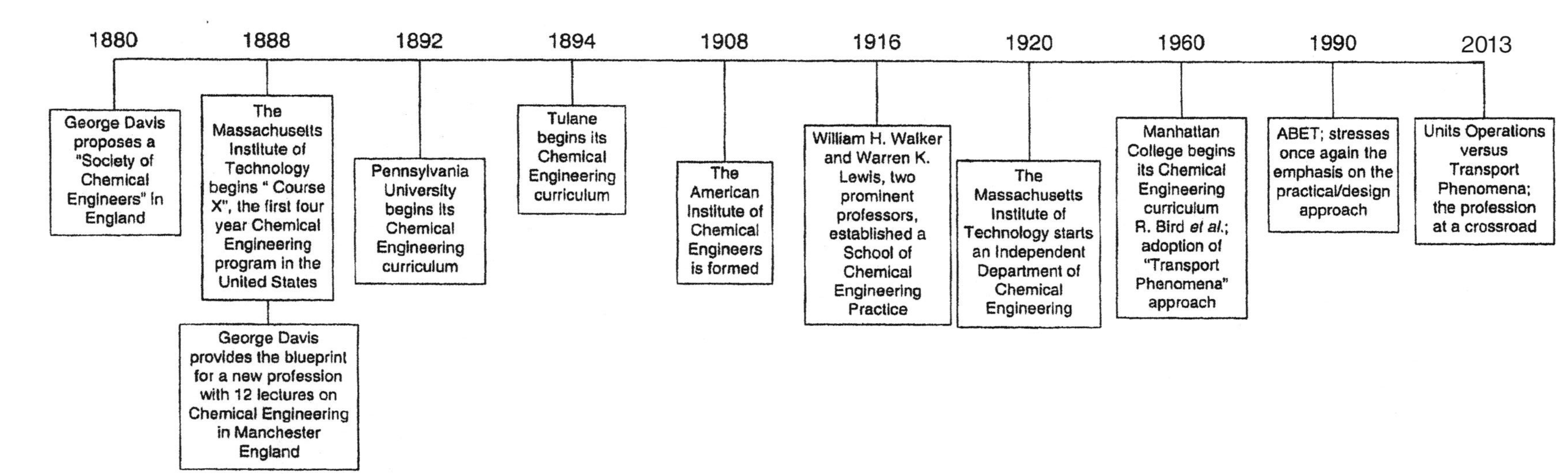

Source: N. Serino, "2005 Chemical Engineering 125th Year Anniversary Calendar," term project, submitted to L. Theodore, 2004.

PART

Introduction to Engineering

Part Outline

1 Basic Calculations and Key Tables

Chapter Outline

1–1 Introduction

This first chapter provides a review of basic calculations plus eight key tables of interest to the practicing chemical engineer. Eight topics receive treatment: units and dimensions, conversion of units, the gravitational constant g_c, significant figures and scientific notation, dimensionless groups, mathematics tables, tables of atomic weights, and water-steam tables. The reader is directed to the literature in the reference section of this chapter for additional information on these eight topics.

Sections 1–2 through 1–5 present basic calculations and tables. Section 1–6 discusses dimensionless groups and provides a table on dimensionless numbers. Section 1–7 contains four tables on mathematical operations. Section 1–8 lists the atomic weights of all the elements. Section 1–9 contains three tables providing enthalpy values for water.

1–2 Units and Dimensions

The units used in this text are consistent with those adopted by the engineering profession in the United States. For engineering work, Système International (SI) and English units are generally used, although efforts have been in place for years to obtain universal adoption of SI units for all engineering and science applications. The SI units have the advantage of being based on the decimal base (10) system, which allows for the convenient conversion of units within the system.

There are other systems of units. Some of the more common of these that have appeared in numerous references are shown in Table 1–1; however, English engineering units are primarily used in this text. Tables 1–2 and 1–3, also drawn from an expansive literature, present units for both the English and SI systems, respectively; Table 1–4 lists the multiples and prefixes for SI units.

Table 1–1

Common Systems of Units

System	Length	Time	Mass	Force	Energy	Temperature
SI	meter	second	kilogram	newton	joule	kelvin, degree Celsius (centigrade)
cgs	centimeter	second	gram	dyne	erg, joule, or calorie	kelvin, degree Celsius (centigrade)
American engineering	foot	second	pound	pound (force)	British thermal unit, horsepower-hour	degree Rankine, degree Fahrenheit
British engineering	foot	second	slug	pound (force)	British thermal unit, horsepower-hour	degree Rankine, degree Fahrenheit

Table 1–2

English Engineering Units

Physical quantity	Name of unit	Unit representation
Length	foot	ft
Time	second	s
Mass	pound (mass)	lb
Temperature	degree Rankine (absolute)	°R
Temperature (alternative)	degree Fahrenheit	°F
Moles	pound-mole	lbmol
Energy	British thermal unit	Btu
Energy (alternative)	horsepower · hour	hp · h
Force	pound (force)	lb_f
Acceleration	feet per second square	ft/s^2
Velocity	feet per second	ft/s
Volume	cubic foot	ft^3
Area	square foot	ft^2
Frequency	cycles per second, hertz	cycles/s, Hz
Power	horsepower, Btu per second	hp, Btu/s
Heat capacity	British thermal unit per (pound mass · degree Rankine)	Btu/(lb · °R)
Density	pounds (mass) per cubic foot	lb/ft^3
Pressure	pounds (force) per cubic foot	lb_f/ft^3
	pounds (force) per square inch	psi (lb_f/in^2)
	pounds (force) per square foot	psf (lb_f/ft^2)
	atmospheres	atm
	bar	bar

Table 1–3

SI Units

Physical unit	Name of unit	Symbol for unit
Length	meter	m
Mass	kilogram, gram	kg, g
Time	second	s
Temperature	kelvin (absolute)	K
Temperature (alternative)	degree Celsius	°C
Moles	gram-mole	gmol
Energy	joule	J, $(kg \cdot m^2)/s^2$, N · m
Force	newton	N, $(kg \cdot m)/s^2$, J/m
Acceleration	meters per square second	m/s^2
Pressure	pascal, newtons per square meter	Pa, N/m^2
Pressure (alternative)	bar	bar
Velocity	meters per second	m/s
Volume	cubic meter, liters	m^3, L
Area	square meter	m^2
Frequency	hertz	Hz, cycles/s
Power	watt	W, $kg \cdot m^2/s^3$, J/s
Heat capacity	joules per kilogram · kelvin	J/(kg · K)
Density	kilograms per cubic meter	kg/m^3
Angular velocity	radians per second	rad/s

Table 1–4

SI Multiples and Prefixes

Multiple and submultiple		Prefix	Symbol
100 000 000 000	10^{12}	tera	T
100 000 000	10^{9}	giga	G
100 000	10^{6}	mega	M
1 000	10^{3}	kilo	k
100	10^{2}	hecto	h
10	10^{1}	deka	da
Base unit 1	10^{0}	none	none
0.1	10^{-1}	deci	d
0.01	10^{-2}	centi	c
0.001	10^{-3}	milli	m
0.000 001	10^{-6}	micro	μ
0.000 000 001	10^{-9}	nano	n
0.000 000 000 001	10^{-12}	pico	p
0.000 000 000 000 001	10^{-15}	femto	f
0.000 000 000 000 000 001	10^{18}	atto	a

1–3 Unit Conversion Factors

Converting a measurement from one unit to another can be accomplished conveniently by using *unit conversion factors*; these factors are obtained from a simple equation that relates the two units numerically. For example, from

$$12 \text{ inches (in)} = 1 \text{ foot (ft)} \tag{1–1}$$

one obtains the following conversion factor:

$$(12 \text{ in}/1 \text{ ft}) = 1 \tag{1–2}$$

Since this factor is equal to unity, multiplying some quantity (e.g., 18 ft) by this factor cannot alter its value. Hence

$$18 \text{ ft } (12 \text{ in}/1 \text{ ft}) = 216 \text{ in} \tag{1–3}$$

Note that in Eq. (1–3), the units of *feet* on the left-hand side cancel out, leaving only the desired units of *inches*.

Physical equations must be dimensionally consistent. For the equality to hold, each additive term in the equation must have the same dimensions. This condition can and should be checked when solving engineering problems.

In this text, care is exercised in maintaining not only the dimensional formulas of all terms but also the dimensional homogeneity of each equation. Equations will generally be developed in terms of specific units rather than general dimensions, such as feet rather than length. This approach should aid the reader in attaching physical significance to the equations presented in these chapters. Conversion constants and selected common abbreviations are provided in Tables 1–5 and 1–6, respectively.

Table 1–5

Conversion Constants (SI)

To convert from	to	multiply by
	Length	
m	cm	100
m	mm	1000
m	micro meters (μm)	10^6
m	angstroms (Å)	10^{10}
m	nanometers (nm)	10^9
m	in	39.37
m	ft	3.281
m	mi	6.214×10^{-4}
ft	in	12
ft	m	0.3048
ft	cm	30.48
ft	mi	1.894×10^{-4}
	Mass	
kg	g	1000
kg	lb	2.205
kg	oz	35.24
kg	ton	2.268×10^{-4}
kg	grains (gr)	1.543×10^4
lb	oz	16
lb	ton	5×10^{-4}
lb	g	453.6
lb	kg	0.4536
lb	gr	7000
	Time	
s	min	0.01667
s	h	2.78×10^{-4}
s	day	1.157×10^{-7}
s	wk	1.653×10^{-6}
s	yr	3.171×10^{-8}
	Force	
N	$(kg \cdot m)/s^2$	1.0
N	dynes (dyn)	10^5
N	$(g \cdot cm)/s^2$	10^5
N	lb_f	0.2248
N	$lb \cdot ft/s^2$	7.233
lb_f	N	4.448
lb_f	dyn	4.448×10^5
lb_f	$(g \cdot cm)/s^2$	4.448×10^5
lb_f	$(lb \cdot ft)/s^2$	32.17

Continued

Table 1–5

Conversion Constants (SI) (*Continued*)

To convert from	to	multiply by
	Pressure	
atm	N/m^2 (Pa)	1.013×10^5
atm	kPa	101.3
atm	bars	1.013
atm	dyn/cm^2	1.013×10^6
atm	lb_f/in^2 (psi)	14.696
atm	mm Hg at 0°C (torr)	760
atm	in Hg at 0°C	29.92
atm	ft H_2O at 4°C	33.9
atm	in H_2O at 4°C	406.8
psi	atm	6.80×10^{-2}
psi	mm Hg at 0°C (torr)	51.71
psi	in H_2O at 4°C	27.7
in H_2O at 4°C	atm	2.458×10^{-3}
in H_2O at 4°C	psi	0.0361
in H_2O at 4°C	mm Hg at 0°C (torr)	1.868
	Volume	
m^3	L	1000
m^3	cm^3 (cc, mL)	10^6
m^3	ft^3	35.31
m^3	gal (U.S.)	264.2
ft^3	qt	1057
ft^3	in^3	1728
ft^3	gal (U.S.)	7.48
ft^3	m^3	0.02832
ft^3	L	28.32
	Energy	
J	$N \cdot m$	1.0
J	erg	10^7
J	$dyne \cdot cm$	10^7
J	kWh	2.778×10^7
J	cal	0.2390
J	$ft \cdot lb_f$	0.7376
J	Btu	9.486×10^{-4}
cal	J	4.186
cal	Btu	3.974×10^{-3}
cal	$ft \cdot lb_f$	3.088
Btu	$ft \cdot lb_f$	778
Btu	$hp \cdot h$	3.929×10^{-4}
Btu	cal	252
Btu	$kW \cdot h$	2.93×10^{-4}
$ft \cdot lb_f$	cal	0.3239

Table 1–5
(Continued)

To convert from	to	multiply by
	Energy	
$ft \cdot lb_f$	J	1.356
$ft \cdot lb_f$	Btu	1.285×10^{-3}
	Power	
W	J/s	1.0
W	cal/s	0.2390
W	$(ft \cdot lb_f)/s$	0.7376
W	kW	10^{-3}
kW	Btu/s	0.949
kW	hp	1.341
hp	$(ft \cdot lb_f)/s$	550
hp	kW	0.7457
hp	cal/s	178.2
hp	Btu/s	0.707
	Concentration	
$\mu g/m^3$	lb/ft^3	6.243×10^{-11}
$\mu g/m^3$	lb/gal	8.346×10^{-12}
$\mu g/m^3$	g/ft^3	4.370×10^{-7}
g/ft^3	$\mu g/m^3$	2.288×10^{6}
g/ft^3	g/m^3	2.288
lb/ft^3	$\mu g/m^3$	1.602×10^{10}
lb/ft^3	$\mu g/L$	1.602×10^{7}
lb/ft^3	lb/gal	7.48
	Viscosity	
P	$g/(cm \cdot s)$	1.0
P	centipoise (cP)	100
P	$kg/(m \cdot h)$	360
P	$lb/(ft \cdot s)$	6.72×10^{-2}
P	$lb/(ft \cdot h)$	241.9
P	$lb/(m \cdot s)$	5.6×10^{-3}
$lb/(ft \cdot s)$	P	14.88
$lb/(ft \cdot s)$	$g/(cm \cdot s)$	14.88
$lb/(ft \cdot s)$	$kg/(m \cdot h)$	5.357×10^{3}
$lb/(ft \cdot s)$	$lb/(ft \cdot h)$	3600
	Heat Capacity	
$cal/(g \cdot °C)$	$Btu/(lb \cdot °F)$	1.0
$cal/(g \cdot °C)$	$kcal/(kg \cdot °C)$	1.0
$cal/(g \cdot °C)$	$cal/(gmol \cdot °C)$	Molecular weight
$cal/(gmol \cdot °C)$	$Btu/(lbmol \cdot °F)$	1.0
$J/g \cdot °C$	$Btu/(lb \cdot °F)$	0.2389
$Btu/(lb \cdot °F)$	$cal/(g \cdot °C)$	1.0
$Btu/(lb \cdot °F)$	$J/(g \cdot °C)$	4.186
$Btu/(lb \cdot °F)$	$Btu/(lbmol \cdot °F)$	Molecular weight

Table 1–6

Common Abbreviations in Engineering Practice

Å	Angstrom unit of length	liq	liquid
abs	absolute	L	liter
amb	ambient	log	logarithm (common, base 10)
atm	atmospheric	ln	logarithm (natural)
at. wt	atomic weight	m, M	meter
bbl	barrel	μm	micrometer
Btu	British thermal unit	mks system	meter·kilogram·second system
cal	calorie	mph (mi/h)	miles per hour
cg	centigram	mg	milligram
cm	centimeter	mL	milliliter
cgs system	centimeter · gram · second system	mm	millimeter
conc	concentrated, concentration	mμm	millimicrometer
cc, cm^3	cubic centimeter	min	minute
cu ft, ft^3	cubic feet	mol wt, MW, M	molecular weight
cfh (ft^3/h)	cubic feet per hour	oz	ounce
cfm (ft^3/min)	cubic feet per minute	ppb	parts per billion
cfs (ft^3/s)	cubic feet per second	pphm	pans per hundred million
m^3, M^3	cubic meter	ppm	parts per million
°	degree	lb	pound
°C	degree Celsius, degree Centigrade	psi (lb$_f$/in^2)	pounds per square inch
°F	degree Fahrenheit	psia (lb$_f$/in^2 absolute)	pounds per square inch absolute
°R	degree Reamur, degree Rankine	psig (lb$_f$/in^2 gauge)	pounds per square inch gauge
ft	foot	rpm (r/min)	revolutions per minute
ft·lb	foot-pound	s	second
fpm (ft/min)	feet per minute	sp gr	specific gravity
fps (ft/s)	feet per second	sp ht	specific heat
fps system	foot · pound · second system	sp wt	specific weight
gr	grain	sq	square
g	gram	scf (standard ft^3)	standard cubic foot
h	hour	scfm (standard ft^3/min)	standard cubic feet per minute
in	inch	STP	standard temperature and pressure
k	degree kekvin		
kcal	kilocalorie	t	time, temperature (on occasion)
kg	kilogram	temp	temperature
km	kilometer	wt	weight

Now consider the example of calculating the perimeter P of a rectangle with length L and height H. Mathematically, this may be expressed as $P = 2L + 2H$. This is a simple mathematical equation. However, it applies only when P, L, and H are in the same units. There are many other examples that can be provided.

Terms in equations must also be consistent from a "magnitude" viewpoint. Differential terms *cannot* be equated with finite or integral terms. Care should also be exercised in solving differential equations. In order to solve differential equations to obtain a description of the pressure, temperature, composition, and other variables of a system, it is necessary to specify boundary and/or initial conditions (BC a/o IC) for the system. This information arises from a description of the problem or the physical system. The number of boundary conditions (BCs) that must be specified is the sum of the highest-order derivative for each independent differential equation. A value of the solution on the boundary of the system is one type of boundary condition. The number of ICs that must be specified is the highest-order derivative of time appearing in the differential equation. Thus, the value for the solution at time equal to zero constitutes an initial condition.

With reference to the above (as applied to mass transfer), the equation

$$\frac{dC_A}{dz^2} = 0 \qquad (C_A = \text{concentration}) \tag{1–4}$$

requires 2 BCs (in terms of the position variable z). The equation

$$\frac{dC_A}{dt} = 0 \quad (t = \text{time}) \tag{1–5}$$

requires 1 IC. Finally, the equation

$$\frac{\partial C_A}{\partial t} = D\frac{\partial^2 C_A}{\partial y^2} \tag{1–6}$$

requires 1 IC and 2 BCs (in terms of the position variable y).

1–4 The Gravitational Constant, g_c

Momentum finds application in many fluid flow studies. The *momentum* of a system is defined as the product of the mass and velocity of the system:

$$\text{Momentum} = \text{mass} \times \text{velocity} \tag{1–7}$$

A commonly employed set of units for momentum are therefore (lb · ft)/s. The units of the time rate of change of momentum (hereafter referred to as *rate of momentum*) are simply the units of momentum divided by time:

$$\text{Rate of momentum} \equiv \frac{\text{lb} \cdot \text{ft}}{\text{s}^2} \tag{1–8}$$

These units can be converted to units of pound force (lb_f) by multiplying by an appropriate constant. As noted earlier, a *conversion constant* is a term that is used to obtain units in a more convenient form; all conversion constants have magnitude and units in the term, but can also be shown to be equal to 1.0 (unity) with no units (i.e., be dimensionless).

A defining equation is

$$1\ \text{lb}_\text{f} = 32.2\frac{\text{lb} \cdot \text{ft}}{\text{s}^2} \tag{1–9}$$

If this equation is divided by lb_f, one obtains

$$1.0 = 32.2\frac{\text{lb} \cdot \text{ft}}{\text{lb}_\text{f} \cdot \text{s}^2} \tag{1–10}$$

This serves to define the conversion constant g_c. If the rate of momentum is divided by g_c as 32.2 (lb·ft)/(lb$_f$·s) (this operation is equivalent to dividing by 1.0), the following units result:

$$\text{Rate of momentum} \equiv \left(\frac{\text{lb}\cdot\text{ft}}{\text{s}^2}\right)\left(\frac{\text{lb}_\text{f}\cdot\text{s}^2}{\text{lb}\cdot\text{ft}}\right)$$

$$\equiv \text{lb}_\text{f} \quad (1\text{–}11)$$

One can conclude from this dimensional analysis that a force is equivalent to a rate of momentum.

1–5 Signficant Figures And Scientific Notation[1]

Significant figures provide an indication of the precision with which a quantity is measured or known. The last digit represents, in a qualitative sense, some degree of doubt. For example, a measurement of 8.32 inch implies that the actual quantity is somewhere between 8.315 and 8.325 inch. This applies to calculated and measured quantities; quantities that are known exactly (e.g., pure integers) have an infinite number of significant figures.

The significant digits of a number are the digits from the first nonzero digit on the left to either (1) the last digit (whether it is nonzero or zero) on the right if there is a decimal point or (2) the last nonzero digit of the number if there is no decimal point. For example,

2.50	has 3 significant figures
250	has 2 significant figures
250.	has 3 significant figures
250.0	has 4 significant figures
25.040	has 5 significant figures
.025	has 2 significant figures
.0250	has 3 significant figures
.02504	has 4 significant figures

Whenever quantities are combined by multiplication and/or division, the number of significant figures in the result should equal the lowest number of significant figures of any of the quantities. In long calculations, the final result should be rounded off to the correct number of significant figures. When quantities are combined by addition and/or subtraction, the final result cannot be more precise than any of the quantities added or subtracted. Therefore, the position (relative to the decimal point) of the last significant digit in the number that has the lowest degree of precision is the position of the last permissible significant digit in the result. For example, the sum of 3702., 370, 0.037, 4, and 37. should be reported as 4110 (without a decimal). The least precise of the five numbers is 370, which has its last significant digit in the *tens* position. The answer should also have its last significant digit in the *tens* position.

Unfortunately, engineers and scientists are rarely concerned with significant figures in their calculations. However, it is recommended that the reader attempt to follow the calculation procedure described in this section.

In the process of performing engineering calculations, very large and very small numbers are often encountered in practice. A convenient way to represent these numbers is to use *scientific notation*. Generally, a number represented in scientific notation is the product of a number (< 10 but > or = 1) and 10 raised to an integer power. For example,

$$25{,}050{,}000{,}000 = 2.505 \times 10^{10}$$

$$0.000002505 = 2.505 \times 10^{-6}$$

A positive feature of using scientific notation is that only the significant figures need be presented in the number.

1–6 Dimensionless Groups

To scale up (or scale down) a process, it is necessary to establish geometric and dynamic similarities between the model and the prototype. These two similarities are discussed below.

Geometric similarity implies using the same geometry of equipment. A circular pipe prototype should be modeled by a tube in the model. Geometric similarity establishes the scale of the model or prototype design. A 1/10th scale model indicates that the characteristic dimension of the model is 1/10th that of the prototype.

Dynamic similarity implies that certain important dimensionless numbers must be the same in the model and the prototype. For example, for a flowing fluid being heated in a tube of an exchanger, it has been shown that the friction factor f is a function of the dimensionless Reynolds number (Re). By selecting the operating conditions such that Re in the model equals the Re in the prototype, then the friction factor in the prototype will equal the friction factor in the model.[2]

Dimensionless numbers, as well as dimensionless groups, play an important role in the analysis of many engineering problems. Some of the key dimensionless numbers that are encountered in fluid flow practice are listed in Table 1–7.[3] The definition (and units) of the various terms and symbols is provided in Chapter 9. Dimensionless numbers in other chemical engineering subjects, many of which are discussed later in this book, also are available in the literature.[3–6]

Table 1–7 Dimensionless Numbers in Fluid Flow

Parameter	Definition	Importance	Qualitative ratio
Cavitation number	$\text{Ca} = \dfrac{P - p'}{\rho v^2/2}$	Cavitation	$\dfrac{\text{Pressure}}{\text{Inertia}}$
Eckert number	$\text{Ec} = \dfrac{v^2}{C_p\,\Delta T}$	Energy dissipation	$\dfrac{\text{Kinetic energy}}{\text{Inertia}}$
Euler number	$\text{Eu} = \dfrac{\Delta P}{\rho v^2/2}$	Pressure drop	$\dfrac{\text{Pressure}}{\text{Inertia}}$
Froude number	$\text{Fr} = \dfrac{v^2}{gL}$	Free surface flow	$\dfrac{\text{Inertia}}{\text{Gravity}}$
Mach number	$\text{Ma} = \dfrac{v}{c}$	Compressible flow, sonic velocity	$\dfrac{\text{Flow speed}}{\text{Sound speed}}$
Poiseuille number	$\text{Po} = \dfrac{D^2\,\Delta P}{\mu\,Lv}$	Laminar flow in pipes, slow flow, creeping flow	$\dfrac{\text{Pressure}}{\text{Viscous forces}}$
Relative roughness	$\dfrac{k}{D}, \dfrac{\epsilon}{D}$	Turbulent flow, rough walls	$\dfrac{\text{Wall roughness}}{\text{Body length}}$
Reynolds number	$\text{Re} = \dfrac{\rho v D}{\mu} = \dfrac{vD}{\nu}$	Various fluid flow applications	$\dfrac{\text{Inertia forces}}{\text{Viscous forces}}$
Strouhal number	$\text{St} = \dfrac{\varpi L}{v}$	Oscillating flow	$\dfrac{\text{Oscillation speed}}{\text{Mean speed}}$
Weber number	$\text{We} = \dfrac{\rho v^2 L}{\sigma}$	Surface forces effect, mixing effects	$\dfrac{\text{Inertia}}{\text{Surface tension}}$

Note: p' = vapor pressure, C_p = heat capacity.
Source: Ref. 2.

1–7 Mathematics Tables

This section contains five tables that one would classify in the mathematics area: a table of integrals (Table 1–8), a table of derivatives (Table 1–9), standard normal probability tables (Tables 1–10 and 1–11), and a table of Student's *t* distribution (Table 1–12). The reader is referred to the literature for a more expansive treatment of the above.[7–9]

Table 1–8

Analytical Integrals

$$\int_0^x a\,dx = ax$$

$$\int_0^x \left[\frac{1}{(1-x)}\right] dx = \ln\left[\frac{1}{1-x}\right]$$

$$\int_0^x \left[\frac{1}{(1-x)^2}\right] dx = \left[\frac{1}{1-x}\right]$$

$$\int_0^x \left[\frac{1}{(1-x)^2}\right] dx = \left[\frac{1}{1-x}\right]$$

$$\int \left[\frac{1}{x}\right] dx = \ln x$$

$$\int e^x dx = e^x$$

$$\int e^{ax} dx = \frac{e^{ax}}{a}$$

$$\int \sin x\; dx = -\cos x$$

$$\int \cos x\; dx = \sin x$$

$$\int_0^x \left[\frac{1}{1+\varepsilon x}\right] dx = \left(\frac{1}{\varepsilon}\right) \ln\,(1+\varepsilon x)$$

$$\int_0^x \left[\frac{1+\varepsilon x}{1-x}\right] dx = (1+\varepsilon)\ln\left(\frac{1}{1-x}\right) - \varepsilon x$$

$$\int_0^x \left[\frac{1+\varepsilon x}{(1-x)^2}\right] dx = \frac{(1+\varepsilon)x}{(1-x)} - (\varepsilon)\ln\left(\frac{1}{1-x}\right)$$

$$\int_0^x \left[\frac{(1+\varepsilon x)^2}{(1-x)^2}\right] dx = 2\varepsilon\,(1+\varepsilon)\ln\,(1-x) + \varepsilon^2 x + \frac{(1+\varepsilon)^2\,(x)}{(1-x)}$$

$$\int_0^x \left(\frac{1}{[(1-x)\,(\theta_B - x)]}\right) dx = \frac{1}{\theta_B - 1}\ln\frac{\theta_B - x}{\theta_B(1-x)} \quad \text{for} \quad \theta_B \neq 0$$

$$\int_0^x \frac{1}{ax^2+bx+c}\,dx = -\left(\frac{2}{2ax+b}\right) + \left(\frac{2}{b}\right) \quad \text{for} \quad b^2 = 4ac$$

$$\int_0^x \frac{1}{ax^2+bx+c}\,dx = \frac{1}{a(p-q)}\ln\left(\frac{q\,(x-p)}{p(x-q)}\right) \quad \text{for} \quad b^2 > 4ac$$

where p and q are roots of the equation $ax^2+bx+c=0$, i.e., $p,\,q = \dfrac{-b \pm \sqrt{b^2-4ac}}{2a}$

Table 1–8

(Continued)

$$\int_0^x \left[\frac{a+bx}{c+gx}\right] dx = \frac{bx}{g} + \frac{ag-bc}{g^2} \ln|c+gx|$$

$$\int_0^x \left[\frac{x}{ax+b}\right] dx = \frac{x}{a} + \frac{b}{a^2} \ln(ax+b)$$

$$\int_0^x \frac{x^2}{ax+b} dx = \left[\frac{(ax+b)^2}{2a^3}\right] - \left[\frac{2b}{a^2}(ax+b)\right] + \left[\left(\frac{b^2}{a^3}\right) \ln(ax+b)\right]$$

$$\int_0^x \frac{1}{x(ax+b)} dx = \frac{1}{b} \ln\left(\frac{x}{ax+b}\right)$$

Note: There are numerous other integrals. There are no limits on some of the integrals.

Table 1–9

Analytical Derivatives

$$\frac{d(a)}{dx} = 0 \qquad (a = \text{constant})$$

$$\frac{d(x)}{dx} = 1$$

$$\frac{d(ax)}{dx} = a$$

$$\frac{d(ax^n)}{dx} = nax^{n-1}$$

$$\frac{d(au)}{dx} = a\left(\frac{du}{dx}\right) \qquad [u = f(x)]$$

$$\frac{d(uv)}{dx} = u\left(\frac{dv}{dx}\right) + v\left(\frac{du}{dx}\right) \qquad [u = f(x)], [v = f(x)]$$

$$\frac{d\left(\frac{u}{v}\right)}{dx} = \left(\frac{1}{v}\right)\left(\frac{du}{dx}\right) - \left(\frac{du}{dx}\right)\left(\frac{dv}{dx}\right)$$

$$\frac{d(u^n)}{dx} = nu^{n-1}\left(\frac{du}{dx}\right)$$

$$\frac{d\left(u^{\frac{1}{2}}\right)}{dx} = \left(\frac{1}{2u^{\frac{1}{2}}}\right)\left(\frac{du}{dx}\right)$$

$$\frac{d\left(\frac{1}{u}\right)}{dx} = -\frac{1}{u^2}\left(\frac{du}{dx}\right)$$

$$\frac{d\left(\frac{1}{u^n}\right)}{dx} = -\frac{n}{u^{nn+1}}\left(\frac{du}{dx}\right)$$

Note: There are numerous other derivatives.

Source: Ref. 8.

Table 1–10

Standard Normal Probability (P) in Right-Hand Tail (Area under Curve for Specified Values of z_0)

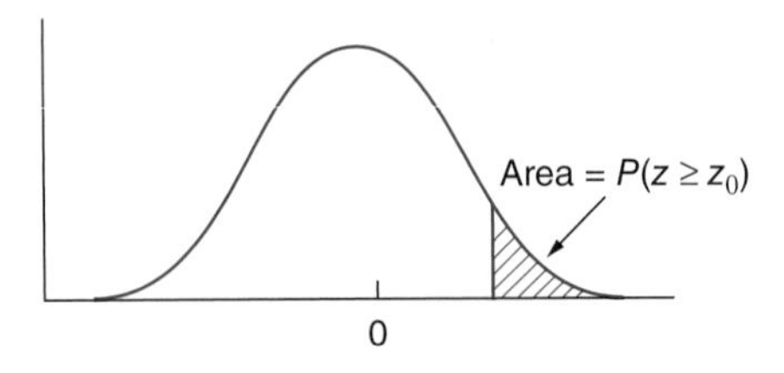

The Standard Normal Distribution

z	0.00	0.01	0.02	0.03	0.04	0.05	0.06	0.07	0.08	0.09
0	0.5	0.496	0.492	0.488	0.484	0.48	0.476	0.472	0.468	0.464
0.1	0.46	0.456	0.452	0.448	0.444	0.44	0.436	0.433	0.429	0.425
0.2	0.421	0.417	0.413	0.409	0.405	0.401	0.397	0.394	0.39	0.386
0.3	0.382	0.378	0.374	0.371	0.367	0.363	0.359	0.356	0.352	0.348
0.4	0.345	0.341	0.337	0.334	0.33	0.326	0.323	0.319	0.316	0.312
0.5	0.309	0.305	0.302	0.298	0.295	0.291	0.288	0.284	0.281	0.278
0.6	0.274	0.271	0.268	0.264	0.261	0.258	0.255	0.251	0.248	0.245
0.7	0.242	0.239	0.236	0.233	0.23	0.227	0.224	0.221	0.218	0.215
0.8	0.212	0.209	0.206	0.203	0.2	0.198	0.195	0.192	0.189	0.187
0.9	0.184	0.181	0.179	0.176	0.174	0.171	0.169	0.166	0.164	0.161
1	0.159	0.156	0.154	0.152	0.149	0.147	0.145	0.142	0.14	0.138
1.1	0.136	0.133	0.131	0.129	0.127	0.125	0.123	0.121	0.119	0.117
1.2	0.115	0.113	0.111	0.109	0.107	0.106	0.104	0.102	0.1	0.099
1.3	0.097	0.095	0.093	0.092	0.09	0.089	0.087	0.085	0.084	0.082
1.4	0.081	0.079	0.078	0.076	0.075	0.074	0.072	0.071	0.069	0.068
1.5	0.067	0.066	0.064	0.063	0.062	0.061	0.059	0.058	0.057	0.056
1.6	0.055	0.054	0.053	0.052	0.051	0.049	0.048	0.047	0.046	0.046
1.7	0.045	0.044	0.043	0.042	0.041	0.04	0.039	0.038	0.038	0.037
1.8	0.036	0.035	0.034	0.034	0.033	0.032	0.031	0.031	0.03	0.029
1.9	0.029	0.028	0.027	0.027	0.026	0.026	0.025	0.024	0.024	0.023
2	0.023	0.022	0.022	0.021	0.021	0.02	0.02	0.019	0.019	0.018
2.1	0.018	0.017	0.017	0.017	0.016	0.016	0.015	0.015	0.015	0.014
2.2	0.014	0.014	0.013	0.013	0.013	0.012	0.012	0.012	0.011	0.011
2.3	0.011	0.01	0.01	0.01	0.01	0.009	0.009	0.009	0.009	0.008
2.4	0.008	0.008	0.008	0.008	0.007	0.007	0.007	0.007	0.007	0.006
2.5	0.006	0.006	0.006	0.006	0.006	0.005	0.005	0.005	0.005	0.005
2.6	0.005	0.005	0.005	0.005	0.005	0.004	0.004	0.004	0.004	0.004
2.7	0.003	0.003	0.003	0.003	0.003	0.003	0.003	0.003	0.003	0.003
2.8	0.003	0.002	0.002	0.002	0.002	0.002	0.002	0.002	0.002	0.002
2.9	0.002	0.002	0.002	0.002	0.002	0.002	0.002	0.002	0.002	0.002

Sources: Refs. 9 and 10.

Table 1–11

Areas under a Standard Normal Curve between 0 and z

z	0.00	0.01	0.02	0.03	0.04	0.05	0.06	0.07	0.08	0.09
0	0.00	0.0040	0.0080	0.0120	0.0160	0.0199	0.0239	0.0279	0.0319	0.0359
0.1	0.0000	0.0438	0.0478	0.0517	0.0557	0.0596	0.0636	0.0675	0.0714	0.0753
0.2	0.0398	0.0832	0.0871	0.0910	0.0948	0.0987	0.1026	0.1064	0.1103	0.1141
0.3	0.0793	0.1217	0.1255	0.1293	0.1331	0.1368	0.1406	0.1443	0.1480	0.1517
0.4	0.1179	0.1591	0.1628	0.1664	0.1700	0.1736	0.1772	0.1808	0.1844	0.1879
0.5	0.1554	0.1950	0.1985	0.2019	0.2054	0.2088	0.2123	0.2157	0.2190	0.2224
0.6	0.1915	0.2291	0.2324	0.2357	0.2389	0.2422	0.2454	0.2486	0.2517	0.2549
0.7	0.2257	0.261 1	0.2642	0.2673	0.2704	0.2734	0.2764	0.2794	0.2823	0.2852
0.8	0.2580	0.2910	0.2939	0.2967	0.2995	0.3023	0.3051	0.3078	0.3106	0.3133
0.9	0.2881	0.3186	0.3212	0.3238	0.3264	0.3289	0.3315	0.3340	0.3365	0.3389
1	0.3159	0.3438	0.3461	0.3485	0.3508	0.3531	0.3554	0.3577	0.3599	0.3621
1.1	0.3413	0.3665	0.3686	0.3708	0.3729	0.3749	0.3770	0.3790	0.3810	0.3830
1.2	0.3643	0.3869	0.3888	0.3907	0.3925	0.3944	0.3962	0.3980	0.3997	0.4015
1.3	0.3849	0.4049	0.4066	0.4082	0.4099	0.4115	0.4131	0.4147	0.4162	0.4177
1.4	0.4032	0.4207	0.4222	0.4236	0.4251	0.4265	0.4279	0.4292	0.4306	0.4319
1.5	0.4192	0.4345	0.4357	0.4370	0.4382	0.4394	0.4406	0.4418	0.4429	0.4441
1.6	0.4332	0.4463	0.4474	0.4484	0.4495	0.4505	0.4515	0.4525	0.4535	0.4545
1.7	0.4452	0.4564	0.4573	0.4582	0.4591	0.4599	0.4608	0.4616	0.4625	0.4633
1.8	0.4554	0.4649	0.4656	0.4664	0.4671	0.4678	0.4686	0.4693	0.4699	0.4706
1.9	0.4641	0.4719	0.4726	0.4732	0.4738	0.4744	0.4750	0.4756	0.4761	0.4767
2	0.4713	0.4778	0.4783	0.4788	0.4793	0.4798	0.4803	0.4808	0.4812	0.4817
2.1	0.4772	0.4826	0.4830	0.4834	0.4838	0.4842	0.4846	0.4850	0.4854	0.4857
2.2	0.4821	0.4864	0.4868	0.4871	0.4875	0.4878	0.4881	0.4884	0.4887	0.4890
2.3	0.4861	0.4896	0.4898	0.4901	0.4904	0.4906	0.4909	0.4911	0.4913	0.4916
2.4	0.4893	0.4920	0.4922	0.4925	0.4927	0.4929	0.4931	0.4932	0.4934	0.4936
2.5	0.4918	0.4940	0.4941	0.4943	0.4945	0.4946	0.4948	0.4949	0.4951	0.4952
2.6	0.4938	0.4955	0.4956	0.4957	0.4959	0.4960	0.4961	0.4962	0.4963	0.4964
2.7	0.4953	0.4966	0.4967	0.4968	0.4969	0.4970	0.4971	0.4972	0.4973	0.4974
2.8	0.4965	0.4975	0.4976	0.4977	0.4977	0.4978	0.4979	0.4979	0.4980	0.4981
2.9	0.4974	0.4982	0.4982	0.4983	0.4984	0.4984	0.4985	0.4985	0.4986	0.4986
3.0	0.4981	0.4987	0.4987	0.4988	0.4988	0.4989	0.4989	0.4989	0.4990	0.4990

Sources: Refs. 9 and 10.

1–8 Table of Atomic Weights

The atomic weights of the elements are presented in Table 1–13. Note that

1. Atomic weights apply to naturally occurring isotopic compositions and are based on an atomic mass of $^{12}C = 12$.
2. A value given in parentheses for radioactive elements is the atomic mass number of the isotope with the longest known half-life.
3. Geologically exceptional samples are known in which radium has an isotopic composition outside the limits for normal material.

Note that physical and chemical properties receive treatment in the next chapter. Additional information on elements is available throughout the engineering and science literature.[7]

Table 1–12

Student's t Distribution

	Values of t					
ν	α = 0.1	α = 0.05	α = 0.025	α = 0.01	α = 0.005	ν
1	3.077684	6.313752	12.70620	31.82052	63.65674	1
2	1.885618	2.919986	4.30265	6.96456	9.92484	2
3	1.637744	2.353363	3.18245	4.54070	5.84091	3
4	1.533206	2.131847	2.77645	3.74695	4.60409	4
5	1.475884	2.015048	2.57058	3.36493	4.03214	5
6	1.439756	1.94318	2.44691	3.14267	3.70743	6
7	1.414924	1.894579	2.36462	2.99795	3.49948	7
8	1.396815	1.859548	2.30600	2.89646	3.35539	8
9	1.383029	1.833113	2.26216	2.82144	3.24984	9
10	1.372184	1.812461	2.22814	2.76377	3.16927	10
11	0.363430	1.795885	2.20099	2.71808	3.10581	11
12	0.356217	1.782288	2.17881	2.68100	3.05454	12
13	0.350171	1.770933	2.16037	2.65031	3.01228	13
14	0.345030	1.761310	2.14479	2.62449	2.97684	14
15	0.340606	1.753050	2.13145	2.60248	2.94671	15
16	0.336757	1.745884	2.11991	2.58349	2.92078	16
17	0.333379	1.739607	2.10982	2.56693	2.89823	17
18	0.330391	1.734064	2.10092	2.55238	2.87844	18
19	0.327728	1.729133	2.09302	2.53948	2.86093	19
20	0.325341	1.724718	2.08596	2.52798	2.84534	20
21	0.321237	1.717144	2.07387	2.50832	2.81876	21
22	0.317836	1.710882	2.06390	2.49216	2.79694	22
23	0.314972	1.705618	2.05553	2.47863	2.77871	23
24	0.312527	1.701131	2.04841	2.46714	2.76326	24
25	0.281552	1.644854	1.95996	2.32635	2.57583	25
26	0.363430	1.795885	12.70620	31.82052	63.65674	26
27	0.356217	1.782288	4.30265	6.96456	9.92484	27
28	0.350171	1.770933	3.18245	4.54070	5.84091	28

Note: ν = degrees of freedom; α = confidence degree; degree of confidence.

Sources: Refs. 9 and 10.

1–9 Water (Steam) Tables

The term *phase* for a pure substance refers to a state of matter that is gas, liquid, or solid. Latent enthalpy (heat) effects are associated with phase changes. These phase changes involve no change in temperature but there is a transfer of energy to and from the substance. There are three possible latent effects, as detailed below: (1) vapor-liquid, (2) liquid-solid, and (3) vapor-solid.

Vapor-liquid changes are referred to as *condensation* when the vapor is condensing and *vaporization* when liquid is vaporizing. *Liquid-solid* changes are referred to as *melting* when the solid melts to liquid and *freezing* when a liquid solidifies. *Vapor-solid* changes are referred to as *sublimation*. One should also note that there are enthalpy effects associated with a phase change of a solid to another solid form; however, this enthalpy effect is small compared to the other effects mentioned above. Specific volume, enthalpy, and entropy data for saturated steam, superheated steam, and steam-ice are provided in Tables 1–14, 1–15, and 1–16, respectively.[11,12] Other physical properties are addressed in the next chapter.

Table 1–13

Atomic Weights of the Elements

Element	Symbol	Atomic Weight	Element	Symbol	Atomic Weight	Element	Symbol	Atomic Weight
Actinium	Ac	227.0278	Hafnium	Hf	178.49	Promethium	Pm	(145)
Aluminum	Al	26.9815	Helium	He	4.0026	Protactinium	Pa	231.0359
Americium	Am	(243)	Holmium	Ho	164.93	Radium	Ra	226.0254*
Antimony	Sb	121.75	Hydrogen	H	1.00797	Radon	Rn	(222)
Argon	Ar	39.948	Indium	In	114.82	Rhenium	Re	186.2
Arsenic	As	74.9216	Iodine	I	126.9044	Rhodium	Rh	102.905
Astatine	At	(210)	Iridium	IT	192.2	Rubidium	Rb	84.57
Barium	Ba	137.34	Iron	Fe	55.847	Ruthenium	Ru	101.07
Berkelium	Bk	(247)	Krypton	Kr	83.8	Samarium	Sm	150.35
Beryllium	Be	9.0122	Lanthanum	La	138.91	Scandium	Sc	44.956
Bismuth	Bi	208.98	Lawrencium	Lr	(260)	Selenium	Se	78.96
Boron	B	10.811	Lead	Pb	207.19	Silicon	Si	28.086
Bromine	Br	79.904	Lithium	Li	6.939	Silver	Ag	107.868
Cadmium	Cd	112.4	Lutetium	Lu	174.97	Sodium	Na	22.9898
Calcium	Ca	40.08	Magnesium	Mg	24.312	Strontium	Sr	87.62
Californium	Cf	(251)	Manganese	Mn	54.938	Sulfur	S	32.064
Carbon	C	12.01115	Mendelevium	Md	(258)	Tantalum	Ta	180.948
Cerium	Ce	140.12	Mercury	Hg	200.59	Technetium	Tc	(98)
Cesium	Cs	132.905	Molybdenum	Mo	95.94	Tellurium	Te	127.6
Chlorine	Cl	35.453	Neodymium	Nd	144.24	Terbium	Tb	158.924
Chromium	Or	51.996	Neon	Nc	20.183	Thallium	Tl	204.37
Cobalt	Co	58.9332	Neptunium	Np	237.0482	Thorium	Th	232.038
Copper	Cu	63.546	Nickel	Ni	58.71	Thulium	Tm	168.934
Curium	Cm	(247)	Niobium	Nb	92.906	Tin	Sn	118.69
Dysprosium	Dy	162.5	Nitrogen	N	14.0067	Titanium	Ti	47.9
Einsteinium	Es	(252)	Nobelium	No	(259)	Tungsten	W	183.85
Erbium	Er	167.26	Osmium	Os	190.2	Uranium	U	238.03
Europium	Eu	151.96	Oxygen	O	15.9994	Vanadium	V	50.942
Fermium	Fm	(257)	Palladium	Pd	106.4	Xenon	Xe	131.3
Fluorine	F	18.9984	Phosphorus	P	30.9738	Ytterbium	Yb	173.04
Francium	Fr	(223)	Platinum	Pt	195.09	Yttrium	Y	88.905
Gadolinium	Gd	157.25	Plutonium	Pu	(244)	Zinc	Zn	65.37
Gallium	Ga	69.72	Polonium	Po	(209)	Zirconium	Zr	91.22
Germanium	Ge	72.59	Potassium	K	39.102			
Gold	Au	196.967	Praseodymium	Pr	140.907			

Table 1–14

Saturated Steam

Temperature T, °F	Absolute pressure P, lb_f/in^2	Specific volume, ft^3/lb; v			Enthalpy, Btu/lb; h			Entropy, Btu/lb·°R; s		
		Saturated liquid	Evaporation difference	Saturated vapor	Saturated liquid	Evaporation difference	Saturated vapor	Saturated liquid	Evaporation difference	Saturated vapor
32	0.08854	0.01602	3306	3306	0	1075.8	1075.8	0	2.1877	2.1877
35	0.09995	0.01602	2947	2947	3.02	1074.1	1077.1	0.0061	2.1709	2.177
40	0.1217	0.01602	2444	2444	8.05	1071.3	1079.3	0.0162	2.1435	2.1597
45	0.14752	0.01602	2036.4	2036.4	13.06	1068.4	1081.5	0.0262	2.1167	2.1429
50	0.17811	0.01603	1703.2	1703.2	18.07	1065.6	1083.7	0.0361	2.0903	2.1264
60	0.2563	0.01604	1206.6	1206.7	28.06	1059.9	1088	0.0555	2.0393	2.0948
70	0.3631	0.01606	867.8	867.9	38.04	1054.3	1092.3	0.0745	1.9902	2.0647
80	0.5069	0.01608	633.1	633.1	48.02	1048.6	1096.6	0.0932	1.9428	2.036
90	0.6982	0.0161	468	468	57.99	1042.9	1100.9	0.1115	1.8972	2.0087
100	0.9492	0.01613	350.3	350.4	67.97	1037.2	1105.2	0.1295	1.8531	0.9826
110	0.2748	0.01617	265.3	265.4	77.94	1031.6	1109.5	0.1471	0.8106	0.9577
120	0.6924	0.0162	203.25	203.27	87.92	1025.8	1113.7	0.1645	0.7694	0.9339
130	0.2225	0.01625	157.32	157.34	97.9	1020	1117.9	0.1816	0.7296	0.9112
140	0.8886	0.01629	122.99	123.01	107.89	1014.1	1122.0	0.1984	0.691	0.8894
150	0.718	0.01634	97.06	97.07	117.89	1008.2	1126.1	0.2149	0.6537	0.8685
160	0.741	0.01639	77.27	77.29	127.89	1002.3	1130.2	0.2311	0.6174	0.8485
170	0.992	0.01645	62.04	62.06	137.9	996.3	1134.2	0.2472	1.5822	0.8293
180	0.51	0.01651	50.21	50.23	147.92	990.2	1138.1	0.263	1.548	1.8109
190	0.339	0.01657	40.94	40.96	157.95	984.1	1142	0.2785	1.5147	1.7932
200	11.526	0.01663	33.62	33.64	167.99	977.9	1145.9	0.2938	1.4824	1.7762
210	14.123	0.0167	27.8	27.82	178.05	971.6	1149.7	0.309	1.4508	1.7598
212	14.696	0.01672	26.78	26.8	180.07	970.3	1150.4	0.312	1.4446	1.7566
220	17.186	0.01677	23.13	23.15	188.13	965.2	1153.4	0.3239	1.4201	1.744
230	20.78	0.01684	19.365	19.382.	198.23	958.8	1157	0.3387	1.3901	1.7288
240	24.969	0.01692	16.306	16.323	208.34	952.2	1160.5	0.3531	1.3609	1.714
250	29.825	0.017	13.804	13.82	218.48	945.5	1164	0.3675	1.3323	1.6998

260	35.429	0.01709	11.746	11.763	228.64	938.7	1167.3	0.3817	1.3043	1.686
270	41.858	0.01717	10.044	10.061	238.84	931.8	1170.6	0.3958	1.2769	1.6727
280	49.203	0.01726	8.628	8.645	249.06	924.7	1173.8	0.4096	1.2501	1.6597
290	57.556	0.01735	7.444	7.461	259.31	917.5	1176.8	0.4234	1.2238	1.6472
300	67.013	0.01745	6.449	6.466	269.59	910.1	1179.7	0.4369	1.198	1.635
310	77.68	0.01755	5.609	5.626	279.92	902.6	1182.5	0.4504	1.1727	1.6231
320	89.66	0.01765	4.896	4.914	290.28	894.9	1185.2	0.4637	1.1478	1.6115
330	103.06	0.01776	4.289	4.307	300.68	887	1187.7	0.4769	1.1233	1.6002
340	118.01	0.01787	3.77	3.788	311.13	879	1190.1	0.49	1.0992	1.5891
350	134.63	0.01799	3.324	3.342	321.63	870.7	1192.3	0.5029	1.0754	1.5783
360	153.04	0.01811	2.939	2.957	332.18	862.2	1194.4	0.5158	1.0519	0.5677
370	173.37	0.01823	2.606	2.625	342.79	853.5	1196.3	0.5286	1.0287	1.5573
380	195.77	0.01836	2.317	2.335	353.45	844.6	1198.1	0.5413	1.0059	1.5471
390	220.37	0.0185	2.0651	2.0836	364.17	835.4	1199.6	0.5539	0.9832	1.5371
400	247.31	0.01864	1.8447	1.8633	374.97	826	1201	0.5664	0.9608	1.5272
410	276.75	0.01878	1.6512	1.67	385.83	816.3	1202.1	0.5788	0.9386	1.5174
420	308.83	0.01894	1.4811	1.5	396.77	806.3	1203.1	0.5912	0.9166	1.5078
430	343.72	0.0191	1.3308	1.3499	407.79	796	1203.8	0.6035	0.8947	1.4982
440	381.59	0.01926	1.1979	1.2171	418.9	785.4	1204.3	0.6158	0.873	1.4887
450	422.6	0.0194	1.0799	1.0993	430.1	774.5	1204.6	0.628	0.8513	1.4793
460	466.9	0.0196	0.9748	0.9944	441.4	763.2	1204.6	0.6402	0.8298	1.47
470	514.7	0.0198	0.8811	0.9009	452.8	751.5	1204.3	0.6523	0.8083	1.4606
480	566.1	0.02	0.7972	0.8172	464.4	739.4	1203.7	0.6645	0.7868	1.4513
490	621.4	0.0202	0.7221	0.7423	476	726.8	1202.8	0.6766	0.7653	1.4419
520	812.4	0.0209	0.5385	0.5594	511.9	686.4	1198.2	0.713	0.7006	1.4136
540	962.5	0.0215	0.4434	0.4649	536.6	656.6	1193.2	0.7374	0.6568	1.3942
560	1133.1	0.0221	0.3647	0.3868	562.2	624.2	1186.4	0.7621	0.6121	1.3742
580	1325.8	0.0228	0.2989	0.3217	588.9	588.4	1177.3	0.7872	0.5659	1.3532
600	1542.9	0.0236	0.2432	0.2668	617	548.5	1165.5	0.8131	0.5176	1.3307
620	1786.6	0.0247	0.1955	0.2201	646.7	503.6	1150.3	0.8398	0.4664	1.3062
640	2059.7	0.026	0.1538	0.1798	678.6	452	1130.5	0.8679	0.411	1.2789
660	2365.4	0.0278	0.1165	0.1442	714.2	390.2	1104.4	0.8987	0.3485	1.2472
680	2708.1	0.0305	0.0810	0.1115	757.3	309.9	1067.2	0.9351	0.2719	1.2071
700	3093.7	0.0369	0.0392	0.0761	823.3	172.1	995.4	0.9905	0.1484	1.1389

Sources: Refs. 11 and 12.

Table 1–15

Superheated Steam

Absolute pressure, lb_f/in^2 (saturated temperature, °F)		Temperature, °F												
		200	220	300	350	400	450	500	550	600	700	800	900	1000
	v	392.6	404.5	452.3	482.2	512.0	541.8	571.6	601.4	631.2	690.8	750.4	809.9	869.5
1	*l*	1150.4	1159.5	1195.8	1218.7	1241.7	1264.9	1288.3	1312.0	1335.7	1383.6	1432.7	1482.6	1533.4
(101.74)	*s*	2.0512	2.0647	2.1153	2.1444	2.1720	2.1983	2.2233	2.2468	2.2702	2.3137	2.3542	2.3923	2.4283
	v	78.16	80.59	90.25	96.26	102.26	108.24	114.22	120.19	126.16	138.1	150.03	161.95	173.87
5	*l*	1148.8	1158.1	1195.0	1218.1	1241.2	1264.5	1288.0	1311.7	1335.4	1383.6	1432.7	1482.6	1533.4
(162.24)	*s*	1.8718	1.8857	1.9370	1.9664	1.9942	2.0205	2.0456	2.0692	2.0927	2.1361	2.1776	2.2148	2.2509
	v	38.85	40.09	45.00	48.03	51.04	54.05	54.05	60.04	63.03	69.01	74.98	80.95	86.92
10	*l*	1146.6	1156.2	1193.9	1217.2	1240.6	1264.0	1264.0	1311.3	1335.1	1383.4	1432.5	1482.4	1533.2
(193.21)	*s*	1.7927	1.8071	1.8595	1.8892	1.9172	1.9436	1.9436	1.9924	2.0160	2.0596	2.1002	2.1383	2.1744
	v		27.15	30.53	32.62	34.68	36.73	36.73	40.82	42.86	46.94	51.00	55.07	59.13
14.696	*l*		1154.4	1192.8	1216.4	1239.9	1263.5	1263.5	1310.9	1334.8	1383.2	1432.3	1482.3	1533.1
(212.00)	*s*		1.7624	1.8160	1.8460	1.8743	1.9008	1.9008	1.9498	1.9734	2.0170	2.0576	2.0958	2.1319
	v			22.36	23.91	25.43	26.95	26.95	29.97	31.47	34.47	37.46	40.45	43.44
20	*l*			1191.6	1215.6	1239.2	1262.9	1262.9	1310.5	1334.4	1382.9	1432.1	1482.1	1533.0
(227.96)	*s*			1.7808	1.8112	1.8396	1.8664	1.8664	1.916	1.9392	1.9829	2.0235	2.0618	2.0978
	v			11.040	11.843	12.628	13.401	13.401	14.93	15.688	17.198	18.702	20.20	21.70
40	*l*			1186.8	1211.9	1236.5	1260.7	1260.7	1308.9	1333.1	1381.9	1431.3	1481.4	1532.4
(267.25)	*s*			1.6994	1.7314	1.7608	1.7881	1.7881	1.8384	1.8619	1.9058	1.9467	1.9850	2.0214
	v			7.259	7.818	8.357	8.884	8.884	9.916	10.427	11.441	12.449	13.452	14.454
60	*l*			1181.6	1208.2	1233.6	1258.5	1258.5	1307.4	1331.8	1380.9	1430.5	1480.8	1531.9
(292.71)	*s*			1.6492	1.6830	1.7135	1.7416	1.7678	1.7926	1.8162	1.8605	1.9015	1.9400	1.9762
	v				5.803	6.220	6.624	7.020	7.410	7.797	8.562	9.322	10.077	10.830
80	*l*				1204.3	1230.7	1256.1	1281.1	1305.8	1330.5	1379.9	1429.7	1480.1	1531.3
(312.03)	*s*				1.6475	1.6791	1.7078	1.7346	1.7598	1.7836	1.8281	1.8694	1.9079	1.9442
	v				4.592	4.937	5.268	5.589	5.905	6.218	6.835	7.446	8.052	8.656
100	*l*				1200.1	1227.6	1253.7	1279.1	1304.2	1329.1	1378.9	1428.9	1479.5	1530.8
(327.81)	*s*				1.6188	1.6518	1.6813	1.7085	1.7339	1.7581	1.8029	1.8443	1.8829	1.9193

	v	3.783	4.081	4.363	4.636	4.902	5.165	5.683	6.195	6.702	7.207
120	*l*	1195.7	1224.4	1251.3	1277.2	1302.5	1327.7	1377.8	1428.1	1478.8	1530.2
(341.25)	*s*	1.5944	1.6287	1.6591	16869	1.7127	1.7370	1.7822	1.8237	1.8625	1.8990
	v		3.468	3.715	3.954	4.186	4.413	4.861	5.301	5.738	6.172
140	*l*		1221.1	1248.7	1275.2	1300.9	1326.4	1376.8	1427.3	1478.2	1529.7
(353.02)	*s*		1.6087	1.6399	1.6683	1.6945	1.7190	1.7645	1.8063	1.8451	1.8817
	v		3.008	3.230	3.443	3.648	3.849	4.244	4.631	5.015	5.396
160	*l*		1217.6	1246.1	1273.1	1299.3	1325.0	1375.7	1426.4	1477.5	1529.1
(363.53)	*s*		1.5908	1.6230	1.6519	1.6785	1.7033	1.7491	1.7911	1.8301	1.8667
	v		2.649	2.852	3.044	3.229	3.411	3.764	4.110	4.452	4.792
180	*l*		1214.0	1243.5	1271.0	1297.6	1323.5	1374.7	1425.6	1476.8	1528.6
(373.06)	*s*		1.5745	1.6077	1.6373	1.6642	1.6894	1.7355	1.7776	1.8167	1.8534
	v		2.361	2.549	2.726	2.895	3.060	3.380	3.693	4.002	4.309
200	*l*		1210.3	1240.7	1268.9	1295.8	1322.1	1373.6	1424.8	1476.2	1528.0
(381.79)	*s*		1.5594	1.5937	1.6240	1.6513	1.6767	1.7232	1.7655	1.8048	1.8415
	v		2.125	2.301	2.465	2.621	2.772	3.066	3.352	3.634	3.913
220	*l*		1206.5	1237.9	1266.7	1294.1	1320.7	1372.6	1424.0	1475.5	1527.5
(389.86)	*s*		1.5453	1.5808	1.6117	1.6395	1.6652	1.7120	1.7545	1.7939	1.8308
	v		1.9276	2.094	2.247	2.393	2.533	2.804	3.068	3.327	3.584
240	*l*		1202.5	1234.9	1264.5	1292.4	1319.2	1371.5	1423.2	1474.8	1526.9
(397.37)	*s*		1.5319	1.5686	1.6003	1.6286	1.6546	1.7017	1.7444	1.7839	1.8209
	v			1.9183	2.063	2.199	2.330	2.582	2.827	3.067	3.305
260	*l*			1232.0	1262.3	1290.5	1317.7	1370.4	1422.3	1474.2	1526.3
(404.42)	*s*			1.5573	1.5897	1.6184	1.6447	1.6922	1.7352	1.7748	1.8118
	v			1.7674	1.9047	2.033	2.156	2.392	2.621	2.845	3.066
280	*l*			1228.9	1260.0	1288.7	1316.2	1369.4	1421.5	1473.5	1525.8
(411.05)	*s*			1.5464	1.5796	1.6087	1.6354	1.6834	1.7265	1.7662	1.8033
	v			1.6364	1.7675	1.8891	2.005	2.227	2.442	2.652	2.859
300	*l*			1225.8	1257.6	1286.8	1314.7	1368.3	1420.6	1472.8	1525.2
(417.33)	*s*			1.5360	1.5701	1.5998	1.6268	1.6751	1.7184	1.7582	1.7954
	v			1.3734	1.4923	1.6010	1.7036	1.8980	2.084	2.266	2.445
350	*l*			1217.7	1251.5	1282.1	1310.9	1365.5	1418.5	1471.1	1523.8
(431.72)	s			1.5119	1.5481	1.5792	1.6070	1.6563	1.7002	1.7403	1.7777
	v			1.1744	1.2851	1.3843	1.4770	1.6508	1.8161	1.9767	2.134
400	*l*			1208.8	1245.1	1277.2	1306.9	1362.7	1416.4	1469.4	1522.4
(444.59)	*s*			1.4892	1.5281	1.5607	1.5894	1.6398	1.6842	1.7247	1.7623

Table 1–16

Saturated Steam-Ice

		Specific volume, ft^3/lb		Enthalpy, Btu/lb			Entropy, Btu/(lb · °R)		
Temperature T, °F	Absolute pressure P, lb_f/in^2	Saturated ice v_i	Saturated steam $v_g \times 10^{-3}$	Saturated ice h_i	Sublimation difference h_{sub}	Saturated steam h_g	Saturated ice s_i	Sublimation difference s_{sub}	Saturated steam s_g
32	0.0885	0.01747	3.306	−143.35	1219.1	1075.8	−0.2916	2.4793	2.1877
30	0.0808	0.01747	3.609	−144.35	1219.3	1074.9	−0.2936	2.4897	2.1961
20	0.0505	0.01745	5.658	−149.31	1219.9	1070.6	−0.3038	2.5425	2.2387
10	0.0309	0.01744	9.05	−154.17	1220.4	1066.2	−0.3141	2.5977	2.2836
0	0.0185	0.01742	14.77	−158.93	1220.7	1061.8	−0.3241	2.6546	2.3305
−10	0.0108	0.01741	24.67	−163.59	1221.0	1057.4	−0.3346	2.7143	2.3797
−20	0.0062	0.01739	42.2	−168.16	1221.2	1053.0	−0.3448	2.7764	2.4316
−30	0.0035	0.01738	74.1	−172.63	1221.2	1048.6	−0.3551	2.8411	2.4860

References

1. J. Reynolds, J. Jeris, and L. Theodore, *Handbook of Chemical and Environmental Engineering Calculations*, Wiley, Hoboken, N.J., 2004.
2. P. Abulencia and L. Theodore, *Fluid Flow for the Practicing Chemical Engineer*, Wiley, Hoboken, N.J., 2009.
3. D. Green and R. Perry, eds., *Perry's Chemical Chemical Engineers' Handbook, 8th ed.*, McGraw-Hill, N.Y., 2008.
4. L. Theodore, *Heat Transfer Applications for the Practicing Engineer*, Wiley, Hoboken, N.J., 2011.
5. L. Theodore and F. Ricci, *Mass Transfer Operations for the Practicing Engineer*, Wiley, Hoboken, N.J., 2010.
6. L. Theodore, *Chemical Reactor Analyses and Applications for the Chemical Engineer*, Wiley, Hoboken, N.J., 2012.
7. D. Green and R. Perry, eds., *Perry's Chemical Engineers' Handbook, 7th ed.*, McGraw-Hill, N.Y., 1998.
8. M. Spiegel, *Mathematical Handbook of Formulas and Tables*, Shaum's Outline Series, McGraw-Hill, N.Y., 1968.
9. S. Shaefer and L. Theodore, *Probability and Statistics in Environmental Science*, CRC Press/Taylor & Francis Group, Boca Raton, Fla., 2007.
10. Adapted from http://www.statsoft/textbook/sttable.html
11. J. Keenan and F. Keyes, *Thermodynamic Properties of Steam*, John Wiley & Sons, Hoboken, N.J., 1936.
12. L.Theodore, F. Ricci, and T. Vanvliet, *Thermodynamics for the Practicing Engineer*, Wiley, Hoboken, N.J., 2009.

2 Process Variables

Chapter Outline

2–1 Introduction

Several years ago, the author originally considered the title "State, Physical, and Chemical Properties" for this type of chapter; however, since these three properties have been used interchangeably and have come to mean different things to engineers and scientists, it was decided to simply employ the title "Process Variables." The three aforementioned properties were therefore integrated into this all-purpose chapter and eliminated the need for differentiating between the three.

This chapter provides a review of some basic concepts from physics, chemistry, and engineering in preparation for material that is covered in later chapters. Because some of these topics are unrelated to each other, this chapter admittedly lacks the cohesiveness that chapters covering a single topic might have. This is usually the case when basic material from such widely differing areas of knowledge as physics, chemistry, and engineering is employed. Although these topics are widely divergent and covered with varying degrees of thoroughness, all of them will find later use in this text. Readers who seek additional information on these review topics are directed to the literature in the reference section of this chapter.

Every compound has a unique set of properties that allows one to recognize and distinguish it from other compounds. These properties can be grouped into two main categories: physical and chemical. *Physical properties* can be measured without changing the identity and composition of the substance. Key properties include viscosity, density, surface tension, melting point, and boiling point. *Chemical properties* may be altered via chemical reaction to form other compounds or substances. Key chemical properties include upper and lower flammability limits, enthalpy of reaction, and autoignition temperature. These properties may be further divided into two categories: intensive and extensive. *Intensive properties* are not a function of the quantity of the substance, while *extensive properties* depend on the quantity of the substance.

Regarding the topic of property estimation, it should be noted that the remainder of the chapter is devoted to a review of the following *process variables*: temperature, pressure, moles and molecular weights, mass and volume, viscosity, surface tension, heat capacity and thermal conductivity, diffusion coefficients, pH, Reynolds number, the ideal gas law, and (finally) property estimation. The reader is referred to any of a host of references in the literature for additional information.

2–2 Temperature

Whether in the gaseous, liquid, or solid state, all molecules possess some degree of kinetic energy; that is, they are in constant motion—vibrating, rotating, or translating. The kinetic energies of individual molecules cannot be measured, but the combined effect of these energies in a very large number of molecules can. This measurable quantity is known as *temperature*; it is a macroscopic concept only and as such does not exist on the molecular level.

Temperature can be measured in many ways; the most common method makes use of the expansion of mercury (usually encased inside a glass capillary tube) with increasing temperature. (In many thermal applications, however, thermocouples or thermistors are more commonly employed.) The two most commonly used temperature scales are the Celsius (or centigrade) and Fahrenheit scales. The Celsius scale is based on the boiling and freezing points of water at 1 atm (atmosphere) pressure; to the former, a value of 100°C is assigned, and to the latter, a value of 0°C. On the older Fahrenheit scale, these temperatures correspond to 212°F and 32°F, respectively. Equations (2–1) and (2–2) show the conversion from one scale to the other:

$$°F = 1.8(°C) + 32 \quad (2\text{–}1)$$

$$°C = (°F - 32)/1.8 \quad (2\text{–}2)$$

where °F = a temperature on the Fahrenheit scale
°C = a temperature on the Celsius scale

Experiments with gases at low to moderate pressures (up to a few atmospheres) have shown that, if the pressure is kept constant, the volume of a gas and its temperature are linearly related via Charles' law (see Sec. 2–12) and that a decrease of 0.3663% or (1/273) of the initial volume is experienced for every temperature drop of 1°C. These experiments were not extended to very low temperatures, but if this linear relationship were extrapolated, the volume of the gas would *theoretically* be zero at a temperature of approximately −273°C or −460°F. This temperature has become known as *absolute zero* and is the basis for the definition of two *absolute* temperature scales. (An *absolute* scale is one which does not allow *negative* values.) These absolute temperature scales are the Kelvin (K) and Rankine (°R) scales; the former is defined by shifting the Celsius scale by 273°C so that 0 K is equal to – 273°C; Equation (2–3) shows this relationship:

$$K = °C + 273 \quad (2\text{–}3)$$

The Rankine scale is defined by shifting the Fahrenheit scale 460°, so that

$$°R = °F + 460 \quad (2\text{–}4)$$

2–3 Pressure

Molecules in the gaseous state possess a high degree of translational kinetic energy, which means that they are able to move quite freely throughout the body of the gas. If the gas is in a container of some type, the molecules are constantly bombarding the walls of the container. The macroscopic effect of this bombardment by a tremendous number of molecules—enough to make the effect measurable—is called *pressure*. The natural units of pressure are force per unit area. In the example of the gas in a container, the *unit area* is a portion of the inside solid surface of the container wall and the *force*, measured *perpendicularly* to the unit area, is the result of the molecules hitting the unit area and giving up momentum during the sudden change of direction.

Various methods are used to express a pressure measurement. Some of them are natural units, i.e, based on a force per unit area, e.g., pound (force) per square inch (abbreviated lb_f/in^2 or psi) or dynes per square centimeter (dyn/cm^2). Others are based on a fluid height, such as inches of water (in H_2O) or millimeters of mercury (mm Hg); units such as these are convenient when the pressure is indicated by a difference between two levels of a liquid as in a *manometer* or *barometer*. *Barometric pressure* and *atmospheric pressure* are synonymous and measure the ambient air pressure. *Standard barometric pressure* is the average atmospheric pressure at sea level, 45° north latitude at 32°F. It is used to define another unit of pressure called the atmosphere (atm). Standard barometric pressure is 1 atm and is equivalent to 14.696 psi or 29.921 in Hg. As one might expect, barometric pressure varies with both weather and altitude.

Measurements of pressure by most gauges indicate the difference in pressure either above or below that of the atmosphere surrounding the gauge. *Gauge pressure* is that pressure indicated by such a device. If the pressure in the system measured by the gauge is greater than the pressure prevailing in the atmosphere, the gauge pressure is expressed positively. The gauge pressure is a negative quantity if lower than atmospheric pressure. The term *vacuum* designates a negative gauge pressure. Gauge pressures are often identified by the letter g after the pressure unit; for example, psig (pounds per square inch gauge) is a gauge pressure in psi units.

Since gauge pressure is the pressure relative to the prevailing atmospheric pressure, the sum of the two provides the *absolute pressure*, indicated by the letter a after the unit [e.g., psia (pounds per square inch absolute)]:

$$P = P_a + P_g \tag{2–5}$$

where P = absolute pressure, (psia)
P_a = atmospheric pressure, (psia)
P_g = gauge pressure, (psig)

The absolute pressure scale is absolute in the same sense that the absolute temperature scale is absolute, i.e., a pressure of zero psia is the lowest possible pressure theoretically achievable—a perfect vacuum.

2–4 Molecular Weights and Moles

An atom consists of protons and neutrons in a nucleus surrounded by electrons. An electron has such a small mass relative to that of the proton and neutron that the weight of the atom (called the *atomic weight*) is approximately equal to the sum of the weights of the particles in its nucleus. Atomic weight may be expressed in *atomic mass units* (amu) *per atom* or *in grams per gram·atom.* One gram·atom contains 6.02×10^{23} atoms (Avogadro's number). The atomic weights of the elements are listed in Table 1–7 in the previous chapter.

The *molecular weight* (MW) of a compound is the sum of the atomic weights of the atoms that make up the molecule. Atomic mass units per molecule (amu/molecule) or grams per gram·mole [g/(gmol)] are used for molecular weight. One gram·mole (gmol) contains an Avogadro number of molecules. For the English or engineering system, a pound·mole (lbmol) contains $454 \times 6.023 \times 10^{23}$ molecules.

Molal units are used extensively in chemical engineering calculations as they greatly simplify material balances where chemical reactions are occurring. For mixtures of substances (gases, liquids, or solids), it is also convenient to express compositions in mole fractions or mole percentages instead of mass fractions. The mole fraction is the ratio of the number of moles of one component to the total number of moles in the mixture. Equations (2–6) to (2–9) express these relationships:

$$\text{Moles of A} = \frac{\text{mass of A}}{\text{molecular weight of } A}$$

$$n_A = \frac{m_A}{(\text{MW})_A} \tag{2–6}$$

$$\text{Mole fraction of A} = \frac{\text{moles of A}}{\text{total moles}}$$

$$y_A = \frac{n_A}{n} \tag{2–7}$$

$$\text{Mass fraction of A} = \frac{\text{mass of A}}{\text{total mass}}$$

$$w_A = \frac{m_A}{m} \tag{2–8}$$

$$\text{Volume fraction of A} = \frac{\text{volume of A}}{\text{total volume}}$$

$$v_A = \frac{V_A}{V} \tag{2–9}$$

The reader should note that, in general, mass fraction (or percent) is *not* equal to mole fraction (or percent).

2–5 Mass and Volume

The *density* (ρ) of a substance is the ratio of its mass to its volume and may be expressed in engineering units of pounds per cubic foot (lb/ft^3), kilograms per cubic meter (kg/m^3), etc. For solids, density can be easily determined by placing a known mass of the substance in a liquid and determining the displaced volume. The density of a liquid can be measured by weighing a known volume of the liquid in a volumetric flask. For gases, the ideal gas law (see Sec. 2–12) can be used to calculate the density from the pressure, temperature, and molecular weight of the gas.

Densities of pure solids and liquids are relatively independent of temperature and pressure and can be found in standard reference books.[1,2] The *specific volume* ν of a substance is its volume per unit mass (ft^3/lb, m^3/kg, etc.) and is, therefore, the inverse of its density.

The *specific gravity* (SG) is the ratio of the density of a substance to the density of a reference substance at a specific condition:

$$\text{SG} = \frac{\rho}{\rho_{\text{ref}}} \tag{2–10}$$

The reference most commonly used for *solids* and *liquids* is water at its maximum density, which occurs at 4°C; this reference density value is 1.000 g/cm^3, 1000 kg/m^3, or 62.43 lb/ft^3. Note that, since the specific gravity is a ratio of two densities, it is dimensionless. Therefore, any set of units may be employed for the two densities as long as they are *consistent*. The specific gravity of *gases* is used only rarely; when it is, air at the same conditions of temperature and pressure as the gas is usually employed as the reference substance.

Another dimensionless quantity related to density is the American Petroleum Institute (API) gravity, which is often employed to describe the densities of fuel oils. The relationship between the API scale and specific gravity is

$$\text{Degrees API, } ^\circ\text{API} = \frac{141.5}{\text{SG}(60/60^\circ\text{F})} - 131.5 \tag{2–11}$$

where SG(60/60°F) = specific gravity of the liquid at 60°F using water at 60°F as the reference.

Petroleum refining is a major industry. Petroleum products serve as an important fuel for the power industry, and petroleum derivatives are the starting point for many syntheses in the chemical industry. Petroleum is also a mixture of a large number of chemical compounds. The common petroleum fractions derived from crude oil and their approximate °API values are listed in Table 2–1.

Table 2–1

°API Values for Crude Oil Components

Components from crude oil	Approximate °API
Light ends and gases	114
Gasoline	75
Naphtha	60
Kerosene	45
Distillate	35
Gas oil	28
Lube oil	18–30
Fuel oil (residue)	25–35

2–6 Viscosity

Viscosity is a property associated with a fluid's resistance to flow and accounts for energy losses which result from shear stresses that occur between different portions of the fluid that are moving at different velocities. It finds wide application in fluid flow applications. The *absolute viscosity* (μ) has units of mass per length × time; the fundamental unit is the *poise* (P), which is defined as 1 g/(cm · s). This unit is inconveniently large for many practical purposes and viscosities are frequently given in *centipoises* (0.01 poise), which is abbreviated cP. The viscosity of pure water at 68.6°F is 1.00 cP. In English units, absolute viscosity is expressed either as pounds (mass) per foot · second [lb/(ft · s)] or pounds per foot · hour (lb/ft · h). The absolute viscosity depends primarily on temperature and to a lesser degree on pressure. The *kinematic viscosity* (ν) *is* the absolute viscosity divided by the density of the fluid and is useful in certain fluid flow problems; the units for this quantity are length squared per time, e.g., square feet per second (ft^2/s) or square meters per hour (m^2/h). A kinematic viscosity of 1 cm^2/s is called a *stoke*, denoted as S. For pure water at 70°F, ν = 0.983 cS (centistokes). Because fluid viscosity changes rapidly with temperature, a numerical value of viscosity has no significance unless the temperature is specified.

Liquid viscosity is usually measured by the amount of time it takes for a given volume of liquid to flow-through an orifice. The *Saybolt universal viscometer* is the most widely used device in the United States for determining the viscosity of fuel oils and liquids. It should be stressed that Saybolt viscosities, which are expressed in *Saybolt second units* (SSU), are not even approximately proportional to absolute viscosities except in the range above 200 SSU; hence, converting units from Saybolt seconds to other units requires the use of special conversion tables. As the time of flow decreases, the deviation becomes more marked. In any event, viscosity is an important property because of potential flow problems that might arise with viscous liquids.

The viscosities of air at atmospheric pressure and water are presented in Tables 2–2 and 2–3, respectively, as a function of temperature. Viscosities of other substances are available in the literature.[1] Additional information is available in Chap. 7.

Table 2–2

Viscosity of Air at 1 Atmosphere Pressure

T(°C)	Viscosity, Micropoise (μP)
0.0	1170.8
18	182.7
40	190.4
54	195.8
74	210.2
229	263.8

Table 2–3

Viscosity of Water

T, °C	Viscosity, centipoise (cP)
0	1.792
5	1.519
10	1.308
15	1.140
20	1.000
25	0.894
30	0.801
35	0.723
40	0.656
50	0.594
60	0.469
70	0.406
80	0.357
90	0.317
100	0.284

2–7 Surface Tension

A liquid forms an interface with another fluid. At the surface, the molecules are more densely packed than those within the fluid. This results in surface tension effects and interfacial phenomena. The surface tension coefficient σ is the force per unit length of the circumference of the interface, or the energy per unit area of the interface area. Surface tension values are available in the literature.[3]

Surface tension causes a *contact angle* to appear when a liquid interface is in contact with a solid surface. If the contact angle $\theta < 90°$, the liquid is termed *wetting*. If $\theta > 90°$, it is a *nonwetting* liquid. Surface tension causes a fluid interface to rise (or fall) in a capillary tube. The capillary rise is obtained by equating the vertical component of the surface tension force to the weight of a given height. Additional details are provided by Abulencia and Theodore.[3]

2–8 Heat Capacity and Thermal Conductivity

The *heat capacity* of a substance is defined as the quantity of heat required to raise the temperature of that substance by one degree on a unit mass (or mole) basis. The term *specific heat* is frequently used in place of *heat capacity*. This is not strictly correct because specific heat has been defined traditionally as the ratio of the heat capacity of a substance to the heat capacity of water. However, since the heat capacity of water is approximately 1 cal/(g · °C) or 1 Btu/(lb · °F), the term *specific heat* has come to imply heat capacity.

For gases, the addition of heat to cause the 1° temperature rise may be accomplished at either constant pressure or constant volume. Since the amounts of heat necessary are different for the two cases, subscripts are used to identify which heat capacity is being used—c_p for constant pressure and c_v for constant volume. For liquids and solids, this distinction does not have to be made since there is little difference between the two. Values of heat capacity are available in the literature.[4, 5]

Heat capacities are often used on a *molar* basis instead of a *mass* basis, in which case the units become cal/(gmol · °C) or Btu/(lbmol · °F). To distinguish between the two bases, uppercase letters (C_p, C_V) are used in this text to represent the molar-based heat capacities, and lowercase letters (c_p, c_V) are used for the mass-based heat capacities or specific heats.

Heat capacities are functions of both the temperature and pressure, although the effect of pressure is generally small and is neglected in almost all chemical engineering calculations. The effect of temperature on C_p can be described by

$$C_p = \alpha + \beta T + \gamma T^2 \tag{2–12}$$

or

$$C_p = \alpha + \beta T + cT^2 \tag{2–13}$$

Values for α, β, γ, and a, b, c, as well as average heat capacity information are provided in the literature.[4,5] Additional information is available in Chap. 8.

Experience has shown that when a temperature difference exists across a solid body, energy in the form of heat will transfer from the high-temperature region to the low-temperature region until thermal equilibrium (same temperature) is reached. This mode of heat transfer where vibrating molecules pass along kinetic energy through the solid is called *conduction.* Liquids and gases may also transport heat in this fashion. The property of *thermal conductivity* provides a measure of how fast (or how easily) heat flows through a substance. It is defined as the amount of heat that flows in unit time through a unit surface area of unit thickness as a result of a unit difference in temperature. Typical units for conductivity are (Btu · ft)/(h · ft^2 · °F) or Btu/(h · ft · °F). Values are available in the literature.[3,6] Additional information is available in Chap. 10.

2–9 Diffusion Coefficient

The diffusion coefficient finds application in numerous mass transfer operations (see also Chap. 11). This coefficient, or diffusivity D_{AB} of component A in solution B, which is a measure of its diffusive mobility, is defined as the ratio of its flux J_A to its concentration gradient and is given by

$$J_A = -D_{AB} \frac{\partial C_A}{\partial z} \tag{2–14}$$

This is Fick's first law[7] written for the z direction in rectangular (cartesian) coordinates. The concentration gradient term represents the variation of the concentration C_A in the z direction. The negative sign accounts for diffusion occurring from high to low concentrations. The diffusivity is a characteristic of the component and its environment (temperature, pressure, concentration, etc.). This equation is analogous to the flux equations defined for momentum transfer[3] (in terms of the previously defined viscosity) and for heat transfer[6] (in terms of the thermal conductivity). The diffusivity is usually expressed with units of (length)^2/time or moles/(time · area). This coefficient, as well as Fick's law, will receive additional treatment in Chap. 11.[8]

2–10 pH

An important chemical property of an aqueous solution is its *pH*, which measures the acidity or basicity of the solution. In a neutral solution, such as pure water, the hydrogen (H^+) and hydroxyl (OH^-) ion concentrations are equal. At ordinary temperatures, this concentration is

$$C_{H^+} = C_{OH^-} = 10^{-7} (\text{g} \cdot \text{ion})/\text{L} \tag{2–15}$$

where C_{H^+} = hydrogen ion concentration
C_{OH^-} = hydroxyl ion concentration

The unit g · ion denotes gram · ion, which represents an Avogadro number of ions. In all aqueous solutions, whether neutral, basic, or acidic, a chemical equilibrium or balance is established between these two concentrations, so that

$$K_{eq} = C_{H^+}C_{OH^-} = 10^{-14} \quad (2\text{–}16)$$

where K_{eq} = equilibrium constant

The numerical value for K_{eq} given in Eq. (2–16) holds for room temperature and only when the concentrations are expressed in gram · ion per liter [(g · ion)/L)]. In acid solutions, C_{H^+} is $> C_{OH^-}$, in basic solutions, C_{OH^-} predominates.

The pH is a property that is a direct measure of the hydrogen ion concentration and is defined by

$$\text{pH} = -\log C_{H^+} \quad (2\text{–}17)$$

Thus, an acidic solution is characterized by a pH below 7 (the lower the pH, the higher the acidity), a basic solution by a pH above 7, and a neutral solution by a pH of 7. It should be pointed out that Eq. (2–17) is not the exact definition of pH but is a close approximation to it. Strictly speaking, the *activity* of the hydrogen ion a_{H^+}, and not the ion concentration belongs in Eq. (2–17). The reader is directed to the literature[4,5] for a discussion of chemical activities.

2-11 Reynolds Number

The most commonly used dimensionless number employed in chemical engineering practice is the Reynolds number. The Reynolds number (Re), is a dimensionless number (as noted in Chap. 1) that indicates whether a moving fluid is flowing in the laminar or turbulent mode. *Laminar* flow is characteristic of fluids flowing slowly enough so that there are no eddies (whirlpools) or macroscopic mixing of different portions of the fluid. (*Note:* In any fluid, there is always *molecular* mixing due to the thermal activity of the molecules; this is distinct from *macroscopic* mixing due to the swirling motion of different portions of the fluid.) In laminar flow, a fluid can be imagined to flow like a deck of cards, with adjacent layers sliding past one another. *Turbulent* flow is characterized by eddies and macroscopic currents. In practice, moving gases are generally in the turbulent region. For flow in a pipe, a Reynolds number above 2100 is an indication of turbulent flow. (See also Chap. 9.)

The Reynolds number is dependent on the fluid velocity, density, viscosity, and some characteristic *length* of the system or conduit; this characteristic length is the inside diameter for pipes:

$$\text{Re} = \frac{Dv\rho}{\mu} = \frac{Dv}{\nu} \quad (2\text{–}18)$$

where Re = Reynolds number
D = inside diameter of the pipe, ft
v = fluid velocity, ft/s
ρ = fluid density, lb/ft^3
μ = fluid viscosity, lb/(ft · s)
ν = fluid kinematic viscosity, $\text{ft}^2\text{/s}$

Any consistent set of units may be used with Eq. (2–18).

2–12 The Ideal Gas Law

The most commonly used law in chemical engineering practice is the ideal gas law. The two precursors of the ideal gas law were *Boyle's* and *Charles'* laws. Boyle found that the volume of a given mass of gas is inversely proportional to the *absolute* pressure if the temperature is kept constant:

$$P_1V_1 = P_2V_2 \tag{2–19}$$

where V_1 = volume of gas at pressure P_1 and temperature T
V_2 = volume of gas at pressure P_2 and temperature T

Charles found that the volume of a given mass of gas varies directly with the *absolute* temperature at constant pressure:

$$\frac{V_1}{T_1} = \frac{V_2}{T_2} \tag{2–20}$$

where V_1 = volume of gas at pressure P and absolute temperature T_1
V_2 = volume of gas at pressure P and absolute temperature T_2

Boyle's and Charles' laws may be combined into a single equation in which neither temperature nor pressure need be held constant:

$$\frac{P_1V_1}{T_1} = \frac{P_2V_2}{T_2} \tag{2–21}$$

For Eq. (2–21) to hold, the mass of gas must be constant as the conditions change from (P_1, T_1) to (P_2, T_2). This equation indicates that for a given mass of a specific gas, PV/T has a constant value. Since, at the same temperature and pressure, volume and mass must be directly proportional, this statement may be extended to

$$\frac{PV}{mT} = C \tag{2–22}$$

where m = mass of a specific gas
C = constant that depends on the gas

Note that volume terms may be replaced by volume rate (or volumetric flow rate), q.

Experiments with different gases showed that Eq. (2–22) could be expressed in a far more generalized form. If the number of moles n *is* used in place of the mass m, the constant is the same for all gases:

$$\frac{PV}{nT} = R \tag{2–23}$$

where R = universal gas constant

Equation (2–23) is defined as the ideal gas law. Numerically, the value of R depends on the units used for P, V, T, and n (see Table 2–4). In this text, gases are generally assumed to approximate ideal-gas behavior. As is usually the case in engineering practice, the ideal-gas law is assumed to be valid unless otherwise stated. If a case is encountered in practice where the gas behaves in a *nonideal* fashion, e.g., a high-molecular-weight gas (such as a chlorinated organic) at elevated pressures, one of the many *real gas* correlations should be employed.[4,5]

Table 2–4

Values of R in Various Units

R	Temperature scale	Units of V	Units of n	Units of P	Unit of PV (energy)
10.73	°R	ft^3	lbmol	psfa	—
0.7302	°R	ft^3	lbmol	atm	—
555.0	°R	ft^3	lbmol	mm Hg	—
297.0	°R	ft^3	lbmol	in H_2O	—
0.7398	°R	ft^3	lbmol	bar	—
1545.0	°R	ft^3	lbmol	lb_f/ft^2 absolute	—
24.75	°R	ft^3	lbmol	ft H_2O	—
1.9872	°R	—	lbmol	—	Btu
0.0007805	°R	—	lbmol	—	hp.h
0.0005819	°R	—	lbmol	—	kW·h
1.314	K	ft^3		atm	—
0.08205	K	L	gmol	atm	—
0.08314	K	L	gmol	bar	—
8314	K	L	gmol	Pa	—
8.314	K	m^3	gmol	Pa	—
82.057	K	cm^3	gmol	atm	—
1.9872	K	—	gmol	—	cal
8.314	K		gmol	—	J

Other useful forms of the ideal-gas law are shown in Eqs. (2–24) and (2–25). Equation (2–24) applies to gas volume flow rate rather than to a gas volume confined in a container:

$$Pq = nRT \tag{2–24}$$

where q = gas volumetric flow rate, ft^3/h
P = absolute pressure, (psia)
n = molar flow rate, (lbmol/h)
T = absolute temperature, °R
R = 10.73 psia · ft^3/(lbmol · °R)

Equation (2–25) combines n and V from Eq. (2–23) to express the law in terms of the density:

$$P(\mathrm{MW}) = \rho RT \tag{2–25}$$

where MW = molecular weight of gas, lb/(lbmol)
ρ = density of gas, lb/ft^3

Some chemical engineering calculations require the aforementioned deviations from ideality to be included in the analysis. Many of the nonideal correlations involve the critical temperature T_c, the critical pressure P_c, and a term defined as the acentric factor, ω. An abbreviated list of these properties is available in the literature.[4,5] These reduced quantities find wide application in thermodynamic analyses of nonideal systems (see also Chap. 8).

The critical temperature and pressure are employed in the calculation of the reduced temperature T_r, and the reduced pressure P_r, as provided in Eqs. (2–26) and (2–27):

$$T_r = T/T_c \tag{2–26}$$

$$P_r = P/P_c \tag{2–27}$$

Both reduced properties are dimensionless and play important roles in non-ideal gas behavior.

Many physical and chemical properities of elements and compounds can be estimated from models (equations) that are based on the reduced temperature and pressure of the substance in question. These reduced properites have also served as the basis for many equations that are employed in practice to describe non-deal-gas (and liquid) behavior. Although a rigorous treatment of this material is beyond the scope of this book, information is available in the literature.[4,5] Highlights of this topic are presented below.

No real gas conforms exactly to the ideal gas law, but it can be used as an excellent approximation for most gases at pressures about or less than 5 atm and near ambient temperatures. One approach to account for deviations from ideality is to include a *correction factor* Z, which is defined as the *compressibility coefficient* or *compressibility factor*. The ideal gas law is then modified to the following form:

$$PV = ZnRT \tag{2–28}$$

Note that Z approaches 1.0 as P approaches 0.0. For an ideal gas, Z is exactly unity. This equation may also be written as

$$Pv = ZRT \tag{2–29}$$

where v is now the *specific* molar volume (not the total volume) with units of volume/mole.

Regarding gas mixtures, the ideal gas law can be applied directly for ideal gas mixtures. However, the molecular weight of the mixture is based on a mole fraction average of the n components:

$$\overline{\mathrm{MW}} = \sum_{i=1}^{n} (y_i)(\mathrm{MW}_i) \tag{2–30}$$

One approach to account for deviations from ideality is to assume the aforementioned compressibility coefficient for the mixture $\overline{Z}_m$ is a linear mole fraction combination of the individual component Z values:

$$\overline{Z}_m = \sum_{i=1}^{n} y_i Z_i \tag{2–31}$$

Furthermore, Kay[9] has shown that the deviations arising in using this approach can be deduced by employing *pseudocritical* values for T_c and P_c where

$$\overline{T}_c = \sum_{i=1}^{n} y_i T_{ci} \tag{2–32a}$$

$$\overline{P}_c = \sum_{i=1}^{n} y_i P_{ci} \tag{2–32b}$$

In lieu of other information, Theodore[10] suggests employing Eq. (2–33) for the pseudocritical value of ω:

$$\overline{\omega} = \sum_{i=1}^{n} y_i \omega_i \tag{2–33}$$

These pseudocritical values—$\overline{T}_c, \overline{P}_c$, and $\bar{\omega}$—are then employed in the appropriate pure component equation of state. This approach has been defined by some as *Kay's rule*, an approach that has been unfortunately been abandoned in recent years.

2–13 Vapor Pressure, Partial Pressure, and Partial Volume

Vapor pressure is an important property of liquids, and, to a much lesser extent, of solids. If a liquid is allowed to evaporate in a confined space, the pressure in the vapor space increases as the amount of vapor increases. If there is sufficient liquid present, a point is eventually reached when the pressure in the vapor space is exactly equal to the pressure exerted by the liquid at its own surface. At this point, a *dynamic equilibrium* exists in which vaporization and condensation take place at equal rates and the pressure in the vapor space remains constant. The pressure exerted at equilibrium is equal to the vapor pressure of the liquid. The magnitude of this pressure for a given liquid depends on the temperature, but not on the amount of liquid present. Solids, like liquids, also exert a vapor pressure. Evaporation of solids (*sublimation*) is noticeable only for those solids with appreciable vapor pressures. This is reviewed in more detail in Chap. 11.

Mixtures of gases are more often encountered than single or pure gases in chemical engineering practice. The ideal gas law is based on the *number of* molecules present in the gas volume; the *kind* of molecules is not a significant factor, only the number. This law applies equally well to mixtures and pure gases alike. Dalton and Amagat both applied the ideal gas law to mixtures of gases.

Since pressure is caused by gas molecules colliding within the containing walls, it seems reasonable that the total pressure of a gas mixture is made up of pressure contributions consisting of the component gases. These pressure contributions are called *partial pressures.* Dalton defined the partial pressure of a component as the pressure that would be exerted if the same mass of the component gas occupied the same total volume *alone* at the same temperature as the mixture. The sum of these partial pressures then equals the total pressure:

$$P = p_A + p_B + p_C + \ldots + p_n = \sum_{i=1}^{n} p_i \tag{2–34}$$

where P = total pressure
n = number of components
p_i = partial pressure of component i

Equation (2–34) is known as *Dalton's law*. Applying the ideal gas law to one component (A) only, one obtains

$$p_A V = n_A RT \tag{2–35}$$

where n_A is the number of moles of component A. Eliminating R, T, and V between Eqs. (2–23) and (2–35) leads to

$$\frac{p_A}{P} = \frac{n_A}{n} = y_A$$

or

$$p_A = y_A P \tag{2–36}$$

where y_A = mole fraction of component A.

Amagat's law is similar to Dalton's. Instead of considering the total pressure to consist of partial pressures where each component occupies the total container volume, Amagat considered the total volume to consist of the partial volumes in which each component is at (or is exerting) the total pressure. The definition of the *partial volume* is therefore the volume occupied by a component gas alone at the same temperature and pressure as the mixture. For this case:

$$V = V_A + V_B + V_C + \cdots + V_n = \sum_{i=1}^{n} V_i \tag{2–37}$$

Applying Eq. (2–23) as before, one obtains

$$\frac{V_A}{V} = \frac{n_A}{n} = y_A \tag{2–38}$$

where V_A = partial volume of component *A*.

It is common in engineering practice to describe low concentrations of components in gaseous mixtures in parts per million (ppm) by volume. Since partial volumes are proportional to mole fractions, it is necessary only to multiply the mole fraction of the component by 1 million to obtain the concentration in parts per million. [For liquids and solids, parts per million (ppm) is also used to express concentration, although it is usually on a *mass* basis rather than a *volume* basis. The terms ppmv and ppmw are sometimes used to distinguish between the volume and mass bases, respectively.]

2–14 Property Estimation

This last section provides references describing procedures that will allow the practitioner to estimate key physical and chemical properties of materials. Although the scientific community has traditionally resorted to experimental methods to accurately determine the aforementioned properties, that option may very well not be available when dealing with new materials and/or unique applications.

Predictive methods, albeit traditional ones, may be the only option available to obtain a first estimate of these properties. It should be noted that significant errors may be involved since extrapolating (or extending) satisfactory estimation procedures at the macroscale level may not always be reasonable. Notwithstanding these concerns, procedures for estimating some of the key physical and chemical properties given below are available in the literature.[4,11,12]

1. Vapor pressure
2. Latent enthalpy
3. Critical properties
4. Viscosity
5. Thermal conductivity
6. Heat capacity

References 4 and 11 are somewhat complementary, but each provides extensive information on this topic. This includes equations and procedures on several other properties not listed above. The interested reader should check these references for more details.

Is property estimation important? Absolutely. As indicated above, there are times and situations when experimental procedures cannot be implemented. For this scenario, one can turn to theoretical and semitheoretical methods and equations to obtain first estimates of important property information for some studies.

One could reasonably argue that the present procedures available to estimate the properties of materials are based on questionable approaches. Nonetheless, the traditional methods available for property estimation either may be applicable or may suggest alternative theoretical approaches.

Finally, it should be noted that the periodic law correlates properties of elements. Several centuries ago, scientists and engineers came to realize that all matter is composed of a rather limited number of basic building blocks, and the desire to discover all the fundamental units/parts became apparent. As data on the properties of elements became available, a pattern in the physical and chemical properties grew discernible, and because the pattern repeated itself in a rather well-organized fashion, it became known as the *periodic law*. This law is one of the premiere generalizations of science that has proven extremely useful in both predicting and correlating physical and chemical properties. The law essentially states that the properties of the chemical elements are not arbitrary, but depend on the structure of the atom and vary systematically, that is, periodically, with the atomic number. The main feature is that elements and compounds exhibit structural, physical, and chemical properties that are remarkably similar, and these properties establish the periodic relations among them.

References

1. D. Green and R. Perry, ed., *Perry's Chemical Engineers' Handbook*, 8th ed., McGraw-Hill, N.Y., 2008.
2. S. Maron and C. Prutton, *Principles of Physical Chemistry*, 4th ed., Macmillian, N.Y., 1970.
3. P. Abulencia and L. Theodore, *Fluid Flow for the Practicing Chemical Engineer*, Wiley, Hoboken, N.J., 2009.
4. J. Smith, H. Van Ness, and M. Abbott, *Introduction to Chemical Engineering Thermodynamics*, 6th ed., McGraw-Hill, N.Y., 2005.
5. L. Theodore, F. Ricci, and T. Van Vliet, *Thermodynamics for the Practicing Engineer*, Wiley, Hoboken, N.J., 2009.
6. L. Theodore, *Heat Transfer Applications for the Practicing Engineer*, Wiley, Hoboken, N.J., 2009.
7. A. Fick (article title unknown), *Pogg. Ann.*, **XCIV**, 59, 1855.
8. L. Theodore and F. Ricci, *Mass Transfer Operations for the Practicing Engineer*, Wiley, Hoboken, N.J., 2010.
9. W. Kay, "Density of Hydrocarbon Gases and Vapors," *Ind. Eng. Chem.*, **28**, 1014, N.Y., 1936.
10. L. Theodore, personal notes, East Willinston, N.Y., 1975.
11. N. Chopey, *Handbook of Chemical Engineering Calculations*, 2nd ed., McGraw-Hill, N.Y., 1994.
12. R. Bird, W. Stewart, and E. Lightfoot, *Transport Phenomena*, 2nd ed., Wiley, Hoboken, N.J., 2002.

3 Numerical Methods and Optimization

Chapter Outline

3–1 Introduction

This third chapter is concerned primarily with numerical methods. This subject was taught in the past as a means of providing chemical engineers (and scientists) with ways to solve complicated mathematical expressions that they could not otherwise solve. However, with the advent of computers, these solutions can now be readily obtained. A brief overview of numerical methods is given to provide the chemical engineer with some insight into what many of the currently used software packages (MathCad, Mathematica, MatLab, etc.) are actually doing.

Chemical engineers or scientists learn early in their careers how to use equations and mathematical methods to obtain exact answers to a large range of relatively simple problems. However, these techniques are seldom adequate for solving real-world problems, although the reader should note that one rarely needs exact answers in technical practice. Most real-world engineering and science applications are inexact because they have been generated from data or parameters that are measured, and thus represent only approximations. What one is likely to require and/or desire in a realistic situation is either an exact answer or one having reasonable accuracy from an engineering perspective.

As noted in the previous paragraph, the solution to a chemical engineering (or scientific) problem usually requires an answer to an equation or equations, and the answer(s) may be approximate or exact. Obviously an exact answer is preferred, but because of the complexity of some equations, exact solutions may not be attainable. Furthermore, an answer that is precise may not be necessary. For this instance, one may resort to another method that has come to be defined as a *numerical method*. Unlike the exact solution, which is continuous and in closed form, numerical methods provide an inexact (but often reasonably accurate) solution.

The numerical method referred to above often involves a stepwise procedure that ultimately leads to an answer and a solution to a particular problem. The method usually requires a large number of calculations and is therefore ideally suited for digital computation.

High-speed computing equipments (today's computers) have had a tremendous impact on engineering design, scientific computation, and data processing. The ability of computers to handle large quantities of data and to perform the mathematical operations described above at tremendous speeds permits the analysis of many more applications and more engineering variables than could possibly be handled on the slide rule—the trademark of the chemical engineers (including the author) of yesteryear. Scientific calculations previously estimated in lifetimes of computation time are currently generated in seconds and, on many occasions, microseconds, and in some rare instances, nanoseconds.[1]

As noted above, this chapter is concerned with numerical methods. This subject was taught in the past as a means of providing chemical engineers with ways to solve complicated mathematical expressions that they could not solve otherwise. A brief overview of the numerical methods below is given to provide the chemical engineer (as well as other engineers) with some insight into what many of the currently used software packages are actually doing. The author has not attempted to cover all the topics of numerical methods. Topics that traditionally fall in the domain of this subject include:

Differentiation

Integration

Simultaneous linear algebraic equations

Nonlinear algebraic equations

Ordinary differential equation(s)

Partial differential equation(s)

Optimization

Since a detailed treatment of each of these topics is beyond the scope of this text, the reader is referred to the literature[2–4] for more extensive analysis and additional information. The remainder of this chapter briefly examines the topics listed above.

3–2 Differentiation

Several differentiation methods are available for generating expressions for a derivative. Consider the problem of determining the benzene concentration time gradient dC/dt[5] at $t = 4.0$ s; refer to Table 3–1.

Method 1 This method involves the selection of any three data points and calculating the slope m of the two extreme points. This slope is approximately equal to the slope at the point lying in the middle. The value obtained is the "equivalent" of the derivative at that point 4. Using data points from 3.0 to 5.0, one obtains

$$\text{Slope} = m = \frac{C_5 - C_3}{t_5 - t_3}$$

$$= \frac{1.63 - 2.70}{5.0 - 3.0} = -0.535$$

Method 2 This method involves determining the average of two slopes. Using the same points chosen above, two slopes can be calculated, one for points 3 and 4 and the other for points 4 and 5. Adding the two results and dividing them by 2 will provide an approximation of the derivative at point 4. For the points used in this method, the results are:

$$m_1 = \text{slope}_1 = \frac{C_4 - C_3}{t_4 - t_3}$$

$$= \frac{2.01 - 2.70}{4.0 - 3.0} = -0.69$$

$$m_2 = \text{slope}_2 = \frac{C_5 - C_4}{t_5 - t_4}$$

$$= \frac{1.63 - 2.01}{5.0 - 4.0} = -0.38$$

$$m_{\text{avg}} = \text{slope}_{\text{avg}} = \frac{-0.69 + (-0.38)}{2} = -0.535$$

Table 3–1
Concentration-Time Data

Time, s	Concentration of benzene, mg/L
0.0	7.46
1.0	5.41
2.0	3.80
3.0	2.70
4.0	2.01
5.0	1.63
6.0	1.34
7.0	1.17

Method 3 This method consists of selecting any three data points (in this case the same points chosen before) and fitting a curve to it. The equation for the curve is obtained by employing a second-order equation and solving it with the three data points.

$$C = 0.155t^2 - 1.775t + 6.63$$

The derivative of the equation is then calculated and evaluated at any point. Here, point 4 is used:

$$\frac{dC}{dt} = 0.31t - 1.775$$

Evaluated at $t = 4.0$ s:

$$\frac{dC}{dt} = 0.31(4.0) - 1.775 = -0.535$$

Method 4 This method uses *the method of least squares*[6] (see also Chap. 27). In this case, all the data points are used to generate a second-order polynomial equation. This equation is then differentiated and evaluated at the point where the value of the derivative is required. For example, Microsoft Excel can be employed to generate the regression equation (see also Chap. 27). Once all the coefficients are known, the equation has only to be analytically differentiated:

$$C = 0.1626t^2 - 1.9905t + 7.3108$$

$$\frac{dC}{dt} = 0.3252t - 1.9905$$

Evaluated at $t = 4.0$ s:

$$\frac{dC}{dt} = 0.3252(4.0) - 1.9905 = -0.6897$$

Methods 5 and 6 These two methods are somewhat similar. They are based on five data points used to generate coefficients. For this development, represent C and t by f and x (as it appeared in the literature[7]), respectively. Method 5 employs five data points to generate a five coefficient (fourth-order) model using an equation of the form $f = A + Bx + Cx^2 + Dx^3 + Ex^4$. This method is known as *interpolating*. A set of equations is used to evaluate numerical derivatives from the interpolating polynomial. The describing equations are listed below:

$$f'(x_0) = \frac{-25f_0 + 48f_1 - 36f_2 + 16f_3 - 3f_4}{12h} \tag{3–1}$$

$$f'(x_1) = \frac{-3f_0 - 10f_1 + 18f_2 - 6f_3 + 3f_4}{12h} \tag{3–2}$$

$$f'(x_i) = \frac{f_{i-2} - 8f_{i-1} + 8f_{i+1} - f_{i+2}}{12h} \tag{3–3}$$

$$f'(x_{n-1}) = \frac{-f_{n-4} + 6f_{n-3} - 18f_{n-2} + 10f_{n-1} + 3f_n}{12h} \tag{3–4}$$

$$f'(x_n) = \frac{-3f_{n-4} - 16f_{n-3} + 36f_{n-2} - 48f_{n-1} + 25f_n}{12h} \tag{3–5}$$

where $h = x_{i+1} - x_i$
F_i = function evaluated at i

For example, the equation obtained for "the five data set" from 1.0 to 5.0 s, i.e., $t = 1.0$, 2.0, 3.0, 4.0 and 5.0 s, using the equations given above is

$$f(x) = -0.0012x^4 + 0.002x^3 + 0.2616x^2 - 2.34x + 7.467$$

All these equations are evaluated for each value of x and $f(x)$. The value obtained for point 4.0 is −0.5448. Method 6 also employs five data points, but only three coefficients are generated for a second-order polynomial equation of the form $f = A + Bx + Cx^2$. Another set of equations are used to evaluate the derivative at each point using this method. The describing equations are provided below:

$$f'(x_0) = \frac{(-54f_0 + 13f_1 + 40f_2 + 27f_3 - 26f_4)}{70h} \tag{3–6}$$

$$f'(x_1) = \frac{(-34f_0 + 3f_1 + 20f_2 + 17f_3 - 6f_4)}{70h} \tag{3–7}$$

$$f'(x_i) = \frac{(-2f_{i-2} - f_{i-1} + f_{i+1} + 2f_{i+2})}{70h} \tag{3–8}$$

$$f'(x_{n-1}) = \frac{(6f_{n-4} - 17f_{n-3} - 20f_{n-2} - 3f_{n-1} + 34f_n)}{70h} \tag{3–9}$$

$$f'(x_n) = \frac{(26f_{n-4} - 27f_{n-3} - 40f_{n-2} - 13f_{n-1} + 54f_n)}{70h} \tag{3–10}$$

At point 4.0, the solution for the derivative using this method is −0.6897.

Comparing all six values obtained for the derivative at $t = 4.0$ s, one can conclude that the answers are in close proximity to each other. It is important to note that these are approximate values and that they vary depending on the approach and the number of data points used to generate the equations.

Some useful *analytical* derivatives in engineering calculations are provided in Table 1–9.

3–3 Numerical Integration

Numerous chemical engineering and science problems require the solution of integral equations. In a general sense, the problem is to evaluate the function on the right-hand side (RHS) of Eq. (3–11):

$$I = \int_a^b f(x)dx \tag{3–11}$$

where I is the value of the integral. There are two key methods employed in their solution: analytical and numerical. If $f(x)$ is a simple function, it may be integrated analytically. For example, if $f(x) = x^2$, then

$$I = \int_a^b x^2 dx = \frac{1}{3}(b^3 - a^3) \tag{3–12}$$

If, however, if $f(x)$ is a function too complex to integrate analytically {e.g., $\log[\tanh(e^{x^3-2})]$}, one may resort to any of the many numerical methods available. Two simple numerical integration methods that are commonly employed in chemical engineering practice are the trapezoidal rule and Simpson's rule. These are described below.

Trapezoidal Rule

In order to use the trapezoidal rule to evaluate the integral with I given by Eq. (3–11) as

$$I = \int_a^b f(x)dx \tag{3–11}$$

one may use the equation

$$I = \frac{h}{2}[y_0 + 2y_1 + 2y_2 + \cdots + 2y_{n-1} + y_n] \tag{3–13}$$

where h *is* the incremental change in x; i.e., Δx, and y_i are the values of $f(x)$ at x_i, i.e., $f(x_i)$. Thus,

$$y_0 = f(x_0) = f(x = a)$$
$$y_n = f(x_n) = f(x = b)$$
$$h = \frac{b/a}{n}$$

This method is known as the *trapezoidal rule* because it approximates the area under the function $f(x)$—which is generally curved—with a two-point trapezoidal rule calculation. The error associated with this rule is illustrated in Fig. 3–1.

There is an alternative available for improving the accuracy of this calculation—the interval (a — b) can be subdivided into smaller intervals. The trapezoidal rule can then be applied repeatedly in turn over each subdivision.

Simpson's Rule

A higher-degree interpolating polynomial scheme can be employed for more accurate results. One of the more popular integration approaches is Simpson's rule. For Simpson's 3-point (or one-third) rule, one may use the equation

$$I = \frac{h}{3}[y_a + 4y_{(b+a)/2} + y_b] \tag{3–14}$$

For the general form of Simpson's rule (where n is an even integer), the equation is

$$I = \frac{h}{3}(y_0 + 4y_1 + 4y_2 + \cdots + 4y_{n-1} + y_n) \tag{3–15}$$

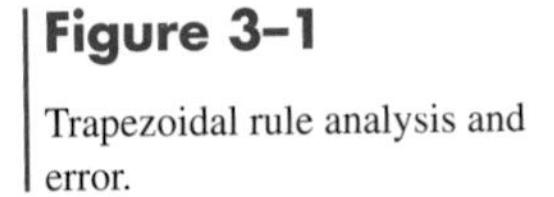

Figure 3–1

Trapezoidal rule analysis and error.

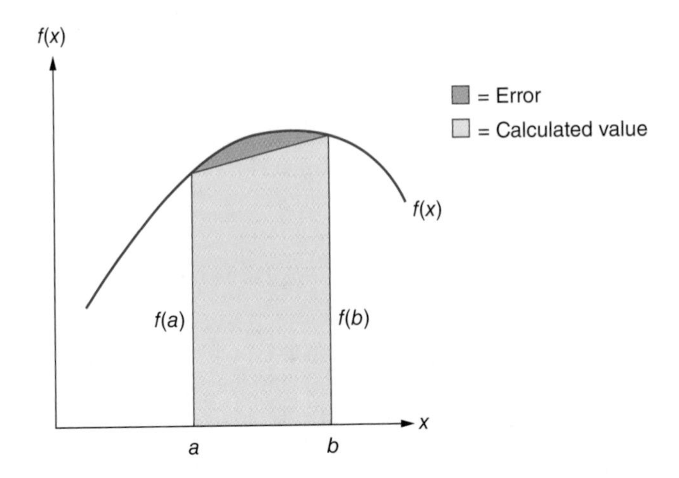

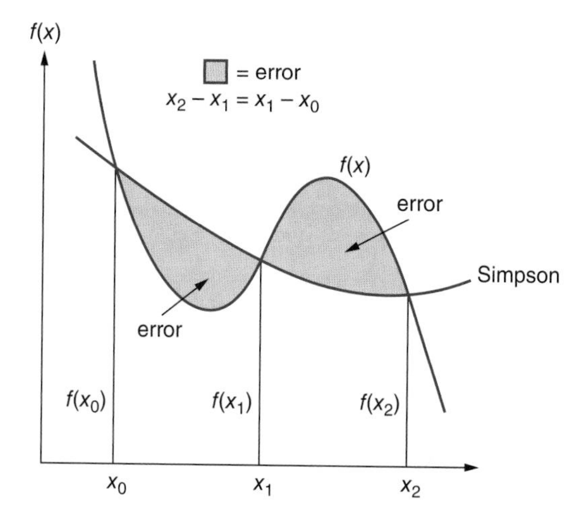

Figure 3–2

Simpson's rule analysis and error.

This method also generates an error, although it is usually smaller than that associated with the trapezoidal rule. A diagrammatic representation of the error for a 3-point calculation is provided in Fig. 3–2.

The reader should note that the trapezoid rule is often the quickest but least accurate way to perform a numerical integration by hand. However, if the step size is decreased, the answer should converge—subject to roundoff errors—to the analytical solution. The results of each numerical integration must be added together to obtain the final answer for smaller step sizes.

Some useful *analytical* integrals in engineering calculations are provided in Table 1–8.[8]

3–4 Simultaneous Linear Algebraic Equations

The chemical engineer often encounters problems that contain not only more than two or three simultaneous algebraic equations but also those that can be nonlinear. There is, therefore, an obvious need for systematic methods of solving simultaneous linear and simultaneous nonlinear equations.[9] This section addresses the linear sets of equations; information on nonlinear sets is available in the literature.[10]

Consider the following set of n equations:

$$\begin{aligned} a_{11}x_1 + a_{12}x_2 + \cdots + a_{1n}x_n &= y_1 \\ a_{21}x_1 + a_{22}x_2 + \cdots + a_{2n}x_n &= y_2 \\ &\vdots \\ a_{n1}x_1 + a_{n2}x_2 + \cdots + a_{nn}x_n &= y_n \end{aligned} \tag{3–16}$$

where a is the coefficient of the variable x and y is a constant. This set is considered to be linear as long as none of the x terms are nonlinear, for example, x_2^2 or $\ln x_1$. Thus, a linear system requires that all terms in x be linear. The system of linear algebraic equations may be set in matrix form:

$$\begin{bmatrix} a_{11} & a_{12} & \cdots & a_{1n} \\ a_{12} & a_{22} & \cdots & a_{2n} \\ \cdots & \cdots & \cdots & \cdots \\ a_{1n} & a_{n2} & \cdots & a_{nn} \end{bmatrix} \begin{bmatrix} x_1 \\ x_2 \\ \cdots \\ x_n \end{bmatrix} = \begin{bmatrix} y_1 \\ y_2 \\ \cdots \\ y_n \end{bmatrix} \tag{3–17}$$

However, it is often more convenient to represent Eq. (3–17) in the *augmented matrix* form provided in the following equation:

$$\begin{bmatrix} a_{11} & a_{12} & \cdots & a_{1n} & y_1 \\ a_{12} & a_{22} & \cdots & a_{2n} & y_2 \\ \cdots & \cdots & \cdots & \cdots & \cdots \\ a_{1n} & a_{n2} & \cdots & a_{nn} & y_n \end{bmatrix} \tag{3–18}$$

Methods of solution available for solving these linear sets of equations include:

1. Gauss–Jordan reduction
2. Gauss elimination
3. Gauss–Seidel approach
4. Cramer's rule
5. Cholesky's methods

Only methods 1 to 3 are discussed in this section.

Gauss–Jordan Reduction

Carnahan and Wilkes[9] provide an example that solved the following two simultaneous equations using the Gauss–Jordan reduction method:

$$3x_1 + 4x_2 = 29 \tag{3–19}$$

$$6x_1 + 10x_2 = 68 \tag{3–20}$$

The four-step procedure is provided below:

1. Divide Eq. (3–19) through by the coefficient of x_1

$$x_1 + \frac{4}{3}x_2 = \frac{29}{3} \tag{3–21}$$

2. Subtract a suitable multiple (6, in this case) of Eq. (3–21) from Eq. (3–20), so that x_1 is *eliminated*. Equation (3–21) remains intact so that what remains is

$$\frac{34}{3} - \frac{5}{3}x_2 + \frac{4}{3}x_2 = \frac{29}{3} \tag{3–22}$$

$$2x_2 = 10 \tag{3–23}$$

3. Divide Eq. (3–23) by the coefficient of x, i.e., that is, solve Eq. (3–23)

$$x_2 = 5 \tag{3–24}$$

4. Subtract a suitable factor of Eq. (3–24) from Eq. (3–22) so that x_2 is eliminated. When $(4/3)x_2 = 20/3$ is subtracted from Eq. (3–22), one obtains

$$x_1 = 9 \tag{3–25}$$

Gauss Elimination

Gauss elimination is another method used to solve linear sets of equations. This method utilizes the augmented matrix as described in Eq. (3–18). The goal with Gauss elimination is to rearrange the augmented matrix into a *triangle form* where all the elements below the diagonal are zero. This is accomplished in much the same way as in Gauss–Jordan reduction. The procedure employed follows. Start with the first equation in the set. This is known as the *pivot* equation and will not change throughout the procedure. Once the matrix is in triangle form, back substitution can be used to solve for the variables.

Gauss elimination is useful for systems that contain fewer than 30 equations. Systems larger than 30 equations become subject to *roundoff error* where numbers are truncated by computers performing the calculations.

Gauss–Seidel Approach

Another approach to solving an equation or series/sets of equations is to make an informed or educated guess. If the first assumed value(s) does not work, the value is updated. By carefully noting the influence of these guesses on each variable, one can approach these answers or correct set of values for a system of equations. The reader should note that when this type of iterative procedure is employed, a poor initial guess does not prevent the correct solution from ultimately being obtained.

Ketter and Prawler[3] provide several excellent illustrative examples.

3–5 Nonlinear Algebraic Equations

The subject of the solution to a nonlinear algebraic equation is considered in this section. Although several algorithms are available in the literature, the presentation will focus on the *Newton–Raphson* (NR) method of evaluating the root(s) of a nonlinear algebraic equation.

The solution to the equation

$$f(x) = 0 \tag{3–26}$$

is obtained by guessing a value for x, e.g., (x_{old}) that will satisfy this equation. This value is continuously updated (x_{new}) using the equation (the prime represents a derivative)

$$x_{new} = x_{old} - \frac{f(x_{old})}{f'(x_{new})} \tag{3–27}$$

until either little or no change in ($x_{new} - x_{old}$) is obtained. One can also express this operation graphically (see Fig. 3–3). Noting that

$$f'(x_{old}) = \frac{df(x)}{dx} \approx \frac{\Delta f(x)}{\Delta x} = \frac{f(x_{old}) - 0}{x_{old} - x_{new}} \tag{3–28}$$

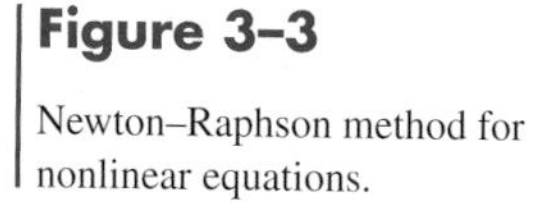

Figure 3–3

Newton–Raphson method for nonlinear equations.

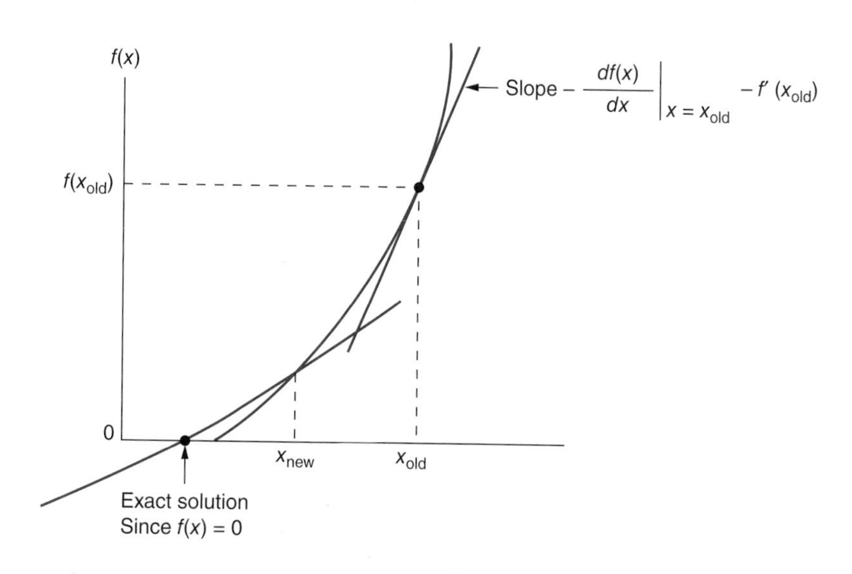

one may rearrange Eq. (3–28) to yield Eq. (3–29) below. The x_{new} then becomes the x_{old} in the next calculation.

This method is also referred to as *Newton's method of tangents* (NMT) and is a widely used method for improving a first approximation to a root to the aforementioned equation of the form $f(x) = 0$. This development can be rewritten in subscripted form to (perhaps) better accommodate a computer calculation. Thus

$$f'(x_n) = \frac{f(x_n)}{x_n - x_{n+1}} \tag{3–29}$$

from which

$$x_{n+1} = x_n - \frac{f(x_n)}{f'(x_n)} \tag{3–30}$$

The term x_{n+l} is again the improved estimate of x_n—the solution to the equation $f(x) = 0$. The value of the function and the value of the derivative of the function are determined at $x = x_n$ using this procedure, and the new approximation to the root x_{n+1} is obtained. The same procedure is repeated, with the new value, to obtain a still better approximation of the root. This continues until successive values for the approximate root differs by less than a prescribed small value ε which controls the *allowable error* (or *tolerance*) in the root. Relative to the previous estimate, ε is given by

$$\varepsilon = \frac{x_{n+1} - x_n}{x_n} \tag{3–31}$$

Despite its popularity, the method suffers for two reasons. First, an analytical expression for the derivative, specifically, $f'(x_n)$, is required. In addition to the problem of having to compute an analytical derivative value at each iteration, one would expect Newton's method to converge fairly rapidly to a root in the majority of cases. However, as is common with some numerical methods, it may fail occasionally in certain instances. A possible initial oscillation followed by a displacement away from a root can occur. Note, however, that the method would have converged if the initial guess had been somewhat closer to the exact root. Thus, the first guess may be critical to the success of the calculation.

3–6 Ordinary Differential Equations

The *Runge–Kutta* (RK) method is one of the most widely used techniques in chemical engineering practice for solving first-order differential equations. For the equation

$$\frac{dy}{dx} = f(x, y) \tag{3–32}$$

the solution takes the form

$$y_{n+1} = y_n + \frac{h}{6}(D_1 + 2D_2 + 2D_3 + D_4) \tag{3–33}$$

where

$$\begin{aligned} D_1 &= hf(x, y) \\ D_2 &= hf\left(x_n + \frac{h}{2}, y_n + \frac{D_1}{2}\right) \\ D_3 &= hf\left(x_n + \frac{h}{2}, y_n + \frac{D_2}{2}\right) \\ D_4 &= hf(x_n + h,\ y_n + D_3) \end{aligned} \tag{3–34}$$

The term h represents the increment in x. The term y_n is the solution to the equation at x_n, and y_{n+1} is the solution to the equation at x_{n+1} where $x_{n+1} = x_n + h$. Thus, the RK method provides a straightforward means for developing expressions for Δy, namely, $y_{n+1} - y_n$, in terms of the function $f(x, y)$ at various "locations" along the interval in question.

For a simple equation of the form

$$\frac{dC}{dt} = a + bC \tag{3–35}$$

where at $t = 0$, $C = C_0$, the RK algorithm given above becomes (for $t = h$)

$$C_1 = C_0 + \frac{h}{6}(D_1 + 2D_2 + 2D_3 + D_4) \tag{3–36}$$

where

$$\begin{aligned} D_1 &= hf(x, y) = h(a + bC_0) \\ D_2 &= hf\left(x_n + \frac{h}{2}, y_n + \frac{D_1}{2}\right) = h\left[a + b\left(C_0 + \frac{D_1}{2}\right)\right] \\ D_3 &= hf\left(x_n + \frac{h}{2}, y_n + \frac{D_2}{2}\right) = h\left[a + b\left(\frac{C_0 + D_2}{2}\right)\right] \\ D_4 &= hf(x_n + h,\ y_n + D_3) = h[a + b(C_0 + D_3)] \end{aligned} \tag{3–37}$$

The same procedure is repeated to obtain values for C_2 at $t = 2h$, C_3 at $t = 3h$, and so on.

The RK method can also be used if the function in question also contains the independent variable. Consider the following equation:

$$\frac{dC}{dt} = f_1(C, t) \tag{3–38}$$

For this situation, one obtains

$$C_1 = C_0 + \frac{h}{6}(D_1 + 2D_2 + 2D_3 + D_4) \tag{3–39}$$

with

$$\begin{aligned} D_1 &= hf(C, t) \\ D_2 &= hf\left(C_0 + \frac{D_1}{2}, t_0 + \frac{h}{2}\right) \\ D_3 &= hf\left(C_0 + \frac{D_2}{2}, t_0 + \frac{h}{2}\right) \\ D_4 &= hf(C_0 + D_3,\ t_0 + h) \end{aligned} \tag{3–40}$$

For example, if

$$\frac{dC}{dt} = 5C - e^{-ct} \tag{3–41}$$

then

$$D_2 = h\left\{5\left(C_0 + \frac{D_1}{2}\right) - e^{[-(C_0 + \frac{D_1}{2})(t_0 + \frac{h}{2})]}\right\} \tag{3–42}$$

Situations may arise when there is a need to simultaneously solve *more* than one ordinary differential equation (ODE). In a more general case, one could have n dependent variables $y_1, y_2, \ldots, y_n$ with each related to a single independent variable x by the following system of n simultaneous first-order ODEs:

$$\begin{aligned} \frac{dy_1}{dx} &= f_1(x, y_1, y_2, \ldots, y_n) \\ \frac{dy_2}{dx} &= f_2(x, y_1, y_2, \ldots, y_n) \\ &\vdots \\ \frac{dy_n}{dx} &= f_n(x, y_1, y_2, \ldots, y_n) \end{aligned} \tag{3–43}$$

Note that the equations in Eq. (3–43) are interrelated, i.e., they are dependent on each other. This is illustrated in the following two equations:

$$\frac{dC}{dt} = -Ae^{-E/RT}C = f(C, t) \tag{3–44}$$

$$\frac{dT}{dt} = -kC\frac{\Delta H}{\rho C_P} = g(C, t) \tag{3–45}$$

or in a more general sense

$$\frac{dy}{dx} = f(x, y, z); \qquad \text{(e.g., } xyz\text{)} \tag{3–46}$$

$$\frac{dz}{dt} = g(x, y, z); \qquad \text{(e.g., } x^2y^2e^{-z}\text{)} \tag{3–47}$$

The RK algorithm for Eqs. (3–46) and (3–47) is

$$y_1 = y_0 + \frac{1}{6}(RY_1 + 2RY_2 + 2RY_3 + RY_4) \tag{3–48}$$

$$z_1 = z_0 + \frac{1}{6}(RZ_1 + 2RZ_2 + 2RZ_3 + RZ_4) \tag{3–49}$$

where $y_1 - y_0 = \Delta y$, $z_1 - z_0 = \Delta z$, $h = \Delta x$ and

$$\begin{aligned} RY_1 &= h \times f(x_0, y_0, z_0) \\ RZ_1 &= h \times g(x_0, y_0, z_0) \\ RY_2 &= h \times f\left(\frac{x_0 + h}{2}, y_0 + \frac{RY_1}{2}, z_0 + \frac{RZ_1}{2}\right) \\ RZ_2 &= h \times g\left(\frac{x_0 + h}{2}, y_0 + \frac{RY_1}{2}, z_0 + \frac{RZ_1}{2}\right) \\ RY_3 &= h \times f\left(\frac{x_0 + h}{2}, y_0 + \frac{RY_2}{2}, z_0 + \frac{RZ_2}{2}\right) \\ RZ_3 &= h \times g\left(\frac{x_0 + h}{2}, y_0 + \frac{RY_2}{2}, z_0 + \frac{RZ_2}{2}\right) \\ RY_4 &= h \times f(x_0 + h, y_0 + RY_3, z_0 + RZ_3) \\ RZ_4 &= h \times g(x_0 + h, y_0 + RY_3, z_0 + RZ_3) \end{aligned} \tag{3–50}$$

Although the RK approach (and other similar methods) has traditionally been employed to solve first-order ODEs, it can also treat higher ODEs. The procedure requires reducing an nth-order ODE to n first-order ODE. For example, if the equation is of the form[11]

$$\frac{d^2y}{dx^2} = f(y,x) \tag{3–51}$$

set

$$z = \frac{dy}{dx} \tag{3–52}$$

so that

$$\frac{dz}{dx} = \frac{d^2y}{dx^2} \tag{3–53}$$

The second-order equation in Eq. (3–53) has now been reduced to the two first-order ODEs provided in Eq. (3–54):

$$\begin{aligned} \frac{d^2y}{dx^2} &= \frac{dz}{dx} = f(y,x) \\ \frac{dy}{dx} &= z \end{aligned} \tag{3–54}$$

The procedure expressed in Eqs. (3–48) and (3–49) can be applied to generate a solution to Eq. (3–51). Note, however, that the first derivative (i.e., dy/dx or its estimate) is required at the start of the calculation. Extending the procedure to higher-order equations is left as an exercise for the reader.

The selection of increment size remains a variable to the practicing chemical engineer. Few numerical analysis methods provided in the literature are concerned with error analysis. In general, *roundoff and numerical errors* appear as demonstrated in Fig. 3–4. In the limit, when the increment $\rightarrow 0$, one approaches an analytical solution. However, the number of calculations correspondingly increases the error ε, which increases exponentially as the increment $\rightarrow 0$. Note that selecting the increment size that will minimize the error is rarely a problem in practice; in addition, computing it is also rarely a concern.

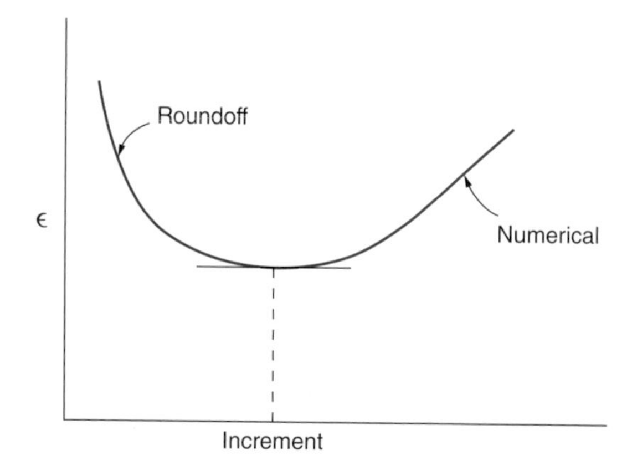

Figure 3–4

Error analysis in numerical calculations.

3–7 Partial Differential Equations

Many practical problems in chemical engineering practice involve at least two independent variables; thus, the dependent variable is defined in terms of (or is a function of) more than one independent variable. The derivatives describing these independent variables are defined as *partial* derivatives. Differential equations containing partial derivatives are referred to as *partial differential equations* (PDEs).

It has been said that "the solution of a partial differential equation is essentially a guessing game." In other words, one cannot expect to be provided with a formal method that will yield exact solutions for all partial differential equations.[10] Fortunately, numerical methods for solving these equations were developed during the mid to late twentieth century, and are routinely used today.

The three main PDEs encountered in chemical engineering practice are briefly introduced below, employing T (e.g., the temperature as the dependent variable) with t (time) and x, y, z (position) as the independent variables (note that any dependent variable, such as, pressure or concentration, could have been selected).

Parabolic equation:

$$\frac{\partial T}{\partial t} = \alpha \frac{\partial^2 T}{\partial z^2} \qquad (3\text{–}55)$$

Elliptical equation:

$$\frac{\partial^2 T}{\partial x^2} + \frac{\partial^2 T}{\partial y^2} = 0 \qquad (3\text{–}56)$$

Hyperbolic equation:

$$\frac{\partial^2 T}{\partial t^2} = \alpha \frac{\partial^2 T}{\partial x^2} \qquad (3\text{–}57)$$

The preferred numerical method of solution involves finite differencing. Only the parabolic and elliptical equations are considered below.

Parabolic PDE

Examples of parabolic PDEs include

$$\frac{\partial T}{\partial t} = \alpha \frac{\partial^2 T}{\partial x^2} \qquad (3\text{–}58)$$

and (the two-dimensional)

$$\frac{\partial T}{\partial t} = \alpha \left[\frac{\partial^2 T}{\partial x^2} + \frac{\partial^2 T}{\partial y^2} \right] \qquad (3\text{–}59)$$

Ketter and Prawler,[3] as well as many others, have reviewed the finite-difference approach to solving Eq. (3–58). This is detailed below.

Consider the (t, x) grid provided in Fig. 3–5. The partial derivatives may be replaced by

$$\frac{\partial T}{\partial t} \cong \frac{\Delta T}{\Delta t} = \frac{-T_4 + T_2}{2(\Delta t)} = \frac{-T_4 + T_2}{2k}; \qquad \Delta t = k \qquad (3\text{–}60)$$

and

$$\frac{\partial^2 T}{\partial x^2} \cong \frac{\Delta}{\Delta x}\left(\frac{\Delta T}{\Delta x}\right) = \frac{T_3 - 2T_0 - T_1}{h^2}; \qquad \Delta x = h \qquad (3\text{–}61)$$

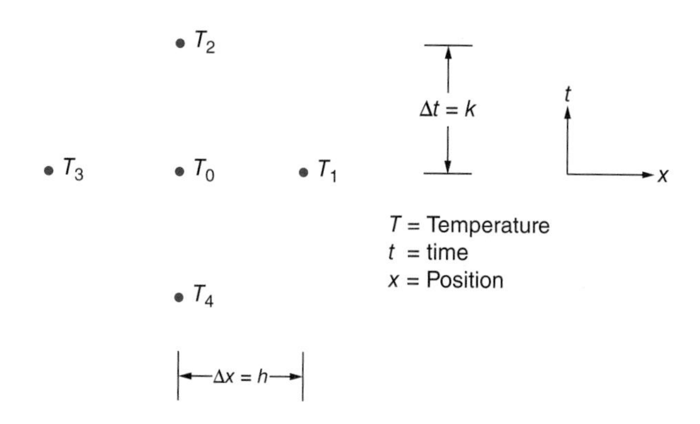

Figure 3–5

Partial differentiation equation parabolic grid.

Substituting Eqs. (3–60) and (3–61) into Eq. (3–58) leads to

$$\frac{-T_4 + T_2}{2(\Delta t)} = \frac{T_3 - 2T_0 - T_1}{h^2}; \qquad \Delta x = h \tag{3–62}$$

Solving for T_2, one obtains

$$T_2 = T_4 + 2r(T_3 - 2T_0 - T_1) \tag{3–63}$$

where $r = k/h^2$

Thus, one may calculate T_2 if T_0, T_1, T_3, and T_4 are known. Unfortunately, *stability* and *error* problems can arise employing the above approach. These can be removed by replacing the *central* difference term in Eq. (3–60) by a *forward* difference term:

$$\frac{\partial T}{\partial t} \cong \frac{\Delta T}{\Delta t} = \frac{-T_0 + T_2}{(\Delta t)} = \frac{-T_0 + T_2}{k} \tag{3–64}$$

With this substitution, Eq. (3–63) becomes

$$T_2 = T_0 + r(T_3 - 2T_0 + T_1) \tag{3–65}$$

It can be shown that the problem associated with the central difference derivative is removed if $r \leq 0.5$.

Elliptical PDE

For this equation, examine the grid in Fig. 3–6. Using finite differences to replace the derivatives Eq. (3–56) ultimately leads to

$$T_0 = \frac{1}{4}(T_1 + T_2 + T_3 + T_4); \qquad \Delta x = \Delta y \tag{3–66}$$

In effect, each T value calculated reduces to the average of its four nearest neighbors in the square grid. This difference equation may then be written at each interior grid point, resulting in a linear system of N equations, where N is the number of grid points. The system can then be solved by one of several methods provided in the literature.[12,13]

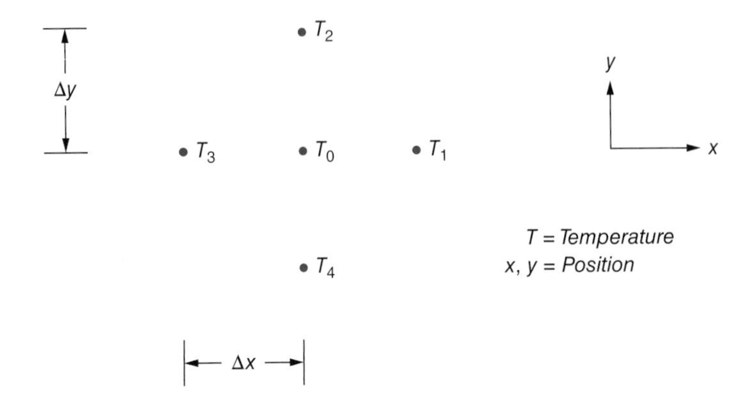

Figure 3–6

Partial differential equation elliptic grid.

Another solution method involves applying the *Monte Carlo* approach, requiring the use of random numbers.[12,13] Consider the squares shown in Fig. 3–7. If the describing equation for the variation of T within the grid structure is

$$\frac{\partial^2 T}{\partial x^2} + \frac{\partial^2 T}{\partial y^2} = 0 \tag{3–67}$$

with specified boundary conditions (BCs) for $T(x, y)$ of $T(0, y)$, $T(a, y)$, $T(x, 0)$, and $T(x, a)$, one may employ the following approach:

1. Proceed to calculate T at point 1 (i.e., T_1).
2. Generate a random number between 00 and 99.
3. If the number is between 00 and 24, move to the left. For 25 to 49, 50 to 74, 75 to 99, move upward, to the right, and downward, respectively.
4. If the move in step 3 results in a new position that is at the outer surface (boundary), terminate the first calculation for point 1 and record the T value of the boundary at that new position. However, if the move results in a new position that is not at a

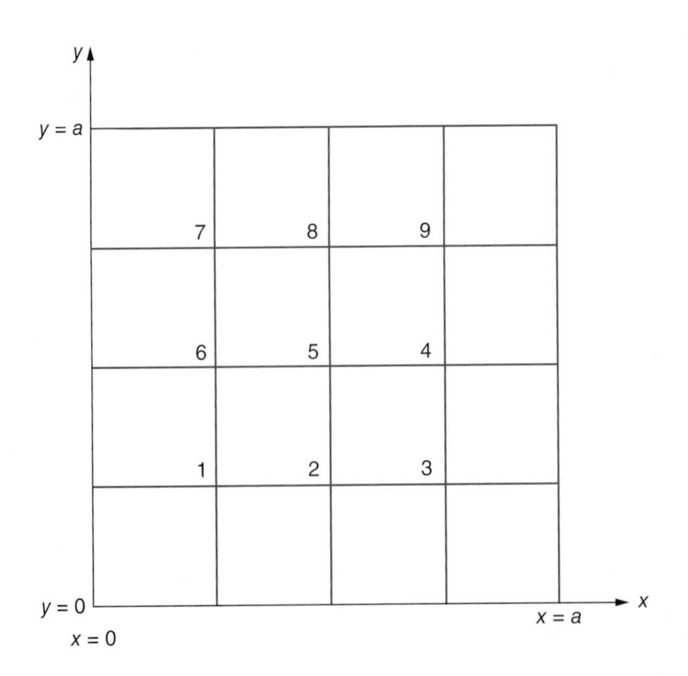

Figure 3–7

Monte Carlo grid approach.

boundary, and is still at one of the other eight interval grid points, repeat steps 2 and 3. This process is continued until an outer surface or boundary is reached.

5. Repeat steps 2 to 4 numerous times, for example, 1000 times.
6. After completing step 5, sum all the T values obtained and divide this value by the number of times steps (2 to 4) have been repeated. The resulting value provides a reasonable estimate of T_1.
7. Return to step 1 and repeat the calculation for the remaining eight grid points.

This method of solution is not limited to square systems. In addition, the author[14] has applied this method of solution to numerous real-world applications.

3–8 Optimization

Optimization has come to mean different things to different people. However, one might offer the following generic definition for many chemical engineers: "Optimization is concerned with determining the 'best' solution to a given problem." Alternatively, a dictionary would offer something to the effect: "to make the most of ... develop or realize to the utmost extent ... often the most efficient or optimum use of." This process or operation in chemical engineering practice is required in the solution of many problems and involves the maximization or minimization of a given function.

A significant number of optimization problems face the practicing chemical engineer. The optimal design of industrial processes as well as process equipment has long been of concern to the practicing chemical engineer, and indeed, for some, might be regarded as a definition of the function and goal of applied engineering. The practical attainment of an optimum design is generally a result of factors that include mathematical analysis, empirical information, and both the subjective and objective experience of the chemical engineer.

In a general sense, optimization applications can be divided into four categories:

1. The number of independent variables involved
2. Whether the optimization is "constrained"
3. Time-independent systems
4. Time-dependent systems

In addition, if no unknown factors are present, the system is defined as *deterministic*, while a system containing experimental errors and/or other random factors is defined as *stochastic*.

Formal optimization techniques have as their goal for the development of procedures for the attainment of an optimum in a system which can be characterized mathematically. The mathematical characterization may be (1) partial or complete, (2) approximate or exact, and/or (3) empirical or theoretical. The resulting optimum may be a final implementable design or a guide to practical design and a criterion for judging practical designs. In either case, the optimization techniques should serve as an important part of the total effort in the design of the units, structure, and control of not only equipment but also industrial system process.

Optimization is qualitatively viewed by many as a tool in decision making. It often aids in the selection of values that allow the chemical engineer to better solve a problem. In its most elementary and basic form, one may say (as noted above) that optimization is concerned with the determination of the "best" solution to a given problem. This process is required in the solution of many general problems in chemical engineering and applied science such as in the maximization (or minimization) of a given function(s), in the selection of a control variable to facilitate the realization of a desired condition, in the scheduling of a series of operations or events to control completion dates of a given project, and in the development of optimal layouts of organizational units within a given design space.[4]

The optimization problem has been described succinctly by Aris[15] as "getting the best you can out of a given situation." Problems amenable to solution by mathematical optimization techniques generally have one or more independent variables whose values must be chosen to yield a viable solution and measure of "goodness" available to distinguish between the many viable solutions generated by different choices of these variables. Mathematical optimization techniques are also used for guiding the problem solver to that choice of variables that maximizes the goodness measure (e.g., profit) or that minimizes some "badness" measure (e.g., cost).

One of the most important areas for the application of mathematical optimization techniques is in engineering design. Applications include:

1. Generation of best functional representations (e.g., curve fitting)
2. Design of optimal control systems
3. Determining the optimal height (or length) of a mass transfer unit
4. Determining the optimal diameter of a unit
5. Finding the best equipment materials of construction
6. Generating operating schedules
7. Selecting operating conditions

Once a particular piece of equipment or process scheme has been selected, it is common practice to optimize the process from a capital cost and OM (operation and maintenance) standpoint (see also Chap. 20). There are many optimization procedures available, most of them too detailed for meaningful application in this text. These sophisticated optimization techniques, some of which are also routinely used in the design of conventional chemical and petrochemical plants, invariably involve computer calculations. However, use of these techniques is not always warranted.

One simple optimization procedure that is recommended by the author is the *perturbation study*. This involves a systematic change (or *perturbation*) of variables, one by one, in an attempt to locate the optimum design in terms of both cost and operation. To be practical, this often means that the chemical engineer must limit the number of variables by assigning constant values to those (process) variables that are known beforehand to play an insignificant role. Reasonable guesses or simple or shortcut mathematical methods can further simplify the procedure. Much information can be gathered from this type of study since it usually identifies those variables that significantly impact the solution, such as on the overall performance of equipment or process, and helps identify the major contributors effecting the optimization calculations.

Additional details on optimization are presented in Chaps. 15, 24, and 28. More detailed and sophisticated optimization procedures are available, most of which are described in the literature.[15,16] Illustrative examples can also be found in the literature.[13,17]

References

1. M. Moyle, *Introduction to Computers for Engineers*, Wiley, Sons, Hoboken, N.J., 1967.
2. B. Carnahan and J. Wilkes, *Digital Computing and Numerical Methods*, Wiley, Hoboken, N.J., 1973.
3. R. Ketter and S. Prawler, *Modern Methods of Engineering Computations*, McGraw-Hill, N.Y., 1969.
4. J. Reynolds, J. Jeris, and L. Theodore, *Handbook of Chemical and Environmental Engineering Calculations*, Wiley, Hoboken, N.J., 2004.

5. L. Perez, homework assignment submitted to L. Theodore, Manhattan College, Bronx, N.Y., 2003.
6. S. Shaefer and L. Theodore, *Probability and Statistics Applications for Environmental Science*, CRC Press/Taylor & Francis Group, Boca Raton, Fla., 2007.
7. F. Lavery, "The Perils of Differentiating Engineering Data Numerically," *Chem. Eng.*, N.Y., 1979.
8. L. Theodore; class notes, Manhattan College, Bronx, N.Y., 1991.
9. B. Carnahan and J. Wilkes, *Digital Computing and Numerical Methods*, Wiley, Hoboken, N.J., 1973.
10. J. Reynolds, class notes (with permission), Manhattan College, Bronx, N.Y., 2001.
11. L. Theodore, class notes, Manhattan College, Bronx, N.Y., 1971.
12. L. Theodore, personal notes, East Williston, N.Y., 1965.
13. L. Theodore, *Heat Transfer for the Practicing Engineer*, Wiley, Hoboken, N.J., 2011.
14. L. Theodore, personal notes, East Williston, N.Y., 1985.
15. R. Aris, *Discrete Dynamic Programming*, Blaisdell, N.Y., 1964.
16. J. Santoleri, J. Reynolds, and L. Theodore, *Introduction to Hazardous Waste Incineration*, 2nd ed., Wiley, Hoboken, N.J., 2000.
17. L. Theodore, personal notes, East Williston, N.Y., 2011.

4 Oral and Written Communication

Chapter Outline

4–1 Introduction

Communication has been defined by some as an act of expressing ideas, especially in speech and writing, by others as an act of transmitting ideas or information; and still others as an exchange of information or messages by speech, writing, and so on. The word *oral* implies something uttered by mouth, spoken, or involving the use of speech, while the word *written* is defined as an expression recorded in a readable format, such as books or other literary material, or an idea that is "put into writing or written form." What does the above mean? Communication is important, particularly in today's high-technology and Internet environment.

There is an old saying that a graduate from an Ivy League university can't count and an MIT (Massachusetts Institute of Technology) graduate can't read or write. But today, successful chemical engineers must be able to express themselves in both oral and written communication. Technical ability is of little use if the chemical engineer cannot transmit ideas to others. In addition, the major contact one has with other administrators of an organization who can determine raise(s) and promotion(s) is through written communication. Oral and written communications are therefore important not only in chemical engineering but also in all fields of engineering, science, and management.

During the author's tenure as director of the graduate program at a local metropolitan (New York City) institution, the program's consulter group from industry regularly bemoaned the communication skills of many of their chemical engineers.[1] This problem has become exacerbated in recent years since most practicing chemical engineers spend nearly three-fourths of their time in communication-related activities. Also keep in mind that most new chemical engineers are required to work with others in a group. An inability to communicate effectively with the members or partners of the group could be disastrous. The technology explosion has only added to this dilemma. It is important for the chemical engineer to understand that the ability to communicate successfully orally and in writing can help ensure a happy, productive, and lucrative career.

This chapter addresses the following topics in an attempt to emphasize this point: technical jargon, sources of information, oral presentation, technical writing, and that first job.

4–2 Technical Jargon

Many readers classify themselves as "professionals." A good number of them would also classify themselves as "technical." Here is what the author once wrote about technical professionals communicating with the average citizen.[2] "Be honest, fresh, and open . . . technical language and jargon are useful as professional shorthand, but they are usually barriers to successful communication with the public." This is good advice when dealing with the public but not with another professional. All bets are then off and many revert to a language that resides solely in the domain of the professional. The author[2] became aware of this unique phenomenon following the submission of a report to the U.S. Department of Justice (DOJ) and the preparation of a proposal to the National Science Foundation (NSF). Although words can mean different things to different people, some key sentences and phrases employed by the gifted few are listed below; each is followed by a layperson's pragmatic definition and/or explanation. (As the reader will soon discover, this is a tongue-in-cheek presentation):

- Congratulations continue to pour in—the only thing being poured is the Dewars to help me forget this mess.
- The data are as consistent as one could hope for—the scatter was mind-boggling.
- It is with pride that we inform you that the report is already out and, unfortunately, we are culpable.

- The evidence is irrefutable—no, it isn't.
- I can say with certainty—my brother-in-law Vito (on my wife's side) mentioned it to me last night in a discotheque.
- We can say with certainty—my cohorts were there also.
- A foundation director concluded—she read it in the New York tabloids.
- An informed local official concluded—he's got an IQ to match his age.
- An informed state official concluded—her IQ is in the 1 to 10 range.
- An informed federal official concluded—he doesn't have an IQ.
- An informed career federal official concluded—his IQ approaches negative infinity during any thought process.
- There may be negative international repercussions—only if our findings get out.
- Definitely worthy of NSF support—it is a totally useless project that can in no way serve humankind.
- There is universal agreement—that's what my favorite brother-in-law Biff (on my sister's side) told me.
- I can assure you—if I can, I have a bridge to sell you.
- The results are correct to at least one order of magnitude—they're totally screwed up.
- The results clearly suggest numerous applications—we can't think of one.
- The final results provide conclusive evidence—we've decided not to publish.
- A reliable source has just informed me—someone named Pepe mentioned it to me in a bar last night; incidentally we were both inebriated.
- There is wide-range agreement—no one believes a word of it, including my colleague Lefty.
- The results appear to indicate a trend—unfortunately, we can't figure out what the trend is.
- The results appear to indicate a definite trend—we are totally confused at this point in time.
- We intend to pursue this aspect of the investigation vigorously—only if someone puts up the moola (i.e., finances it).
- Our new process will require minimum maintenance—the system is essentially impossible to maintain.
- On the basis of my limited experience—I'm looking for clues.
- On the basis of my experience—I'm desperately looking for clues.
- On the basis of my extensive experience—I'm clueless.
- Getting started on the study was no problem—we simply assumed the subject in question to be a sphere.
- The data seem to suggest—it's anybody's guess.
- If I were you—thank God I'm not.
- Thanks are due—I had to mention his name since dinner at his restaurant last week was on the house.
- The final results of the project will find wide application—not if anybody has any sense.
- Thanks are due the NSF—I have a policy of thanking anyone who gives me money on a no-strings-attached basis.

Unfortunately, this type of jargon is widespread in some technical literature; it is also in common use in oral communication. The bottom line is that technical jargon has no place in professional technical activities.

4–3 Sources of Information

The field of chemical engineering encompasses many diverse fields of activity. It is therefore important that means be available to place widely scattered information in the hands of the chemical engineer. One need only examine the open-ended problems in Chap. 30 to appreciate the need for the ready availability of a vast variety of information. Such problems are almost always approached by checking on and reviewing all sources of information. Most of this information can be classified in the following source categories:

Traditional

Chemical engineering

New

Personal experience

Each topic is discussed below.

Traditional Sources

The library is the major repository of traditional information. To efficiently use the library, one must become familiar with its classification system. In general, chemical engineers using the technical section of a library seek information on a certain subject and not material by a particular author or in a specific book. For this reason, classification systems are based on subject matter. The question confronting the user then is how to find and/or obtain information on a particular subject. Two basic subject-matter-related classification systems are employed in U.S. libraries: the Dewey decimal system and the Library of Congress system.

These systems have changed somewhat with the advent of computers and the Internet. A library (particularly a technical one) usually contains all or some of the following:

General books (fiction, etc.)

Reference books

Handbooks

Journals

Transactions

Encyclopedias

Periodicals

Dictionaries

Trade literature

Catalogs

Chemical Engineering Sources

Chemical engineering technical literature includes textbooks, handbooks, periodicals, magazines, journals, dictionaries, encyclopedias, and industrial catalogs. There are a host of textbooks, including not only the classic works of Walker et al.,[3] Kern,[4] Treybal's first edition on mass transfer operations,[5] Shreve,[6] and Happel[7] but also (more recently) the work of Bird et al.[8] The three major handbooks include those by Perry and Green,[9] Kirk and Othmer,[10] and Albright.[11] Finally, in terms of journals and magazines, one notes *Chemical Engineer Progress* (CEP), the *American Institute of Chemical Engineers* (*AIChE*) Journal (formally the *Transactions*), and *Chemical Engineering* (*CE*).

New Sources

The four key new sources[12] include:

The Internet

Google

Wikipedia

Web sites

A brief introduction to each is provided below.

The granddaddy of them all is the Internet. The Internet serves as the electronic backbone that connects all the computers, storages devices, servers, Web sites, etc. In addition to technical information, the Internet is regularly used for online shopping, generating map searches, accessing libraries and databases, and so on.

Google is a comprehensive search engine. It provides information for the Internet. It has developed software, e.g., Google Docs, that allows collaboration on document preparation, Google Images and Google Earth provide three-dimensional (3D) satellite information that enables user to access the entire planet. There are other search engines, including Yahoo!, Bing, and Dogpile.

Wikipedia (derived from the phrase "*w*hat *I k*now *i*s . . .") is essentially an online site that provides free and open information. It is also a collaborative encyclopedia where one can obtain or apply information to a host of topics. There are many other similar sites.

There are multitudes of websites. Virtually every university, business, government (national, state, local), and organization has one. The USEPA and DOE have one. Individuals also have one or more Web sites. Even your humble author has one: *www.theodorenewsletter.com*; this site provides information on education, sports, economics, politics, and a host of other topics.

Note that much of the above will almost certainly have changed between the times when this section was written (March 2013) and when it reaches the reading audience.

Personal Experience

Another information source—and a frequently underrated one—is the chemical engineer's personal experience, personal files, and company experience files. Although the traditional, chemical engineering, and new sources described above are important, fields other than chemical engineering or the particular field with which the chemical engineer is directly involved often provide the most useful information. In fact, most equipment designs, process changes, and new-plant designs are based on either prior experience, company files, or both.

4-4 Oral Presentations

Making a successful oral presentation did not come easy for the author. There was at time when any public speaking engagement was a frightening and nerve-racking experience. It took nearly 30 years before he could walk into a classroom (to give a lecture) and feel completely relaxed. Today, the author feels *nearly* completely at ease when serving as a luncheon or an after-dinner speaker.

The chemical engineer who is a total introvert will not become a fluent speaker overnight. Those who are aware of their difficulties in speaking in front of a crowd or with interviewers are strongly advised to take a speech course. The objective of attending courses or seminars on presenting oral reports is to provide some experience in technical speaking, and also provide the opportunity to observe techniques used by other speakers. People learn social behavior by mimicking others. One should observe others as they speak and analyze the positive and negative aspects of their techniques in order to improve one's own style.

There is much standard advice given on public speaking. Here are some hints. Know the subject. Outline the talk instead of writing and reading it. Look directly at one person in the audience, then another, then another. Prepare good visual aids. Stick to the time limit. Practice in front of a friend or a colleague who can inform you of any annoying tendencies you might have (mumbling or using speech danglers such as "uh..." and "well . . .," clinging to the podium, fussing with your hair, etc.). Most professional organizations require the use of visual aids for a speech. A microsoft PowerPoint presentation is preferred. One also has the option of using additional props (membranes, packing, adsorbents, etc.) if this is appropriate to the topic. Finally, in the words of the author,[12] "Tell them what you are going to tell them, then tell them, and then tell them what you told them."

Here are some suggestions that one may employ in preparing visual aids (or the equivalent):

1. The first slide should contain the title of the speech and the name(s) of the speaker(s). Acknowledgments may also be included.
2. The second slide should give an *overview* of the speech.
3. The penultimate slide should describe important conclusions.
4. The final slide should provide a *summary* of what was presented (occasionally, this slide may be dedicated to *acknowledgments*.)
5. Prepare each slide with a border.
6. Never photocopy books or typewritten material.
7. Limit the amount of information on any one slide; the listening audience should be able to read it.
8. Finally, prepare slides that are informative, lively, and interesting.

Here are a baker's dozen suggestions that one may employ in preparing and presenting the speech:

1. Dress accordingly.
2. Express gratitude to the person who introduced you and invited you to speak.
3. Try to memorize your opening remarks.
4. Begin your speech on a positive note.
5. Speak in a cheerful manner and in your own individual and innovative style.
6. Demonstrate enthusiasm.
7. As noted earlier, maintain good eye contact with the audience.
8. Be aware of your standing position in relation to the screen and the audience; stand near the screen, on one side.
9. If possible, avoid continuous reading of notes and cue cards.
10. Maintain a constant decibel level during the presentation.
11. Remain standing during the presentation, including the question period.

12. Do *not* begin the answer to a question with: "Well. . . ."
13. Voice delivery should be kept in mind. All sentences should not end in an *up note* as if in a question and do not use the word *like* with every other sentence.

Keep in mind that oral talks are generally judged on overall professionalism of the presentation with emphasis on the following points:

1. Quality (technical content) of verbal presentation
2. Accuracy, completeness, and organization of material presented
3. Quality of visual aids (quality of preparation)
4. Effective speaking technique (professionalism)
5. Poise and technical competence in answering questions (if applicable)

Finally, don't be afraid to keep it light. An occasional joke or comment may enhance the presentation.

Skilled oral communication is an art to be learned and developed through hard work and practice. Probably every college and university now includes courses that require practice in oral technical presentations. Realizing the importance of communication from an economic perspective, many companies are now providing training courses for their employees in this area.

In addition to technical presentations, engineers must be able to communicate with their subordinates to get things done, with their supervisors to get projects and budgets approved, and with the public to maintain their company's good image. As noted earlier, the author feels that training in oral presentations should be an ongoing activity during the formative years of the chemical engineer. No hard-and-fast rules exist for how the training should be implemented. However, practicing an oral presentation with a friend or a colleague is the best advice that can be offered.

4–5 Technical Writing

The reader should be aware of the style differences between nontechnical and technical writing. Technical documents are written in a factual, "dry" style emphasizing quantitative analysis, whereas nontechnical passages contain numerous adjectives and modifiers. Similarly, engineering texts, by contrast, contain numerous color illustrations and graphics. The contrast in style between the two is reminiscent of the contrast between a black-and-white "how to" video, and a high-budget movie. Generally, engineering texts use a pragmatic and mechanistic approach, while the nontechnical books have a policy and philosophical bent, or what some define as the liberal arts approach. Further, one generally transmits information, while the other entertains. In any event, this section is concerned with technical writing, a topic that should be of interest to the chemical engineer.

Technical writing really isn't that difficult; it is *not* a talent that only a handful of people are born with. That said, there are a few basic rules that can transform one's writing from a confusing garbled mess into something that will impress readers:

1. If applicable, know who the reader is.
2. It helps if the subject has not been written about before. If it has, improve what information is available by editing, rewriting, expanding, and updating.
3. There should be an element of interest to the reader(s).
4. Prepare an outline; this should include an appropriate title, objective, introduction, background material, results, conclusions, and recommendations.
5. Improve the foundation of the outline by filling it in with notes and sentences.

6. Keep related ideas together and establish a logical flow from paragraph to paragraph and section to section.
7. The abstract or executive summary (for technical reports) is the most important part of the writing; spend a significant amount of time here. It is the only material that is read by the majority of readers. The reader (perhaps your immediate supervisor) can then decide whether to go deeper into your writing material. Note that the abstract should contain a brief summary of the report *without* referring to the main body of the report.
8. Sentences should contain little or no unnecessary words; paragraphs should contain little or no unnecessary sentences. One certainly would have no unnecessary lines in a diagram or unnecessary components in a machine.
9. The swan song of the successful writer is "revision, revision, revision." Start early, e.g., weeks before it is due. Get colleagues, friends, or your wife (the author's usual enabler) to review and critique your material.
10. Proofread—and keep proofreading.
11. Some key words that one can employ are *analyses*, *contradicts*, *demonstrates*, *details*, *displays*, *expresses*, *illustrates*, *implies*, *importance*, *indicates*, *reflects*, *signifies*, *suggests*, *therefore*, and *thus*. However, be wary of the comments and materials listed in Sec. 4–2.
12. Some key words one should try *not* to employ are *because*, *did*, *done*, *in spite of*, *made sure*, *might*, *perhaps*, *occasionally*, and *owing to the fact*.

In the final analysis, the more one reads and writes, the easier it becomes. It is like tying shoes; it's a little hard at first, but once you have mastered it, you will always know how.

In essay form, one could summarize the above as follows. The primary objective of any writing is to inform others. In English classes, one may be rewarded for using long, flowery sentences and verbose, meandering paragraphs. Technical writing has to be concise and succinct. First-person words such as *I* and *we* are generally not recommended; other personal pronouns such as *you*, *he*, *she*, or *they* should be minimized. The only possible exception is in the recommendations section(s). This section can be written in the first person since it reflects the author's personal option.

Obviously, grammatically correct written English is required throughout any writing. Each paragraph needs opening and closing sentences with supporting sentences in between. The abstract and conclusions should each be one or two or three paragraphs long. One should be brief and informative with a strong opening sentence. Longer sections of the report also need an opening paragraph and a closing paragraph and the body in these sections must unfold in a logical and informative manner. Finally, a paragraph should have at least three sentences, but lengthy, unwieldy paragraphs should be avoided.

Most of the following items are required in a written technical report:

Cover

Title page

Abstract

Table of contents

List of tables

List of figures

Background and theory, with equipment description (if applicable)

Results and discussion

Conclusions

Recommendations (optional section)

Nomenclature

References

Appendix

As noted earlier, the abstract (or executive summary) is the most important section of the report. The abstract must briefly state the objectives of the study, e.g., what experiment (if applicable) was done, how it was done, what calculations were made, and the most important results. The purpose of the abstract is to immediately acquaint the reader with the contents of the report. Do not provide procedural details or refer to figures that are discussed in the body of the report. Do not write out equations; just mention what types of equations were used. Summarize important numerical results if determining these values is one of the objectives. Avoid or eliminate details in the abstract of the report. The abstract requires the most careful writing of any section of the report. Although it appears first, it should be the last section to be written. Its length is usually one to three paragraphs, or approximately one-half of a page. In some cases, an *extended abstract* or *executive summary* is prepared which can extend to two or three pages.

4-6 On That First Job

This section was drawn from an article written by the author[13] with the above title. An edited version of that article follows.

Many of the younger generation in the chemical engineering profession are now rapidly approaching crunch time regarding employment. In effect, it's job time.

Over the years, the author's students have often asked for advice on employment and careers. His response to them centers on four subject areas:

1. What are you looking for?
2. What is the company looking for?
3. What do you need to get the job?
4. What is needed to succeed?

Well, what about success? Over the years, the author has been privileged to maintain close social and professional ties with successful graduates in the chemical engineering industry. In thinking about what character traits likely contributed to their success, he found the following to be the most common:

1. *Communication Abilities*. A leader must be an exceptional communicator both orally and in writing.
2. *Appearance*. Leaders must dress appropriately, generally in business attire, and present themselves with a pleasant, confident demeanor.
3. *Self-Awareness*. Leaders have an ability to recognize and understand their moods, emotions, and drives, as well as how they affect others, are able to laugh at themselves, and are not bothered by what others might say about them.
4. *Action-Oriented*. Perhaps most important, leaders are doers and have an ability to make things happen, even when the odds are stacked against them.

Interestingly, technical ability (or the equivalent) and grade-point average (GPA) correlate weakly with successful leadership.

The author concludes this section with a tale that appeared in *EM Magazine* in December 2000, authored by a former student, Anthony Buonicore. The moral of the tale may register with a few of the readers.

One stormy night many years ago, an elderly man and his wife entered the lobby of a small hotel in Philadelphia. Trying to get out of the rain, the couple approached the front desk hoping to get shelter for the night.

"Could you possibly give us a room here?" the man asked. The clerk, a friendly man with a winning smile, explained that there were three conventions in town.

All of our rooms are taken," the clerk said. "But I can't send a nice couple like you into the rain at one o'clock in the morning. Would you perhaps be willing to sleep in my room? It's not exactly a suite, but it will be good enough to make you folks comfortable for the night."

When the couple declined, the young man pressed on. "Don't worry about me; I'll make out just fine," the clerk told them. So the couple agreed.

As he paid the bill the next morning, the elderly man said to the clerk; "You are the kind of manager who should be the boss of the best hotel in the country." The clerk looked at them and smiled. As they drove away, the elderly couple agreed that the helpful clerk was indeed exceptional.

Two years passed. The clerk had almost forgotten the incident when he received a letter from the old man. It recalled the stormy night and enclosed a round-trip ticket to New York, asking the young man to pay them a visit.

The old man met him at a street corner in New York City. He then pointed to a great new building, a palace of reddish stone, with turrets, and watchtowers thrusting up to the sky. "That," said the old man, "is the hotel I have just built for you to manage." "You must be joking," the clerk said. "I can assure you I am not," said the old man.

The old man's name was William Waldorf-Astoria, and the magnificent structure he built was the Waldorf-Astoria Hotel. The young clerk who became its first manager was George C. Boldt. The clerk never foresaw the chain of events that would lead him to become the manager of one of the world's most glamorous hotels.

References

1. L. Theodore, personal notes, East Williston, N.Y., 1980–1984.
2. L. Theodore, "On Technical Jargon," in "As I See It" Column, *The Roslyn News*, East Williston, N.Y., 2000.
3. W. Walker, W. Lewis, W. McAdams, and E. Gilliland, *Principles of Chemical Engineering*, McGraw-Hill, N.Y., 1923.
4. D. Kern, *Process Heat Transfer*, McGraw-Hill, N.Y., 1950.
5. R. Treybal, *Mass Transfer Operations*, McGraw-Hill, N.Y., 1955.
6. R. Shreve, *The Chemical Process Industries*, McGraw-Hill, N.Y., 1942.
7. J. Happel, *Chemical Process Economics*, Wiley, N.Y., 1958.
8. R. Bird, W. Stewart, and E. Lightoot, *Transport Phenomena*, 2nd ed., Wiley, N.Y., 1960.
9. R. Perry and D. Green, *Perry's Chemical Engineers' Handbook*, 8th ed., McGraw-Hill, N.Y., 2008.
10. R. Kirk and D. Othmer, *Encyclopedia of Chemical Technology*, 4th ed., Wiley, N.Y., 2001.
11. L. Albright, *Albright's Chemical Engineering Handbook*," CRC Press/Taylor & Francis Group, Boca Raton, Fla., 2008.
12. L. Theodore, personal notes, East Williston, N.Y., 2012.
13. L. Theodore, "On That First Job," *The Williston Times*, East Williston, N.Y., May 17, 2002.
14. A. J. Buonicore, *Industry Insight*, EM (Air & Water Management Association), Pittsburgh, Pa., Dec. 2000.

5 Second Language(s)

Chapter Outline

5-1 Introduction

The author spent his early childhood (ages 2 to 5 years) in Manhattan's Hell's Kitchen with English as a second language (ESL). Greek was spoken at home. Later, the author moved to a predominantly Puerto Rican–Cuban neighborhood and learned some Spanish, particularly the off-colored or colloquial words. Did these other languages help the author in his professional career? The answer is definitely "not really." But this is not the mid-twentieth century; it is the twenty-first century, and times have changed.

Today, "possessing" a second language—or several other languages—is considered by many to be important to the U.S. engineering and science professional. The second language of choice appears to be Mandarin. For example, Manhattan College's chemical engineering program recently instituted a series of lectures concerned with providing their students with an introduction to Mandarin. Is this a good choice? Not according to some,[1] who have argued that it is English that will remain or become the second language for the rest of the world.

Although voice is not discussed in this chapter, the chemical engineer should be knowledgeable of this topic since it can adversely affect one's language qualifications. In English, for example, there are the "valley girl" inflections that result in sentences ending on an "up" note as though they are questions. There is also the excessive use of the terms word "you know" and "like." In addition, there is the "well, ..." first-word response to a question.

There is no doubt that language and communication are at the heart of the human experience. The United States may ultimately be required to educate chemical engineers who are equipped linguistically and culturally to communicate successfully abroad, and envision a future in which all engineers will develop and maintain proficiency in at least one other language. Regarding timing, attention has focused on the importance of learning a second language as early as possible. Some research indicates that younger children learn language better than do older children and adults. In addition to developing a lifelong ability to communicate with people from other countries and backgrounds, other benefits that have been reported include improved overall school performance and superior problem-solving skills. Therefore, the educational process should consider whether the teaching of a second language should begin earlier rather than later in a chemical engineer's career.

Ignorance of foreign languages and foreign cultures can threaten the security of the United States as well as its ability to compete in the global marketplace. The U.S. education system has, in more recent years, placed less value on speaking languages other than English or on understanding cultures other than one's own. However, second languages have reemerged as a concern, primarily following events that presented immediate and direct threats to the country's future, such as the attacks on September 11, 2001. These actions have caused government to reflect on the expertise of its personnel and to focus attention on the need for more and better language skills in certain parts of the world. Language skills and cultural expertise are also urgently needed to address changing economic challenges in an increasingly global marketplace. Thus, renewed interest in second languages will ultimately have an impact on the following stakeholders:

1. Professors
2. Students
3. Parents
4. Industry administrators
5. Government administrators
6. Community leaders

Ultimately, second language(s) will not only affect the global economy but also build stronger international relationships while avoiding international conflicts.

The following topics are addressed in this chapter: early history; questions about the learning process; second-language factors; the choice of language(s); and barriers. Finally, the reader should note that the actual teaching of a (second) language is not an objective of the chapter.

5–2 Early History

The term a *Second-language acquisition* (SLA), derives from two separate endeavors: (1) adult language teaching and (2) first-language acquisition for children. Both are briefly discussed below.

Behavioral theory was employed in endeavor (1). It essentially was based on the premise that if event A occurs, event B follows. For example, if test mice learn that approaching a piece of cheese induces an electric shock, the mice do not approach the cheese. Alternatively, if a shock is followed by a piece of cheese in an obscure location, the mice immediately migrate to that location.

World War II, the Cold War, the Korean War, and other wars demonstrated a need to develop a language teaching process. It was based on a similar approach that featured language drilling. In effect, the students (in this case armed forces personnel) were required to learn certain language operations, and were subsequently rewarded in some manner or form. The end result was the *audiolingual method* (ALM), which became the approach employed in academia. The approach emphasized memorizing dialogue and drills involving rehearsed statements. Interestingly, so-called spontaneous speech was discouraged since it caused problems in attempts to correct mistakes.

During this period, studies on first-language acquisition for children were initiated. Although the analysis of these studies are beyond the scope of this chapter, most people generally now agree that language is an integral part of human behavior and knowledge, i.e., it is a part of one's genetic makeup. This is what accounts for the rapid learning process of children.

5–3 Questions Regarding the Learning Process

The two key questions regarding the learning process (for another language) are:

1. What is the best method for teaching and learning languages?
2. How does acquisition of a second language occur?

These two basic questions generate a number of other questions:

1. What do learners (engineers) do when acquiring a second language?
2. What stages do they go through?
3. What does their second language look like?
4. What kind of errors do they make?
5. What factors affect acquisition of this second language?

These questions have produced a subset of other questions:

1. What mechanics do learners use to master a second language?
2. Are errors that learners produce related to the mechanics used to produce a second language?
3. How does fluency subsequently develop?

Answers to these questions can determine which instructional efforts further second-language acquisition and which do not.

The interested reader is referred to the literature[1] not only for answers to these questions but also for developmental material. Obviously this is an area requiring detailed analysis and is beyond the scope of this chapter.

5–4 Second-Language Factors

As noted earlier, a second language is a language learned after the first language. For some, an individual's first language may not be their dominant language, i.e., the one they use most or are most comfortable with. For others, the earliest language may be lost; this a process is known as *language attrition*. This can happen when young children move, with or without their families because of immigration or international adoption to a new language environment.

Several factors impact the individual who moves to an area where another language is spoken:

1. The person's age
2. Similarities and differences between the second and first languages
3. Theories of second language acquisition (SLA)
4. Foreign language(s)

Each of these factors is briefly discussed below.[2]

Age

There is some question as to the difference(s) between first and second languages. For some, it is the age when the person learned the second language. Some use the term *second language* to represent a language consciously acquired or used by its speaker after puberty, noting that in most cases, people never achieve the same level of fluency and comprehension in their second languages as in their first language. Some claim the age of 6 or 7 to be the cutoff point for bilinguals to achieve native-like proficiency; after that age, their language would have a sufficient number of errors to set them apart from a first-language group. There is a hypothesis that phonetic accents develop during puberty. This claim is somewhat substantiated by the fact that the chemical processes in the brain are more geared toward language and social communication. Following puberty, the ability for learning a language without an accent has been rerouted to function in another area of the brain, perhaps in the frontal lobe area promoting cognitive function, or in the neural system of hormones allocated for reproduction and sexual organ growth.

Others have claimed that individuals who are exposed to a second language at an early age obtain better proficiency than do those who learn the second language as an adult. However, this theory does not apply to rate of learning. Adults acquire a second language faster than do younger children, i.e., adults and older children are fast learners, particularly when it comes to the initial stage of second-language education.

Although fluency is enhanced during an early learning period, it can cause someone learning a foreign language at an early age to develop a "weak [sense of] identification," or, a sense of dual nationality or ethnicity. This can impact a child's relations, attitudes, and behaviors, and ability to become absorbed into a foreign culture. These factors can negatively impact one's perspective of one's native country.

Similarities and Differences between the Second and First Languages

Error correction does not seem to directly influence learning a second language. Instruction may affect the rate of learning, but the learning stages remain the same. However, error correction remains a controversial topic, with many differing schools of thought.

Approaches have varied from the need to correct all errors at all costs, with little thought to the students' feelings of self-esteem, to the notion that students learn better when their teachers help them recognize and correct their own errors. This latter concept has been attributed to students' active participation in the corrective process.

Learners of a first or second language possess knowledge that goes beyond the input they received; in other words, the whole is greater than the parts. Learners of a language are able to construct correct utterances (e.g., phrases, sentences, questions) that they never saw or heard earlier in the learning process.

Success in first- and second-language learning can be measured in one of two ways: likelihood or quality. First-language learners are generally successful in both endeavors. It is inevitable that most people will learn a first-language and with a few exceptions, will be successful. For second-language learners, on the other hand, success is not guaranteed. As noted above, second-language learners rarely achieve complete native-like control of the new language.

Achieving success in learning a second language can seem like a daunting task. Studies have indicated why some students are more successful than others. It is generally agreed that a good language learner uses positive learning strategies and is an active learner who is constantly searching for meaning. In addition, good language learners often are willing to practice and use the language in real communication; they also monitor the learning process, have a strong drive to communicate, and have good listening skills.

Theories of Second-Language Acquisition

The distinction between acquiring and learning a language is a natural process.[2,3] Learning a language requires a conscious effort, whereas *acquiring* a language often results when the student needs to partake in natural communicative situations. In the learning process, error correction is present, as is the study of grammatical rules isolated from the natural language.

Not all second-language educators agree on the above distinction; however, research in SLA focuses on developing the knowledge and use of a language by children and adults who already know at least one other language. Also, knowledge of the SLA process helps educational policy makers set more realistic goals for programs for both foreign language courses and the learning of the majority language by minority-language children and adults.[2,4] The acquisition process has been influenced by both linguistic and psychological theories. One of the dominant linguistic theories hypothesizes that a device or module of sorts in the human brain contains innate knowledge. Many psychological theories, on the other hand, hypothesize that cognitive mechanisms are responsible for much of human learning.

Foreign Language(s)[2]

In pedagogy and sociolinguistics, there is often a distinction between a second language and a foreign language. The latter is learned for use in an area where that language is not generally spoken. Arguably, English in countries such as India, Pakistan Bangladesh, and the Philippines as well as Scandinavian countries and The Netherlands can be considered a second language for many of its speakers, because they learn it young, speak it fluently, and use it regularly. In addition, it is the official language of the courts, government and business in southern Asia.

The same rule of thumb applies for French in the Arab Maghreb Union, except for Libya. Similar to English in the Scandinavian countries and The Netherlands, French is nominally not an official language in any of these Arabic-speaking countries. In the post-Soviet states such as Uzbekistan, Kyrgyzstan, and Kazakhstan, Russian can be considered a second language.

5–5 The Choice of Language[5]

The bilingual chemical engineer may be confronted with a problem requiring a choice of language during either a social or professional meeting. Not all bilinguals have the opportunity to use both their languages on a regular basis. When a bilingual lives in a largely single-language environment, there may be little choice about language use from day to day. However, in environments where two or more languages are widely spoken, bilinguals may use both their languages on a daily or frequent basis. When bilinguals use both languages, language choice is rarely haphazard or arbitrary. If the other person is already known to the bilingual, such as with a family member, friend, business colleague, or fellow engineer, a relationship has usually been established through one language. If both are bilingual, they have the option of changing to the other language (e.g., to include others in the conversation).

If the other individual is not known, a bilingual should attempt to quickly pick up clues as to which language to use. Clues such as dress, appearance, age, accent, and command of a language may suggest to the bilingual which language would be appropriate to use. For example, in bilingual areas of Canada and the United States, employees dealing with the general public may glance at a person's name on their records to help them decide which language to use. For instance, a person named Pierre Rouleau in Quebec or Maria García in southern Texas might be addressed first in French or Spanish, respectively, rather than in English.

An individual's own attitudes and preferences can also influence their choice of language. In a minority-for majority-language situation, older people may prefer to speak the minority language. Younger people, such as second-generation immigrants, may reject the minority language in favor of the majority language because of its higher status and more fashionable image. In situations where the native language is perceived to be under threat, some bilinguals may avoid speaking the majority or dominant language to assert and reinforce the status of the other language. For example, French-Canadians in Quebec sometimes refuse to speak English in shops and offices to emphasize the status of French. The degree of contact with the majority-language community can also be a factor in language choice. Finally, most minority languages are confined to private and domestic roles.

5–6 Barriers

This last section is concerned with the barriers that presently exist to learning other languages. The following six comments summarize this problem. The reader should note that the list below does *not* address the disadvantages associated with other language acquisition(s).

1. The United States is failing to study in sufficient numbers many of the languages essential to meeting the challenges of a new era.
2. Opportunities are lacking in many low-income, minority, and urban school districts.
3. Other-language instruction is offered in only one-quarter of urban public schools, compared with about two-thirds of suburban private schools.
4. Heavily populated minority school districts anticipate decreases in programed instruction time needed to learn foreign languages.
5. African-American, Hispanic, and American-Indian students earn fewer credits in foreign languages than do their Caucasian peers.
6. Instruction spread over two years is woefully inadequate for high school students to develop any usable level of proficiency.

It should be noted in closing this chapter that there are many who believe that the world's international language will continue to be English. English dominance may be ensured by the difficulties encountered by someone trying to learn Chinese, for example. Grappling with different tone-related meanings of the term *maÓ,* which can mean horse, mother, scold, or hemp, is a nasty business if you aren't born immersed in Chinese. The basics of English, however, are more easily learned. Most languages challenge the learner with either tables of prefixes and suffixes (as in Spanish) or tones (as in Chinese). English has no tones, and gets by with conjugation as simple as "I walk," "you walk," or "he walks." English is, in effect, a user-friendly language.[6]

References

1. B. Van Patten, *From Input to Output*, McGraw-Hill, N.Y., 2003.
2. Second Language, *Wikipedia—the Free Encyclopedia*, 2012.
3. Krashen et al., "Age, Rate and Eventual Attainment in Second Language Acquisition," TESOL Quart. [Online] **13** (4) Dec. 1979.
4. Source unknown
5. C. Baker, *Foundations of Bilingual Education and Bilingualism,* 4th ed., Multilingual Matters, Clevedon, U.K., 2006.
6. J. McWhorter, "English is Here to Stay," *The Daily Beast*, Jan. 17, 2011.

Chemical Engineering Processes

Chapter Outline

6–1 Introduction

The chemical engineer has employed two perspectives in describing a chemical process: unit operations and unit processes. Although both have influenced chemical engineering education, the unit operations approach has survived the test of time. However, this chapter's major objective is to introduce the reader to the latter approach i.e., unit processes, and to list and in some cases briefly define these processes. These unit processes are those processes with which chemical engineers are involved. Both subject areas and perspectives are briefly described below.

The mold for what is now called *chemical engineering*—as described in Chap. 1—was fashioned at the 1922 meeting of the American Institute of Chemical Engineers (AIChE) when A. D. Little's committee presented a report on chemical engineering education. The 1922 meeting marked the official endorsement of the *unit operations* concept for the first time and saw the approval of a "declaration of independence" for the profession.[1] A key component of this report included the following:

> "Any chemical process, on whatever scale conducted may be resolved into a coordinated series of what may be termed *unit operations*, as pulverizing, mixing, heating, roasting, absorbing, precipitation, crystallizing, filtering, dissolving, and so on. The number of these basic unit operations is not very large and relatively few of them are involved in any particular process .... An ability to cope broadly and adequately with the demands of this (the chemical engineer's) profession can be attained only through the analysis for processes into the unit actions as they are carried out on the commercial scale under the conditions imposed by practice.

The key *unit operations* were ultimately reduced to three: fluid flow, heat transfer, and mass transfer, subject areas that receive treatment in Part II of this book.

Chemical engineers also realize that a process (plant) can be regarded as a collection of individually operated process, i.e., *unit processes*. It is becoming increasingly evident that each separate unit of a plant *usually* influences all others in either a subtle or major way, justifying the unit process method of analysis. This approach is highlighted in this chapter.

Many now believe that *chemistry* deals with the combination of atoms and *physics* with the forces between atoms. Atomic combination involves atomic forces, and it is one of the objects of *physical chemistry* to see how far the chemical interactions that occur between atoms and molecules can be interpreted by studying the forces existing within and between atoms. The study of atomic structure provides information of why atoms combine.

A *chemical reaction* is a process by which atoms or groups of atoms are combined and/or redistributed, resulting in a change in the molecular composition and properties. The products obtained from reactants depend on the condition under which a chemical reaction occurs. The products are the end result of the chemical process and can be categorized in three general areas: inorganic chemicals, organic chemicals, and petrochemicals. Additional details are provided in Chap. 13.

This chapter addresses the following topics: the chemical industry today; inorganic and organic chemicals; petrochemicals; and Shreve's unit processes.

6–2 The Chemical Industry Today

The growth of chemical industries and the training of professional chemical engineers and chemists are both intertwined. A number of universities were established in Germany during the industrial revolution in the early nineteenth century. They drew students from around the world, and other universities soon followed suit. A large group of young chemists was thus trained at approximately the same time when the chemical industry was beginning to utilize new discoveries. This interaction between the universities and the chemical industry resulted in the rapid growth of the (particularly organic) chemical industry and provided Germany with scientific predominance in the field until World War I. Following the war, the German system was introduced into all industrial nations of the world, and chemistry and chemical industries both expanded and progressed rapidly.

The scientific explosion in more recent years has had an enormous influence on society. Processes were developed for synthesizing completely new substances that were either better than the natural ones or could replace them more cheaply. As the complexity of synthesized compounds increased, entirely new products appeared. Plastics and new textiles were developed, energy use increased, and new drugs conquered whole classes of disease.

The progress brought forth by both chemists and chemical engineers in recent years has been spectacular, although the benefits of this progress have included corresponding liabilities. The most obvious risks have come from nuclear weapons and radioactive materials, with their potential for producing cancer(s) in exposed individuals and mutations in their children. In addition, certain pesticides have potentially damaging effects. This led to the emergence of a new industry that is now defined as *environmental engineering*. Mitigating these negative effects is one of the challenges that the science community will have to meet in the future.[2]

The chemical industry is large, accounting for approximately one-sixth of the U.S. gross national product (GNP). It brings numerous new products to the marketplace each year. The current U.S. chemical industry has been defined as a sprawling complex of raw-materials sources, manufacturing plants, and distribution facilities which supplies both the U.S and the world with chemical products, almost all of which were unknown 125 years ago.

As discussed in Sec. 6–1, no clear compartmentalization can be used to define the various industries. One approach is to classify the chemical products as inorganic, organic, and petrochemical. These classifications receive treatment in the next three sections. Another approach is to attempt to list all the various industries. This method of presentation is employed in Sec. 6–6, where the *Shreve list* of chemical industries is presented. There are also several dozen chemical conversion "reactions" that are an integral part of all of the process. A partial list of these chemical conversions is provided in Table 6–1. In any event, it should be noted once again that it is the chemical engineer who participates in all the activities associated with bringing useful products to society.[3]

Table 6–1

Major Chemical Conversions

Acylation	Dehydrogenation	Neutralization
Alcoholysis	Electrolysis	Nitration
Alkylation	Esterification	Oxidation
Aromatization or Cyclization	Fermentation	Polymerization
Calcination	Halogenation	Pyrolysis
Carboxylation	Hydrogenation	Reduction
Combustion	Hydrolysis	Sulfonation
Condensation	Ion exchange	
Dehydration	Isomerization	

6–3 Inorganic Chemicals

Inorganic chemistry is a general chemistry field that is concerned with chemical reactions and properties of all the chemical elements and their compounds, with the exception of hydrocarbons (compounds composed of carbon and hydrogen) and their derivatives.

Inorganic chemistry is too vast a subject to form a convenient unit of study; the term would be of little importance except for the tendency in most engineering and science schools to refer to courses labeled "Inorganic Chemistry" when a better title might be "Elementary Chemistry." The subject matter of such courses includes the elementary laws of chemistry, its symbols and nomenclature, and an introduction to the experimental methods that are important in experimental chemistry. The student is introduced to such fundamental topics as valence, ionization, reactivity, atomic theory, and the kinetic theory of gases. The properties and reactions of substances in aqueous solution also receive attention. Modern inorganic chemistry overlaps parts of many other scientific fields, including biochemistry, metallurgy, mineralogy, and solid-state physics. Finally, the increased understanding of the chemical behavior of the elements and inorganic compounds has led to the discovery of not only a wide variety of new synthesizing techniques but also many new classes of inorganic substances.[2]

Industrial inorganic chemical processes have existed for over 200 years. Many early processes involved simple treatment of natural materials. With time, more complex processes evolved; these often consisted of several steps involving the reaction of several natural raw materials. With the development of the scientific chemical industry, many more inorganic chemicals were discovered and produced. Table 6–2 provides a partial list of the major inorganic chemicals that are produced in the United States today.

Inorganic chemicals are often more easily produced than are their organic chemical counterparts. Inorganic reactions usually proceed rapidly and to near completion, with few, if any, side reactions or by-products. The problem of thermal decomposition is unimportant in many inorganic materials when compared to most heat-sensitive organic compounds.

6–4 Organic Chemicals

The branch of chemistry in which carbon compounds and their reactions are studied is defined as *organic chemistry*. A wide variety of substances, such as drugs, vitamins, plastics, and natural and synthetic fibers, as well as carbohydrates, proteins, and fats consist of organic molecules. The subject of organic chemistry involves (1) determining the structures of organic molecules, (2) studying their various reactions, and (3) developing procedures for the synthesis of organic compounds.

Table 6–2

Key U.S. Inorganic Chemicals

Aluminum chloride	Hydrochloric acid	Potassium hydroxide
Ammonia	Hydrofluoric acid	Sodium
Ammonium nitrate	Hydrogen	Sodium carbonate
Ammonium sulfate	Nitric acid	Sodium chlorate
Calcium carbide	Oxygen	Sodium hydroxide
Calcium phosphate	Phosphoric acid	Sodium phosphates
Carbon dioxide	Phosphorus	Sodium sulfate
Chlorine	Potassium	Sulfuric acid

Organic chemistry has a profound effect on society; it has improved natural materials and has synthesized natural and artificial materials that have, in turn, improved health, increased comfort, and added to the convenience associated with nearly every product manufactured today.

The ability of carbon to form covalent bonds with other carbon atoms in long chains and rings distinguishes carbon from all other elements. No other elements is known to form chains consisting of more than eight similar atoms. This particular property of carbon, and the fact that carbon nearly always forms four bonds to other atoms, accounts for the large number of known compounds. At least 80 percent of the 5 million reported chemical compounds contain carbon.

Regarding sources of organic compounds, *coal tar* was once the only source of aromatic and some heterocyclic compounds. *Petroleum* was the source of aliphatic compounds since it contains such substances as gasoline, kerosene, and lubricating oil. *Natural gas* provides (primarily) methane and ethane. These three categories of natural fossil compounds are still the major sources of organic compounds. When petroleum is not available, a chemical industry can be based on acetylene (if available), which, in turn, can be synthesized from limestone and coal. During World War II (1939 to 1945), Germany was compelled to employ this process when it lost reliable petroleum and natural-gas supplies in northern Africa following the battle of El Alamein.

Covalent organic compounds are distinguished from inorganic salts by the former's low melting points and boiling points. Hydrocarbons have low specific gravities (approximately 0.8 compared to 1.0 for water), but functional groups may increase the densities of organic compounds. Only a few organic compounds, such as carbon tetrachloride, possess specific gravities in excess of unity.

The chemical engineer usually designs organic reactions to be carried out at optimum conditions to produce maximum conversion or yields. One often resorts to catalysts, regardless of whether not the reaction is reversible, and attempts to take advantage of equilibrium considerations. In addition, catalysts are frequently essential when there is a need for rapid chemical reactions.[2]

There are various classes of organic compounds, including alkanes, alkenes, and alkynes. Additional classes of organics include alcohols, aldehydes, amines, carboxylic acids, esters, alkyl halides, ketones, nitrites, and thiols (often referred to as *mercaptans*).

Other atoms, such as chlorine, oxygen, and nitrogen, may be substituted for hydrogen in an alkane, provided the correct number of chemical bonds is allowed, with chlorine forming one bond to other atoms, oxygen forming two bonds to other atoms, and nitrogen three bonds. The chlorine atom in ethyl chloride, the ^{-}OH group in ethyl alcohol, and the $^{-}NH_2$ group in ethyl amines are called *functional* groups. Functional groups determine many of the chemical properties of compounds. Many of the chlorine-bearing compounds are known to be carcinogens or otherwise toxic agents.[2]

The chemical industry produces a wide variety of organic chemical products. Major categories include dyes, pigments, flavor and perfume materials, medicinals, plasticizers, plastics and synthetic resins, synthetic fibers, synthetic rubbers, rubber-processing chemicals, surface-active agents (surfactants; e.g., detergents), pesticides and other agricultural chemicals, and intermediates.

Intermediates, which are defined as chemicals used in the manufacture of other organic chemicals, have the largest annual production of any of the categories listed. A few of the more important intermediates are acetic anhydride, aniline, formaldehyde, phenol, phthalic anhydride, and styrene. Table 6–3 provides a partial list of some of the major organic chemicals produced in the United States.

Table 6–3
Major Organic Chemicals

Acetic acid	Cresols	Penicillin
Acetic anhydride	Dichlorodiphenyltrichloroethane (DDT)	Phenol
Acetone	Dyes	Phthalic anhydride
Acetylsalicylic acid (aspirin)	Ethyl acetate	Pigment
Aniline	Ethyl alcohol	Pyridine
Butyl alcohol	Ethylene glycol	Streptomycin
Carbon disulfide	Formaldehyde	Styrene
Carbon tetrachloride	Methanol	Vitamins
Chlorobenzene	Naphthalene	

6–5 Petrochemicals

The chemical composition of all petroleum is principally hydrocarbons, although a few sulfur-containing and oxygen-containing compounds are usually present. Three broad classes of crude petroleum exist: the paraffin types, the asphaltic types, and the mixed-base types. Petroleum contains gaseous, liquid, and solid elements. The consistency of petroleum varies from liquid as thin as gasoline to liquid so thick that it will barely pour. Small quantities of gaseous compounds are usually dissolved in the mix; when larger quantities of these compounds are present, the petroleum deposit is associated with a deposit of natural gas.

Modern industrial societies use petroleum primarily for transportation purposes and also for energy purposes, primarily to generate electricity. In addition, petroleum and its derivatives are used in the manufacture of medicines, fertilizers, foodstuffs, plastics, building materials, paints, clothing, and electronic applications.

As noted above, petroleum is employed as a fuel to serve as an energy source. The properties of fuels most often used in thermal combustion applications are reviewed next, noting that fuels may be gaseous, liquid, or solid:[4]

1. *Gaseous fuels* are principally natural gas (80–95% methane, with the balance ethane, propane, and small quantities of other gases). Light hydrocarbons obtained from petroleum or coal treatment may also be used.
2. *Liquid fuels* are mainly hydrocarbons obtained by distilling crude oil (petroleum). The various grades of fuel oil, gasoline, shale oil, and various petroleum cuts and residues are considered in this category.
3. *Solid fuels* consist principally of coal (a mixture of carbon, water, noncombustible ash, hydrocarbons, and sulfur). Coke, wood, and solid waste (garbage) may also be considered in this category.

These classifications are not mutually exclusive and necessarily overlap in some areas. Additional details follow.[5]

Natural gas is perhaps the closest approach to an ideal fuel because it is practically free of noncombustible gases or particulate residue. Of the many gaseous fuels, natural gas is the most important. No fuel preparation is necessary because natural gases are easily mixed with air and the combustion reaction proceeds rapidly once the ignition temperature is reached.

Fuel oil may be defined as petroleum or any of its liquid residues remaining after the more volatile constituents have been removed or separated. Thus, the term *fuel oil* may conveniently cover a wide range of petroleum products. It may be applied to crude petroleum, to a light petroleum fraction similar to kerosene or gas oil, or to a heavy residue remaining after distillation. The principal industrial liquid fuels are therefore the by-products of natural petroleum. These fuel oils are marketed in two principal classes: distillates and residuals. The principal industrial boiler fuel is residual oil, known as No. 6 or *Bunker C*. The *residual oil* remains after the more valuable products are removed by distillation. As an industrial and utility fuel, it competes with coal, although its price has fluctuated wildly over the years. It is specified mainly by viscosity. Grades 1 and 2 are sometimes designated as *light* and *medium* domestic fuel oils and are specified mainly by the temperature range of the distillation and specific gravity.

There is no satisfactory definition of *coal*. It is a mixture of organic, chemical, and mineral materials produced by a natural process of growth and decay, accumulation of both vegetable and mineral debris, and accomplished by chemical, biological, bacteriological, and metamorphic action that occurred over millions of years. The characteristics of coal vary considerably with location; some variation in composition is usually encountered even within a given mine.

Coals are classified according to *rank*, which refers to the degree of conversion from one form of coal to another (e.g., lignite to anthracite). The following ranking of coals is employed today: anthracite, bituminous, subbituminous, and lignite. Solid fuels, including coal, consist of free carbon, moisture, hydrocarbons, oxygen (mostly in the form of oxygenated hydrocarbons), small amounts of sulfur and nitrogen, and nonvolatile noncombustible materials designated as ash.

By definition, a *petrochemical* is produced from raw materials derived from petroleum or natural gas. Several refinery streams are often potential sources of raw materials leading to a given finished product. In other cases, a refinery stream may be useful for several purposes, and economic considerations determine the purpose for which the stream may be employed.

Petrochemicals are used in many areas in everyday life. They have many diverse applications, including uses as fertilizers (ammonia), explosives (toluene, TNT, etc.), and synthetic rubber (butadiene, styrene, etc.). Petrochemicals may be aliphatic, aromatic, or inorganic. Many of these petrochemicals are intermediates in the production of other chemicals. A complete list of all the chemicals derived from petroleum would be extensive. Additional information is available in the literature.[3] The principal petrochemicals are listed in Table 6–4.

Table 6–4

Petrochemicals

Acetaldehyde	Carbon tetrachloride	Methyl ethyl ketone (MEK)
Acetic acid	Ethyl alcohol	Percholorethylene
Acetic anhydride	Ethyl chloride	Phenol
Acetone	Ethylene glycol	Phtalic anhydride
Acetylene	Ethylene oxide	Polyethylene
Acrylonitrile	Formaldehyde	Polypropylene
Ammonia	Glycerol	Propyl alcohol
Amyl alcohols	Hydrazine	Styrene
Aniline	Hydrogen peroxide	Vinyl chloride
Butadiene	Isopropyl alcohol	
Butyl alcohols	Methanol	

6–6 Shreve's Unit Processes

The first edition of Shreve's *Chemical Process Industries* was published in 1945. (It was the assigned text for a graduate course—Unit Processes—that the author took at NYU in 1956.) The latest (fifth) edition was published in 1984.[6] The latest edition lists 38 chemical industries (unit processes). It provides the following for each process: costs of reactants; energy requirements; and information related to efficient and profitable operations.

Shreve's chapter numbers and their accompanying titles are provided below:

3. Water Conditioning and Environmental Protection
4. Energy, Fuels, Air Conditioning, and Refrigeration
5. Coal Chemicals
6. Fuel Gases
7. Industrial Gases
8. Industrial Carbon
9. Ceramic Industries
10. Portland Cements, Calcium, and Magnesium Compounds
11. Glass Industries
12. Salt and Miscellaneous Sodium Compounds
13. Chlor-Alkali Industries: Soda Ash, Caustic Soda, and Chlorine
14. Electrolytic Industries
15. Electrothermal Industries
16. Phosphorus Industries
17. Potassium Industries
18. Nitrogen Industries
19. Sulfur and Sulfuric Acid
20. Hydrochloric Acid and Miscellaneous Inorganic Chemicals
21. Nuclear Industries
22. Explosives, Propellants, and Toxic Chemical Agents
23. Photographic Products Industries
24. Surface-Coating Industries
25. Food and Food By-Product Industries
26. Agrichemical Industries
27. Fragrances, Flavors, and Food Additives
28. Oils, Fats, and Waxes
29. Soap and Detergents
30. Sugar and Starch Industries
31. Fermentation Industries
32. Wood-Derived Chemicals
33. Pulp and Paper Industries
34. Plastic Industries
35. Man-made Fiber and Film Industries
36. Rubber Industries
37. Petroleum Processing

38. Petrochemicals

39. Cyclic Intermediates and Dyes

40. Pharmaceutical Industries

On the positive side, Shreve's book provides a wealth of information. On the negative side, it provides only a listing of the major traditional chemical and chemical-related compounds in the United States. Since the latest edition is dated, industries such as nanotechnology, health and hazard risk assessment, cryogenics, aerospace, and communications receive no treatment. Furthermore, certain chapters are incomplete. Chapters 3 and 4 should be devoted to water and energy management, respectively. Chapter 5 should address other fossil fuels such as shale oil and tar sands. Despite these shortcomings, Shreve's book remains the standard for describing processes.

References

1. N. Serino, *2005 Chemical Engineering 125th Year Anniversary Calendar*, term project submitted to L. Theodore, Manhattan College, Bronx, N.Y., 2005.
2. L. Theodore, *Chemical Reaction Analyses and Applications for the Practicing Engineer*, Wiley, Hoboken, N.J., 2012.
3. L. Andersen and L. Wenzel, *Introduction to Chemical Engineering*, McGraw-Hill, N.Y., 1961.
4. K. Skipka and L. Theodore, *U.S. Energy Resources: Past, Present, and Future Management*, CRC Press/Taylor & Francis Group, Boca Raton, Fla., 2013 (in print).
5. J. Santoleri, J. Reynolds, and L. Theodore, *Introduction to Hazardous Waste Incineration*, 2nd ed., Wiley, N.Y., 2000.

PART II Unit Operations

Part Outline

7 The Conservation Laws and Stoichiometry

Chapter Outline

7–1 Introduction

In order to better understand the design as well as the operation and performance of industrial equipment, it is necessary to understand the fundamentals and principles underlying this technology. How can one predict what products will be emitted from effluent streams? At what temperature must a unit be operated to ensure the desired performance? How much energy in the form of heat is given off? Is it economically feasible to recover this heat? Is the design appropriate? The answers to these questions are rooted in the various theories of chemistry, physics, and applied economics. Answers to some of these questions will be provided in the following sections in this chapter and in Chaps. 8 to 15.

The topics covered in this chapter include the general conservation law and the conservation laws for mass, energy, and momentum. The chapter concludes with a section on stoichiometry.

NOTE The bulk of the material in this chapter has been drawn from the original work of Reynolds.[1]

7–2 The Conservation Laws

Mass, energy, and momentum are all conserved. As such, each quantity obeys the general conservation law expressed in the following equation as applied within a system:

$$\begin{Bmatrix}\text{Quantity}\\\text{into}\\\text{system}\end{Bmatrix}-\begin{Bmatrix}\text{quantity}\\\text{out of}\\\text{system}\end{Bmatrix}+\begin{Bmatrix}\text{quantity}\\\text{generated}\\\text{in system}\end{Bmatrix}=\begin{Bmatrix}\text{quantity}\\\text{accumulated}\\\text{in system}\end{Bmatrix} \tag{7–1}$$

Equation (7–1) may also be written on a *time* basis:

$$\begin{Bmatrix}\text{Rate of}\\\text{quantity}\\\text{into}\\\text{system}\end{Bmatrix}-\begin{Bmatrix}\text{rate of}\\\text{quantity}\\\text{out of}\\\text{system}\end{Bmatrix}+\begin{Bmatrix}\text{rate of}\\\text{quantity}\\\text{generated}\\\text{in system}\end{Bmatrix}=\begin{Bmatrix}\text{rate of}\\\text{quantity}\\\text{accumulated}\\\text{in system}\end{Bmatrix} \tag{7–2}$$

The conservation law may be applied at the *macroscopic*, *microscopic*, or *molecular level*. One can best illustrate the differences in these methods with an example. Consider a system in which a fluid is flowing through a cylindrical tube (see Fig. 7–1). One may define the system as the fluid contained within the tube between points 1 and 2 at any given time. If one is interested in determining changes occurring at the inlet and outlet of the system, the

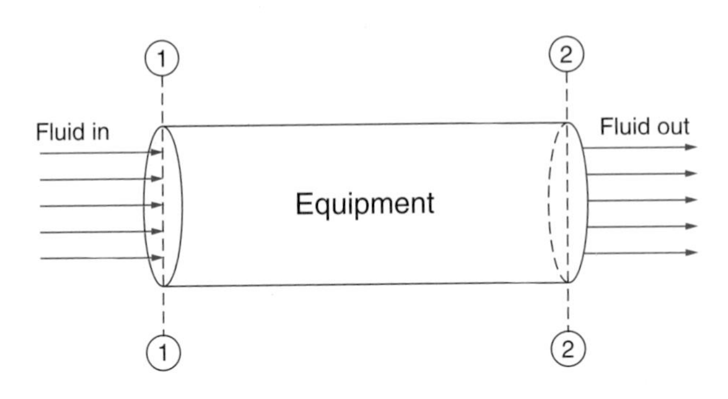

Figure 7–1 Conservation law example

conservation law is applied on a macroscopic level to the entire system. The resultant equation describes the overall changes occurring *to* the system without regard for internal variations *within* the system. This approach is usually applied by the practicing chemical engineer. The microscopic approach is employed when detailed information concerning the behavior *within* the system is required, and this is occasionally requested of and by the chemical engineer or scientist. The conservation law is then applied to a *differential* element within the system which is large compared to an individual molecule, but small compared to the entire system. The resultant equation is then expanded, via an integration, to describe the behavior of the entire system. This is defined by some as the *transport phenomenon approach.* The molecular approach involves the application of the conservation law to individual molecules. This leads to a study of statistical and quantum mechanics, both of which are beyond the scope of this text. In any case, the description of individual molecules at the molecular level is of little value to the practicing chemical engineer. However, the statistical averaging of molecular quantities in either a differential or finite element within a system leads to a more meaningful description of the behavior of the system. The macroscopic approach is primarily adopted and applied in this text, and little to no further reference to microscopic or molecular analyses will be made. This chapter's aim, then, is to express the laws of conservation for mass, energy, and momentum in algebraic or finite-difference form.

It should also be noted that the applied mathematician has developed differential equations describing the detailed behavior of systems by applying the appropriate conservation law to a differential element or shell within the system. Equations were derived with each new application. The chemical engineer later removed the need for these tedious and error-prone derivations by developing a general set of equations that could be used to describe systems. These came to be defined as the aforementioned *transport equations.*[1,2] Needless to say, these transport equations have proved an asset in describing the behavior of some systems, operations, and processes. This text departs from the approach of developing these differential equations, even though this method has a great deal to commend it. Experience has indicated that the chemical engineer possessing a working knowledge of the conservation laws is likely to obtain a more integrated and unified picture of chemical engineering by developing the equations in algebraic form, i.e., macroscopic form.

7–3 Conservation of Mass[1]

The *conservation law* for mass can be applied to any process, equipment, or system. The general form of this law is given by the following two equations:

$$(\text{Mass in}) - (\text{mass out}) + (\text{mass generated}) = (\text{mass accumulated}) \tag{7–3}$$

or on a time-rate basis by

$$\begin{pmatrix}\text{Rate of}\\ \text{mass in}\end{pmatrix} - \begin{pmatrix}\text{rate of}\\ \text{mass out}\end{pmatrix} + \begin{pmatrix}\text{rate of mass}\\ \text{generated}\end{pmatrix} = \begin{pmatrix}\text{rate of mass}\\ \text{accumulated}\end{pmatrix} \tag{7–4}$$

The law *of conservation of mass* states that mass can be neither created nor destroyed. *Nuclear reactions*, in which interchanges between mass and energy are known to occur, provide a notable exception to this law. Even in chemical reactions, a certain amount of mass–energy interchange takes place. However, in normal chemical engineering applications, nuclear reactions do not occur, and the mass–energy exchange in chemical reactions is so minuscule that it is not worth taking into account.

The *material (mass) balance* is a direct result of the aforementioned law of conservation of mass. The *complete balance equation* is discussed in the later sections; only a simple application of it is presented here. Before applying a material balance, the process to which the balance is applied must be defined. All processes take place within certain limits of space and time, and these limits determine what is known as the *system*. Everything outside the system is called the *surroundings,* and the limits of the system, often perceived as an imaginary wall or envelope around the system, are referred to collectively as the *system boundary.* Exactly what constitutes the system is often arbitrary in the sense that one can decide what the system is. The system does not have to include a whole chemical plant or an entire process that produces a given chemical. It may also be redefined during the course of an application or a problem solution. A small part of the process may be isolated and defined as the system for the purpose of one set of calculations, and then another part may be chosen for another set.

Some definitions pertinent to systems and material balances are listed below:

1. *Open system*: a system in which material crosses the system boundary during the process.
2. *Closed system*: a system in which no material enters or leaves during the process.
3. *Batch process*: a process that starts with a definite amount of feed material in the system and converts that material to products. Batch processes take place in closed systems.
4. *Continuous process*: a process in which inputs and outputs flow continuously during the course of the operation; continuous processes take place in *open* systems.
5. *Steady state*: a condition in which the values of all variables, including temperature, pressure, composition, and flow rate, do not vary with time during the course of a continuous process; note that this definition does not imply that the process variables (e.g., temperature) at all points in the system must have the same value, but that the value at a given point in the system must not change with time.

The simplest form of the material balance provided in Eqs. (7–3) and (7–4) is

$$\text{Input} = \text{output} \tag{7–5}$$

This form applies to batch processes in which no chemical reaction is involved. The *input* refers to the initial amount of material in the system (expressed as either mass or number of moles), and the *output* refers to the product in the system at the end of the process. When the system contains a mixture of chemicals, the material balance may be applied to the total mixture or to each individual chemical species.

Suppose, for example, that 50 lb of solid silver nitrate is to be dissolved in 1000 lb of water to produce a silver nitrate solution. Three different material balances may be applied to this system: one for the silver nitrate, one for the water, and one for the total amount of material. This third balance is referred to as a total *material balance.* One of the three equations resulting from the three balances will not be very useful in the calculations since it provides no new information not already contained in the other two equations. In other words, the third balance is not *independent* of the other two. This can be shown by adding the two equations for the silver nitrate and water balances; the resulting equation will be that for the total material balance.

The simple form of the material balance given above also applies to continuous *processes* in which no chemical reaction occurs, as long as the process is at steady state. In this case, *input* refers to the rate at which material enters the system [in such units as lb/h, kg/min, (lbmol)/day] and *output*, to the rate at which material exits the system.

The material balance can be more complicated. If the process is at not-steady state, the input and output flow rates of each chemical need not be the same. If, for example, the output is less than the input rate, the chemical might be building up or accumulating within the system boundaries with time.

The occurrence of a chemical reaction poses another complication. Suppose that inside the system, a reaction is taking place in which a chemical is either partially or completely consumed. The mass of that chemical leaving the system would obviously be less than that coming in. Or, suppose that the chemical is being generated rather than consumed by the reaction. The output of that chemical would be greater than the input.

The complete balance factors in all of these possible occurrences. The complete balance equation may then be written as [see also Eqs. (7–3) and (7–4)]:

$$I + G = O + C + A \tag{7–6}$$

where I = Input—the amount (mass or moles) or flow rate (mass or moles) entering the system

G = Generation—the amount produced or rate of production by a chemical reaction inside the system

O = Output—the amount or flow rate exiting the system

C = Consumption—the amount consumed or the rate of consumption by a chemical reaction inside the system

A = Accumulation—the amount accumulating or rate of accumulation in the system

In the absence of a chemical reaction, one obtains

$$G = C = \text{zero} \tag{7–7}$$

If the reaction is steady state, then

$$A = \text{zero} \tag{7–8}$$

As indicated above, the material balance may be applied to the total amount of material or the amount of any one chemical involved in the process. If a mixture of oxygen and hydrogen enters the system and a mixture of hydrogen, oxygen, and water vapor exits the system, the material balance equation may be applied to the oxygen alone, to the hydrogen alone, to the water vapor alone, or to the total mixture.

In the general balance equation, the A or accumulation term was defined as the amount accumulating or the rate of accumulation in the system. It is possible to have a situation where a chemical in the system is being *depleted* for some reason other than a chemical reaction. This could occur during a non-steady-state continuous process. Rather than introducing another term in the equation to handle this case, one simply assigns a negative value to the A term.

In real-world applications, a commercial chemical process is rarely limited to the use of a single unit or piece of equipment. As mentioned earlier, when more than one unit is involved, the choice of the system need not involve the entire process. Remember that the choice of the system is arbitrary; it is that portion of the entire process that is chosen for isolation, study, and performance of calculations. Indeed, in the course of solving a problem, one may find it necessary to choose a number of different systems to obtain all the information needed to arrive at a complete solution.

Figure 7–2 illustrates a hypothetical process involving two units: a mixer M and a flash separator F. The mixer has two feed streams, A and B. The output of the mixer is stream C, which is the input stream for the flash unit. Streams D and E are the product streams from the flash. The dashed-line enveloping around the mixer, the flash, and the

Figure 7–2

Flow diagram for a multiunit system.

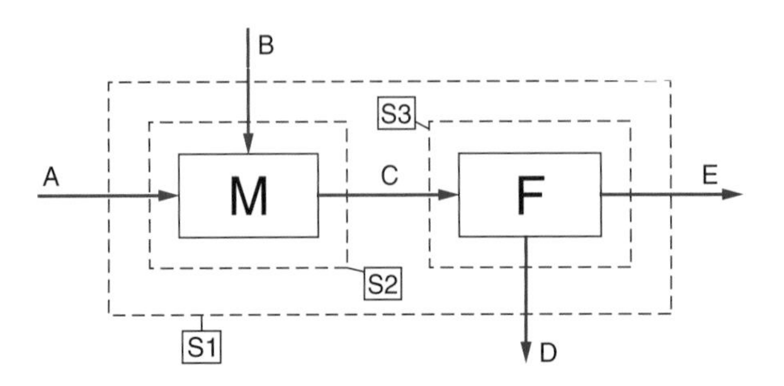

entire process (labeled S2, S3, and S1, respectively) represents various systems that one could choose for the material balances. (*Note*: In some flow diagrams, a mixer is represented simply as a point instead of a box, i.e., as two arrows merging together to represent the streams to be mixed, and one output arrow representing the mixture.) Balances on the S1 envelope are referred to as *overall balances.* In solving a problem based on this flowsheet, all three systems or envelopes could be used. However, some of the balance equations would yield redundant information; in other words—and as noted earlier—not all of the equations would be independent. When two units or pieces of equipment are involved (counting the mixer as a piece of equipment even though it may simply consist of two pipelines coming together), the number of *independent* systems or envelopes that can be used is *two.* It doesn't matter which two; S1-S2, S2-S3, or S1-S3 are all acceptable sets. If there were 10 pieces of equipment, the number of independent systems would be 10—one for each piece of equipment.

Once all the material balances have been applied to a flowchart and, as a result, all streams have been fully defined in terms of amounts or flow rates and compositions, the flow diagram is considered to be *balanced.* It is then possible to multiply all stream amounts or flow rates by the same constant, and the flowchart will remain in balance. This is called *scale-up.* Note that multiplication of the stream amounts or stream flow rates does not affect the stream compositions. If the stream compositions change, the flow diagram will no longer be in balance.

For example, consider a simple process in which 10 lb of sodium chloride is mixed with 190 lb of water to produce 200 lb of solution. This process is depicted in the flow diagram in Fig. 7–3. If all three streams are now multiplied by 2, one will have 20 lb of sodium chloride mixed with 380 lb of water to form 400 lb of solution. The flowchart is still balanced. Note that the fractions of NaCl (1) and water in the output stream have not been changed.

The scale-up constant need not be a pure number. Suppose that one multiplies the mixture described above by a constant, 500 kg/lb. One would still have a balanced flowchart with 100,000 kg of solution produced. The flowchart may also be altered from batch to continuous operation by scaling. In *batch* operation, streams are defined in terms of amounts, whereas in *continuous* operation, streams are defined by flow rates. In this example, one could multiply all streams by the constant, (1000 lb/h)/lb. This would lead to 10,000 lb/h and 190,000 lb/h for the input sodium chloride and water streams, respectively, and 200,000 lb/h for the product stream.

Figure 7–3

Flow diagram for a mixing operation.

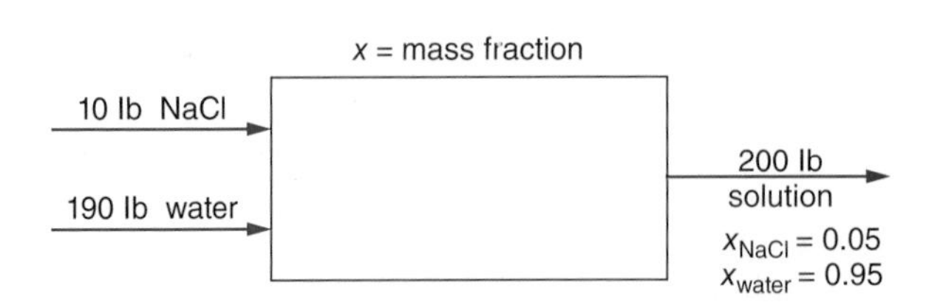

A word of warning regarding the term *constant* is in order. The multiplier constant used for the scale-up must be truly independent of all streams and the material in those streams. Determining whether the constant is a valid one is not always obvious. Suppose, in the last example, that one used (1000 lbmol/h)/lb as the multiplier. Now, there is a problem. This multiplier does vary from stream to stream since $(1000\ lbmol_{NaCl}/h)/lb_{NaCl}$ is not the same as $(1000\ lbmol_{water}/h)/lb_{water}$ because the molecular weight (which is used to convert lb to lbmol) of NaCl is not the same as that of water.

The fact that balanced flowcharts can always be scaled allows the chemical engineer, at the start of calculations, to assume, for the sake of convenience, any amount of material for a given stream. After applying material balances, the process can—as noted—be easily be scaled up or down, or changed from batch to continuous operation.

The assumed amount of material is called a *basis.* If the amount or flow rate of a stream is given in a problem statement, it is usually convenient (but not necessary) to choose this quantity as a basis. If the amount or flow rate is not given, any convenient basis, such as 100 lb or 100 lbmol of one of the streams or a components of one of the streams, can be chosen.

7-4 The Conservation Law for Energy[1]

The *law of conservation of energy,* which, like the law of conservation of mass, holds for all processes that do not involve nuclear reactions, states that energy can be neither created nor destroyed. As a result, the energy level of the system can change only when energy crosses the system boundary:

$$\Delta(\text{energy level of system}) = \text{energy crossing boundary} \qquad (7\text{–}9)$$

NOTE The symbol Δ means *change in* or *difference*.

Energy crossing the boundary can be classified in one of two different ways: heat Q or work W. Heat is energy moving between the system and the surroundings by virtue of a temperature difference driving force. Heat flows from high temperature to low temperature. The entire system is not necessarily at the same temperature; nor are the surroundings. If a portion of the system is at a higher temperature than a portion of the surroundings and, as a result, energy is transferred from the system to the surroundings, that energy is classified as *heat*. If part of the system is at a higher temperature than another part of the system and energy is transferred between the two parts, that energy is *not* classified as heat because it is not crossing the boundary. *Work* is also energy moving between the system and surroundings, but the driving force here is something other than a temperature difference, e.g., a mechanical force, a pressure difference, gravity, a voltage difference, or a magnetic field. Note that the definition of work is a *force acting through a distance*. All of the examples of driving forces just cited can be shown to provide a force capable of acting through a distance.

The energy level of the system has three contributions: kinetic energy, potential energy, and internal energy. Any body in motion possesses kinetic energy. If the system is moving as a *whole,* its kinetic energy E_k is proportional to the mass of the system and the square of the velocity at its center of gravity. The phrase *as a whole* indicates that motion inside the system relative to the system's center of gravity does not contribute to the E_k term, but rather to the internal energy or U term. The terms *external kinetic energy* and *internal kinetic energy* are sometimes used here. An example would be a moving railroad tank car carrying propane gas. (The propane gas is the system.) The center of gravity of the propane gas is moving at the same velocity as that of the train, and this constitutes the system's external kinetic energy. The gas molecules are also moving in random directions relative to the center of gravity, and this constitutes the system's internal energy due to motion inside the system, i.e., internal kinetic energy. The potential energy E_p involves any energy the system as a whole possesses by virtue of its position (more precisely, the position of its center of gravity) in some force field, (e.g., gravity, centrifugal, electrical) that gives the system the potential for accomplishing work. Again, the phrase *as a whole* is used to differentiate between *external* potential energy E_p and *internal* potential energy. *Internal potential energy* refers to potential energy due to force fields inside the system. For example, the electrostatic force fields (bonding) between atoms and molecules provide those particles with the potential for work. The internal energy U is the sum of all internal kinetic and internal potential energy contributions.

The *law of conservation of energy,* which is also called the *first law of thermodynamics,* may now be written as

$$\Delta(U + E_k + E_p) = Q + W \tag{7–10}$$

or equivalently as

$$\Delta U + \Delta E_k + \Delta E_p = Q + W \tag{7–11}$$

It is important to note the sign convention for Q and W defined in Eqs. (7–10) and (7–11). Since any Δ term is always defined as the final minus the initial state, both the heat and work terms must be positive when they cause the system to gain energy, i.e., when they represent energy flowing from the surroundings to the system. Conversely, when the heat and work terms cause the system to lose energy; i.e., when energy flows from the system to the surroundings, the Δ terms are negative in sign. This sign convention is not universal, and the reader must carefully check what sign convention is being used by a particular author when referring to the literature. For example, work is often defined in some texts as positive when the system does work on the surroundings.[3, 5]

When dealing with open systems, it is convenient to divide the work term into two contributions: flow work W_f and shaft work W_s. *Shaft work* is work done on the system by a moving part within the system. The vanes of a centrifugal pump and the piston of a reciprocating compressor are examples. An example of shaft work done by the system (which would be negative by the above mentioned convention) is the turning of a turbine blade by a moving fluid. *Flow work* is work that occurs at every inlet to and outlet from the system. A portion of fluid about to enter a system through a pipeline is *pushing* the fluid immediately ahead of it. The fluid being pushed is part of the system since it has already crossed the boundary. This means that the surroundings (the outside fluid) is performing work on the system. The work done by fluid entering the system in pushing a volume of fluid V_{in} across the system boundary is expressed by

$$W_{in} = P_{in} V_{in} \tag{7–12}$$

At the outlet, the fluid is pushing the fluid ahead of it that is already out. This time, the system is working on the surroundings, and therefore this work W_{out} is negative:

$$W_{out} = P_{out} V_{out} \tag{7–13}$$

The net *flow work* is the sum of the inlet work performed by all inlet streams minus the sum of the outlet work performed by all outlet streams:

$$W_f = P_{in} V_{in} - P_{out} V_{out} \tag{7–14}$$

or

$$W_f = -\Delta(PV) \tag{7–15}$$

Substituting for W_{in} in the conservation of energy or energy balance equation, one may now write that equation as

$$\Delta U + \Delta E_k + \Delta E_p = Q + W_s - \Delta(PV) \tag{7–16}$$

or equivalently

$$\Delta(U + PV) + \Delta E_k + \Delta E_p = Q + W_s \tag{7–17}$$

The *enthalpy is* defined as

$$H = U + PV \tag{7–18}$$

The significance of the *internal energy U* was discussed earlier. The significance of the *enthalpy H* is difficult to pinpoint other than the fact that it is a convenient mathematical grouping of *U*, *P* and *V*. At any rate, this definition allows one to write the energy balance equation for an open *steady-state system* as

$$\Delta H + \Delta E_k + \Delta E_p = Q + W_s \tag{7–19}$$

This equation may also be applied to *closed systems* when the process is carried out at *constant pressure.* For a closed system, the only type of work, outside of shaft work, that can be performed is work done by virtue of an expansion or contraction of the system. This is called *pressure-volume work,* W_{PV}, and is given by

$$W_{PV} = -\int_{V_1}^{V_2} P_{\text{surroundings}}\, dV \tag{7–20}$$

The minus sign is required because of the sign convention employed. Work done by the system, e.g., during expansion, must be negative. For a constant pressure process, this becomes

$$\begin{aligned} W_{PV} &= -\int_{V_1}^{V_2} P_{\text{surroundings}}\, dV \\ &= -PV \\ &= -\Delta(PV) \end{aligned} \tag{7–21}$$

Note that the expression for W_{PV} is the same as that for W_f used earlier. This gives rise to almost the same equation as Eq. (7–11):

$$\Delta H + \Delta E_k + \Delta E_p = Q + W_s \tag{7–22}$$

This equation can also be applied to a closed system *at constant pressure.* In most chemical engineering applications, the ΔE_k and ΔE_p terms are negligible, and many textbooks present the energy balance equations with these two terms omitted.

Before leaving the first law of thermodynamics, certain important terms are defined as follows:

1. *Isothermal* means constant temperature.
2. *Isobaric* means constant pressure.

3. *Isochoric* is constant volume
4. *Adiabatic* specifies no transfer of heat to or from systems.

These topics are revisited in Chaps. 8 and 10.

7–5 The Conservation Law for Momentum

The reader is referred to Chap. 9 for a detailed treatment of this topic. Subject matter is also available in the literature.[6,7]

7–6 Stoichiometry[1]

When chemicals react, they do so according to a strict composition ratio. When oxygen and hydrogen combine to form water, the ratio of the amount of oxygen to the amount of hydrogen consumed is always 7.94 by mass and 0.500 by moles. The term *stoichiometry* refers to this phenomenon, which is sometimes called the *chemical law of combining weights.* The reaction equation for the combining of hydrogen and oxygen is

$$2H_2 + O_2 \rightarrow 2H_2O \tag{7–23}$$

In chemical reactions, atoms are neither generated nor consumed, merely rearranged with different bonding partners. The manipulation of the coefficients of a reaction equation, so that the number of atoms of each element on the left is equal to that on the right, is referred to as *balancing* the equation. Once the equation is balanced, the whole-number molar ratio that must exist between any two components of the reaction can be determined simply by observation; these are known as *stoichiometric ratios.* There are three such ratios (not counting the reciprocals) for the above reaction [Eq. (7–23)]:

2 mol H_2 consumed per mol O_2 consumed
mol H_2O generated per mol H_2 consumed
mol H_2O generated per mol O_2 consumed

The unit mole represents either the gmol or the lbmol. Using molecular weights, these stoichiometric ratios (which are molar ratios) may easily be converted to mass ratios. For example, the first ratio above may be converted to a mass ratio by using the molecular weights of H_2 (2.016) and O_2 (31.999) as follows:

$$(2 \text{ gmol } H_2 \text{ consumed}) [2.016 \text{ g/(gmol)} = 4.032 \text{ g } H_2 \text{ consumed}$$

$$(1 \text{ gmol } O_2 \text{ consumed}) [31.999 \text{ g/(gmol)} = 31.999 \text{ g } O_2 \text{ consumed}$$

The mass ratio between the hydrogen and oxygen consumed is therefore

$$4.032/31.999 = 0.126 \text{ g } H_2 \text{ consumed per g } O_2 \text{ consumed}$$

These molar and mass ratios are used in material balances to find the amounts or flow rates of components involved in chemical reactions.

Multiplying a balanced reaction equation through by a constant does nothing to alter its meaning. The reaction used as an example above is often written as

$$H_2 + \frac{1}{2}O_2 \rightarrow H_2O \tag{7–24}$$

Occasionally, however, care must be exercised because the solution to the problem depends on the manner or form in which the reaction is written. This is the case with chemical equilibrium problems and problems involving thermochemical reaction equations. These are addressed in the next chapter.

Three different types of material balance may be expressed when a chemical reaction is involved: the molecular balance, the atomic balance, and the extent-of-reaction *balance.* It is a matter of convenience which of the three types is used. Each is briefly discussed below.

The *molecular balance* is the same as that described earlier. Assuming a steady-state continuous reaction, the accumulation term A is zero, and, for all components involved in the reaction, the balance equation becomes

$$I + G = O + C \tag{7–25}$$

If a total material balance is performed, this form of the balance equation must be used if the amounts or flow rates are expressed in terms of moles, e.g., lbmol or gmol/h, since the total number of moles can change during a chemical reaction. If, however, the amounts or flow rates are given in terms of mass, such as kg or lb/h, the G and C terms may be dropped, since mass cannot be lost or gained in a chemical reaction. Thus

$$I = O \tag{7–26}$$

In general, however, when a chemical reaction is involved, it is usually more convenient to express amounts and flow rates using moles rather than mass.

A material balance that is not based on the chemicals (or molecules), but rather on the atoms that make up the molecules, is referred to as an *atomic balance.* Since atoms are neither created nor destroyed in a chemical reaction, the G and C terms equal zero and the balance once again becomes

$$I = O \tag{7–26}$$

As an example, consider the combination of hydrogen and oxygen to form water:

$$2H_2 + O_2 \rightarrow 2H_2O \tag{7–23}$$

As the reaction progresses, O_2 and H_2 molecules (or moles) are consumed while H_2O molecules (or moles) are generated. On the other hand, the number of oxygen atoms (or moles of oxygen atoms) and the number of hydrogen atoms (or moles of hydrogen atoms) do not change. Care must also be taken to distinguish between molecular oxygen and *atomic* oxygen. If, in the above reaction, [Eq.(7–23)] one starts out with 1000 lbmol of O_2 (oxygen molecules), one is also starting out with 2000 lbmol of O (oxygen atoms).

The term *extent-of-reaction balance* is derived from the quantitative aspects of determining the amounts of chemicals involved in the reaction in terms of how much of a particular reactant has been consumed or how much of a particular product has been generated. As an example, consider the formation of ammonia from hydrogen and nitrogen:

$$N_2 + 3H_2 \rightarrow 2NH_3 \tag{7–27}$$

Given 10 mol of hydrogen and 10 mol of nitrogen, and letting z represent the amount of nitrogen consumed at the end of the reaction, the output amounts of all three components are $(10 - z)$ for the nitrogen, $(10 - 3z)$ for the hydrogen, and $2z$ for the ammonia. The following is a convenient way of representing the extent-of-reaction balance:

$$\begin{array}{lcccccc} \text{Input} \rightarrow & 10 & & 10 & & 0 & \\ & N_2 & + & 3H_2 & \rightarrow & 2NH_3 & (7\text{–}28) \\ \text{Output} \rightarrow & 10 - z & & 10 - 3z & & 2z & \end{array}$$

Which of the three types of balance—molecular, atomic or extent of reaction—is used to solve a particular problem is mainly a matter of convenience and ease of solution. In solving some problems, the use of one type of balance yields the result more quickly and easily than using the other types. In many instances, the best procedure is to combine the types of balances in the same solution, ensuring that the final set of balances is an *independent* set. Needless to say, the reader should become familiar with all three types of balance.

When methane is burned completely, the stoichiometric equation for the reaction is

$$CH_4 + 2O_2 \rightarrow CO_2 + 2H_2O \tag{7–29}$$

The stoichiometric ratio of the oxygen to the methane is 0.5 mol methane consumed per/mol oxygen consumed. Starting out with 1 mol of methane and 3 mol of oxygen in a reaction vessel, only 2 mol of oxygen would be used up, leaving an excess of one mole. In this case, the oxygen is called the *excess reactant* and methane is the *limiting reactant*. The *limiting reactant* is defined in percent excess terms and is given by

$$\%\ \text{excess} = \frac{n - n_s}{n_s} \times 100\% \tag{7–30}$$

where n = number of moles of excess reactant at start of reaction

n_s = stoichiometric number of moles of excess reactant (i.e., number of moles needed to react completely with limiting reactant)

In the example above, the stoichiometric amount of oxygen is 2 mol since that is the amount that would react with the 1 mol of methane. The *excess* amount of oxygen is 1 mol, which is a *percentage excess* of 50% or a *fractional excess* of 0.50.[8]

As another example using the same reaction equation, suppose one started with 10 mol of methane and 10 mol of oxygen in the reaction vessel. If the reaction goes to completion, all 10 mol of the oxygen would disappear, as would 5 mol of the methane. In this case, oxygen is the *limiting* reactant and the methane is the *excess*. The percent excess of the methane is 100%. The reader is left with the exercise of determining how 100% was arrived at.

When the amounts of reactants are present initially in stoichiometric proportion, there is no excess reactant because none of the reactants will remain if the reaction goes to completion. In the burning of methane, starting amounts of 10 mol of methane and 20 mol of oxygen will cause both reactants to be completely depleted.

In practice, few chemical reactions proceed to completion. There are two reasons for this.

1. All reactions take time to occur. In some cases, such as an explosion, the reaction time is measured in fractions of a second. Many other reactions, however, are quite slow and as the reaction progresses, the reaction rate gets slower and slower. In these cases, it is not economically feasible to wait for the reaction to finish before removing the desired products.

2. Many industrial reactions are reversible (in fact, *all* reactions are reversible to some extent), which means that, while the reactants are combining to form products, the products are also reacting to produce the reactants. If enough time is allowed during a reversible reaction, an equilibrium point is reached where the amounts of reactants and products will no longer be changing. Obviously, this condition will also prevent the *forward reaction* (reactants to products) from reaching completion.

Although a reaction may never be completed, the definitions of *excess* and *limiting* reactants given earlier still apply. Note that both definitions are based on the assumption that the reaction goes to completion, even though this may not actually be the case. How far a reaction goes to completion is measured for *each* reactant i by the *fractional conversion* f_i which is defined as

$$f_i = \frac{\text{amount } i \text{ reacted}}{\text{amount } i \text{ initially present}} \tag{7–31}$$

for batch processes, and

$$f_i = \frac{\text{rate } i \text{ reacted}}{\text{rate } i \text{ input}} \tag{7–32}$$

for continuous processes. The *percentage conversion* is also used and is given by $f_i \times 100\%$. Note that each reactant will usually have a different value of f_i for the same process.

Suppose, in the reaction for the oxidation of sulfur dioxide to sulfur trioxide, that

$$2SO_2 + O_2 \rightarrow 2SO_3 \tag{7–33}$$

where 100 mol of SO_2 and 100 mol of O_2 are present initially in the reaction vessel and after a time lapse, 50 mol of the SO_2 have disappeared. Applying the preceding definition, the fractional conversion of the SO_2 is 50 mol reacted per 100 mol initially present or 0.50. The fractional conversion of the O_2, on the other hand, is 25 mol reacted per 100 mol initially present or 0.25.[8]

A *recycle stream* is often used with a reactor to increase the fractional conversion of a desired product. Suppose, for example,that a 100 lbmol/h stream of pure chemical A is introduced to a reactor in which the reaction

$$A \rightarrow B \tag{7–34}$$

is occurring. If the conversion of A is 50%, the product stream will contain 50 lbmol/h of A and 50 lbmol/h of B. The 50% is referred to as a *single-pass conversion.* If chemical A in the product stream could be separated from chemical B, it could (and often should) be recycled and mixed with the incoming feed stream of 100 lbmol/h of chemical A. This would obviously increase the amount of desirable product B that is obtained from a given amount of the raw material A. In effect, the recycling would lead to a higher *conversion* of A. This increased conversion resulting from recycling is referred to as the *overall conversion.* The distinction between single-pass and overall conversion is shown in the following two equations:

$$f_{\text{overall}} = \frac{A_{\text{input to process}} - A_{\text{output from process}}}{A_{\text{input to process}}} \tag{7–35}$$

$$f_{\text{single-pass}} = \frac{A_{\text{input to reactor}} - A_{\text{output from reactor}}}{A_{\text{input to reactor}}} \tag{7–36}$$

The term *process* in the first definition refers to that part of the flowchart that includes the recycle stream.

Recycling is also used as a method of pollution prevention (see also Chap. 25). If chemical A in the above example is a potential pollutant, decreasing the amount of A in the waste stream provides an added incentive for recycling. The reader is referred to the literature[9,11] for a detailed discussion of recycling as a pollution prevention technique.

Finally, consider the following combustion reaction. The balanced reaction for the combustion of 1.0 lbmol of chlorobenzene (C_6H_5Cl) with stoichiometric oxygen is

$$C_6H_5Cl + 7O_2 \rightarrow 6CO_2 + 2H_2O + HCl \tag{7–37}$$

Table 7–1

Chlorobenzene Stoichiometric Calculations

	C_6H_5Cl	$7O_2$	$[26.3N_2]$	$\rightarrow$	$6CO_2$	$2H_2O$	HCl	$[26.3N_2]$
Moles	1	7	26.3	$\neq$	6	2	1	26.3
MW	112.5	32	28		44	18	36.5	28
Mass	112.5	224	736.4	=	264	36	36.5	736.4

If air, not oxygen, is employed in the process, the reaction with air becomes

$$C_6H_5Cl + 7O_2 + [26.3N_2] \rightarrow 6CO_2 + 2H_2O + HCl + [26.3N_2] \quad (7\text{–}38)$$

where the nitrogen in the air has been retained in brackets on both sides of the equation since it does not participate in the combustion reaction.

The moles and masses involved in this reaction based on the stoichiometric combustion of 1.0 lbmol of C_6H_5Cl are given in Table 7–1.

Thus,

Initial mass = 1072.9	Initial number of moles = 34.3
Final mass = 1072.9	Final number of moles = 35.3

Note that in accordance with the conservation law for mass, the initial and final masses balance. The number of moles, as is typically the case in chemical reaction-combustion calculations, do not balance. The concentrations of the various species may also be calculated. For example,

% CO_2 by weight	=	(264/1072.9)100% = 24.61%
% CO_2 by mol (or volume)	=	(6/35.3)100% = 17.0%
% CO_2 by weight (dry basis)	=	(264/1036.9)100% = 25.46%
% CO_2 by mol (dry basis)	=	(6/33.3)100% = 18.0%

The air requirement for this reaction is 33.3 lb.mol. A Theodore Tutorial, This is stoichiometric or 0% excess air (EA). For 100% EA (100% above stoichiometric), one would use (33.3)(2.0) or 66.6 lbmol of air. For 50% EA one would use (33.3)(1.5) or 50 lbmol of air; for this condition, 16.7 lbmol excess or additional air is employed. If 100% excess air is employed in the combustion of 1.0 lbmol of C_6H_5Cl, the reaction would become

$$C_6H_5Cl + 14O_2 + [52.6N_2] \rightarrow 6CO_2 + 2H_2O + HCl + 7O_2[52.6N_2] \quad (7\text{–}39)$$

References

1. J. Reynolds, *Material and Energy Balances*, A Theodore Tutorial, Theodore Tutorials, East Williston, N.Y., originally published by the U.S. Environmental Protection Agency Air Pollution Training Institute (USEPA/APTI), Research Triangle Park, N.C., 1992.
2. L. Theodore, *Transport Phenomena for Engineers*, Theodore Tutorials, East Williston, N.Y., originally published by International Textbook Co., Scranton, Pa., 1971.
3. I. Farag and J. Reynolds, *Heat Transfer*, A Theodore Tutorial, Theodore Tutorials, East Williston, N.Y., a text originally published by the USEPA/APTI, Research Triangle Park, N.C., 1996.
4. L. Theodore and J. Reynolds, *Thermodynamics*, A Theodore Tutorial, Theodore Tutorials, East Williston, N. Y., originally published by the USEPA/APTI, Research Triangle Park, N.C. 1991.

5. L. Theodore, F. Ricci, and T. VanVliet, *Thermodynamics for the Practicing Engineer*, Wiley, Hoboken, N.J., 2009.
6. I. Farag, *Fluid Flow*, A Theodore Tutorial, Theodore Tutorials, East Williston, N.Y., originally published by the USEPA/APTI, Research Triangle Park, N.C., 1994.
7. P. Abulencia and L. Theodore, *Fluid Flow for the Practicing Chemical Engineer*, Wiley, Hoboken, N.J., 2009.
8. L. Theodore, *Chemical Reactor Design and Application for the Practicing Engineer*, Wiley, Hoboken, N.J., 2012.
9. M. K. Theodore and L. Theodore, *Introduction to Environmental Management*, CRC Press/Taylor & Francis Group, Boca Raton, Fla., 2010.
10. R. Dupont, L. Theodore, and K. Ganesan, *Pollution Prevention*, CRC Press/Taylor & Francis Group, Boca Raton, Fla., 2000.
11. R. Dupont, K. Ganesan, and L. Theodore, *Pollution Prevention, Sustainability, Industrial Toxicology, and Green Science and Engineering*, CRC Press/Taylor & Francis Group, Boca Raton, Fla., 2014 (in preparation).
12. J. Santoleri, J. Reynolds, and L. Theodore, *Introduction to Hazardous Waste Incineration*, 2nd ed., Wiley, Hoboken, N.J., 2000.

8 Thermodynamics

Chapter Outline

8–1 Introduction

Thermodynamics was once defined by Webster as "the science which deals with the intertransformation of heat and work." The fundamental principles of thermodynamics are contained in the first, second, and third laws of thermodynamics. These principles have been defined as "pure" thermodynamics. These laws were developed and extensively tested in the latter half of the nineteenth century and are essentially based on experience. (The third law was developed during the twentieth century.)

Practically all thermodynamics, in the ordinary meaning of the term is *applied thermodynamics* in that it is essentially the application of these three laws, coupled with certain facts and principles of mathematics, physics, and chemistry, to problems in chemical engineering. The fundamental laws are of such generality that it is not surprising that these laws find application in other disciplines, including physics, chemistry, and mechanical engineering.

One of the more important equations in thermodynamics is the *Gibbs phase rule* (GPR), which is applied in several chemical engineering and science areas, including phase equilibrium. The reader is introduced to the term *degrees of freedom* before proceeding to the GPR. The degrees of freedom F or the *variance* of a system is defined as the *smallest number* of independent variables (such as pressure, temperature, concentration) that must be specified in order to completely define (the remaining variables of) the system. The significance of the degrees of freedom of a system may be drawn from the following example. In order to specify the density of a pure gaseous (vapor) stream, it is necessary to state both the temperature and pressure to which this density corresponds. For example, the density of steam (see also steam tables) has a particular value at 200°C and 1 atm pressure. Obtaining a value of the density at 200°C without mention of pressure does not clearly define the state of the steam for at 200°C, the steam may exist at many other possible pressures. Similarly, mention of the pressure without the temperature leaves ambiguity. Therefore, two variables must be specified for the complete description of the state of the steam. Also, this phase, when present alone in a system, possesses 2 degrees of freedom (F), or the system is said to be *bivariant*.

There is a definite relation in a system between the number of degrees of freedom, the number of components, and the number of phases present. This relationship was first established by J. Willard Gibbs in 1876. This relation, known as the *Gibbs phase rule*, is a principle of the widest generality. It is one of the most often used rules in thermodynamic analyses, particularly in the representation of equilibrium conditions existing in heterogeneous systems. Theodore et al.[1] provide a mathematical derivation of the phase rule. The equation is as follows:

$$F = C - P + 2 \tag{8–1}$$

Equation (8–1) is the celebrated phase rule of Gibbs. The F term is the aforementioned number of degrees of freedom in a system and provides the number of variables whose values must be specified (arbitrarily) before the state of the system can be completely and unambiguously characterized. According to this rule, the number of degrees of freedom of a system is therefore related to the number of components and the number of phases present.

Theodore et al. also provide a qualitative review of the second law.[1] As described in Chap. 7, the first law of thermodynamics is a conservation law for energy transformations.

---

*The bulk of the material presented in this chapter has been drawn from Theodore and Reynolds.[2]

Regardless of the types of energy involved in a processes (thermal, mechanical, electrical, elastic, magnetic, etc.), the energy change of a system is equal to the difference between energy input and energy output. The first law also allows free *convertibility* from one form of energy to another, as long as the overall energy quantity is conserved. Thus, this law places no restriction on the conversion of work into heat, or on its counterpart: the conversion of heat into work.

The unrestricted conversion of work into heat is well known to most chemical engineers. Frictional effects are frequently associated with mechanical forms of work that result in a temperature rise of the bodies in contact. However, the transformation of heat into work is of greater concern to the practicing chemical engineer. In nations with a partially developed or developing technological society, the ability to produce energy in the form of work takes on more significant importance. Work transformations are necessary to transport people and goods, drive machinery, pump liquids, compress gases, and provide energy input to so many other processes that are taken for granted in highly developed societies such as those in the United States. Much of the work input in these societies is available in the form of electrical energy, which can then be converted to rotational mechanical work. Although some of this electrical energy (work) is produced by hydroelectric power plants, by far the greatest part of it is obtained from the combustion of fossil or nuclear fuels. These fuels allow the chemical engineer to produce a relatively high-temperature gas or liquid stream that acts as a thermal (energy) source for the production of work. Hence, the study of the conversion of heat to work is extremely important, especially in the light of developing shortages of fossil and nuclear fuels, along with the accompanying environmental problems, particularly with questionable concerns regarding global warming.

The brief discussion of energy conversion devices above leads to an important second-law consideration: that energy has *quality* as well as quantity. Because work is 100 percent convertible to heat whereas the reverse situation is *not* true, work is a *more valuable* form of energy than heat. Although it is not as obvious, it can also be shown through second-law arguments that work has a quality-related value, which varies according to the temperature at which heat is discharged from a system. The higher the temperature at which heat transfer occurs, the greater the potential for energy transformation into work. Thus, thermal energy stored at higher temperatures is generally more useful to society than that available at lower temperatures. While there is an immense quantity of energy stored in the oceans, for example, its present *availability* to society for performing useful tasks is essentially nonexistent.

The choice of topics to be reviewed in this chapter was initially an area of concern. After some deliberation, it was decided to provide an introduction to five areas that many have included (at one time or another) in this broad engineering subject: enthalpy effects, chemical reaction enthalpy effects, second-law calculations, phase equilibrium, and chemical reaction equilibrium.

8–2 Enthalpy Effects

There are many different types of enthalpy effects, including the following:

1. Sensible (temperature)
2. Latent (phase)
3. Dilution (with water), e.g., HCl with H_2O
4. Solution (nonaqueous), e.g., HCl with a solvent other than H_2O
5. Reaction (chemical)

This section is concerned with effects 1 to 5.

Sensible Enthalpy Effects

Sensible enthalpy effects are associated with temperature. This subsection provides methods that can be employed to calculate these changes. These methods include the use of

1. Enthalpy values
2. Average heat capacity values
3. Heat capacity as a function of temperature

Detailed calculations on effects 1 to 3 are provided by Theodore et al.[1,2] Two expressions for heat capacity are considered in topic 3 employing a, b, c constants and α, β, γ constants.

If enthalpy values are available, the enthalpy change is given by

$$\Delta h = h_2 - h_1 \qquad \text{(mass basis)} \tag{8–2}$$

$$\Delta H = H_2 - H_1 \qquad \text{(mole basis)} \tag{8–3}$$

If average molar heat capacity data are available, Eq. 8–5 may be integrated to yield

$$\Delta H = \bar{C}_p \Delta T \tag{8–4}$$

where $\bar{C}_p$ is the average molar value of C_p in the temperature range ΔT. Average molar heat capacity data are provided in the literature.[1–4]

A more rigorous approach to enthalpy calculations can be provided if heat capacity variation with temperature is available. If the heat capacity is a function of the temperature, the enthalpy change is written in differential form:

$$dH = C_p \, dT \tag{8–5}$$

If the temperature variation of the heat capacity is given by

$$C_p = \alpha + \beta T + \gamma T^2 \tag{8–6}$$

then Eq. (8–5) may be integrated directly between some reference or standard temperature (T_0) and the final temperature (T_1):

$$\Delta H = H_1 - H_0 \tag{8–7}$$

$$\Delta H = \alpha(T_1 - T_0) + \frac{\beta}{2}(T_1^2 - T_0^2) + \frac{\gamma}{3}(T_1^3 - T_0^3) \tag{8–8}$$

Equation (8–5) may also be integrated if the heat capacity is a function of temperature of the form

$$C_p = a + bT + cT^{-2} \tag{8–9}$$

The enthalpy change is then given by

$$\Delta H = a(T_1 - T_0) + \frac{b}{2}(T_1^2 - T_0^2) + c(T_1^{-1} - T_0^{-1}) \tag{8–10}$$

Tabulated values of α, β, γ and a, b, c for a host of compounds (including some chlorinated organics are available in the literature.[2]

Latent Enthalpy Effects

The term *phase*, when used to describe a pure substance, refers to a state of matter that is a gas, liquid, or solid. *Latent* enthalpy effects are associated with *phase* changes. These phase changes involve no change in temperature, but there is a transfer of energy to and from the substance. There are three possible latent effects: (1) vapor-liquid, (2) liquid-solid, and (3) vapor-solid.

Vapor-liquid changes are referred to as *condensation* when the vapor is condensing and *vaporization* when liquid is vaporizing. *Liquid-solid* changes are referred to as *melting* when the solid melts to liquid and *freezing* when a liquid solidifies. *Vapor-solid* changes are referred to as *sublimation*. Also, there are enthalpy effects associated with a phase change of a solid to a solid of another form; however, this enthalpy effect is small compared to the other effects mentioned above.

When a pure substance is vaporized ...

$$\Delta H = T\Delta v \frac{dp'}{dt} \tag{8–11}$$

where ΔH = latent enthalpy of vaporization or condensation, energy/mass
Δv = specific volume change accompanying the phase change at temperature T, volume/mass
(dp'/dT) = rate of change of vapor pressure with temperature, pressure/degrees absolute

The term (dp'/dT) is the slope of the vapor pressure–temperature curve at the temperature in question.

Equation (8–11) can be integrated to give

$$\ln\left(\frac{p_2'}{p_1'}\right) = -\frac{\Delta H(\text{at } T_{av})}{R}\left(\frac{1}{T_2} - \frac{1}{T_1}\right) \tag{8–12}$$

where $T_{av} = (T_2 - T_1)/2$.

Vapor pressure data is available in the literature.[5] However, two equations can be used in lieu of vapor pressure information: the Clapeyron equation and the Antoine equation. The two-constant (A, B) Clapeyron equation is given by

$$\ln p' = A - \frac{B}{T} \tag{8–13}$$

where p' and T are the vapor pressure and temperature, respectively, with appropriate units specified. The three-constant (A, B, C) Antoine equation is given by

$$\ln p' = A - \frac{B}{T + C} \tag{8–14}$$

Note that for both equations, the units of p' and T must be specified for given values of A, B, and (possibly) C. Values for the Clapeyron equation coefficients (A and B) are available in the literature.[1–3,5]

Enthalpy of Mixing

When two or more pure substances are mixed to form a solution, a heat effect invariably results. The *heat* (or *enthalpy*) *of mixing* is defined as the enthalpy change that occurs when two or more pure substances at a reference state are mixed at constant temperature and pressure to form a solution. When the mixing occurs in the presence of water (H_2O), the process is normally defined as a *dilution* effect. If mixing occurs in the presence of a solvent other than H_2O, it is defined as a *solution* effect. For example, the mixing of HCl with H_2O is considered a dilution process, while the mixing of HCl with H_2SO_4 is a solution process.

Enthalpy concentration diagrams offer a convenient way to calculate enthalpy of mixing effects and temperature changes associated with this type of process. These diagrams, for a two-component mixture, are graphs of the enthalpy of a binary solution plotted as a function of composition (mole fraction or weight fraction of one component) with the temperature as a parameter. For an ideal solution, isotherms on an enthalpy–concentration diagram would be straight lines. For a real solution, the actual isotherm is usually displaced vertically from the ideal-solution isotherm at a given point by the value of ΔH at that point, where ΔH is the enthalpy of mixing. This indicates that heat must be evolved whenever the pure components at the given temperature are mixed to form a solution at the same temperature. Such a system is said to be *exothermic*. For an *endothermic* system, the heats of solution are positive; i.e., solution at constant temperature is accompanied by the absorption of heat. Organic mixtures often fit this description. These enthalpy concentration diagrams are particularly useful in solving problems involving *adiabatic* mixing.

The reader should note that there is also an enthalpy effect when gases or solids are dissolved in liquids. This effect is normally small but can be significant in some applications, particularly with acid gases in water (see also Chap. 11). This type of calculation should be included in some absorber systems when temperature changes through the unit are significant.[6]

When gases or solids are dissolved in liquids, the enthalpy effect is referred to as the *enthalpy of solution*, and is normally based on the solution of 1.0 mol of solute (gas or solid). If component 1 is the solute, x is normally defined as the mole fraction of solute in the solution. With ΔH defined as the enthalpy effect of solution, the term $\Delta H/x_1$ represents the enthalpy effect per mole of solute so that

$$\overline{\Delta H} = \frac{\Delta H}{x_1} \tag{8–15}$$

where $\overline{\Delta H}$ is the enthalpy of solution based on a mole of solute. An important property of this system is that the enthalpy at infinite dilution $\overline{\Delta H}_\infty$ represents the value of $\overline{\Delta H}$ as n approaches infinity (∞). For a HCl(g) – H_2O(l) system at 25°C, $\Delta H_\infty = -75$ kJ/gmol HCl. In effect, this represents the quantity of energy liberated when 1 gmol of HCl at 25°C is mixed with an infinite number of gmoles of H_2O at 25°C to produce a final solution at 25°C. Data on enthalpy of mixing at different dilutions is understandably limited. Selected values for dilution with water were employed to develop the following model:[6]

$$\overline{\Delta H} = \Delta H_\infty [1 - e^{-an}];\ n = \text{mol } H_2O/\text{mol solute} \tag{8–16}$$

where coefficient $a = 0.4375$ at approximately 25°C. In lieu of data for other systems, it is recommended that Eq. (8–16) be used with ΔH_∞ as at the value for the system in question.

Chemical Reaction Enthalpy Effects

The equivalence of mass and energy needs to be addressed qualitatively. This relationship is important only in nuclear reactions, the details of which are beyond the scope of this text. The energy-related effects discussed in this section arise because of the rearrangement of electrons outside the nucleus of the atom. However, it is the nucleus of the atom that undergoes rearrangement in a nuclear reaction, releasing a significant quantity of energy; this process occurs with a miniscule loss of mass. The classic Einstein equation relates E to mass, as provided in Eq. (8–17).

$$\Delta E = (\Delta m)c^2 \tag{8–17}$$

where ΔE = energy change
Δm = decrease in mass
c = velocity of light

The standard enthalpy (heat) of reaction can be calculated from standard enthalpy of formation data. To simplify the presentation that follows, examine the author's favorite equation:

$$a(\text{A}) + b(\text{B}) \rightarrow c(\text{C}) + d(\text{D}) \tag{8–18}$$

If this reaction is assumed to occur in a standard (or reference) state, then the standard enthalpy of reaction ΔH^0, is given by

$$\Delta H^0 = c(\Delta H_f^0)_C + d(\Delta H_f^0)_D - a(\Delta H_f^0)_A - b(\Delta H_f^0)_B \tag{8–19}$$

where $(\Delta H_f^0)_i$ is the standard enthalpy of formation of species i. Thus, the (standard) enthalpy of a reaction is obtained by taking the difference between the (standard) enthalpy of formation of products and reactants. If the (standard) enthalpy of reaction or formation is negative (exothermic), as is the case with most combustion reactions, then energy is liberated due to the chemical reaction. Energy is absorbed if ΔH^0 is positive (endothermic).

Tables of enthalpies of formation and reaction are available in the literature (particularly thermodynamics texts and reference books) for a wide variety of compounds.[1–5] It is important to note that these are valueless unless the stoichiometric equation and the state of the reactants and products are included.

Theodore et al.[1,2] provide equations to describe the effect of temperature on the enthalpy of reaction. For heat capacity data in α, β, γ form, one obtains

$$\Delta H_T^0 = \Delta H_{298}^0 + \Delta\alpha(T - 298) + \frac{1}{2}\Delta\beta(T^2 - 298^2) + \frac{1}{3}\Delta\gamma(T^3 - 298^3) \tag{8–20}$$

and for the reaction presented in Eq. (8–18)

$$\begin{aligned}\Delta\alpha &= c\alpha_C + d\alpha_D - a\alpha_A - b\alpha_B \\ \Delta\beta &= c\beta_C + d\beta_D - a\beta_A - b\beta_B \\ \Delta\gamma &= c\gamma_C + d\gamma_D - a\gamma_A - b\gamma_B\end{aligned} \tag{8–21}$$

For heat capacity data in a, b, c form, one obtains

$$\Delta H_T^0 = \Delta H_{298}^0 + \Delta a(T - 298) + \frac{1}{2}\Delta b(T^2 - 298^2) + \Delta c(T^{-1} - 298^{-1}) \tag{8–22}$$

The two other terms that have been used are the gross (or higher) heating value and the net (or lower) heating value.

The gross heating value (HV_G) represents the enthalpy change or heat released when a gas is stoichiometrically combusted at 60°F, with the final (flue) products at 60°F and any water present in the liquid state. Stoichiometric combustion requires that no oxygen be present in the flue gas combustion of the hydrocarbon. The net heating value, (HV_N) is similar to HV_G except the water is in the vapor state. The net heating value is also known as the *lower heating value,* and the gross heating value is known as the *higher heating value*. Note that (HV_G) is greater than (HV_N); the difference is simply the heat of vaporization of any water present.

8–3 Second-Law Calculations[2]

The law of conservation of energy has already been defined as the *first law of thermodynamics.* Its application allows calculations of energy relationships associated with all kinds of processes. The "limiting" law is called the *second law of thermodynamics* (SLT). Applications involve calculations for maximum power outputs from a power plant and equilibrium yields in chemical reactions. In principle, this law states that water cannot flow uphill and heat cannot flow from a cold to a hot body of its own accord. Other defining statements for this law that have appeared in the literature are provided here:

1. Any process, the sole net result of which is the transfer of heat from a lower temperature level to a higher one, is impossible.
2. No apparatus, equipment, or process can operate in such a way that its only effect (on system and surroundings) is to convert heat taken in completely into work.
3. It is impossible to convert the heat taken into a system completely into work in a cyclical process.

The second law also serves to define another important thermodynamic function called *entropy*, which is normally denoted as S. The change in S for a reversible adiabatic process is always zero:

$$\Delta S = 0 \tag{8–23}$$

For liquids and solids, the entropy change for a system undergoing a temperature change from T_1 to T_2 is given by

$$\Delta S = C_p \ln\frac{T_2}{T_1} \qquad (C_p = \text{constant}) \tag{8–24}$$

The entropy change of an ideal gas undergoing a change of state from P_1 to P_2 at a constant temperature T is given by

$$\Delta S_T = R \ln\frac{P_1}{P_2}, \quad \text{Btu/(lbmol}\cdot{}^\circ\text{R)} \tag{8–25}$$

The entropy change of one mole of an ideal gas undergoing a change of state from T_1 to T_2 at a constant pressure is given by

$$\Delta S_p = C_p \ \ln\frac{T_2}{T_1} \qquad [C_p(\text{gas}) = \text{constant}] \tag{8–26}$$

Correspondingly, the entropy change for an ideal gas undergoing a change from (P_1, T_1) to (P_2, T_2) is

$$\Delta S = R \ \ln\frac{P_1}{P_2} + C_p \ \ln\frac{T_2}{T_1} \tag{8–27}$$

Some fundamental facts relative to the entropy concept are discussed below. The entropy change of a system may be positive (+), negative (–), or zero (0); the entropy change of the surroundings during this process may likewise be positive, negative, or zero. However, the total entropy change ΔS_T must be equal to or greater than zero:

$$\Delta S_T \geq 0 \tag{8–28}$$

The equality sign applies if the change occurs *reversibly* and *adiabatically*.

The *third law of thermodynamics* is concerned with the absolute values of entropy. By definition, the entropy of all pure crystalline materials at absolute zero temperature is exactly zero. Note, however, that chemical engineers are usually concerned with *changes* in thermodynamic properties, including entropy.

8–4 Phase Equilibrium

Relationships governing the equilibrium distribution of a substance between two phases, particularly gas and liquid phases, are the principal subject matter of phase equilibrium thermodynamics. These relationships form the basis of calculational procedures that are employed in the design and the prediction of the performance of several mass transfer processes.[7]

The term *phase,* for a pure substance, indicates a state of matter, i.e., solid, liquid, or gas. For mixtures, however, a more stringent connotation must be used, since a totally liquid or solid system may contain more than one phase (e.g., a mixture of oil and water). A phase is characterized by uniformity or homogeneity; the same composition and properties must exist throughout the phase region. At most temperatures and pressures, a pure substance normally exists as a single phase. At certain temperatures and pressures, two or perhaps even three phases can coexist in equilibrium.

Chemical engineering calculations rarely involve single (pure) components. Phase equilibria for multicomponent systems are considerably more complex, mainly because of the addition of composition variables. For example, in a ternary or three-component system, the mole fractions of two of the components are pertinent variables along with temperature and pressure. In a single-component system, dynamic equilibrium between two phases is achieved when the rate of molecular transfer from one phase to the second one equals that in the opposite direction. In multicomponent systems, the equilibrium requirement is more stringent—the rate of transfer of each component must be the same in both directions.

Generally, phase equilibrium examines the physical properties of various classes of mixtures and analyzes how different components affect each other within those mixtures. There are three classes of two-phase mixtures:

1. Vapor-liquid
2. Vapor-solid
3. Liquid-solid

A *vapor-liquid-solid* phase equilibrium occurs at the triple point, but this class has little to no industrial applications.

Because of its importance to chemical engineers, the development to follow will address solely vapor–liquid equilibrium. A *vapor–liquid* phase equilibrium example involving raw data is the psychrometric or humidity chart. A *humidity chart* is used to describe the properties of moist air and to calculate moisture content in air. The ordinate of the chart is the absolute humidity H, which is defined as the mass of water vapor per mass of bone-dry air. (Some charts base the ordinate on moles instead of mass.)

The most important equilibrium phase relationship is that between a liquid and a vapor. Raoult's and Henry's laws theoretically describe liquid-vapor behavior and under certain conditions are applicable in practice. Raoult's law is sometimes useful for mixtures of components of similar structure. It states that the partial pressure of any component in the vapor phase is equal to the product of the vapor pressure of the pure component and the mole fraction of that component in the liquid:

$$p_i = p_i' x_i \tag{8–29}$$

where p_i = partial pressure of component i in vapor
p_i' = vapor pressure of pure i at same temperature
x_i = mole fraction of component i in liquid

This expression may be applied to all components. If the gas phase is ideal, this equation becomes

$$y_i = \frac{p_i'}{P} x_i \tag{8–30}$$

where y_i = mole fraction component i in vapor
P = total system pressure

Thus, the mole fraction of water vapor in a gas that is saturated, i.e., in equilibrium contact with pure water ($x = 1.0$) is given simply by the ratio of the vapor pressure, of water at the system temperature divided by the system pressure.[1] This finds application in many quench, distillation, and absorber calculations.[7]

Unfortunately, relatively few mixtures follow Raoult's law. Henry's law is a more empirical relationship used for representing data on many systems. Here

$$p_i = H_i x_i \tag{8–31}$$

where H_i is Henry's law constant for component i (in units of pressure). If the gas behaves ideally, Eq. (8–31) may be written as

$$y_i = m_i x_i \tag{8–32}$$

where m_i is a constant (dimensionless). This is a more convenient form of Henry's law to work with. The constant m_i (or H_i) has been determined experimentally for a large number of compounds and is usually valid at low concentrations. In word equation form, Henry's law states that the partial pressure of a solute in equilibrium in a solution is proportional to its mole fraction in the limit of zero concentration.

Henry's law may be assumed to apply for most dilute solutions. This law finds widespread use in absorber calculations since the concentration of the solute in some process gas streams is often dilute. This greatly simplifies the study and design of absorbers.[7] One should note, however, that Henry's law constant is a strong function of temperature.

In some engineering applications, mixtures of condensable vapors and noncondensable gases must be handled. A common example is water vapor and air; a mixture of organic vapors and air is another such example that often appears in air pollution applications. Condensers can be used to recover and/or control organic emissions to the atmosphere by lowering the temperature of the gaseous stream, although an increase in pressure will produce the same result. The calculation for this process can be accomplished using the *phase equilibrium constant* (K_i). This constant has been referred to in industry as *a componential split factor* since it provides the ratio of the mole fractions of a component in two equilibrium phases. The defining equation is

$$K_i = \frac{y_i}{x_i} \tag{8–33}$$

where K_i is the phase equilibrium constant for component i (dimensionless). This equilibrium constant is a function of the temperature, pressure, the other properties of in the system, and the mole fractions of these components. However, as a first approximation, K_i is generally treated as a function only of the temperature and pressure. For ideal-gas conditions, K_i may be approximated by

$$K_i = \frac{p_i'}{P} \tag{8–34}$$

where p' is the vapor pressure.

Many of the phase equilibrium calculations involve hydrocarbons. Fortunately, most mixtures of hydrocarbons approach ideal-gas behavior over a fairly wide range of temperature and pressure. Values for K_i for a large number of hydrocarbons are provided in two de Priester nomographs. These two nomographs or charts were originally developed by de Priester in 1953.[8] Both have withstood the test of time and continue to be used in some phase equilibrium calculations.

Bubble-point (BP) and dew-point (DP) calculations are centered around the equation:

$$y_i = K_i x_i \tag{8–35}$$

It is important to understand how and why a vapor is converted into a liquid and when a liquid is converted into a vapor. Four key definitions involving this conversion process follow. The *bubble-point temperature* (BPT) is the temperature at which the first bubble of vapor forms when the liquid is heated slowly at a constant pressure. The *bubble-point pressure* (BPP) is the pressure at which the first bubble of vapor forms when the pressure above a liquid is reduced at a constant temperature. The *dew-point temperature* (DPT) is the temperature at which the first drop of liquid forms when the vapor is cooled at a constant pressure, while the *dew-point pressure* (DPP) is the pressure at which the first drop of liquid appears when the pressure of a vapor is increased at a constant temperature.

In a bubble-point calculation, x_i values are normally known and the y_i values can be calculated from

$$\sum y_i = \sum (K_i x_i) = 1.0 \tag{8–36}$$

A trial-and-error calculation on either the temperature or pressure is dictated in order to determine the BPT or BPP, respectively. In a dew-point calculation (DPT or DPP), the procedure is reversed and

$$\sum x_i = \sum \frac{y_i}{K_i} = 1.0 \tag{8–37}$$

Summarizing, if the system pressure P is above the BPP, the state is all liquid. If the system pressure is below the DPP, the state is all vapor. If the system pressure between the BPP and DPP, the state is a mixture of vapor and liquid.

If the system is ideal, Raoult's law presented above can also be used to perform these calculations. If the system is not ideal, other equations are available to account for the deviation from ideality. This information can be better understood by providing either a *P–x, y* (pressure–mole fraction liquid, mole fraction vapor) or *T–x, y* (temperature) diagram for a binary system under study. A *T–x, y* diagram is a graph of equilibrium temperatures versus the liquid and vapor mole fraction of one of the species of the binary system. The diagrams are made up of two curves, a liquid curve and a vapor curve, with a two-phase region existing between these two curves. The development to follow will focus on the generation and interpretation of these diagrams in the remainder of the section.

As noted above, vapor-liquid equilibrium (VLE) relationships for a binary mixture are often provided as the *P–x, y* diagram or a *T–x, y* diagram, or both, and it is these two types of diagrams that are of interest to the chemical engineer. VLE data can be generated assuming Raoult's law applies. VLE diagrams can also be developed for two-component systems where the liquid phase is not assumed to be ideal.[1–3,5,6] Once again, the same two types of diagrams are of interest: *P–x,y* and *T–x,y*. Procedures for generating these figures are available in the literature.[1,3]

The remainder of this section on phase equilibrium deals exclusively with VLE applications. The de Priester charts are a valuable source of VLE data for many hydrocarbons that approach ideal behavior. However, the de Priester chart data are based on experimental data. The fact that these compounds approach ideal behavior allows the data to be presented in a simple form, i.e., as a function solely of temperature and pressure.

Although Raoult's law was included in the early analysis, nonideal deviations can be accounted for by two theoretical models that have been verified by rather extensive experimental data: The two models are Wilson's method[9] and the nonrandom two-liquid (NRTL) model.[10] Details follow.

In the case where liquid solutions cannot be considered ideal, Raoult's law will give highly inaccurate results. For these nonideal liquid solutions, various alternatives are available; these are considered in most standard thermodynamics texts and will be treated to some extent below. One important case will be mentioned because it is frequently encountered in distillation, namely, where the liquid phase is not an ideal solution but the pressure is low enough to permit the vapor phase to behave as an ideal gas. In this case, the deviations from ideality are localized in the liquid and treatment is possible by quantitatively considering deviations from Raoult's law. These deviations are taken into account by incorporating a *correction* (or finagle) factor γ into Raoult's law. The purpose of γ, defined as the *activity coefficient*, is to account for the departure of the liquid phase from ideal solution behavior. It is introduced into the Raoult's law equation (for component *i*) as follows:

$$y_i P = p_i = \gamma_i x_i p_i' \tag{8–38}$$

This activity coefficient is a function of liquid phase composition and temperature. Phase-equilibria problems of the above mentioned type are often effectively reduced to evaluating γ.

Vapor-liquid equilibrium calculations performed with Eq. (8–38) are slightly more complex than those made with Raoult's law. The key equation becomes

$$P = \sum p_i = \sum (y_i P) = \sum (\gamma_i x_i p_i') \tag{8–39}$$

When applied to a two-component (A-B) system, Eq. (8–39) becomes

$$P = y_A P + y_B P = \gamma_A x_A p_A' + \gamma_B x_B p_B' \tag{8–40}$$

so that

$$y_A P = p_A = \gamma_A x_A p_A' \tag{8–41}$$

$$y_B = p_B = \gamma_B x_B p_B' \tag{8–42}$$

Methods (available in the literature) to determine the activity coefficient(s) now follow.

Theoretical developments (often referred to as *molecular thermodynamics*) of liquid-solution behavior are often based on the concept of *local composition*. Within a liquid solution, local compositions, different from the overall mixture composition, are presumed to account for the short-range order and nonrandom molecular orientations that result from differences in molecular size and intermolecular forces. The concept was introduced by G. M. Wilson in 1964 with the publication of a model of solution behavior, since known as the *Wilson equation*.[9] The success of this equation in the correlations of VLE data prompted the devolvement of several alternative local-composition models. Perhaps the most notable of these is the nonrandom two-liquid (NRTL) equation of Renon and Prausnitz.[10]

8–5 Chemical Reaction Equilibrium

With regard to chemical reactions, there are two important questions that are of concern to the chemical engineer: (1) how *far* will the reaction go and (2) how *fast* will the reaction be? Chemical thermodynamics provides the answer to the first question; however, it tells nothing about the second. Reaction rates fall within the domain of chemical reaction kinetics. To illustrate the difference and importance of both questions in an engineering analysis of a chemical reaction, consider the following process. Substance A, which costs 1 cent/ton, can be converted to B, which is worth $1 million/lb, by the reaction A ↔ B. Chemical thermodynamics will provide information on the maximum amount of B that can be formed. If 99.99% of A can be converted to B, the reaction would then appear to be economically feasible from a *thermodynamic* perspective. However, a *kinetic* analysis might indicate that the reaction is so slow that, for all practical purposes, its rate is vanishingly small. For example, it might take 10^6 years to obtain a 10^{-6} percent conversion of A. The reaction is then economically unfeasible. Thus, it can be seen that both equilibrium and kinetic effects must be considered in an overall engineering analysis of a chemical reaction.[2] This topic is further discussed in Chap. 13.

A rigorous, detailed presentation of this topic is beyond the scope of this text. However, a superficial treatment is presented in the hope that it may provide at least a qualitative introduction to chemical reaction equilibrium. As will be shown in Chap. 13, if a chemical reaction is conducted in which reactants go to products, the products will be formed at a rate governed (in part) by the concentration of the reactants and conditions such as temperature and pressure. Eventually, as the reactants form products and the products react to form reactants, the *net* rate of reaction must equal zero. At this point, equilibrium will have been achieved.

Chemical reaction equilibrium calculations are structured around a thermodynamic term referred to as *free energy*. This so-called energy (G) is a thermodynamic property that cannot be easily defined without some basic grounding in thermodynamics. It will not be defined here, and the interested reader is directed to the literature[2] for further development of this topic. Note that free energy has the same units as enthalpy and may be used on a mole or total mass basis.

Consider the *equilibrium* reaction

$$a(\mathrm{A}) + b(\mathrm{B}) = c(\mathrm{C}) + d(\mathrm{D}) \tag{8–43}$$

For this reaction

$$\Delta G^0_{298} = c(\Delta G^0_f)_C + d(\Delta G^0_f)_D - a(\Delta G^0_f)_A - b(\Delta G^0_f)_B \tag{8–44}$$

The standard free energy of reaction ΔG^0 may be calculated from standard free energy of formation data, in a manner similar to that for the standard enthalpy of reaction. The following equation is used to calculate the chemical reaction equilibrium constant K at a temperature T:

$$\Delta G^0_T = -RT \ln(K_T) \tag{8–45}$$

The effect of temperature on the standard free energy of reaction ΔG_T^0, and the chemical reaction equilibrium constant K are described below. Some readers may choose to review Sec. 8–2 for more information on heat capacity variations with temperature and (standard) enthalpy of reaction details. If the molar heat capacity for each chemical species taking part in the reaction is known and can be expressed as a power series in T (kelvins), then

$$C_P = \alpha + \beta T + \gamma T^2 \tag{8–46}$$

If $\Delta\alpha$ and so on represents the difference between the values for the products and the reactants multiplied by their respective stoichiometric coefficients, then one can show that

$$\ln K_T = -\frac{\Delta H_0}{RT} - \frac{\Delta\alpha}{R}\ln T + \frac{\Delta\beta}{2R}T - \frac{\Delta\gamma}{6R}T^2 + I \tag{8–47}$$

and

$$\Delta G_T^0 = -\Delta H_0 - (\Delta\alpha)T\ln T - \frac{\Delta\beta}{2}T^2 - \frac{\Delta\gamma}{6}T^3 - IRT \tag{8–48}$$

Two unknowns, ΔH_0 and I, appear in Eqs. (8–47) and (8–48). The term ΔH_0 may be evaluated from the equation

$$\Delta H_T^0 = \Delta H_0 + (\Delta\alpha)T + \frac{\Delta\beta}{2}T^2 + \frac{\Delta\gamma}{3}T^3 \tag{8–49}$$

if ΔH_T^0 is given (or available) at a specified temperature (usually 25°C). The term I can be calculated if ΔG_f^0 or K is given at a specified temperature.

If the heat capacities are in the form

$$C_P = a + bT + cT^{-2} \tag{8–50}$$

then

$$\ln K_T = -\frac{\Delta H_0}{RT} + \frac{\Delta a}{R}\ln T + \frac{\Delta b}{2R}T + \frac{\Delta c}{2R}T^{-2} + I \tag{8–51}$$

and

$$\Delta G_T^0 = \Delta H_0 - (\Delta a)T\ln T - \frac{\Delta b}{2}T^2 - \frac{\Delta c}{2}T - IRT \tag{8–52}$$

with

$$\Delta H_T = \Delta H_0 + (\Delta a)T + \frac{\Delta b}{2}T^2 - \Delta c T^{-1} \tag{8–53}$$

Once the chemical reaction equilibrium constant (for a particular reaction) has been determined, one can proceed to estimate the quantities of the participating species at equilibrium. The problem that remains is to relate K to understandable physical quantities. For gas phase reactions, the term K may be approximately represented in terms of the partial pressures of the components involved. This functional relationship for the hypothetical reaction

$$a(\mathrm{A}) + b(\mathrm{B}) = c(\mathrm{C}) + d(\mathrm{D}) \tag{8–43}$$

is given by

$$K = \frac{p_C^c p_D^d}{p_A^a p_B^b} \tag{8–54}$$

where p_A = partial pressure of component A, and so on.

Assuming that a K value is available or calculable, Eq. (8–54) may be used to estimate the partial pressures of the participating components at equilibrium. To determine these quantities, one must account for the amounts reacted or produced. Thus, if x is the amount of C formed in the above reaction

$$\begin{aligned} p_C &= p_C(\text{initial}) + x \\ p_D &= p_D(\text{initial}) + \frac{d}{c}x \\ p_A &= p_A(\text{initial}) - \frac{a}{c}x \\ p_B &= p_B(\text{initial}) - \frac{b}{c}x \end{aligned} \tag{8–55}$$

This approach is valid for an ideal gas at constant temperature where x is small.[1,2] A more rigorous approach is provided in the literature.[3,5]

References

1. L. Theodore, F. Ricci, and T. VanVliet, *Thermodynamics for the Practicing Engineer*, Wiley, Hoboken, N.J., 2009.
2. L. Theodore and J. Reynolds, *Thermodynamics*, A Theodore Tutorial, Theodore Tutorials, East Williston, N.Y., originally published by the USEPA/APTI, Research Triangle Park, N.C., 1991.
3. J. Smith, H. Van Ness, and M. Abbott, *Chemical Engineering Thermodynamics*, 6th ed., McGraw-Hill, N.Y., 2001.
4. K. Pitzer, *Thermodynamics*, 3rd ed., McGraw-Hill, N.Y., 1995.
5. R. Perry and D. Green, eds., *Perry's Chemical Engineers' Handbook*, 8th ed., McGraw-Hill, N.Y., 2008.
6. L. Theodore, personal notes, East Williston, N.Y., 1990.
7. L. Theodore and F. Ricci, *Mass Transfer Operations for the Practicing Engineer*, Wiley, Hoboken, N.J., 2010.
8. C. DePriester, *Chem. Eng. Prog. Symp. Ser.*, **49**(7), 42, N.Y., 1953.
9. G. Wilson, J. *Am. Chem. Soc.*, **86**, 27–130, Washington D.C., 1964.
10. H. Renon and J. Prausnitz, *AIChE J.*, **14**, 135–144, N.Y., 1968.

9 Fluid Flow

Chapter Outline

9–1 Introduction

The topic of this chapter is introduced by examining the units of some of the pertinent quantities that will be encountered below. As noted in Part I, the momentum of a system is defined as the product of the mass and velocity of the system:

$$\text{Momentum} = (\text{mass})(\text{velocity}) \tag{9–1}$$

The engineering units (one set) for momentum are therefore (lb · ft)/s. The units of the time rate of change of momentum (hereafter referred to as *rate of momentum*) are simply the units of momentum divided by time:

$$\text{Rate of momentum} = \frac{\text{lb}\cdot\text{ft}}{\text{s}^2} \tag{9–2}$$

These units can be converted to lb_f if multiplied by an appropriate conversion constant. The conversion constant in this case is

$$g_c = 32.2\frac{(\text{lb}\cdot\text{ft})}{(\text{lb}_\text{f}\cdot\text{s}^2)} \tag{9–3}$$

This serves to define the conversion constant g_c, which is also equal to unity with no units. If the rate of momentum is divided by g_c as 32.2 $(\text{lb}\cdot\text{ft})/(\text{lb}_\text{f}\cdot\text{s}^2)$, the following units result:

$$\text{Rate of momentum} = \left(\frac{\text{lb}\cdot\text{ft}}{\text{s}^2}\right)\left(\frac{\text{lb}_\text{f}\cdot\text{s}^2}{\text{lb}\cdot\text{ft}}\right) \equiv \text{lb}_\text{f} \tag{9–4}$$

One may conclude from this dimensional analysis that a force is equivalent to a rate of momentum. This development is also discussed in Chap. 1.

When a fluid flows past a stationary solid wall, the fluid adheres to the wall at the interface between the solid and fluid. Therefore, the local velocity v of the fluid at the interface is zero. The velocity of the fluid is finite at some distance y normal to and displaced from the wall. Therefore, there is a velocity variation from point to point in the flowing fluid. This causes a velocity field, in which the velocity is a function of the distance y normal from the wall: $v = f(y)$. If $y = 0$ at the wall, then $v = 0$, and v increases with y. Thus, rate of change of velocity with respect to distance is the *velocity gradient*:

$$\frac{dv}{dy} = \frac{\Delta v}{\Delta y} \tag{9–5}$$

This *velocity derivative* (or *gradient*) is also called the *strain rate*, *shear rate*, *time of shear*, or *rate of deformation*. The term is important in the classification of real fluids. The relationships between shear stress and strain rate are presented in diagrams called *rheograms*.

As noted in Chap. 2, *viscosity* is a fluid property that describes the resistance of the fluid to flow. A high-viscosity fluid such as molasses, has a higher resistance to flow than does a lower-resistance fluid such as water. In other words, it takes more energy to pump 1 gal/min of molasses than to pump 1 gal/min of water in the same pipe.

It has been shown that when a fluid is sheared with a shear stress τ, its strain rate (or deformation rate) is proportional to the shear stress. The proportionality constant is termed the *fluid viscosity* μ. For a fluid sheared between two long parallel plates, the local velocity (at any height y) varies from zero at the fixed plate to the velocity V at the upper moving plate. The derivative of the local velocity (v) with respect to the height y, (dv/dy), is termed the aforementioned *velocity gradient, strain rate* or *deformation rate*. The shear stress is related to dv/dy by the equation

$$\tau = \mu \frac{dv}{dy} \tag{9–6}$$

This is Newton's law of viscosity. Fluids obeying Newton's law are termed *Newtonian fluids*. The viscosity μ is frequently referred to as the *absolute* viscosity or the *dynamic* viscosity to avoid confusion with the *kinematic* viscosity ν which, is defined as the ratio of the dynamic viscosity μ to the density ρ. The viscosity is a fluid property listed in many engineering books.

Fluids can be classified according to their viscosity. An imaginary fluid of zero viscosity is called a *Pascal fluid*. The flow of a Pascal fluid is termed *inviscid* (or non-viscous) flow. Viscous fluids are classified on the basis of their rheological (viscous) properties:

Newtonian fluids—fluids that obey Newton's law of viscosity, i.e., fluids in which the shear stress is linearly proportional to the velocity gradient. All gases are considered Newtonian fluids. Examples of Newtonian liquid are water, benzene, ethyl alcohol, hexane, and sugar solutions. Nearly all liquids of a simple chemical formula are considered Newtonian fluids.

Nonnewtonian fluids—fluids that do not obey Newton's law of viscosity. Generally, they are complex mixtures, such as polymer solutions and slurries.

Nonnewtonian fluids are classified into three types:

Time-independent fluids—fluids in which the viscous properties do not vary with time.

Time-dependent fluids—fluids in which the viscous properties vary with time.

Viscoelastic or memory fluids—fluids with elastic properties which allow them to spring back after the release of a shear force. Examples include egg white and rubber cement.

The steady–state behavior of most dilatant and pseudoplastic fluids can be represented by the Ostwald–deWaele power-law equation:

$$\tau = K\left(\frac{dv}{dy}\right)^n \tag{9–7}$$

where τ = shear stress, Pa or lb_f/ft^2
K = point consistency coefficient, $Pa \cdot s^n$ or $kg/(m \cdot s)^{(2-n)}$
dv/dy = velocity gradient, deformation rate, or strain rate, s^{-1} (reciprocal seconds)
n = flow behavior index (<1 for pseudoplastics, >1 for dilatant fluids)

The apparent viscosity η is obtained from the equations

$$\tau = \eta\left(\frac{dv}{dy}\right) = K\left(\frac{dv}{dy}\right)^n \tag{9–8}$$

$$\eta = K\left(\frac{dv}{dy}\right)^{n-1} \tag{9–9}$$

The term η has the units of lb/(ft·s) in the engineering system, or kg/(m·s) in the SI system of units. For a Newtonian fluid, $\tau = \mu(dv/dy)$, so that

$$n = 1$$
$$K = \mu$$
$$\eta = \mu = \text{a constant independent of the strain rate} \qquad (9\text{–}10)$$

The remainder of the chapter addresses the following topics: basic laws, fluid flow equations, prime movers, fluid-particle applications, flow-through porous media, and fluidization.

NOTE The bulk of the material presented in this chapter has been drawn from Farag.[1]

9–2 Basic Laws

Momentum transfer was introduced in Sec. 9–1 by reviewing the units and dimensions of momentum, time rate of change of momentum, and force. The *phenomenological* law governing the transfer of momentum by molecular diffusion (Newton's second law) was briefly discussed. In addition to molecular diffusion, momentum (and energy) may also be transferred by bulk motion. Since bulk motion involves transfer of mass from one point in a system to another point in the same system, the conservation law for mass was also discussed earlier. These serve as an excellent warm-up for the discussion of the equation of motion (equation of momentum transfer or conservation law for momentum) that receives treatment in this section.

Momentum Balances

A momentum balance (also termed the *impulse-momentum principle*) is important in flow problems where forces need to be determined. This analysis is inherently more complicated than those presented earlier (i.e., forces possess both magnitude and direction), because the force F and momentum M are *vectors*. To describe force and momentum vectors, both direction and magnitude must be specified; for mass and energy, only the magnitude is required.[2]

Newton's law is applied in order to derive the linear momentum balance equation. Newton's law states that the sum of all forces equals the rate of change of linear momentum:

$$\sum F = \frac{d}{dt}\left(\frac{mv}{g_c}\right) = \frac{dM}{dt} = \dot{M} \qquad (9\text{–}11)$$

Here $\dot{M}$ is the rate (with respect to time) of linear momentum, while m and v represent the mass and velocity, respectively. Newton's law in this form must be applied in a specified direction (e.g., horizontal or vertical). The product mv is defined as *linear momentum*. When this is applied to a fluid entering or leaving a control volume, the following terms may be defined:

$\dot{M}_{\text{out}}$ = momentum rate of the fluid leaving the control volume
$\dot{M}_{\text{in}}$ = momentum rate of the fluid entering the control volume

Equation (9–11) may be now rewritten in finite form:

$$\sum F = \dot{M}_{\text{out}} - \dot{M}_{\text{in}} \qquad (9\text{–}12)$$

(*Note*: This topic did *not* receive treatment in Chap. 7; the reader was referred to this chapter at that time.)

This balance essentially means that for steady-state flow, the force on the fluid equals the net rate of outflow of momentum across the control surface. These force rate of momentum terms are the contributing effects that are employed in development of the conservation law for momentum.

Total Energy Equation[3]

The conservation law for energy finds application in many chemical process units such as heat exchangers, reactors, and distillation columns, where shaft work plus kinetic and potential energy changes are negligible compared with heat flows and either internal energy or enthalpy changes. Energy balances on such units therefore reduce to $Q = \Delta E$ (closed nonflow system) or $\dot{Q} = \Delta \dot{H}$ (open flow system).

Another important class of operations is one for which heat flows and internal energy changes are secondary in importance to kinetic and potential energy changes and shaft work. Most of these operations involve the flow of fluids to, from, and between tanks, reservoirs, wells, and process units. Accounting for energy flows in such processes is most conveniently accomplished with mechanical energy balances.[3] Details of this approach follow.

Consider the steady-state flow of a fluid in the process depicted in Fig. 9–1. The mass entering at location 1 carries with it a certain amount of energy, existing in various forms. Thus, because of its elevation, z_1 ft above any arbitrarily chosen horizontal reference plane, for example, $z = 0$, it possesses a potential energy $(g/g_c)z_1$ (which can be recovered by allowing the fluid to fall from the height at location 1 to that of the reference point). Because of its velocity v_1, the mass also possesses and carries with it into location 1 of the system an amount of kinetic energy $v_1^2/2g_c$. It also possesses its so-called internal energy E_1 because of its temperature. Furthermore, the mass of fluid in question entering at point 1 is forced into the section by the pressure of the fluid behind it, and this form of flow energy must also be included. The amount of this energy is given by the force exerted by the flowing fluid times the distance through which it acts; this force is clearly given by the pressure per unit area P_1 times the area S_1 of the cross-section. The distance through which the force acts is the volume V_1 of the fluid divided by the cross-sectional area S_1. Since the work is the force times the distance, that is, $(P_1S_1)(V_1/S_1) = P_1V_1$, the energy associated with this term is, in turn, the product of the pressure times the volume of the fluid. This was referred to earlier as *flow work.*

Two additional energy terms need to be included in this analysis. Both involve energy exchange in the form of heat Q and work W between the fluid and the surroundings. In the discussion to follow, it will be assumed (consistent with the notation recently adopted by the scientific community; see also Chap. 8) that any energy in the form of heat or work *added* to the system is treated as a *positive* term.[1]

Figure 9–1

Process flow diagram.

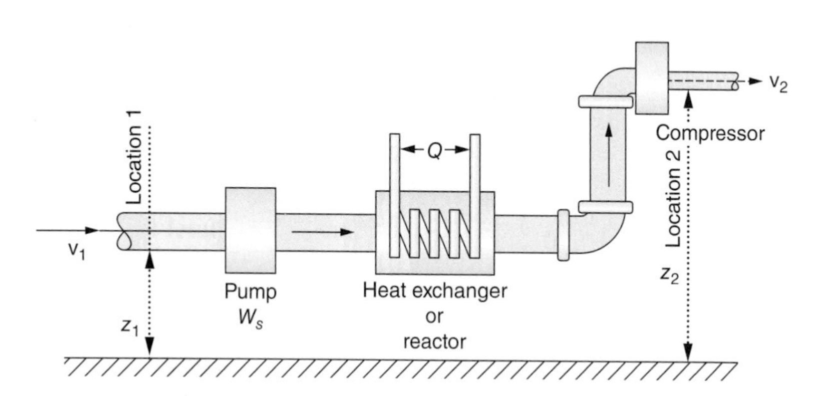

Applying the conservation law of energy mandates that all forms of energy entering the system equal that of those leaving. Expressing all terms in consistent units (e.g., energy per unit mass of fluid flowing) results in the total energy balance presented in the following equation:

$$P_1V_1 + \frac{v_1^2}{2g_c} + \frac{g}{g_c}z_1 + E_1 + Q + W_s = P_2V_2 + \frac{v_2^2}{2g_c} + \frac{g}{g_c}z_2 + E_2 \quad (9\text{–}13)$$

As written, each term in Eq. (9–13) is associated with a mechanincal energy effect. For this reason, it has also been defined as a form of the *mechanical energy balance equation* [see later discussion and Eq. (9–19)] and may be considered a special application of the conservation law for energy. Also note that, as written, the volume term V (by necessity) is the specific volume, i.e., the volume per unit mass. In terms of the density (mass per volume), Eq. (9–13) becomes

$$\frac{P_1}{\rho} + \frac{v_1^2}{2g_c} + \frac{g}{g_c}z_1 + E_1 + Q + W_s = \frac{P_2}{\rho} + \frac{v_2^2}{2g_c} + \frac{g}{g_c}z_2 + E_2 \quad (9\text{–}14)$$

and

$$V = \frac{1}{\rho} \quad (9\text{–}15)$$

Note once again that Q and W_s can be written on a time-rate basis in this equation by simply dividing by the mass flow rate through the system; the equation then dimensionally reduces to an energy per mass balance, such as $(\text{ft}\cdot\text{lb}_f)/\text{lb}$.

Four points are in order here before leaving this discussion:

1. The term Q should represent the total net heat added to the fluid. But, in this analysis it includes only the heat passing into the fluid across the walls of the conduit walls from an external source. This excludes heat generated by friction by the fluid within the unit. However, this effect can normally be safely neglected in most engineering fluid flow applications.

2. The work W_s, similar to Q, must pass through the retaining walls. While it could conceivably enter in other ways, it is supplied in most applications by some form of moving mechanism, such as a pump, or a fan, and is often referred to as *shaft work.*

3. Variations in the local point velocity over the cross-sections of a conduit may have to be considered. By definition, the kinetic energy of a small element of fluid having *local* velocity v is $v^2/2g$. If the local velocities at all points in the cross-section were uniform, the *velocity* v would be equal to the average velocity, $\bar{V}$, and $\bar{V}^2/2g$ would be the correct value of the kinetic energy. Ordinarily, there is a velocity profile across the conduit, and hence the use of $V^2/2g$ can introduce an error, the magnitude of which depends on the nature of the velocity profile and the shape of the cross-section. As will be discussed later, parabolic flow usually occurs when the velocity is low, with a corresponding low value of the kinetic energy. This can be included in the development of the describing equations by including a coefficient α in the numerator of each kinetic energy term, e.g., $\alpha v^2/2g$. If the velocity is approximately uniform, the error in using $\bar{V}^2/2g$ is not serious. Since the error tends to cancel because of the appearance of $\bar{V}^2/2g$ terms on each side of the energy balance equations, it is customary to use $\bar{V}^2/2g$ as the kinetic energy. When the velocity distribution is parabolic, it can be shown that the correct value of the kinetic energy is $\bar{V}^2/g$, not $\bar{V}^2/2g$.

4. The internal energy term E corresponds to the thermodynamic definition provided earlier. For convenience, the sum of E and PV may be treated as the single function defined in Chap. 8 as the enthalpy H:

$$H = E + PV \quad (9\text{–}16)$$

This also is a property of the fluid, uniquely determined by point conditions. Similar to the case for E, its absolute value is arbitrary; differences in value are often given above a specified reference.

With these considerations, Eq. (9–13) becomes:

$$\frac{v_1^2}{2g_c} + \frac{g}{g_c} z_1 + H_1 + Q + W_s = \frac{v_2^2}{2g_c} + \frac{g}{g_c} z_2 + H_2 \tag{9–17}$$

or simply

$$\frac{\Delta v^2}{2g_c} + \frac{g}{g_c} \Delta z + \Delta H = Q + W_s \tag{9–18}$$

As noted in the presentation of Eq. (9–14), each term is dimensional with units of energy and mass. If this equation is multiplied by the fluid flow rate, i.e., mass · time, the units of each term are converted to energy · time. In the absence of both kinetic and potential energy effects, the above equation reduces to the energy equation provided in Eq. (7–19). Also note that Δ, the difference term, refers to a difference between the value at station 2 (the usual designation for the outlet) minus that at station 1 (the inlet).

9-3 Key Fluid Flow Equations

As noted, the solutions to many fluid flow problems are based on the aforementioned mechanical energy balance equation. This equation was derived, in part, from the *general* (or total) energy equation presented above; Eq. (9–13) is repeated here:

$$P_1V_1 + \frac{v_1^2}{2g_c} + \frac{g}{g_c} z_1 + E_1 + Q + W_s = P_2V_2 + \frac{v_2^2}{2g_c} + \frac{g}{g_c} z_2 + E_2 \tag{9–13}$$

One may now add some assumptions regarding this equation:

1. Assume adiabatic flow, that is, $Q = 0$.
2. For isothermal, or near-isothermal, flow (valid in most applications), the internal energy is constant, so that $E_1 = E_2$.
3. A term $\sum F$, representing the total friction arising as a result of fluid flow, is added to Eq. (9–13). This is treated as a positive term in Eq. (9–19) later.
4. An efficiency (fractional) term η is combined with the shaft work term W_s. If work is imparted on the system, the term becomes ηW_s; if work is extracted (with an engine or turbine), the term appears as W_s/η. The efficiency term must be included since part of the work added to or extracted from the system is lost as a result of irreversibilities associated with the mechanical device. The notation h_s will be employed later for this term.

Equation (9–13) now becomes

$$\frac{\Delta P}{\rho} + \frac{\Delta v^2}{2g_c} + \frac{g}{g_c} \Delta z - \eta W_s + \sum F = 0 \tag{9–19}$$

This equation is also defined as the mechanical energy balance equation.

The Bernoulli Equation

The *Bernoulli equation* has come to mean different things to different people. One definition of this equation is obtained by neglecting both work and friction effects in Eq. (9–19):

$$\frac{\Delta P}{\rho} + \frac{\Delta v^2}{2g_c} + \frac{g}{g_c} \Delta z = 0 \tag{9–20}$$

Hydrostatics

When a simple fluid is at rest, there is no shear stress and the pressure at any point in the fluid is the same in all directions. The pressure is the same across any horizontal section; it varies only in the vertical direction. A force balance on the element in the vertical direction yields:

$$\frac{dP}{dz} = -\rho g \tag{9–21}$$

This equation is the *hydrostatic* or *barometric* differential equation. This term dP/dz is often used to denote the *pressure gradient*. Equation (9–21) is a first-order ordinary differential equation. It may be integrated by separation of variables:

$$\int dP = -\int \rho g \, dz = -g \int \rho \, dz \tag{9–22}$$

For most engineering applications involving liquids, and many applications involving gases, the density is constant; i.e., the fluid is incompressible. Taking ρ outside the integration sign and integrating between any two stations in the fluid (at station 1 the pressure equals P_1 and the elevation is z_1; station 2 has a pressure of P_2 and elevation z_2), provides pressure–height relationship:

$$P_2 - P_1 = -\rho g(z_2 - z_1) \tag{9–23}$$

Equation (9–23) may be rewritten as

$$\frac{P_1}{\rho g} + z_1 = \frac{P_2}{\rho g} + z_2 = \text{constant} \tag{9–24}$$

The term $P/\rho g$ is the *pressure head* of the fluid, with units of meter (or feet) of fluid. Equation (9–24) states that the sum of the pressure head + potential head is constant in hydrostatic applications. Equation (9–24) is sometimes termed the *hydrostatic Bernoulli equation*. It is useful in calculating the pressure at any liquid depth.

Manometry and Pressure Measurement

Pressure usually is measured by allowing it to act across some area and opposing it with some type of force, such as gravity, a compressed spring, or an electrical force. If the force is gravity, the device is usually a manometer. A very common device used to measure pressure is the *Bourdon tube pressure gauge*, which is a reliable and inexpensive direct-displacement device. It consists of a stiff metal tube bent in a circular shape. One end is fixed, and the other is free to deflect when pressurized. This deflection is measured by a linkage attached to a calibrated dial gauge. Bourdon gauges are available with an accuracy of ±0.1 percent of the full scale. Other pressure gauges measure the pressure by displacement of the sensing element electrically. Among the common methods are (1) capacitance, (2) resistive, and (3) inductive. The interest here is in the manometer.

Consider the open manometer shown in the Fig. 9–2. Assume that P_1 is unknown and P_a is the known atmospheric pressure. The heights z_a, z_1, and z_2 are also known. First apply Bernoulli's hydrostatic equation at points 1 and 2, and again at points 2 and *a*.

$$P_1 - P_2 = -\rho_1 g(z_1 - z_2) \tag{9–25}$$

$$P_2 - P_a = -\rho_2 g(z_2 - z_a) \tag{9–26}$$

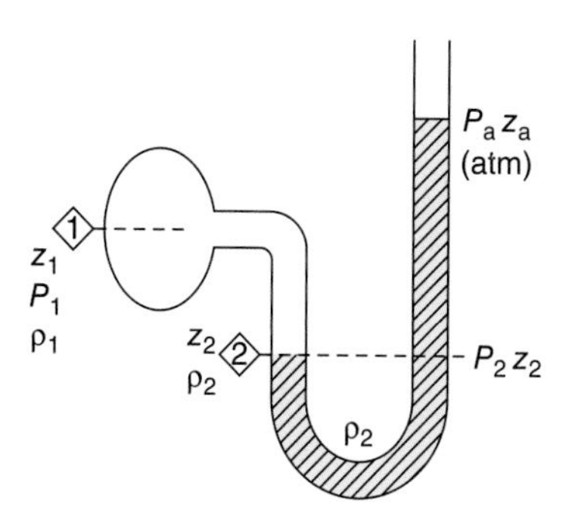

Figure 9–2

Open manometer.

By adding the two equations, one obtains

$$P_1 - P_a = -[\rho_1 g(z_1 - z_2) + \rho_2 g(z_2 - z_a)] \tag{9–27}$$

Reynolds Number

The *Reynolds number* Re is a dimensionless quantity, and is a measure of the relative ratio of inertia to viscous forces in the fluid:

$$\begin{aligned} \text{Re} &= \rho VL/\mu \\ &= VL/\nu \end{aligned} \tag{9–28}$$

where L = a characteristic length
V (or v) = average velocity
ρ = fluid density
μ = dynamic (or absolute) viscosity
ν = kinematic viscosity

In flow-through round pipes and tubes, L is the length and D is the diameter. The Reynolds number provides information on flow behavior. It is particularly useful in scaling up bench-scale or pilot plant data to full-scale applications.

Laminar flow is always encountered at a Reynolds number below 2300 in a circular duct, but it can persist up to higher Reynolds numbers. Very slow flow (in circular ducts) for which Re is less than 1 is termed *creeping flow*. Under ordinary conditions of flow, the flow (in circular ducts) is turbulent at a Reynolds number above approximately 4000. Between 2300 and 4000, where the type of flow may be either laminar or turbulent, predictions are unreliable. The Reynolds numbers at which the fluid flow changes from laminar to transition or to turbulent are termed *critical numbers*. In the case of flow in circular ducts there are two critical Reynolds numbers: 2300 and 4000. In other geometries, different Re criteria exist.

Conduits

Fluids are usually transported in tubes or pipes. Generally speaking, pipes are thick-walled and have a relatively large diameter. Tubes are thin-walled and often are supplied in coils. Pipes are specified in terms of their diameter and wall thickness. The nominal diameters with steel pipe range from ⅛ to 30 in. Standard dimensions of steel pipe are available in the literature[1, 3] and are known as *iron pipe size* (IPS) or *nominal pipe size* (NPS). The pipe wall thickness is indicated by the schedule number. Tube sizes are indicated by the outside diameter (OD). The wall thickness is usually given by the *Birmingham wire gauge* (BWG) number. The smaller the BWG, the heavier is the tube. For example, a

¾-in 16-BWG tube has an OD of 0.75 in, inside diameter (ID) of 0.62 in, wall thickness of 0.065 in, and a weight of 0.476 lb/ft.

Flow in conduits that are not cylindrical (e.g., a rectangular parallelepiped), are treated as if the flow occurs in a pipe. For this situation, a hydraulic radius r_h is defined as follows:

$$r_h = \frac{\text{cross-sectional area perpendicular to flow}}{\text{wetted perimeter}} \tag{9–29}$$

For flow in a circular tube

$$r_h = \frac{(\pi D^2/4)}{\pi D} = \frac{D}{4}$$

$$D = 4r_h \tag{9–30}$$

The hydraulic diameter approach is usually valid for laminar flow and always valid for turbulent flow.

Another important concept is referred to as *calming*, *entrance*, or *transition* length. This is the length of conduit required for a velocity profile to become *fully developed* following some form of disturbance in the conduit. This disturbance can arise because of the presence of a valve, a bend in the line, an expansion in the line, or another irregularity. This is an important concern when measurements are conducted in the cross-section of the pipe or conduit. An estimate of this *calming* length L_c for laminar flow is

$$\frac{L_c}{D} = 0.05 \text{ Re} \tag{9–31}$$

For turbulent flow, one may employ

$$L_c = 50D \tag{9–32}$$

Mechanical Energy Equation: Modified Form

The following equation was derived in the previous section:

$$\frac{\Delta P}{\rho} + \frac{\Delta v^2}{2g_c} + \frac{g}{g_c}\Delta z - \eta W_s + \sum F = 0 \tag{9–19}$$

This was defined as the *mechanical energy* balance equation. This equation was later rewritten without the work and friction terms:

$$\frac{\Delta P}{\rho} + \frac{\Delta v^2}{2g_c} + \frac{g}{g_c}\Delta z = 0 \tag{9–20}$$

This equation was defined as the basic form of the *Bernoulli equation*. Equation (9–19) can also be written as

$$\frac{P_1}{\rho}\frac{g_c}{g} + \frac{v_1^2}{2g} + z_1 = \frac{P_2}{\rho}\frac{g_c}{g} + \frac{v_1^2}{2g} + z_2 - h_s'\frac{g_c}{g} + h_f'\frac{g_c}{g} \tag{9–33}$$

where h_s' and h_f' effectively replaced ηW_s and ΣF, respectively, in Eq. (9–19). Note that h_s' is a positive term, as is h_f'. The units of both h_s and h_f are $(\text{ft}\cdot\text{lb}_f)/\text{lb}$. However, each term in this equation, as written, has units of feet of flowing fluid (a pressure term). If the equation is multiplied by g/g_c (keeping in mind that g_c is a conversion constant), each term returns to units of energy/mass [e.g., $(\text{ft} \cdot \text{lb}_f)/\text{lb}$]. The result is as follows:

$$\frac{\Delta P}{\rho} + \frac{\Delta v^2}{2g_c} + \Delta z \frac{g}{g_c} - h_s + h_f = 0 \tag{9–34}$$

Note that the prime notation is retained if the units of h'_s and h'_f are given in height of flowing fluid, such as ft H_2O. The choice of units for h_s and h_f is, of course, optional or arbitrary.

The reader should also note that in the process of converting Eq. (9–33) to Eq. (9–34), the Δ term (representing a difference) applies to the outlet minus inlet conditions ($P_2 - P_1$, etc.). One can now examine Eq. (9–34) at the following extreme or limiting conditions:

1. If only the first ($\Delta P/\rho$) and last (h_f) terms are present, an increase in the latter's frictional losses would result in a corresponding decrease in the outlet or discharge pressure term P_2. Thus, a pressure drop results, which is to be expected.
2. If only the first ($\Delta P/\rho$) and fourth (h_s) terms are present, an increase in the latter's input mechanical (shaft) work term would result in a corresponding increase in the pressure term P_2. Thus, a smaller pressure drop results, and again, this is in agreement with what one would expect.
3. If only the latter two terms are present, $h_s = h_f$, and this also agrees with the expected result since both terms are positive and any frictional effect is compensated by the mechanical (shaft) work introduced to the system.

Care should be exercised in interpretation of the term ΔP. Although the notation Δ represents a difference, ΔP can be used to describe the difference between the inlet minus the outlet pressure (i.e., $P_1 - P_2$), or it can describe the difference between the outlet minus the inlet pressure (i.e., $P_2 - P_1$). When a fluid is flowing in the $1 - 2$ direction, the term $P_1 - P_2$ is a positive term and represents a decrease in pressure that is defined as the pressure drop. The term $P_2 - P_1$, however, also represents a pressure change whose difference is negative and is also defined as a pressure drop. One's wording and interpretation of this pressure change is obviously a choice that is left to the user or reader.

As indicated above, the h_f term was included to represent the loss of energy due to friction in the system. The(se) frictional loss(es) can take several forms. An important chemical engineering problem is the calculation of these losses. It has been shown that the fluid can flow in either of two modes: laminar or turbulent. For *laminar* flow, an equation is available from basic theory to calculate friction loss in a pipe. In practice, however, fluids (particularly gases) are rarely moving in laminar flow.

Laminar Flow-through a Circular Tube

Fluid flow in circular tubes (or pipes) is encountered in many applications. The flow is always accompanied by friction. Consequently, there is energy loss, indicating a pressure drop in the direction of flow. A momentum balance is an excellent tool for analyzing laminar flow in circular tubes. Cylindrical coordinates are the natural choice.

Since the two classes of flow are so different, the equations describing frictional resistance will differ for *turbulent* versus laminar flow. On the other hand, it will be shown later that a single equation can be used for both types of flow in such a way as to render unnecessary a preliminary calculation to determine whether the flow is occurring above or below the critical Reynolds number.[4]

One can theoretically derive the h_f term for laminar flow.[2, 5] The equation can be shown to take the form

$$h_f = \frac{32\mu v L}{\rho g_c D^2} \tag{9–35}$$

for a fluid flowing through a straight cylinder of diameter D and length L. A friction factor f that is dimensionless may now be defined as (for laminar flow):

$$f = \frac{16}{\text{Re}} \tag{9–36}$$

so that Eq. (9–35) takes the form

$$h_f = \frac{4fLv^2}{2g_c D} \tag{9–37}$$

Although this equation describes friction loss or the pressure drop across a conduit of length L, it can also be used to provide the pressure drop due to friction per unit length of conduit, $\Delta P/L$, by simply dividing Eq. (9–37) by L.

Another friction factor term exists that differs from that presented in Eq. (9–36). In this other case, f_D is defined as

$$f_D = \frac{64}{\text{Re}} \tag{9–38}$$

The term f_D is used to distinguish the value of Eq. (9–36) from that of Eq. (9–38). In essence, then

$$f_D = 4f \tag{9–39}$$

The f term is defined as the *Fanning friction* factor, while f_D is defined as the *Darcy* or *Moody friction factor.*[6] The importance of selecting the correct friction factors for use in calculations will become more apparent shortly. In general, chemical engineers employ the Fanning friction factor; other engineers prefer the Darcy (or Moody) factor. This text employs the Fanning friction factor.

With reference to Eq. (9–36), one should also note that this is an equation of a straight line with a slope of −1 if f is plotted versus Re on log-log coordinates. Note (once again) that the equation for f applies only to laminar flow, i.e., when Re is <2300 for pipe flow.

Employing Eq. (9–37), one may extend Eq. (9–34) and rewrite it as

$$\frac{\Delta P}{\rho} + \frac{\Delta v^2}{2g_c} + \Delta z\frac{g}{g_c} - h_s + \sum \frac{4fLv^2}{2g_c D} = 0 \tag{9–40}$$

The summation sign has been inserted before the new term for the loss in a straight pipe because this loss may result from flow-through several sections in a series of pipes of various lengths and diameters. The symbols Σh_c and Σh_e, representing the sum of the contraction and expansion losses, respectively, may also be added to the equation as provided in Eq. (9–41) (these effects are discussed later in this chapter):

$$\frac{\Delta P}{\rho} + \frac{\Delta v^2}{2g_c} + \Delta z\frac{g}{g_c} - h_s + \sum \frac{4fLv^2}{2g_c D} + \sum h_c + \sum h_e = 0 \tag{9–41}$$

As noted, flow in a pipe may be laminar or turbulent. Some characteristics of the two types of flow are given in Table 9–1. The treatment of turbulent flow follows.

Table 9–1

Flow Characteristics

Flow Type	Characteristic
Laminar	Flow moves in axial direction
	No motion normal to tube axis
Turbulent	No net flow normal to tube axis
	Strong, local, oscillating motion (or eddy) normal to tube axis
	Eddies cause shear stress to increase over that of laminar flow*
	Friction factor is larger than in laminar flow

*The extra shear stress is called the *Reynolds stress*.

Turbulent Flow-through a Circular Conduit

The Fanning friction factor was defined above and presented in Eq. (9–36):

$$f = \frac{16}{\text{Re}} \tag{9–36}$$

However, this applies *only* to laminar flow. Unlike laminar flow, the friction factor for turbulent flow cannot be derived from (fundamental) principles. Fortunately, extensive experimental data are available, and this permits the numerical evaluation of the friction factor for turbulent flow.

As described earlier, as the Reynolds number is increased above 2300 for flow in pipes, eddies and turbulence begin to develop in the flowing fluid. From Re = 2300 to Re ~4000, the flow becomes more unstable. As the Reynolds number is increased to values above 4000, the turbulent state of the fluid core becomes well developed and the velocity distribution across a diameter of the pipe becomes similar to that of a flattened parabola. This flattened parabola can be approximately expressed as

$$v = v_{\max}\left(\frac{2n^2}{(n+1)(2n+1)}\right) \tag{9–42}$$

where $v_{\max}$ is the centerline (maximum) velocity and n is 7 (the one-seventh power law applies).

It is important to note that almost all the key fluid flow equations presented for laminar flow apply as well to turbulent flow, provided the appropriate friction factor is employed. These key equations [Eqs. (9–37), (9–40), and (9–41)] are again provided below; note once again that v (the average velocity) is given by $q/(\pi D^2/4)$, where q is the volumetric flow rate:

$$h_f = \frac{4fLv^2}{2g_cD} = \frac{32fLq^2}{\pi g_c D^5} \tag{9–37}$$

so that

$$\frac{\Delta P}{\rho} + \frac{\Delta v^2}{2g_c} + \Delta z\frac{g}{g_c} - h_s + \sum\frac{4fLv^2}{2g_cD} = 0 \tag{9–43}$$

$$\frac{\Delta P}{\rho} + \frac{\Delta v^2}{2g_c} + \Delta z\frac{g}{g_c} - h_s + \sum\frac{4fLv^2}{2g_cD} + \sum h_c + \sum h_e = 0 \tag{9–44}$$

Figure 9–3

Fanning friction factor–Reynolds number plot.

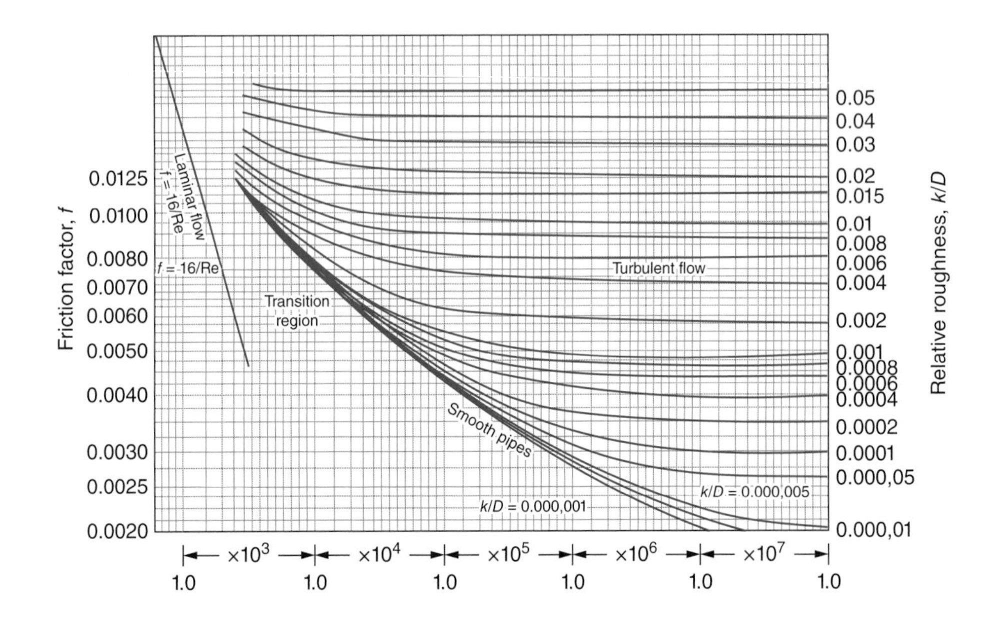

The effect of the Reynolds number on the Fanning friction factor employed above is shown in Fig. 9–3. Note that Eq. (9–36) appears on the far left-hand side of Fig. 9–3.

In the turbulent regime, the "roughness" of the pipe become a consideration. In his original work on the friction factor, Moody[6] defined the term k (or $\in$), as the roughness and the ratio k/D as the relative roughness. Thus, for rough pipes and tubes in turbulent flow

$$f = f(\text{Re}, k/D) \tag{9–45}$$

This equation indicates that the friction factor is a function of *both* the Re and k/D. However, as noted earlier, the dependence on the Reynolds number is a weak one. Moody[6] provided one of the original friction factor charts. His data and results, as applied to the Fanning friction factor, are presented in Fig. 9–3 and cover the laminar, transition, and turbulent flow regimes. It should be noted that the laminar flow friction factor is independent of the relative roughness. Figure 9–3 also contains friction factor data for various relative roughness values. The average roughness values of commercial pipes are listed in Table 9–2.

The reader should note that Moody's original work provided a plot of the Darcy (or Moody) friction factor, not the Fanning friction factor. Thus, his chart has been adjusted

Table 9–2

Average Roughness of Commercial Pipes

	Roughness k	
Material (new)	**ft**	**mm**
Asphalted cast iron	0.0004	0.12
Cast iron	0.00085	0.26
Commercial steel (wrought iron)	0.00015	0.046
Concrete	0.001–0.01	0.3–3.0
Drawn tubing	0.000005	0.0015
Galvanized iron	0.0005	0.15
Glass	Smooth	Smooth
Riveted steel	0.003–0.03	0.9–9.0

to provide the Fanning friction factor; i.e., the plot in Fig. 9–3 is for the Fanning friction factor. Those choosing to work with the Darcy friction factor need only multiply the Fanning friction factor by 4, since

$$f_D = 4f \tag{9–39}$$

In summary, for Re <2300, the flow will always be laminar and the value of f should be taken from the line at the far left in Fig. 9–3. For Re >4000, the flow will practically always be turbulent and the values of f should be read from the lines at the right. Between Re = 2300 and Re = 4000, no accurate calculations can be made because it is generally impossible to predict the flow type in this range. To estimate the friction loss in this range, the figures for turbulent flow should be used, as this would provide an estimate on the high side.[4]

Compressible versus Incompressible Flow

The first step in this flow analysis is to classify the flow. One has to specify one condition from columns 1 and 2 in Table 9–3. Column 1 pertains to variation with time. Steady flow is time-invariant. Column 2 pertains to variation of fluid density. If the density is constant, or its variation is very small (most liquids, and gases with a Mach number <0.3), the flow is incompressible.

The *Mach number* (Ma) is a dimensionless group defined as the ratio of fluid velocity to the speed of sound in the fluid:

$$\text{Ma} = \frac{\text{average velocity}}{\text{speed of sound of fluid}} \tag{9–46}$$

If the Mach number is less than or equal to 0.3 (Ma ≤ 0.3), compressibility effects may be neglected, and one may safely assume incompressible flow. The speed of sound of common liquids is available in the literature.[1,3] The speed of sound C of an ideal gas is

$$C = \sqrt{kRT/\text{MW}} \tag{9–47}$$

where $k = C_p/C_v$
R = universal gas constant = 8314.4 $m^2/(s^2 \cdot K)$
T = absolute temperature, K
MW = molecular weight of fluid

For air, this simplifies to

$$C = 20\sqrt{T(\text{K})} \quad \text{m/s} \tag{9–48}$$

$$C = 49\sqrt{T(°\text{R})} \quad \text{m/s} \tag{9–49}$$

Two-Phase Flow

The simultaneous flow of two phases in pipes (as well as in other conduits) is complicated by the fact that the action of gravity tends to cause settling and "slip" of the heavier phase; as a result, the lighter and heavier phases flow at different velocities in the pipe. The results of this phenomenon vary depending on the classification of the two

Table 9–3
Compressible versus Incompressible Flow

(1)	(2)
Steady	Incompressible
Unsteady	Compressible

phases, the flow regime, and the inclination of the pipe (conduit). As one might suppose, the major industrial application in this area is gas (G) - liquid (L) flow in pipes.

The general subject of flashing and boiling liquids (a two-phase application), is also not considered in this chapter. However, when a saturated liquid flows in a pipeline from a given point at a given pressure to another point at a lower pressure, several processes can take place. As the pressure decreases downstream due to frictional effects, the saturation or boiling temperature decreases, leading to the possible evaporation of a portion of the liquid. The net result is that a one-phase flowing mixture is transformed into a two-phase mixture with a corresponding increase in frictional resistance in the pipe or conduit. Boiling liquids arise when liquids are vaporized in pipelines at approximately constant pressure. Alternatively, the flow of condensing vapors in pipes (or conduits) is complicated by the constant variation of properties of the mixture with changes in pressure, temperature, and fraction condensed. Further, the condensate, which forms on the walls, requires energy for transformation into a spray, and this energy must be obtained from the main vapor stream, resulting in an additional pressure drop. An analytical treatment of these topics is also beyond the scope of this book. However, information is available in the literature.[1]

There are five basic types of two-phase flow systems:

1. Gas (G)—liquid (L) flow principles: generalized approach
2. Gas (turbulent) flow—liquid (turbulent) flow
3. Gas (turbulent) flow—liquid (viscous) flow
4. Gas (viscous) flow—liquid (viscous) flow
5. Gas—solid flow

Abulencia and Theodore[3] provide a detailed analysis of these five systems with a host of illustrative examples.

9–4 Prime Movers

This section is concerned with fluid flow transport. It includes determining power requirements, designing and sizing fans, pumps, and compressors, and reviews the various valves and fittings.

The handling and flow of either gases or liquids is much simpler, cheaper, and less troublesome than that of solids. Consequently, the chemical engineer attempts to transport most quantities in the form of a gas or liquid whenever possible. It is also important to note that throughout this book, the word *fluid* will always be used to include both liquids and gases. In many operations, a solid is handled in a finely subdivided state in a fluid to ensure that it remains in suspension in a fluid.

If a pressure difference is required between two points in a system, a prime mover such as a fan, pump, or compressor is often used to provide the necessary pressure and/or flow impetus. Chemical engineers are often called on to specify prime movers more frequently than any other piece of processing equipment, particularly in the chemical industry. In a general sense, these primary movers are to a process plant what the engine is to one's automobile, or what the heart is to a human. Whether one is processing petrochemicals, caustic soda, acids, etc., the fluid must usually be transferred from one point to another somewhere in the process. At a chemical plant, chemicals must be loaded or unloaded, sent to heat exchangers or cooling/heating coils, transferred from one processing unit to another, or packaged for shipment.

To move material (either a fluid or a slurry) through the various pieces of equipment at a facility or plant (including piping and duct work) requires mechanical energy not only to impart an initial velocity to the material but also, more importantly, to overcome pressure losses that occur throughout the flow path of the moving fluid. This energy may be imparted to the moving stream in at least one of three modes: (1) an increase in stream velocity, (2) an increase in stream pressure, or (3) an increase in stream's height, or some combination of the three. In mode 1, the additional energy takes the form of an increase in the kinetic energy as the bulk stream velocity increases. In mode 2, the internal energy (mainly potential energy, but usually some kinetic as well) of the stream increases. This pressure increase may also cause a stream temperature rise, which represents an internal energy increase. In mode 3, which may be relatively small for some operations, the bulk fluid experiences an increase in potential energy in Earth's gravitational field.

Three devices that convert electrical energy into the mechanical energy that can be applied to various streams are discussed in this section. These devices are fans, which move low-pressure gases; pumps, which move liquids and liquid-solid mixtures such as slurries, suspensions, and sludges; and compressors, which move high-pressure gases. Three general process classifications of prime movers—centrifugal, rotary, and reciprocating—can be selected. Except for special applications, centrifugal units are normally employed. These units are normally rated in terms of four characteristics:

1. *Capacity*—the quantity of fluid discharged per unit time (the mass flow rate).
2. *Pressure increase*—often reported for pumps as head: *head* can be expressed as the energy supplied to the fluid per unit mass and is obtained by dividing the increase in pressure (the pressure change) by the fluid density.
3. *Power*—the energy consumed by the mover per unit time.
4. *Efficiency*—the energy supplied to the fluid divided by the energy supplied to the unit.

Finally, the net effect of most prime movers is to increase the pressure of the fluid.

Fans

The terms *fans* and *blowers* are often used interchangeably, and are not differentiated in the following discussion; thus, any statements about fans apply equally to blowers. Strictly speaking, however, fans are used for low pressure (drop) operation, generally below 2.0 lb_f/in^2 (psi). Blowers are generally employed when generating pressure heads range from 2.0 to 14.7 psi. Higher-pressure operations require compressors.

Fans are usually classified as centrifugal or axial-flow type. In *centrifugal fans*, the gas is introduced into the center of a revolving wheel (the eye) and discharges at right angles to the rotating blades. In *axial-flow fans*, the gas moves directly (forward) through the axis of rotation of the fan blades. Both types are used in industry, but it is the centrifugal fan that is employed at most facilities.

The fan selection procedure for the practicing chemical engineer requires, in part, an examination of the fan curve and the system curve. A fan curve, relating static pressure with flow rate, is provided in Fig. 9–4. Note that each type of fan has its own characteristic curve. Also note that fans are usually tested in the factory or laboratory with open inlets and long smooth straight discharge ducts. Since these conditions are seldom duplicated in the field, actual operation often results in lower efficiency and reduced performance. A system curve is also shown in Fig. 9–4. This curve must be calculated or provided prior to the purchase of a fan in order to obtain the best estimate of the pressure drop across the system through which the fan must deliver the gas. The curve should approach a straight line with an approximate slope of 1.8 on log-log or ln-ln coordinates.) The system pressure (drop), defined as the *resistance* associated *of ducts,* fittings, equipment, contractions, expansions, and so on, is a measure of h_p, the head loss of system.

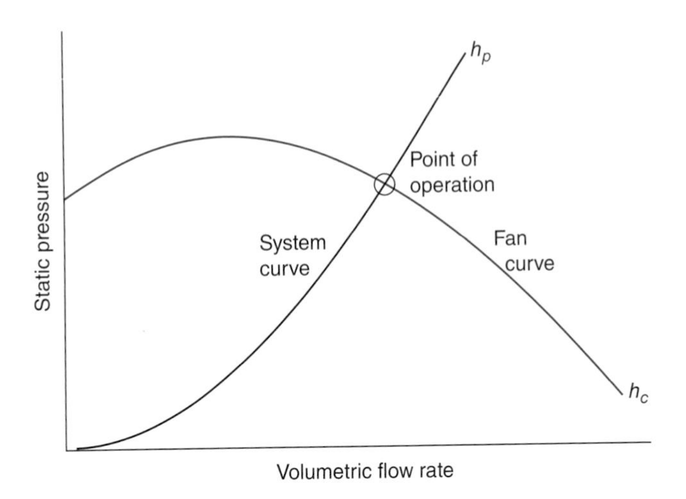

Figure 9–4

System and fan characteristic curves.

Numerous process and equipment variables are classified as part of a fan specification. These include flow rate (acfm), temperature, density, gas stream characteristics, static pressure that needs to be developed, motor type, drive type, materials of construction, fan location, and noise controls.

Gas (or air) horsepower (GHP) and brake horsepower (BHP) are the two terms of interest. These may be calculated from Eqs. (9–50) and (9–51):

$$\text{GHP} = 0.0001575 q_a \Delta P = \frac{q_a \Delta P}{6356} \tag{9–50}$$

$$\text{BHP} = \frac{0.0001575 q_a \Delta P}{\eta_f} \tag{9–51}$$

where q_a = volumetric flow rate, actual cubic feet per minute (actual ft^3/min)
Δp = SP = static pressure head developed, in H_2O
η_f = fractional fan efficiency

Pumps

Pumps are required to transport liquids, liquid-solid mixtures such as slurries and sludges, auxiliary fuel, and other materials. Pumps are also needed to transport water to or from such peripheral devices as boilers, quenchers, and scrubbers. As indicated earlier, pumps may be classified as reciprocating, rotary, or centrifugal. The reciprocating and rotary types are referred to as *positive-displacement* pumps because, unlike the centrifugal type, the liquid or semiliquid flow is broken into small portions as it passes through the pump.

Reciprocating pumps operate by the direct action of a piston on the liquid contained in a cylinder. As the liquid is compressed by the piston, the higher pressure forces it through discharge valves to the pump outlet. As the piston retracts, the next batch of low-pressure liquid is drawn into the cylinder and the cycle is repeated. The piston may be either directly driven by steam or moved by a rotating crankshaft through a crosshead. The rate of liquid delivery is a function of the volume swept out by the piston and the number of strokes per unit time. A fixed volume is delivered for each stroke, but the actual delivery may be less because of both leakage past the piston and failure to fill the cylinder when the piston retracts. The *volumetric efficiency* of the pump is defined as the ratio of the actual volumetric discharge to the pump displacement volume. For well-maintained pumps, the volumetric efficiency should be at least 95 percent.

The pressure increase (head) developed by a centrifugal pump is almost always a function of the discharge rate. In a real pump, as the flow rate increases, the total head delivered by the pump decreases. To obtain the discharge rate for a given system, the chemical engineer has to solve, simultaneously, the system curve of head h_p versus discharge with the pump curve head h_c. As noted earlier, the system h_p is usually obtained by applying the mechanical energy equation (or the equivalent). The pump characteristic curve of h_c is usually supplied by the manufacturer. The point at which the two curves intersect (see Fig. 9–4) yields the operating conditions for the system.

When a centrifugal pump is operating at high flow rates, the high velocities occurring at certain points in the eye of the impeller or at the vane tips cause local pressures to fall below the vapor pressure of the liquid. This is predicted by the Bernoulli equation. Vaporization can occur at these points, forming bubbles that collapse violently on moving to a region of higher pressure or lower velocity. This momentary vaporization and so-called destructive collapse of the bubbles is called *cavitation*. This should be avoided if maximum capacity is to be obtained and damage to the pump prevented. The shock of bubble collapse causes severe pitting of the impeller and creates considerable noise and vibration. Cavitation may be reduced or eliminated by reducing the pumping rate or by slight alterations in the impeller design to enhance streamlining. Note, however, that cavitation seldom occurs at low flow rates on any given pump.[7] Abulencia and Theodore[3] provide additional details plus numerous illustrative examples.

Reciprocating pumps are used for some applications. Reciprocating pumps can potentially deliver the highest pressure of any type of pump (20,000 psig); however, their capacities are relatively small compared to those of the centrifugal pump. Also, because of the nature of the operation of the reciprocating pump, the discharge flow rate tends to be somewhat pulsating, and liquids containing abrasive solids can damage the machined surfaces of the piston and cylinder. Because of its positive-displacement operation, a reciprocating pump can also be used to measure liquid volumetric flow rates.

The *rotary* pump combines rotation of the liquid with positive displacement. The rotating elements mesh with elements of the stationary casing in much the same way that two gears mesh. As the rotating elements come together, a pocket is created that first enlarges, drawing in liquid from the inlet or suction line. As rotation continues, the pocket of liquid is trapped, being forced into the discharge line at a higher pressure.

Pumps can also be set up in *parallel*. This configuration is used to administer systems where higher capacity is desired. Parallel pumps work as two or more single pumps with merging outlets. In addition to creating greater capacity, parallel pumps can be helpful for systems that cannot be shut down for long periods of time, as each pump works independently of the other. The calculations used for parallel pumps are similar to those used for pumps in series. The capacity and volumetric flow rate of the pair of pumps are found in the same way as that of the single pump and series pumps. The total head of the system is simply the greater head produced from each of the pumps; these values are calculated as the head for a single pump. The power and total efficiency are calculated in the same way as those for two pumps in series.

Compressors

Compressors, unlike fans and pumps, find only limited use for specialized applications. They are employed primarily to increase the pressure of a fluid in some types of systems, such as when liquids are broken up into tiny droplets (atomized) before entering a unit. This can be accomplished by using a high-pressure stream of air or steam that impinges on the liquid steam and atomizes it. Pressurization of the air or steam is accomplished through the use of compressors.

Compressors operate much the same way as do pumps and have the same classification category: rotary, reciprocating, and centrifugal. An obvious difference between pump and compressor operations is the large decrease in volume resulting from the compression of a gaseous stream relative to the negligible change in volume caused by the pumping of a liquid stream.

The following equation may be used to calculate compressor power requirements when the compressor operation is adiabatic and the gas (usually air) assumes ideal-gas behavior:[1,3]

$$W_s = \frac{\gamma RT}{\gamma - 1}\left[\left(\frac{P_2}{P_1}\right)^{(\gamma-1)/\gamma} - 1\right] \qquad (9\text{–}52)$$

where W_s = compressor work required per lbmol of air
$R = 1.987$ Btu/(lbmol · °R)
T = air temperature at compressor inlet conditions, °R
P_1, P_2 = air inlet and discharge pressures, consistent units
γ = ratio of heat capacity at constant pressure to that at constant volume (typically 1.3 for air)

Valves and Fittings

As indicated earlier, tubing and other conduits are used for the transportation of gases, liquids, and slurries. These ducts are often connected and may also contain various valves and fittings, including expansion and contraction joints. There are four basic types of connecting conduits:

1. Threaded
2. Bell-and-spigot
3. Flanged
4. Welded

Extensive information on these connections classes is available in the literature.[1,3]

Because of the diversity of types of systems, fluids, and the environments in which valves must operate, a vast array of valve types have been developed. Examples of the common types are the globe valve, gate valve, ball valve, plug valve, pinch valve, butterfly valve, and check valve. Each type of valve has been designed to meet specific needs. Some valves are capable of throttling flow, other valve types can only stop flow, others work well in corrosive systems, and others handle high-pressure fluids.

Valves have two main functions in a pipeline: to control the amount of flow or to stop the flow completely. Of the many different types of valves employed in practice, the most commonly used are the gate valve and the globe valve. The *gate valve* contains a disk that slides at right angles to the flow direction. This type of valve is used primarily for on/off control of a liquid flow. Because small lateral adjustments of the disk can cause significant changes in the flow cross-sectional area, this type of valve is not suitable for accurate adjustment of flow rates. As the fluid passes through the gate valve, only a small amount of turbulence is generated; the direction of flow is not altered and the flow cross-sectional area inside the valve is only slightly less than that of the pipe. As a result, the valve causes only a minor pressure drop. Problems with abrasion and erosion for the disk arise when the valve is used in positions other than fully open or fully closed. Unlike the gate valve, the globe valve—so called because of the spherical shape of the valve body—is designed for more sensitive flow control. In this type of valve, the liquid passes through the valve in a somewhat tortuous or circuitous route. In one form, the seal is a horizontal ring into which a plug with a slightly beveled edge is inserted when the stem is closed. Good control of flow is achieved with this type of valve, but at the expense of a higher pressure loss than with a gate valve.

Each valve type has advantages and disadvantages. Understanding these differences and how they affect the valve's application or operation is necessary for the successful operation of a facility.

A *fitting* is a piece of equipment whose function is one or a combination of the following:

1. The joining of two pieces of straight pipe (e.g., couplings and unions)
2. The changing of pipeline direction (e.g., elbows and Ts)
3. The changing of pipeline diameter (e.g., reducers and bushings)
4. The terminating of a pipeline (e.g., plugs and caps)
5. The joining of two streams (e.g., Ts and Ys)

Each of these is briefly discussed. A *coupling* is a short piece of pipe threaded on the inside and used to connect straight sections of pipe with no change in direction or size. When a coupling is opened, a considerable amount of piping must usually be dismantled. A *union* is also used to connect two straight sections but differs from a coupling in that it can be opened relatively easily without disturbing the rest of the pipeline. An *elbow* is an angle fitting used to change flow direction, usually by 90°, although 45° elbows are also available. In addition, a T component (shaped like the letter T) can be used to change flow direction; this fitting is more often used to combine two streams into one; i.e., when two branches of piping are to be connected at the same point. A *reducer* is a coupling for two pipe sections of different diameters. A *bushing* is also a connector for pipes of different diameters, but unlike the reducer-coupling, is threaded on both the inside and outside. The larger pipe screws onto the outside of the bushing and the smaller pipe screws into the inside of the bushing. *Plugs*, which are threaded on the outside, and *caps*, which are threaded on the inside, are used to terminate a pipeline. Finally, a Y component (shaped like the letter Y) is similar to the T and is used to combine two streams.

If the cross-section of a conduit enlarges gradually so that the flowing fluid velocity does not undergo any disturbances, energy losses are minor and may be neglected. However, if the change is sudden, as in a rapid expansion, additional friction loss can result. For such sudden enlargement/expansion situations, the exist loss can be estimated by

$$h_{f,e} = \frac{v_1^2 - v_2^2}{2g_c} \quad [e = \text{sudden expansion (SE)}] \tag{9–53}$$

where $h_{f,e}$ is the loss in head, v_2 is the velocity at the larger cross-section, and v_1 is the velocity at the smaller cross-section. When the cross-section of the pipe is reduced suddenly (as with a contraction), the loss may be expressed by

$$h_{f,c} = \frac{Kv_2^2}{2g_c} \quad [c = \text{sudden contration, (SC)}] \tag{9–54}$$

where v_2 is the velocity at the smaller cross-section and K is a dimensionless loss coefficient that is a function of the ratio of the two cross-sectional areas. Both of these calculations receive additional treatment in the next subsection.

Calculating Losses of Valves and Fittings

Pipe systems, as mentioned above, include inlets, outlets, bends, and other devices (e.g., valves, fittings) that create a pressure drop. This drop, which results in energy loss, may in some instances be greater than that due to flow in a straight pipe. Thus, these additional so-called minor losses may not be so minor in some applications. Pressure loss data for a wide variety of valves and fittings have been compiled in terms of either the loss (or resistance) coefficient K. Details on these two concepts follow.

The dimensionless loss coefficient is defined as

$$K = \frac{h_{f,m}}{v^2/2g_c} \tag{9–55}$$

The units of $h_{f,m}$ must have units consistent with $v^2/2g_c$ [e.g., (ft·lb$_f$)/lb] for K to be dimensionless. A pipe system may have several valves and fittings. The minor losses of all of them can therefore be expressed in terms of the velocity (dynamic) head, and can all be summed:

$$\sum h_{f,m} = \frac{v^2}{2g_c}\sum K \tag{9–56}$$

Numerical values for the resistance coefficient K for open valves, elbows, and tees (Ts) are available in the literature.[1, 3,8] The listed K values depend on the nominal diameter, the valve or fitting type, and whether it is screwed or flanged. The tabulated valve loss coefficients are for *fully open* valves. To account for *partially open* valves, the following ratio is employed:

$$\frac{K \text{ of partially open valve}}{K \text{ of fully open valve}} \tag{9–57}$$

The loss coefficient K_{SE} due to a sudden expansion (SE) between two different sizes of pipes D_1 and D_2 ($D_1 < D_2$) is given by Eq. (9–58)

$$h_{f,e} = K_{SE}\frac{v_1^2}{2g_c} \tag{9–58}$$

with

$$K_{SE} = \left[1-\left(\frac{D_1}{D_2}\right)^2\right]^2 \tag{9–59}$$

Combing Eqs. (9–58) and (9–59) gives

$$h_{f,e} = \left[1-\left(\frac{D_1}{D_2}\right)^2\right]^2\frac{v_1^2}{2g_c} \tag{9–60}$$

A sudden contraction (SC) generally leads to a higher pressure drop. The loss coefficient K_{SC} can be *approximated* by the equation

$$K_{SC} = 0.42\left[1-\left(\frac{D_1}{D_2}\right)^2\right] \tag{9–61}$$

Note that values of 0.50 rather than 0.42 have been reported in the literature;[11] for this case,

$$K_{SC} = 0.50\left[1-\left(\frac{D_1}{D_2}\right)^2\right] \tag{9–62}$$

Note that the loss coefficient K_{SC} is based on the velocity head of the smaller pipe at exit conditions.

Another approach to describe these losses is through the *equivalent-length concept*. Assume that L_{eq} is the length of straight pipe that produces the same head loss as the device in question. The term L_{eq} is related to the loss coefficient as follows:

$$L_{eq} = K\frac{D}{4f} \quad (9\text{–}63)$$

The L_{eq} in this equation is rarely expressed as actual feet of straight pipe, but rather as a certain number of pipe diameters. Tabulated values are available in the literature.[3,5] It should be noted that either the K or the L_{eq} approach assumes a *zero-length* fitting or device. The resistance of the fitting may, therefore, be taken as the total resistance of the pipe or fitting system less the resistance of the length of straight pipe of the section of flow under study.

9–5 Fluid-Particle Applications

If a particle is initially at rest in a stationary fluid and is then set in motion by application of a constant external force or forces, the resulting motion occurs in two stages: (1) during *acceleration*, when the particle velocity increases from zero to some maximum velocity; and (2) when the particle achieves this maximum velocity and remains constant. During stage 2, the particle (p) is not accelerating. The final, constant, and maximum velocity attained is defined as the *terminal settling velocity* (v) of the particle. Most particles reach their terminal settling velocity almost instantaneously. These velocities are given by the following three equations:

For the Stokes' law range, with Re < 2.0

$$v = \frac{fd_p^2\rho_p}{18\mu} \quad (9\text{–}64)$$

For the intermediate range, with 2.0 < Re < 500:

$$v = \frac{0.153 f^{0.71} d_p^{1.14} \rho_p^{0.71}}{\mu^{0.43}\rho^{0.29}} \quad (9\text{–}65)$$

For Newton's law range, with Re > 500:

$$v = 1.74(fd_p\rho_p\rho)^{0.5} \quad (9\text{–}66)$$

Note that Re for these equations is based on the particle diameter d_p. Keep in mind that f denotes the external force per unit mass of particle. One consistent set of units (English) for Eqs. (9–64) to (9–66) is ft/s² for f, ft for d_p, lb/ft³ for ρ, lb/(ft · s) for μ, and ft/s for v.

When particles approach sizes comparable to the mean free path of other fluid molecules, the medium can no longer be regarded as continuous since particles can fall between the molecules faster than predicted by aerodynamic theory. To allow for this *slip*, Cunningham's correction factor is introduced to Stokes' law

$$v = \frac{gd_p^2\rho_p}{18\mu}C \quad (9\text{–}67)$$

where C is the Cunningham correction factor (CCF) and

$$C = 1 + \frac{2A\lambda}{d_p} \tag{9–68}$$

The term A is $1.257 + 0.40\exp(-1.10d_p/2\lambda)$, and λ is the mean free path of the fluid molecules (6.53×10^{-6} cm for ambient air). The CCF is usually applied to particles equal to or smaller than 1 μm. Applications include nanotechnology[10] and particulate air pollution studies.[11]

There are three industrial separation techniques that exploit the density difference between a liquid and a solid. The driving force in these processes is usually the result of gravity, centrifugal action, and/or buoyant effects. Unlike the discussion above, which primarily treated gas-particle dynamics, this presentation will be key on liquid-solid separation. The separation of solid particles into several fractions on the basis of their terminal velocities is called *hydraulic classification.*

Gravity sedimentation is a process of liquid-solid separation that separates, under the effect of gravity, a feed slurry into an underflow slurry of higher solids concentration and an overflow of substantially clearer liquid. A difference in density between the solids and the suspended liquid is, as indicated, a necessary prerequisite.

An extended topic of interest in centrifugation is *hydrostatic centrifugation equilibrium.* When a contained liquid of constant ρ and μ is rotated around the vertical axis (z axis) at a constant angular speed ω, it is thrown outward from the axis by centrifugal force. The free surface of the liquid develops a three-dimensional paraboloid of revolution, the cross-section of which has a parabolic shape. When the fluid container has been rotated long enough, the whole volume of fluid will rotate at the same angular velocity.

Finally, *flotation* processes are useful for the separation of a variety of species, ranging from molecular and ionic to microorganisms and mineral fines, from one another for the purpose of the extraction of valuable products as well as the cleaning of wastewaters. *Flotation* is a gravity separation process based on the attachment of air or gas bubbles to solid (or liquid) particles that are then carried to the liquid surface where they accumulate as float material and can be skimmed off. The process consists of two stages: the production of suitably small bubbles, and their attachment to the particles. Depending on the method of bubble production, flotation is classified as dissolved air, electrolytic, or dispersed air, with dissolved air primarily employed by industry.[12]

9–6 Flow-through Porous Media

The flow of a fluid through porous media or a packed bed occurs frequently in chemical process applications. Some chemical engineering examples include flow-through a fixed-bed catalytic reactor, flow-through an adsorption tower, and flow-through a filtration unit. An understanding of this type of flow is also important in the study of some particle dynamics applications and fluidization (see also Sec. 9–7).[5]

A *porous medium* is a continuous solid phase with many void spaces called *pores.* Examples include sponges, paper, sand, and concrete. Packed beds of porous materials are used in a number of chemical engineering operations, such as distillation, absorption, filtration, and drying. There are two types of porous media:

1. *Impermeable media*—solid media in which the pores are not interconnected, e.g., foamed polystyrene
2. *Permeable media*—solid media in which the pores are interconnected, e.g., packed columns and catalytic reactors.

In studying fluid flow in porous medium (e.g., packed bed) the following variables are defined:

1. *Porosity*, or *void fraction* ∈ (*e* may also be employed):

$$\in = \frac{\text{volume of voids}}{\text{total volume of system}} \tag{9–69}$$

$$\in = \frac{\text{cross-sectional area of voids}}{\text{total cross-sectional area}} \tag{9–70}$$

2. *Solid fraction*: fraction of total bed volume occupied by solids = 1− ∈ or 1− e.
3. *Empty bed cross-section S*: equal to $\pi d_b^2/4$, where d_b = bed diameter.
4. *Actual bed cross-section for flow*: (or open-bed cross-section):

$$S_{\text{ac}} = \frac{\pi d_b^2 \in}{4} \tag{9–71}$$

5. *Ultimate density*: ρ_s = density of solid material.
6. *Bulk density*: ρ_{bulk} = (solid fraction)(solid density) + (void fraction)(fluid density)

$$\rho_{\text{bulk}} = (1-\in)\rho_s + \in \rho_f \tag{9–72}$$

7. *Superficial velocity* (also termed *empty-tower velocity*): The term is the average velocity of fluid based on an empty cross-section:

$$v_s = \frac{q}{S} = \frac{4q}{\pi d_b^2} \tag{9–73}$$

8. *Interstitial velocity*: v_1 is the actual velocity of fluid through pores:

$$v_1 = \frac{q}{S_{\text{ac}}} = \frac{q}{\pi d_b^2 \in} = \frac{v_s}{\in} \tag{9–74}$$

9. *Standard screen mesh size*: Particle sizes in the range of 3 to ~0.0015 inches are measured with standard screens. The testing sieves are standardized and have square openings. A common screen is the *Tyler standard screen series*; refer to Abulencia and Theodore[3] for additional details
10. *Particle-specific surface* (a_p): surface area per unit volume of the packed-bed particles; expressed in units of m^2/m^3 (or ft^2/ft^3). For a spherical particle:

$$a_p = \frac{S_p}{V_p} = \frac{\pi d_p^2}{(\pi/6)d_p^3} = \frac{6}{d_p} \tag{9–75}$$

11. *Bed-specific surface area* (a_b): surface area per unit bed volume:

$$a_b = a_p(1-\in) = \frac{6(1-\in)}{d_p} \tag{9–76}$$

12. *Effective diameters ($d_{p,e}$) of nonspherical particles*: diameter of a sphere that has the same specific surface area a_p as that of the nonspherical particle.

13. *Hydraulic diameter* (d_h): The fluid will move through pores (or the open spaces between the spaced particles) which are not necessarily circular. The hydraulic diameter d_h is defined as follows:

$$d_h = 4R_h = 4\left(\frac{\text{volume open to flow}}{\text{total wetted surface}}\right) \tag{9–77}$$

$$d_h = 4R_h = 4\frac{(\text{volume of void})(\text{volume of bed})}{(\text{volume of bed})(\text{total surface area})} \tag{9–78}$$

Fluid flow in a porous medium is classified as laminar, transition, or turbulent, based on the porous medium Reynolds number Re_{pm}:

Laminar	$\text{Re}_{\text{pm}} < 10$
Transition	$10 < \text{Re}_{\text{pm}} < 1000$
Turbulent	$\text{Re}_{\text{pm}} > 1000$

$$\text{Re}_{\text{pm}} = \frac{d_p v_e \rho}{(1-\epsilon)(\mu)} \tag{9–79}$$

Experimental measurements indicate that

$$\frac{\Delta P}{L} = 150 v_s \mu \frac{(1-\epsilon)^2}{\epsilon^3} \tag{9–80}$$

which some have defined as the *Blake–Kozeny* equation. This equation generally applies for void fractions less than 0.5 and is valid only in the laminar region where the particle Reynolds number is < 10.[12] For turbulent flow, experimental data indicate that

$$\frac{\Delta P}{L} = 3.50 \frac{1}{d_p} \frac{\rho v_s^2}{2} \frac{1-\epsilon}{\epsilon^3}$$

$$= 1.75 \frac{\rho v_s^2}{d_p} \frac{1-\epsilon}{\epsilon^3} \tag{9–81}$$

which some have defined as the *Burke–Plummer* equation. When the Blake–Kozeny equation for laminar flow and the Burke–Plummer equation for turbulent flow are simply added together, the result is

$$\frac{\Delta P}{L} = \frac{150 \mu v_s}{d_p^2} \frac{(1-\epsilon)^2}{\epsilon^3} + 1.75 \frac{\rho v_s^2}{d_p}\left(\frac{1-\epsilon}{\epsilon^3}\right) \tag{9–82}$$

This may be rewritten in terms of dimensionless groups (numbers):

$$\left(\frac{\Delta P \rho}{G_s^2}\right)\left(\frac{d_p}{L}\right)\left(\frac{\epsilon^3}{1-\epsilon}\right) = 150 \frac{1-\epsilon}{d_p G_s / \mu} + 1.75; \; G_s = \text{superficial velocity flux} \tag{9–83}$$

This is the *Ergun* equation.[13] It may also be written in the following form:

$$\Delta P = 150 \frac{v_s \mu}{g_c} \frac{(1-\epsilon)^2}{\epsilon^3}\left(\frac{L}{d_p^2}\right) + 1.75 \frac{v_s^2(1-\epsilon)}{g_c \epsilon^3}\left(\frac{L}{d_p}\right)\rho \tag{9–84}$$

Filtration

Filtration is one of the most common chemical engineering applications that involve the flow of fluids through packed beds. As carried out industrially, it is similar to the filtration carried out in the chemical laboratory using a filter paper in a funnel. The object is still the separation of a solid from the fluid in which it is carried; the separation is accomplished by allowing (usually by force) the fluid to pass through through a porous filter. The solids are trapped within the pores of the filter and (primarily) build up as a layer on the surface of this filter. The fluid, which may be either gas or liquid, passes through the bed of solids and through the retaining filter. Abulencia and Theodore[3] provide additional details plus design and predictive equations.

As noted above, solid particles are removed from a slurry by passing it through a filtering medium. A *slurry* can be defined as a watery mixture of insoluble matter such as mud, lime or plaster of Paris. The solids are deposited on the filtering medium, which is normally referred to as the *cake*. Filtration may therefore be viewed as an operation in which a heterogeneous mixture of a fluid and particles is separated by a filter medium that permits the flow of the fluid but retains the particles. Therefore, it primarily involves the flow of fluids through porous media. The reader should note once again that the fluid may be either a liquid or gas.

In all filtration processes, the mixture or slurry flows as a result of some driving force: gravity, pressure (or vacuum), or centrifugal force. In each case, the filter medium supports the particles as a porous cake. This cake, supported by the filter medium, retains the solid particles in the slurry with successive layers added to the cake as additional filtrate passes through the cake and medium.

The various procedures for creating the driving force for the fluid, the different methods of cake deposition and removal, and the different means for removal of the filtrate from the cake, result in a great variety of filter equipment. In general, filters may be classified according to the nature of the driving force supporting filtration.

9–7 Fluidization

Fluidization is a process in which a fluid (gas or liquid) transforms fine solids into a fluid-like state. Excellent particle-fluid contact results. Consequently, fluidized beds are used in many applications, including oil cracking, zinc roasting, coal combustion, gas desulfurization, heat exchanges, plastics coating, and fine-powder granulation. Figure 9–5 demonstrates the kinds of contact between solids and a fluid, starting from a packed bed and ending with pneumatic transport.[3] At low fluid velocity one encounters a fixed-bed configuration of height L_m, which is termed the "slumped" bed height. As the velocity increases, fluidization starts; the point at which fluidization starts is termed *onset of fluidization*. The superficial velocity of the fluid at the onset of fluidization is termed the *minimum fluidization velocity* v_{mf}, and the bed height is L_{mf}. As the fluid velocity increases beyond v_{mf}, the bed expands and the bed voidage increases. At low fluidization velocities (fluid velocity > v_{mf}), the operation is termed *dense*, the bed voidage increases, and fluid is termed *dense phase fluidization*.

The variation in porosity (and hence bed height) with the superficial velocity is calculated from the Kozeny–Carman equation, assuming laminar flow and $\rho_f \ll \rho_s$:

$$v = \frac{d_p^2 g}{150\mu_f}(\rho_p - \rho_f)\frac{\epsilon^3}{(1-\epsilon)} \tag{9–85}$$

The equations for minimum fluidization are similar to those presented earlier for a fixed porous bed. For laminar flow conditions ($\mathrm{Re}_p < 10$), the Blake–Kozeny equation is used

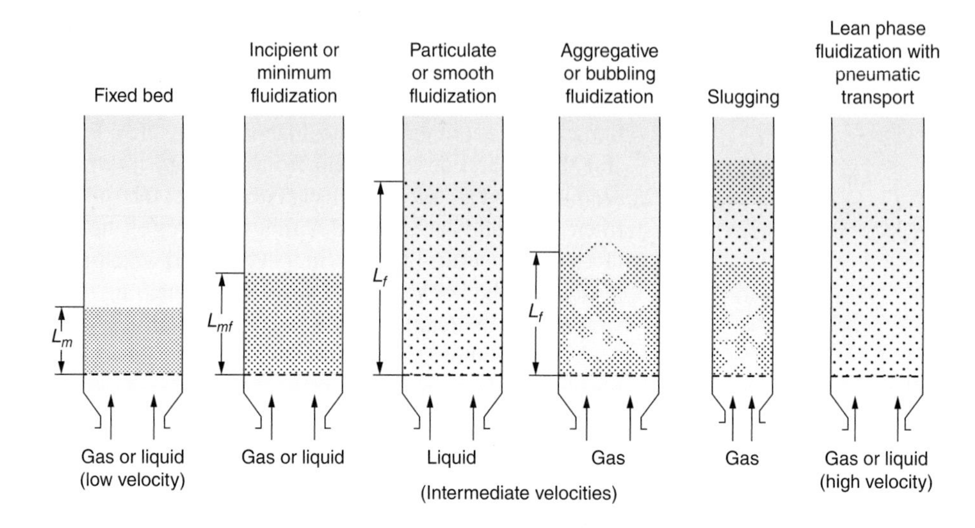

Figure 9.5

Types of particle-fluid contact in a bed.

to express h'_f in terms of the superficial gas velocity at minimum fluidization v_{mf} and other fluid and bed properties:

$$h'_f = (1-\in_{\text{mf}})\left(\frac{\rho_s - \rho_f}{\rho_f}\right)L_{\text{mf}} = 150\frac{v_{\text{mf}}\mu_f}{\rho_f g}\frac{(1-\in_{\text{mf}})^2}{\in_{\text{mf}}^3}\left(\frac{L_{\text{mf}}}{d_p^2}\right) \tag{9–86}$$

or

$$v_{mf} = \frac{1}{150}\left(\frac{\in_{\text{mf}}^3}{1-\in_{\text{mf}}}\right)\frac{g(\rho_s-\rho_f)d_p^2}{\mu_f} \quad (\text{Re}_p < 10) \tag{9–87}$$

The Burke–Plummer equation is used to express the head loss h'_f for turbulent flow conditions ($\text{Re}_p > 1000$):

$$v_{\text{mf}} = \sqrt{\frac{1}{1.75}\in_{\text{mf}}^3\left(\frac{\rho_s-\rho_f}{\rho_f}\right)gd_p} \tag{9–88}$$

A general equation covering the entire range of flow rates but for *various-shaped* particles can be obtained by assuming once again that the laminar and turbulent effects are additive. This result is also referred to as the *Ergun equation*,[13] which takes the form

$$\frac{\Delta P}{L} = \frac{150 v_0 \mu}{g_c \Phi^2 d_p^2}\frac{(1-\in)^2}{\in^3} + \frac{1.75\rho v_0^2}{g_c \Phi^2 d_p^2}\left(\frac{1-\in}{\in^3}\right) \tag{9–89}$$

where Φ is the sphericity or shape factor of the fluidized particles. Typical values for the sphericity of typical fluidized particles range from 0.75 to 1.0. In lieu of any information on Φ, one should employ a value of 1.0, which is typical for spheres, cubes, and cylinders ($L=d$).

Finally, there are two modes of fluidization. When the difference between fluid and solid densities is neglible, or the particles are very small, and therefore the velocity of the flow is low, the bed is fluidized evenly with each particle moving individually through the bed. This is called *smooth* (or *particulate*) *fluidization* and is typical of liquid-solid systems. If the fluid and solid densities differ significantly, or the particles are large, the velocity of the flow must be relatively high. In this case, fluidization is uneven, and the fluid passes through the bed mainly in large bubbles. These bubbles burst at the surface, spraying solid particles above the bed. Here the bed has many of the characteristics of a liquid with the fluid phase acting as a gas bubbling through it. This is called "bubbling" *or aggregative fluidization*, is typical of gas–solid systems, and is due to the large density difference between the solid and the gas. The approximate criterion to estimate the transition from bubbling to smooth fluidization is expressed in terms of the Froude dimensionless number at minimum fluidization. The Froude number is expressed in terms of the minimum fluidization velocity v_{mf}, the particle diameter d_p, and the gravity acceleration g:

$$\mathrm{Fr}_{mf} = \frac{v_{mf}^2}{g d_p}$$

$$< 0.13 \text{ smooth or particulate fluidization}$$

$$> 1.3 \text{ bubbling (or aggregative) fluidization} \quad (9\text{–}90)$$

References

1. I. Farag, *Fluid Flow*, A Theodore Tutorial, Theodore Tutorials (originally published by the USEPA APTI, Research Triangle Park, N.C.), East Williston, N.Y., 1994.
2. L. Theodore, *Transport Phenomena for Engineers*, Theodore Tutorials (originally published by the International Textbook Co.), East Williston, N.Y., 1971.
3. P. Abulencia and L. Theodore, *Fluid Flow for the Practicing Chemical Engineer*, Wiley, Hoboken, N.J., 2009.
4. W. Badger and J. Banchero, *Introduction to Chemical Engineering*, McGraw-Hill, N.Y., 1955.
5. C. Bennett and J. Myers, *Momentum, Heat, and Mass Transfer*, McGraw-Hill, N.Y., 1962.
6. L. Moody, Friction Factors for Dye Flow, *Trans. Am. Soc. Mech. Eng.*, **66**, **67** 1–84, N.Y., 1944.
7. G. Brown and Associates, *Unit Operations*, Wiley, Hoboken, N.J., 1950.
8. D. Green and R. Perry, eds. *Perry's Chemical Engineers' Handbook*, 8th ed., McGraw-Hill, N.Y., 2008.
9. E. Cunningham, *Proc. R. Soc. Lond. Ser.*, **17**, 83, 357, London, 1910.
10. L. Theodore, *Nanotechnology: Basic Calculations for Engineers and Scientists*, Wiley, Hoboken, N.J., 2007.
11. L. Svarousky, "Sedimentation, Centrifugation and Flotation," *Chem. Eng.*, N.Y., July 16, 1979.
12. L. Theodore, *Air Pollution Control Equipment Calculations*, Wiley, Hoboken, N.J., 2008.
13. S. Ergun, "Fluid Flow-through Packed Columns," *Chem. Eng. Prog.*, **48**, 89–94, N.Y., 1952.

10 Heat Transfer

Chapter Outline

10–1 Introduction

A temperature difference between two bodies in close proximity or between two parts of the same body results in heat flow from higher to lower temperatures. There are three different (and classic) mechanisms by which this heat transfer can occur: conduction, convection, and radiation. When the heat transfer is the result of molecular motion (e.g., the vibrational energy of molecules in a solid being passed along from molecule to molecule), the mechanism of transfer is *conduction*. When the heat transfer results from macroscopic motion, such as currents in a fluid, the mechanism is *convection*. When heat is transferred by electromagnetic waves, *radiation* is the mechanism. In most industrial applications, multiple mechanisms are usually involved in the transmission of heat. However, since each mechanism is governed by its own set of physical laws, it is beneficial to discuss them independently of each other.

Topics addressed in this chapter include conduction, convection, and radiation; condensation, boiling, refrigeration, and cryogenics; and, heat exchangers and the heat exchanger dilemma.

NOTE The bulk of the material presented in this chapter was drawn from Farag and Reynolds.[1]

10–2 Conduction

As the temperature of a solid increases, the molecules that make up the solid experience an increase in vibrational kinetic energy. Since every molecule is bonded in some manner to neighboring molecules, this energy can be transmitted through the solid. Thus, heating a wire at one end eventually results in raising the temperature at the other end. This type of heat transfer is defined as *conduction* and is the principle mechanism by which solids transfer heat. Fluids are capable of transporting heat in a similar fashion. Conduction in a stagnant liquid, for example, occurs by the movement not only of vibrational kinetic energy but also of translational kinetic energy as the molecules move throughout the body of the liquid. The ability of a fluid to flow, mix, and form internal currents on a macroscopic level (as opposed to the molecular mixing just described) allows fluids to carry heat energy by *convection* (a topic discussed in the next section) as well.

The rate of heat flow by conduction is given by Fourier's law:[2]

$$q = -kA\frac{dT}{dx} \tag{10–1}$$

where q = heat flow rate, Btu/h
x = direction of heat flow, ft
k = thermal conductivity, Btu/(h · ft ·°F)
A = heat transfer area, a plane perpendicular to x direction, ft^2
T = temperature, °F

Note that English units are employed in this and the development to follow. The negative sign reflects the fact that heat flow is from a high to low temperature, and therefore the sign of the derivative is opposite to that of the direction of the heat flow. If k can be considered constant over a limited temperature range (ΔT), Eq. (10–1) may be integrated to give

$$q = \frac{kA(T_h - T_c)}{x_h - x_c} = \frac{kA\Delta T}{L} \tag{10–2}$$

where T_h = higher temperature at location x_h, °F
T_c = lower temperature at location x_c, °F
$\Delta T = T_h - T_c$
L = distance between locations x_h and x_c, ft

Equation (10–1) is the basis for the definition of thermal conductivity, i.e., the amount of heat (Btu) that flows in a unit of time (1 h) through a unit of surface (1 ft^2) of unit thickness (1 ft) by virtue of a unit difference in temperature (1°F). Values of k employing these units for some insulating solids are given in Table 10–1.

Thermal conductivity k is a property of the material through which the heat is passing and, as such, does vary somewhat with temperature. Equation (10–2) should therefore be employed only for small values of ΔT with an average value used for k. The average value of k can be based on either the average k at the average temperature or the value at the average of the two temperatures.

Equation (10–2) may be written in the form of the general transfer rate equation:

$$\text{Rate} = \frac{\text{driving force}}{\text{resistance}} = \frac{\Delta T}{L/kA} \tag{10–3}$$

Since q in Eq. (10–2) is the heat flow rate and ΔT the driving force, the L/kA term may be considered to be the resistance to heat flow. This approach is useful when heat is flowing by conduction in sequence through different materials.

Consider, for example, a flat incinerator wall containing three different layers: an insulating layer a, a steel plate b, and an outside insulating layer c. The total resistance to heat flow-through the incinerator wall is the sum of the three individual resistances:

$$R = R_a + R_b + R_c \tag{10–4}$$

At steady state, the rate of heat flow-through the wall is, therefore, given by

$$q = \frac{T_a - T_d}{(L_a/k_aA_a) + (L_b/k_bA_b) + (L_c/k_cA_c)} \tag{10–5}$$

where k_a, k_b, k_c = thermal conductivity of each section, Btu/(h · ft ·°F)
A_a, A_b, A_c = area of heat transfer of each section, (ft^2); equal for areas of constant cross section

Table 10–1

Thermal Conductivities of Some Common Insulating Materials

Material	Thermal conductivity k*
Asbestos-cement boards	0.43
Sheets	0.096
Asbestos cement	1.202
Corkboard, 10 lb/ft^3	0.025
Diatomaceous earth (Sil-o-cel)	0.035
Fiber, insulating board	0.028
Glass wool, 1.5 lb/ft^3	0.022
Magnesia, 85%	0.039
Rock wool, 10 lb/ft^3	0.023

*Units for K = Btu/(h · ft · °F).

L_a, L_b, L_c = thickness of each layer, ft
T_a = temperature at inside surface of insulating wall a, °F
T_d = temperature at outside surface of insulating wall d, °F

In the above example, the heat is flowing through a slab of constant cross-section. In many cases of industrial importance, however, this is not the case. For example, in heat flow-through the walls of a cylindrical pipe in a heat exchanger or a rotary kiln, the heat transfer area increases with distance from the center of the cylinder. The heat flow in this case is given by

$$q = \frac{kA_{\text{lm}}\Delta T}{L} \qquad (10\text{–}6)$$

The term A_{lm} in Eq. (10–6) represents the average heat transfer area, or more accurately, the log-mean average heat transfer area. The log-mean average can be calculated by

$$A_{\text{lm}} = \frac{A_2 - A_1}{\ln(A_2/A_1)} \qquad (10\text{–}7)$$

where A_2 = outer surface area of cylinder, ft^2
A_1 = inner surface area of cylinder, ft^2

Chemical engineers spend a good deal of time working with systems that are operating under steady-state conditions. However, many processes are *transient* or unsteady state in nature, and it is necessary to predict how process variables will change with time, as well as how these effects will impact the design and performance of these systems. The prediction of the unsteady-state temperature distribution in solids is an example of one such process. It can be accomplished using transient conduction equations; the energy balance equations can usually be easily solved to calculate the spatial temporal variation of the temperature within the solid.[3–5]

Unsteady-state processes are those in which the heat flow, the temperature, or both, vary with time at a fixed point in space. Batch heat-transfer processes are typical unsteady-state processes. For example, heating reactants in a tank or the start-up of a cold furnace are two unsteady-state applications. Still other common examples include the rate at which heat is conducted through a solid while the temperature of the heat source varies, the daily periodic variations of solar heat on various solids, the quenching of steel in an oil or cold-water bath, cleaning or regeneration processes, or, in general, any process that can be classified as intermittent or in the start-up or shutdown category. A good number of theoretical heat transfer conduction problems are time dependent. These *unsteady* or *transient* situations usually arise when the boundary conditions of a system are changed.

10–3 Convection

When a pot of water (this discussion applies to any liquid) is heated on a stove, the portion of water adjacent to the bottom of the pot is the first to experience a temperature rise. Eventually, the water at the top will experience an increase in temperature and become hotter. Although some of the heat transfer from bottom to top is attributed to conduction through the water, most of it is due to the second mechanism of heat transfer, namely, convection. As the water at the bottom is heated, its density decreases. This results in convection currents as gravity causes the low-density water to move upward and be replaced by the higher-density cooler water from above. As one might suppose, this macroscopic mixing is a far more effective mechanism for moving heat energy through fluids than is conduction. This process is called *natural convection* since no external forces, other than gravity, need be applied to transfer the heat. In most industrial applications, however, it is more economical to increase the mixing action by artificially generating a current by the use of a pump, agitator, or some other mechanical device. This is referred to as *forced convection*.

The flow of heat from a hot fluid to a cooler fluid through a solid wall is a situation often encountered in engineering equipment; examples of such equipment are heat exchangers, condensers, evaporators, boilers, and economizers. The heat absorbed by the cool fluid or given up by the hot fluid may be *sensible* heat, causing a temperature change in the fluid, or it may be *latent* heat, causing a phase change such as vaporization or condensation (see also Chap. 8). The rate of heat transfer between the two streams, assuming no heat loss to the surroundings, may be calculated by the enthalpy change on either fluid:

$$q = \Delta H = \dot{m}_h(h_{h1} - h_{h2}) = \dot{m}_c(h_{c2} - h_{c1}) \qquad (10\text{–}8)$$

where q = rate of heat flow, Btu/h
$\dot{m}_h$ = mass flow rate of hot fluid, lb/h
$\dot{m}_c$ = mass flow rate of cold fluid, lb/h
h_{h1} = enthalpy of entering hot fluid, Btu/lb
h_{h2} = enthalpy of exiting hot fluid, Btu/lb
h_{c1} = enthalpy of entering cold fluid, Btu/lb
h_{c2} = enthalpy of exiting cold fluid, Btu/lb

Equation (10–8) is applicable to the heat exchange between two fluids regardless of whether a phase change is involved.

In a waste heat boiler example, the enthalpy change of the flue gas is calculated from its temperature drop:

$$q = \dot{m}_h(h_{h1} - h_{h2}) = \dot{m}_h c_{ph}(T_{h1} - T_{h2}) \qquad (10\text{–}9)$$

where c_{ph} = heat capacity of hot fluid, Btu/(lb ·°F)
T_{h1} = temperature of entering hot fluid, °F
T_{h2} = temperature of exiting hot fluid, °F

The enthalpy change of the water involves a small amount of sensible heat to bring the water to its boiling point plus a considerable amount of latent heat to vaporize the water. Assuming that all of the water is vaporized and no superheating of the resulting steam occurs, the enthalpy change is given by

$$q = \dot{m}_c(h_{c2} - h_{c1}) = \dot{m}_c c_{pc}(T_{c2} - T_{c1}) + w_c\lambda_c \qquad (10\text{–}10)$$

where c_{pc} = specific heat capacity of cold fluid, Btu/(lb ·°F)
T_{c1} = temperature of entering cold fluid, °F
T_{c2} = boiling point of cold fluid, °F
λ_c = heat of vaporization of cold fluid at T_{c2}, Btu/lb

To design a piece of heat transfer equipment, it is sufficient to know the heat transfer rate calculated by the enthalpy balances described above. This is the rate at which heat travels from the hot fluid, through the tube walls, and into the cold fluid and is employed in the calculation of the contact area. The slower this rate is, for given hot and cold fluid flow rates, the more contact area is required. The rate of heat transfer through a unit of contact area is referred to as the *heat flux* or *heat flux density* and, at any point along the tube length, is given by

$$\frac{dq}{dA} = U(T_h - T_c) \tag{10–11}$$

where dq/dA = local heat flux density, Btu/(h · ft²)
U = local overall heat transfer coefficient, Btu/(h · ft² ·°F)

The use of this overall heat transfer coefficient (U) is a simple yet powerful concept. In most applications it combines both conduction and convection effects, although transfer by radiation can also be included. Methods for calculating the overall heat transfer coefficient are presented later in this section. In actual practice it is not uncommon for vendors to provide a numerical value for U.

With reference to Eq. (10–11), the temperatures T_h and T_c are actually *average values*. When a fluid is being heated or cooled, the temperature will vary throughout the cross-section of the stream. If the fluid is being heated, its temperature will be highest at the tube wall and decrease with increasing distance from the tube wall. The average temperature across the flowing stream's cross-section is therefore, T_c, which is the temperature that would be achieved if the fluid at this cross-section were suddenly mixed to a uniform temperature. If the fluid is being cooled, on the other hand, its temperature will be lowest at the tube wall and increase with increasing distance from the wall.

To apply Eq. (10–11) to an entire heat exchanger, the equation must be integrated. This cannot be accomplished unless the geometry of the exchanger is first defined. For simplicity, one of the simplest geometries will be assumed here: the double-pipe heat exchanger (see later section). This device consists of two concentric pipes. The outer surface of the outer pipe is normally well insulated so that no heat exchange with the surroundings may be assumed. One fluid flows through the center pipe, while the other one flows through the annular channel between the pipes. The fluid flows may be either *concurrent*, when the two fluids flow in the same direction, or *countercurrent*, where the flows are in opposite directions. The countercurrent arrangement is more efficient and is more commonly used.

For a heat exchanger, integration of Eq. (10–11) along the exchanger length, applying several simplifying assumptions, yields

$$q = UA\,\Delta T \tag{10–12}$$

This has come to be defined by some as the *heat transfer equation*. The simplifying assumptions are that U and ΔT do not vary along the length. Since this is not actually the case, both U and ΔT must be regarded as averages of some type. A more careful integration of Eq. (10–11), assuming that only U is constant, would show that the appropriate average for ΔT is the log-mean average:

$$\Delta T_{\text{lm}} = \frac{\Delta T_2 - \Delta T_1}{\ln(\Delta T_2/\Delta T_1)} \tag{10–13}$$

where ΔT_1 and ΔT_2 are the temperature differences between the two fluids at the ends of the exchanger. The area term (A) in Eq. (10–12) denotes the cylindrical contact area between the fluids. However, since a pipe of finite thickness separates the fluids, the cylindrical area (A) must first be defined. Any one of the infinite number of areas between and including the inside and outside surface areas of the pipe may be arbitrarily chosen for this purpose. The usual practice is to use either the inside (A_i) or outside (A_0) surface areas; the outside area is the more commonly used of the two. Since the value of the overall heat transfer coefficient depends on the area chosen, it should be subscripted to correspond to the area on which it is based. Eq. (10–12), based on the outside surface area, now becomes

$$q = U_o A_o \Delta T_{\text{lm}} \tag{10–14}$$

Comparing Eq. (10–14) to the general rate equation [Eq. (10–3)], it can be seen that $(U_o, A_o)^{-1}$ may be regarded as the resistance to heat transfer between the two fluids:

$$R_t = \frac{1}{U_o A_o} \tag{10–15}$$

In practice, the flow of the fluids in a heat exchanger is turbulent, which means that the main bulk of the stream is well mixed and therefore of fairly uniform temperature. Immediately adjacent to the pipe wall, however, is a thin layer of fluid (referred to as the *boundary layer*) that is in laminar flow and is therefore not mixed. In the absence of macroscopic mixing, heat must be transferred through the film by conduction alone, and hence the laminar film presents a considerable contribution to the total resistance (R_t).

The total resistance to heat transfer across the exchanger may therefore be divided into three contributions: the *inside film*, the *tube wall*, and the *outside film*. This is restated mathematically in Eqs. (10–16) and (10–17):

$$R_t = R_i + R_w + R_o \tag{10–16}$$

or

$$\frac{1}{U_o A_o} = \frac{1}{h_i A_i} + \frac{x}{k A_{\text{lm}}} + \frac{1}{h_o A_o} \tag{10–17}$$

where h_i = inside film coefficient, Btu/(h · ft² ·°F)
h_o = inside film coefficient, Btu/(h · ft² ·°F)
A_o = outside surface area of pipe, ft²
A_i = inside surface area of pipe, ft²
A_{lm} = log-mean average surface area of pipe, ft²
x = pipe thickness, ft
k = thermal conductivity of pipe, Btu/(h · ft · °F)

Film coefficients are almost always determined experimentally. Many empirical correlations can be found in the literature for a wide variety of fluids and exchanger geometries. Typical ranges of film coefficients are given in Table 10–2. For fully developed turbulent flow in smooth tubes, the following relationship by Dittus and Boelter[4,6] is recommended for determining the inside film coefficient h_i:

$$\text{Nu} = 0.023\ \text{Re}^{0.8}\ \text{Pr}^{n} \tag{10–18}$$

where Nu = Nusselt number = $h_i D_i / k$ (dimensionless)
Re = Reynolds number = $D_i G/\mu$ (dimensionless)
Pr = Prandtl number = $c_p \mu / k$ (dimensionless)
D_i = inside tube diameter (ft)
k = thermal conductivity of fluid, Btu/(h · ft ·°F)
c_p = specific heat capacity of fluid, (Btu · lb · °F)
μ = viscosity of fluid, lb/(h · ft³)
$G = \dot{m}/A_{cs}$ = fluid mass flux, lb/(h · ft²)
n = constant (see below)
$\dot{m}$ = fluid mass flow rate, lb/h
A_{cs} = inside cross-sectional area for fluid flow, ft²

Table 10–2

Typical Film Coefficients, h*

Fluid	Inside pipes	Outside pipes†,‡
Gases	10–50	1–3 (n), 5–20 (f)
Water (liquid)	200–2000	20–200 (n), 100–1000 (f)
Boiling water	500–5000	300–9000
Condensing steam	—	1000–10,000
Nonviscous liquids	50–500	50–200 (f)
Boiling liquids	—	200–2000
Condensing vapor	—	200–400
Viscous liquids	10–100	20–50 (n), 10–100 (f)
Condensing vapor	—	50–100

*h = Btu/(h · ft^2 ·°F).
†(n) = natural convection.
‡(f) = forced convection.

The properties in Eq. (10–18) are evaluated at the fluid bulk temperature, and the exponent n has the following values:

$$n = 0.4 \quad \text{for heating} \tag{10–19}$$

$$n = 0.3 \quad \text{for cooling} \tag{10–20}$$

Heat transfer by *natural convection* is another mechanism that comes into play. Convective effects, previously described as *forced convection,* are due to the bulk motion of the fluid. The bulk motion is caused by external forces, such as that provided by pumps, fans, compressors, etc., and is essentially independent of "thermal" effects. *Free convection* is another effect that occasionally develops and was briefly discussed in the previous section. This effect is almost always attributed to buoyant forces that are due to density differences within a system. It is treated analytically as another external force term in the momentum equation. The momentum (velocity) and energy (temperature) effects are therefore *interdependent*; consequently, both describing equations must be solved simultaneously. This treatment is beyond the scope of this text but is available in the literature.[5]

If the equipment can be considered a stationary vertical cylinder, the heat transfer coefficient may be approximated by

$$h = 0.29\left(\frac{T_w - T_a}{L}\right)^{0.25} \tag{10–21}$$

where h = natural convection film coefficient, Btu/(h · ft^2 ·°F)
T_w = temperature at wall, °F
T_a = ambient air temperature, °F
L = cylinder length, ft

For stationary horizontal cylinders

$$h = 0.29\left(\frac{T_w - T_a}{D}\right)^{0.25} \tag{10–22}$$

where D = cylinder diameter, ft

After a period of service, thin films of foreign materials such as dirt, scale, or products of corrosion build up on the tube wall surfaces. As shown in Eqs. (10–23) and (10–24), these films R_{fi} and R_{fo} (fouling factor resistance) introduce added resistances to heat flow, and reduce the overall heat transfer coefficient:

$$R_t = R_i + R_{fi} + R_w + R_{fo} + R_o \tag{10–23}$$

For this condition, one obtains

$$\frac{1}{U_o A_o} = \frac{1}{h_i A_i} + \frac{1}{h_{d,i} A_i} + \frac{x}{kA_{\text{lm}}} + \frac{1}{h_{d,o} A_o} + \frac{1}{h_o A_o} \tag{10–24}$$

where $h_{d,i}$ = inside fouling factor, Btu/(h · ft² ·°F)
$h_{d,o}$ = outside fouling factor, Btu/(h · ft² ·°F)

Typical values of fouling factors are provided in the literature.[4–6]

10–4 Radiation

In the heat transfer mechanisms of conduction and convection, the movement of heat energy takes place through a material medium, usually a solid in the case of conduction and a fluid in the case of convection. A transfer medium is not required for the third mechanism, where the energy is carried by electromagnetic radiation. If a piece of steel plate is heated in a furnace until it is glowing red and then placed several inches away from a cold piece of steel plate, the temperature of the cold steel will rise, even if the process takes place in an evacuated container.

Radiation becomes important as a heat transfer mechanism only when the temperature of the source is very high. The driving force for conduction and convection is the temperature difference between the source and the receptor; the actual temperatures have only minor influence since it is the difference in temperatures that count. For these two mechanisms, it rarely matters whether the temperatures are 100°F and 50°F or 500°F and 450°F. Radiation, on the other hand, is strongly influenced by the temperature level; as the temperature increases, the effectiveness of radiation as a heat transfer mechanism increases rapidly. It follows that, at very low temperatures, conduction and convection are the major contributors to the total heat transfer; however, at very high temperatures, radiation is the controlling factor. At temperatures in between, the fraction contributed by radiation depends on such factors as the convection film coefficient and the nature of the radiating surface.[4] To cite two extreme examples, the temperature at which radiation accounts for roughly half of the total heat transmission for large pipes losing heat by natural convection is around room temperature; for fine wires, this temperature is above a red heat.

A perfect or ideal radiator, referred to as a *blackbody,* emits energy at a rate proportional to the fourth power of the absolute temperature of the body. When two bodies exchange heat by radiation, the *net* heat exchange is then proportional to the difference between each T^4 level. This is shown in Eq. (10–25), which is based on the Stefan–Boltzmann law of thermal radiation:

$$q = \sigma FA(T_h^4 - T_c^4) \tag{10–25}$$

where $\sigma = 0.1714 \times 10^{-8}$ Btu/(h · ft² ·°R⁴) (Stefan–Boltzmann constant)
A = area of either surface (chosen arbitrarily) ft²
F = view factor (dimensionless)
T_h = absolute temperature of hotter body, °R
T_c = absolute temperature of colder body, °R

The view factor (F) depends on the geometry of the system, i.e., the surface geometrics of the two bodies plus the spatial relationship between them, and on the surface chosen for A. Values of F are available in the literature for many geometries.[4–6] It is important to note that Eq. (10–25) applies only to blackbodies and is valid only for thermal radiation.

Other types of surfaces besides blackbodies are less capable of radiating energy, although the T^4 law is generally obeyed. The ratio of energy radiating from one of these *gray* (nonblack) bodies to that radiating from the blackbody under the same conditions is defined as the *emissivity* (ε). For *gray* bodies, Eq. (10–25) becomes

$$q = \sigma F F_\varepsilon A(T_h^4 - T_c^4) \tag{10–26}$$

where F_ε = emissivity function dependent on the emissivity of each body and the geometry of the system (dimensionless).

Relationships for F_ε or various geometries are available in the literature.[4–6] In many applications, the system may be satisfactorily approximated by two long concentric cylinders where the energy exchange is between the outside surface of the inner cylinder (A_h) and the inside surface of the outside cylinder (A_c). An example would be heat radiation from the outside surface of a hot pipe to the surrounding air. In this case, the view factor (F_ε) (based on either area) is 1.0 and

$$F_\varepsilon = \frac{1}{\dfrac{1}{\varepsilon_h} + \dfrac{A_h}{A_c}\left(\dfrac{1}{\varepsilon_c} - 1\right)} \tag{10–27}$$

For radiation to the surrounding air, A_h/A_c approaches zero and Eq. (10–27) reduces to

$$F_\varepsilon = \varepsilon_h \tag{10–28}$$

Substitution for F_ε in Eq. (10–26) leads to

$$q = \sigma F \varepsilon_h A(T_h^4 - T_c^4) \tag{10–29}$$

In the development of convection heat transfer in the previous section, the film heat transfer coefficient (h_c) was defined by

$$q_c = h_c A(T_h - T_c) \tag{10–30}$$

where T_h and T_c are the temperatures of the two bodies exchanging heat by convection. In the case of the pipe surface convecting energy to the surrounding air, T_h would be the outside pipe wall temperature and T_c, the average (bulk) temperature for the air. Since radiation heat transfer is often accompanied by convection and the total heat transfer is the sum of both contributions, it is worthwhile to place both processes on a common basis by defining a radiation heat transfer coefficient (h_r) employing the equation

$$q_r = h_r A(T_h - T_c) \tag{10–31}$$

when q_r is the heat transfer by radiation,[4] Btu/h. The total heat transfer is then the sum of the convection and radiation contributions:

$$q = (h_c + h_r)A(T_h - T_c) \tag{10–32}$$

For the example of the hot pipe radiating and convecting to the surrounding air, h_r can be evaluated by solving Eqs. (10–29) and (10–31) simultaneously to obtain

$$h_r = \sigma\varepsilon_h \frac{(T_h^4 - T_c^4)}{T_h - T_c} \tag{10–33}$$

Note that T_h and T_c are *absolute* temperatures. In most instances, the convection heat transfer coefficient is not strongly dependent on temperature; it should be apparent from Eq. (10–33) that this is not the case for the radiation heat transfer coefficient. However, the radiation coefficient is relatively small in comparison to the convection coefficient at the gas temperatures normally employed in many chemical engineering applications.

10–5 Condensation, Boiling, Refrigeration, and Cryogenics

Phase change processes involve changes (sometimes significantly) in density, viscosity, heat capacity, and thermal conductivity of the fluid in question. The heat transfer process and the applicable heat transfer coefficients for boiling and condensation are generally more involved and complicated than those for a single-phase process. It is therefore not surprising that most real-world applications involving boiling and condensation require the use of empirical correlations; information on many of these are available in the literature.[4–6]

Condensation and Boiling

The transfer of heat which accompanies a change of phase is often characterized by high rates. Heat fluxes as high as 50 million Btu/h · ft^2 have been obtained in some boiling systems. This method of transferring heat has become important in rocket technology and nuclear reactor design, where large quantities of heat are usually produced in small or confined spaces. Although condensation rates have not reached a similar magnitude, heat transfer coefficients for condensation as high as 20,000 Btu/(h · ft^2 ·°F) have been reported in the literature.[6]

Phase changes of substances can occur only if heat transfer is involved in the process. The phase change processes of interest that arise include

1. Boiling (or evaporation)
2. Condensation
3. Melting (or thawing)
4. Freezing (or fusion)
5. Sublimation

The corresponding heat-of-transformation phenomena arising during these processes are (see also Chap. 8)

1. Enthalpy of vaporization (or condensation)
2. Enthalpy of fusion (or melting)
3. Enthalpy of sublimation

These common phase change operations are listed in Table 10–3. Other phase changes are possible. During any phase change, it is usual (but not necessary) to have heat transfer without an accompanying change in temperature.

Table 10–3

Phase Change Operations

Process	Description
Boiling	Change from liquid to vapor
Condensation	Change from vapor to liquid
Solidification	Change from liquid to solid
Melting	Change from solid to liquid
Sublimation	Change from solid to vapor

The applications of phase change involving heat transfer are numerous and include utility units where water is boiled, evaporators in refrigeration systems where a refrigerant may be either vaporized or boiled, or both, and condensers that are used to cool vapors to liquids. For example, in a thermal power cycle, pressurized liquid is converted to vapor in a boiler. After expansion in a turbine, the vapor is restored to its liquid state in a condenser. It is then pumped to the boiler to repeat the cycle. Evaporators, in which the boiling process occurs, and condensers are also essential components in vapor-compression refrigeration cycles. Thus, the practicing chemical engineer needs to be familiar with phase change processes.

Applications involving the solidification or melting of materials are also important but not as numerous. Typical examples include the production of ice, freezing of foods, freeze-drying processes, and solidification and melting of metals. The freezing of food and other biological matter usually involves the removal of energy in the form of both sensible heat (enthalpy) and the latent heat of freezing. A large part of biological matter is liquid water. When meat at room temperature is frozen, it is typically placed in a freezer at –30°C, which is considerably lower than the freezing point. The sensible heat to cool any liquids from the initial temperature to the freezing point is first removed, followed by the removal of latent heat λ to accomplish the actual freezing. Once frozen, the substance is often cooled further by removing some sensible heat of the solid substance.

Refrigeration and Cryogenics

In addition to applications for domestic purposes (when a small, portable refrigerator is required), refrigeration and cryogenic units have been used for the storage of materials such as antibiotics, other medical supplies, and specialty foods. Cooling capacities much larger than this are needed in air-conditioning equipment. Some of these units, both small and large, are especially useful in chemical engineering applications that require the accurate control of temperature. Most temperature-controlled enclosures are provided with a unit that can maintain an area or a space below ambient temperature (or at precisely ambient temperature) as required. The use of such devices led to the recognition that cooling units would be well suited to the refrigeration of electronic components and to applications in the field of instrumentation. Such applications usually require small compact refrigerators, with a relatively low cooling capability, where economy of operation is often unimportant.

One of the main financial considerations when dealing with refrigeration and cryogenics is the cost of building and powering the unit. This is a costly element in the process, so it is important to efficiently transfer heat in the refrigeration and cryogenic processes. Since the cost of equipment can be expensive, various factors should be considered when choosing the applicable equipment.

Cryogenics plays a major role in the chemical processing industry. Its importance lies in such processes as the recovery of valuable feedstocks from natural gas streams, upgrading the heat content of fuel gas, purifying many process and waste streams, and producing ethylene, as well as other chemical processes. Cryogenic air separation provides gases (e.g., nitrogen, oxygen, argon) used in the manufacture of metals such as steel, chemical processing and manufacturing industries, electronic industries, enhanced oil recovery, and partial oxidation and coal gasification processes.

Cryogenic liquids have their own applications. Liquid nitrogen is commonly used to freeze food, and liquid cryogenic cooling techniques are used to reclaim rubber tires and scrap metal from old cars. Cryogenic freezing and storage is essential in the preservation of biological materials that include blood, bone marrow, skin, tumor cells, tissue cultures, and animal semen. Magnetic resonance imaging (MRI) also employs cryogenics to cool the highly conductive magnets that are used for these types of noninvasive body diagnostics.[7]

Refrigeration and cryogenics have generated considerable interest among workers in the fields of chemical engineering and science. All refrigeration processes involve work, specifically, the extraction of heat from a body of low temperature and the rejection of this heat to a body willing to accept it. *Refrigeration* generally refers to operations in the temperature range of 120 to 273 K, while cryogenics usually deals with temperatures below 120 K, where gases, including methane, oxygen, argon, nitrogen, hydrogen, and helium can be liquefied. Describing equations and design details are available in the literature.[5]

10–6 Heat Exchangers

Heat exchangers are defined as equipment that affect the transfer of thermal energy in the form of heat from one fluid to another. The simplest exchangers involve the direct mixing of hot and cold fluids. In most industrial exchangers, the fluids are *separated by a wall*. The latter type, referred to by some as a *recuperator*, can range from a simple plane wall between two flowing fluids to more complex configurations involving multiple passes, fins, and/or baffles. Conductive and convective heat transfer principles (see earlier sections) are required to describe and design these units; radiation effects are generally neglected. Kern[4] and Theodore[5] provide extensive predictive and design calculations, most of which are based on Eq. (10–12), the *heat transfer equation*. The presentation in this section focuses on the description of the various heat exchanger equipment (and their classification), not on the log-mean temperature difference driving force ΔT_{lm}, the overall heat transfer coefficient U, and other parameters discussed in Sec. 10–3. However, the design and predictive equation is the *heat transfer equation* provided in Eq. (10–12).

Heat exchangers used in the chemical, petrochemical, petroleum, paper, and power industries encompass a wide variety of designs that are available from many manufacturers. Equipment design practice first requires the selection of safe operable equipment. The selection-design process must also seek a cost-effective balance between initial (capital) installation costs, operating costs, and maintenance costs, topics discussed later in the book in Parts III and IV.

There is a near-infinite variety of heat exchange equipment. These can vary from a simple electric heater in the home to a giant boiler in a utility power plant. A limited number of heat transfer devices likely to be encountered by the practicing chemical engineer have been selected for description in this section. Most of the units transfer heat from one fluid to another fluid, with the heat passing through a solid interface such as a tube wall that separates the two (or more) fluids. The size, shape, and material employed that separates the two fluids is, of course, important. Another potential problem or concern in a heat exchanger is the method of confining one or both of the two fluids involved in the heat transfer process.

Double-Pipe Heat Exchangers

Of the various types of heat exchangers that are employed in industry, perhaps the two most common are the *double-pipe* and the *shell-and-tube models*. Although shell-and-tube heat exchangers (see next subsection) generally provide greater surface area for heat transfer with a more compact design, greater ease of cleaning, and less possibility of leakage, the double-pipe heat exchanger still finds use in practice. It is this latter unit that is discussed in this subsection.

The double-pipe unit consists of two concentric pipes. Each of the two fluids—hot and cold—flow either through the inside of the inner pipe or through the annulus formed between the outside of the inner pipe and the insidc of the outer pipe. Generally, it is more economical (from heat efficiency and design perspectives) for the hot fluid to flow-through the inner pipe and the cold fluid through the annulus, thereby reducing heat losses to the surroundings. To ensure sufficient contacting time, pipes longer than approximately 20 ft are extended by connecting them to *return bends*. The length of pipe is generally kept to a maximum of 20 ft because the weight of the piping may cause the pipes to sag. Sagging may allow the inner pipe to touch the outer pipe, distorting the annulus flow region (as well as the velocity profile and distribution) and thus disturbing proper operation. When two pipes are connected in a U configuration that entails a return bend, the bend is referred to as a "hairpin." In some instances, several hairpins may be connected in series.

Double-pipe heat exchangers have been used in the chemical process for approximately 90 years. The first patent on this unit appeared in 1923. The unique (at that time) design provided the fluid to be heated or cooled and to flow longitudinally and transversely around a tube containing the cooling or heating liquid. The original design has not changed significantly since that time.

Additional information and calculational details are provided by Kern[4] and Theodore.[5]

Shell-and-Tube Heat Exchangers

Shell-and-tube (also referred to as "tube-and-bundle") heat exchangers provide a large heat transfer area economically and practically. The tubes are placed in a bundle, and the ends of the tubes are mounted in tube sheets. The tube bundle is enclosed in a cylindrical shell, through which the second fluid flows. Most shell-and-tube exchangers used in practice are of welded construction. The shells are built from a piece of pipe with flanged ends and necessary branch connections. The shells are made of seamless pipe up to 24 inches in diameter; they consist of bent and welded steel plates of ≥24 inches height. Channel sections are usually of built-up construction, with welding-neck forged-steel flanges, rolled-steel barrels, and welded-in pass partitions. Shell covers are either welded directly to the shell, or are built-up constructions of flanged and dished heads and welding-neck forged-steel flanges. The tube sheets are usually nonferrous castings in which the holes for inserting the tubes have been drilled and reamed before assembly. Baffles are usually employed to control the flow of the fluid outside the tubes and provide turbulence.

There are vast industrial uses of shell-and-tube heat exchangers. These units are used to heat or cool fluids in the chemical process industries through either a single-exchanger or two-phase heat exchanger. In single-phase exchangers, both the tube side and shell-side fluids remain in the same phase with which they enter. In two-phase exchangers (examples include condensers and boilers, the shell-side fluid is usually condensed to a liquid or heated to a gas, while the tube-side fluid remains in the same phase.

Generally, shell-and-tube exchangers are employed when double-pipe exchangers do not provide sufficient area for heat transfer. Shell-and-tube exchangers usually require fewer construction materials and are consequently more economical when compared to double-pipe and/or multiple-double pipe heat exchangers in parallel position. Additional information and calculational details are available in Kern[4] and Theodore.[5]

Heat Exchangers with Fins and Extended Surfaces

One method of increasing the heat transfer rate is to increase the surface area of the heat exchanger. This can be accomplished by mounting metal *fins* on a tube in such a way as to ensure good metallic contact between the base of the fin and the wall of the tube. With this contact, the temperature throughout the fins will approximate that of the temperature of the heating (or cooling) medium. Also, the high thermal conductivity of most metals used in practice reduces the resistance to heat transfer by conduction in the fins. Consequently, the surface will be increased without additional tubes.[8]

Additional metal is often supplied to the outside of ordinary heat transfer components such as pipes, tubes, and walls. These extended surfaces increase the surface available for heat flow and result in an increase of the total transmission of heat. Examples of fin usage include automobile radiators, air conditioning, cooling of electronic components, and heat exchangers used in the chemical process industries. They are primarily employed for heat transfer to gases where film coefficients are very low.

Extended surfaces, or fins, are classified into longitudinal fins, transverse fins, and spine fins. *Longitudinal fins* (also termed *straight fins*) are attached continuously along the length of the surface. They are employed in cases involving gases or viscous liquids. As one might suppose, their primary application is with double-pipe heat exchangers. *Transverse* or *circumferential fins* are positioned approximately perpendicular to the pipe or tube axis and are generally used in the cooling of gases. These fins are applied mainly with shell-and-tube exchangers. Transverse fins may be continuous or discontinuous (segmented). *Annular fins* are examples of continuous transverse fins. *Spine* or *peg fins* employ cones or cylinders, which extend from the heat transfer surface, and are used for either longitudinal flow or cross flow. As noted above, these fins are constructed of highly conductive materials.

Other Heat Exchangers

The following heat exchangers are briefly addressed below: evaporators, waste-heat boilers, condensers, and quenchers. Cooling towers are not discussed; however, descriptive material is available in the literature.[9]

Evaporators

The vaporization of a liquid for the purpose of concentrating a solution is a common mass transfer–heat transfer operation in the chemical process industry. The simplest device is an open pan or kettle that receives heat from a coil or jacket or by direct firing underneath the pan. In practice, the traditional unit is the horizontal-tube evaporator, in which a liquid (to be concentrated) in the shell side of a closed, vertical cylindrical vessel is evaporated by passing steam or another hot gas through a bundle of horizontal tubes contained in the *lower* part of the vessel. The liquid level in the evaporator is usually less than half the height of the vessel, thus permitting the disengagement of entrained liquid from the overhead vapor.[10]

Waste-Heat Boilers

Energy has become too valuable to discard. As a result, waste-heat and/or heat recovery boilers are now common in many process plants. As the chemical processing industries are becoming more competitive, few companies can afford to waste or dump thermal energy. This increased awareness of economic considerations has made waste-heat boilers one of the more important heat exchanger units of the boiler industry. The term *waste-heat boiler* includes units in which steam is generated primarily from the sensible heat of an available hot-flue or hot-gas stream rather than by solely firing fuel. The main purpose of a boiler is to convert a liquid (usually water) into a vapor. In most industrial boilers, the energy required to vaporize the liquid is provided by the direct firing of a fuel in the combustion chamber. The energy is transferred from the burning fuel in the combustion chamber by convection and radiation to the metal wall separating the liquid from the combustion chamber. Conduction then takes place through the metal wall and conduction-convection into the body of the vaporizing liquid. In a waste-heat boiler, no combustion occurs in the boiler itself; the energy for vaporizing the liquids is provided by the sensible heat of hot gases, which are usually product (flue) gases generated by a combustion process occurring either upstream or elsewhere in the system. The waste-heat boilers found at many facilities utilize the flue gases for this purpose. Additional details are available in the literature.[5,11,12]

Condensers[13,14]

Condensation can be accomplished by increasing the pressure, decreasing the temperature (removing heat), or both. In practice, condensers generally operate through extraction of heat. Condensers differ in the means of removing heat and the type of device used. The two different means of condensation are *direct contact* (or contact), where the cooling medium are intimately mixed and combined with vapors and condensate; and *indirect contact* (or surface), where the cooling medium and vapors or condensate are separated by a surface area of some type. (The reader is referred to an earlier section for additional details.) Contact condensers are simpler, are less expensive to install, and require less ancillary equipment and maintenance. The condensate or coolant from a contact condenser has a volume flow 10 to 20 times that of a surface condenser. This condensate usually cannot be reused and may pose a waste disposal problem. Surface condensers form the bulk of the condensers used in industry. Some of the applicable types of surface condensers similar to those described previously are shell-and-tube, double-pipe, spiral-plate, flat-plate, air-cooled, and various extended-surface tubular units.

The choice of a coolant, for this as well as other heat exchange equipment, will usually depend on the particular plant and the efficiency required of the unit. The most common coolant is usually cooling tower or river water. It has been shown[15] that the vapor outlet temperature is critical to the operating efficiency of a condenser. In some instances, use of chilled brine or a boiling refrigerant can achieve a collection efficiency that will be sufficient without additional devices. It is not unusual to specify multi-stage condensation in practice, where condensers, usually two, are connected in series and use cooling mediums with successively lower temperatures. For example, a condenser using cooling-tower water can be used prior to a unit using chilled water or brine, thereby achieving higher thermal efficiency while minimalizing the use of chilled water.

Quenchers[15]

Hot gases must often be cooled before being discharged to the atmosphere or entering another device(s) which normally is not designed for very high temperature operation (>500°F). These gases are usually cooled either by recovering the energy in a (waste-heat) boiler, as discussed in a previous subsection, or by quenching. Both methods may be used in tandem. For example, a waste-heat boiler can reduce an exit gas temperature down to about ~500°F; a water quench can then be used to further reduce the gas temperature to ~200°F, as well as saturate the gas with water. This secondary cooling and saturation can later eliminate the problem of water evaporation and can also alleviate other potential problems.[11]

Although quenching and the use of a waste-heat boiler are the most commonly used methods for gas-cooling applications, several other techniques are used. All methods may be divided into two categories: direct-contact cooling and indirect-contact cooling. *Direct-contact cooling* methods include (1) dilution with ambient air, (2) quenching with water, and (3) contact with high heat capacity solids. Among the *indirect-contact* methods are (1) natural convection and radiation from ductwork, (2) forced-draft heat exchangers, and (3) the aforementioned waste-heat boiler(s).

With the *dilution* method, the hot gaseous effluent is cooled by adding sufficient ambient air that results in a mixture of gases at the desired temperature. The *water quench* method uses the heat of vaporization of water to cool the gases. When water is sprayed into the hot gases under conditions conducive to evaporation, the energy contained in the gases evaporates the water, and this results in a cooling of the gases. The hot exhaust gases may also be quenched using submerged exhaust quenching. This is another technique employed in some applications. In the *solids contact* method, the hot gases are cooled by giving up heat to a bed of ceramic elements. The bed, in turn, is cooled by incoming air to be used elsewhere in the process. As discussed earlier, *natural convection and radiation* occur whenever there is a temperature difference between the gases inside a duct and the atmosphere surrounding it. Cooling hot gases by this method requires the provision of enough heat transfer area to obtain the desired amount of cooling. In *forced-drop heat exchangers,* the hot gases are cooled by forcing cooling fluid past the barrier separating the fluid from the hot gases.

10–7 The Heat Exchanger Dilemma[5,10]

One area where meaningful energy conservation measures can be realized is in the design and specification of process (operating) conditions for heat exchangers. This can be best accomplished by including second-law principles in the analysis. The quantity of heat recovered in an exchanger is not alone in influencing size and cost. As the log-mean temperature difference (LMTD) driving force in the exchanger approaches zero, the "quality" of heat recovered increases.

Most heat exchangers are designed with the requirement or specification that the temperature difference between hot and cold fluids be positive at all times and be at least 20°F. This temperature difference or driving force is referred to by some as the *approach temperature*. However, the corresponding entropy change (see also Chap. 8) is also related to the driving force, with large temperature difference driving forces resulting in large irreversibilites and the associated large entropy changes.

The heat exchanger designer must decide whether to employ (1) a large LMTD driving force, which will result in both a more compact (smaller-area) design [see Eq. (10–12)] and a large entropy increase that is accompanied by the loss of quality energy, or (2) a small LMTD driving force, which will result in both a larger heat exchanger and a smaller entropy change and greater recovery of quality energy. This topic receives extensive treatment by Theodore et al.[5,10]

References

1. L. Farag and J. Reynolds, *Heat Transfer*, A Theodore Tutorial, Theodore Tutorials, East Williston, N.Y., 1996 (originally published by USEPA/APTI, Research Triangle Park, N.C.).
2. J. B. Fourier, *Théorie Analytique de la Chaleur,* Gauthier-Villars, Paris, 1822; German translation by Weinstein, Springer, Berlin, 1884; *Ann. Chim. Et Phys.,* **37**(2), 291, 1828; *Pogg. Ann.*, **13**, 327, 1828.
3. H. Carslaw and J. Jaeger, *Conduction of Heat in Solids,* 2nd ed., Oxford University Press, London, 1959.
4. D. Kern, *Process Heat Transfer,* McGraw-Hill, N.Y., 1950.
5. L. Theodore, *Heat Transfer Applications for the Practicing Engineer*, Wiley, Hoboken, N.J., 2011.
6. C. Bennett and J. Meyers, *Momentum, Heat and Mass Transfer,* McGraw-Hill, N.Y., 1962.
7. R. Kirk and D. Othmer, *Encyclopedia of Chemical Technology,* 4th ed., Vol. 7, *Cryogenics*, Wiley, N.Y., 2001.
8. W. Badger and J. Banchero, *Introduction to Chemical Engineering,* McGraw-Hill, N.Y., 1955.
9. L. Theodore and F. Ricci, *Mass Transfer Operations for the Practicing Engineer,* Wiley, Hoboken, N.J., 2010.

10. L. Theodore, F. Ricci, and T. VanVliet, *Thermodynamics for the Practicing Engineer,* Wiley, Hoboken, N.J., 2009.
11. J. Santoleri, J. Reynolds, and L. Theodore, *Introduction to Hazardous Waste Incineration*, 2nd ed., Wiley, N.Y., 2000.
12. V. Ganapathy, "Size or Check Waste Heat Boilers Quickly," *Hydrocarb. Process.,* 169–170, N.Y., Sept. 1984.
13. W. Connery, "*Condensers,*" in L. Theodore and A. J. Buonicore, eds., *Air Pollution Control Equipment,* Prentice-Hall, Upper Saddle River, N.J., 1982, Chap. 6.
14. L. Theodore, *Air Pollution Control Equipment Calculations,* Wiley, Hoboken, N.J., 2008.
15. W. Montaz, T. Trappi, and J. Seuvert, "Sizing up RTO and RCO Heat Transfer Media," *Pollution Eng.,* 34–38, Dec. 1997.
16. L. Theodore and J. Reynolds, *Hazardous Waste Incineration Manual*, Theodore Tutorials, East Williston, N.Y., originally published by USEPA/APTI, Research Triangle Park, N.C., 1996.

11 Mass Transfer

Chapter Outline

11-1 Introduction

There are various mass transfer operations. The three most commonly encountered in chemical enginnering practice are distillation, absorption, and adsorption. As such, they receive the bulk of the treatment in this chapter. Other operations reviewed include liquid-liquid and liquid-solid extraction, humidification, and drying, and to a lesser degree, crystallization. Crystallization is briefly addressed in the last paragraph of this section.

The topic of stagewise versus continuous operation needs to be addressed before leaving this introduction section. Stagewise operation is considered first. If two insoluble phases are allowed to come into contact so that the various diffusing components of the mixture distribute themselves between the phases, and if the phases are then mechanically separated, the entire operation constitutes one stage. Thus, a *stage* is the unit in which interphase contact occurs and where the phases are separated; and, a single-stage process is naturally one where this operation is conducted once. If a series of stages are arranged so that the phases come into contact and separate once in each succeeding stage, the entire multistage assemblage is called a *cascade* and the phases may move through the cascade in parallel, countercurrent, or cross-flow mode.[1,2]

In order to establish a standard for the measurement of performance, the *ideal*, *theoretical*, or *equilibrium* stage is defined as one where the effluent phases are in equilibrium, so that (any) prolonged contact will result in no additional change of composition in either phase. Thus, at equilibrium, no further net change in composition of the phases is possible for a given set of operating conditions. (In actual equipment in the chemical process industries, it is seldom practical to allow sufficient time, even with thorough mixing, to attain equilibrium.) Therefore, an *actual* stage does not accomplish as large a change in composition as does an equilibrium stage. For this reason, the *fractional stage efficiency* is defined as the ratio of a composition change in an actual stage to that in an equilibrium stage. Stage efficiencies for equipment in the chemical process industry range between a few percent to efficiencies approaching 100 percent. The approach to equilibrium realized in any stage is then defined as the aforementioned fractional stage efficiency.[1,2]

In continuous-contact operation, the phases flow-through the equipment in continuous intimate contact throughout the unit, without repeated physical separation and contact. The nature of the method requires the operation to be either semibatch or steady state, and the resulting change in compositions may be equivalent to that given by a fraction of an ideal stage or by more than one stage. Thus, equilibrium between two phases at any given position in the equipment is rarely completely established.

The essential difference between stagewise and continuous-contact operation may be summarized as follows. In *stagewise operation*, the flow of matter between the phases—that is, the driving force for the mass transfer—is allowed to reduce the concentration difference. If the phases are allowed to remain in contact for long periods, equilibrium can be established after which no further transfer occurs. The rate of transfer and the time (of contact) then determine the stage efficiency realized in any particular application. In *continuous-contact operation*, the departure from equilibrium is deliberately maintained and the transfer between the phases may continue without interruption. Therefore, economics plays a significant role in determining the most suitable method.[1,2]

The remainder of this chapter addresses the following topics: equilibrium and rate principles, absorption, adsorption, distillation, liquid-liquid extraction and leaching, and humidification and drying. There are, of course, other mass transfer operations. One such operation is crystallization, which is discussed in the next paragraph.

In the crystallization process, a crystal usually separates out as a substance of definite composition from a solution of varying composition. Any impurities in the liquid (often referred to as the "mother liquor") are carried in the crystalline product only to the extent that they adhere to the surface or are *occluded* (retained) within the crystals that may have grown together during or after the crystallization operation. The separation of a solid from a solution onto a crystal occurs only if there is a state of imbalance involving a concentration difference driving force. An example is a decrease in chemical potential (or concentration) existing between the bulk of the liquid solution and the crystal interface. This effectively means that the solution must be *supersaturated*. In practice, the four most frequently encountered crystallization processes are (1) cooling, (2) evaporation, (3) cooling and evaporation (also termed *adiabatic evaporation*), and (4) salting out. Process 1 is most commonly employed, provided the solubility of the component being crystallized decreases with decreasing temperature. Theodore and Ricci provide additional details with several illustrative examples.[1]

Other and novel separation processes (not reviewed in this chapter) include:

1. Freeze crystallization
2. Ion exchange
3. Liquid ion exchange
4. Resin adsorption
5. Evaporation foam fractionation
6. Dissociation extraction
7. Electrophoresis
8. Vibrating screens

Details on these operations are also included in the literature.[1] The bulk of the real-world industrial mass transfer operations reside in the category of immiscible phases, and it is primarily these mass transfer processes that are addressed in this chapter.

NOTE The bulk of the material presented in this chapter is drawn from Theodore and Barden.[3]

11-2 Equilibrium and Rate Principles

Equilibrium and rate principles play important roles in designing and predicting the performance of many mass transfer processes. In fact, several mass transfer calculations are based primarily on the application of either equilibrium or rate principles. Distillation is a prime example; virtually every calculation is based on vapor-liquid equilibrium data, e.g., acetone-water, and the rate of transfer of each component into the other.

The term *phase* for a pure substance indicates a state of matter that is solid, liquid, or gas. For mixtures, however, a more stringent connotation must be used, since a totally liquid or solid system may contain more than one phase (e.g., a mixture of oil and water). A phase is characterized by uniformity or homogeneity; the same composition and properties must exist throughout the phase region. At most temperatures and pressures, a pure substance normally exists as a single phase. At certain temperatures and pressures, two or perhaps even three phases can coexist in equilibrium.

Chemical engineering calculations rarely involve single (pure) components. Phase equilibria for multicomponent systems are considerably more complex, mainly because of the addition of composition variables. For example, in a ternary (or three-component) system, the mole fractions of two of the components are pertinent variables along with temperature and pressure. In a single-component system, dynamic equilibrium between two phases is achieved when the rate of molecular transfer from one phase to the second one equals that in the opposite direction. In multicomponent systems, the equilibrium requirement is more stringent since the rate of transfer of each component must be the same in both directions.

The most important equilibrium phase relationship is that between liquid and vapor. Raoult's and Henry's laws were discussed in Chap. 8. They theoretically describe liquid-vapor behavior, and under certain conditions, are applicable in practice. Raoult's law is sometimes useful for mixtures of components of similar structure. It states that the partial pressure of any component in the vapor is equal to the product of the vapor pressure of the pure component and the mole fraction of that component in the liquid. Unfortunately, relatively few mixtures follow Raoult's law. Henry's law is a more empirical relationship that is used for representing data on many systems. Here

$$p_i = H_i x_i \tag{11–1}$$

where H_i is Henry's law constant for component i (in units of pressure). If the gas behaves ideally, Eq. (11–1) may be written as

$$y_i = m_i x_i \tag{11–2}$$

where m_i is a constant (dimensionless). This is a more convenient form of Henry's law to work with. The law may be assumed to apply for most dilute solutions. This law finds widespread use in absorber calculations since the concentration of the solute in some process gas streams is often dilute. This greatly simplifies the study and design of absorbers. One should note, however, that Henry's law constant is a strong function of temperature.

The rate transfer process can be described by the product of three terms: (1) the area available for transfer, (2) the driving force for transfer, and (3) the (reciprocal of the) resistance to the transfer process. In effect, the rate process in equation form is

$$\text{Rate} = \frac{(\text{area})(\text{driving force})}{(\text{resistance})} \tag{11–3}$$

For mass transfer (MT) applications, Eq. (11–3) becomes

$$(\text{Rate of MT}) = \frac{(\text{area available for MT})(\text{driving force for MT})}{(\text{resistance to MT})} \tag{11–4}$$

The *diffusivity*, or *diffusion coefficient* D_{AB} of constituent A in solution B, which is a measure of its diffusive mobility, is defined as the ratio of its flux J_A to its concentration gradient and is given by

$$J_A = -D_{AB}\frac{\partial C_A}{\partial z} \tag{11–5}$$

This is Fick's first law[4] written for the z direction. The concentration gradient term represents the variation of the concentration C_A in the z direction. For certain simplified cases of molecular diffusion in gases, equations can be derived to determine the rate at which mass N_A is being transferred:

$$N_A = \left[\frac{(D_{G_{AB}})(P_t)}{(R)(T)(z)(\bar{p}_{B,M})}\right](p_{A1} - p_{A2}) \tag{11–6}$$

where $\bar{p}_{B,M} = \dfrac{p_{B2} - p_{B1}}{\ln(p_{B2}/p_{B1})}$ = log mean partial pressure difference driving force of component B

$p_{B1} = p - p_{A1}$ = partial pressure of component B at the liquid-vapor interface

$p_{B2} = p - p_{A2}$ = partial pressure of component B at distance z from the interface

p_{A1} = partial pressure of component A at the interface

p_{A2} = partial pressure of component A at distance z from the interface for state 2

R = the ideal gas constant

The bracketed term is an expression for the *mass transfer coefficient* corresponding to the steady-state situation of one component diffusing through a nondiffusing second component. In principle, then, it is not necessary to calculate any other mass transfer coefficients for laminar flow since molecular diffusion prevails, and exact equations are available. However, in general, obtaining such analytically derived expressions is difficult; in most cases encountered, it is impossible since turbulent mass transfer, which becomes quite complex, usually prevails.

Many approaches to the turbulent (convective) mass transfer problem exist. *Film theory* (as applied by Whitman[5]) postulates the existence of an imaginary stagnant film next to the interface whose resistance to mass transfer is equal to the total mass transfer resistance of the system. The difficulty with this theory is in the calculation of the effective film thickness. *Surface renewal theory* assumes that a clump of fluid far from the interface (1) moves to the interface without transferring mass; (2) sits there, stagnant, transferring mass en route; and (3) instantly mixes with the bulk fluid. This theory is somewhat more satisfactory than film theory. *Boundary-layer theory* rests on the solution of a set of simplified differential equations[3] which are approximations to a more nearly correct set of differential equations. *Empirical approaches*, which are merely data correlations, serve for specific cases but are of limited value on extrapolation. Further details regarding any of these approaches are available in the literature[2]. Primary emphasis in this section will be on the more design-applicable approaches.

In the mass transfer operation of gas absorption, two insoluble phases are brought into contact in order to permit the transfer of solute from one phase to the other. (For example, ammonia can be absorbed from an air-ammonia mixture into a water stream with little to no air dissolving in the water.) Concern, then, is with the simultaneous application of the diffusion mechanism for each phase in the combined system. It has already been shown that the rate of diffusion within each phase is dependent on the concentration gradient existing within it. At the same time, the concentration gradients of the two-phase system are indicative of the departure from equilibrium which exists between the phases. This departure from equilibrium provides the driving force for diffusion. In view of Whitman's two-film theory, where it is assumed that at the gas-liquid interface the principal diffusion resistances occurs in a thin film of gas and a thin film of liquid, the rates of diffusion in these two films will define the mass transfer operation.

The diffusion coefficient D in the Fick law equation is inversely proportional to the concentration of the inert material c_B in the liquid film through which the material must diffuse. Replacing D with (k/c_B) yields

$$N_A = \left(\frac{k}{zc_{B,m}}\right)(c_{A,I} - c_{A,L}) \tag{11–7}$$

where $c_{B,m}$ is the log-mean concentration difference of the inert material across the film, k is a proportionality constant, and N is the amount of material transferred per unit area per unit time, or moles per area $\cdot$ time. The practical application of Eq. (11–7) is based on the assumption that z, the film thickness, is a constant, or rather that it represents an effective average value throughout the length of the contact path. Also, $c_{B,m}$ is considered as constant since many mass transfer processes involve fairly dilute mixtures and solutions. Eq. (11–7) reduces to

$$N_A = k_L(c_{A,I} - c_{A,L}) \tag{11–8}$$

or

$$N_A = k_G(p_{A,G} - p_{A,I}) \tag{11–9}$$

where k_L k_G = individual liquid and gas mass transfer coefficients, respectively
$c_{A,I}$ = interfacial (surface) concentration of component A
$c_{A,L}$ = bulk liquid concentration of component A
$p_{A,I}$ = interfacial partial pressure of component A
$p_{A,L}$ = bulk liquid partial pressure of component A

Equation (11–8) expresses the transfer of N molecules of solute (A) through the liquid film under a concentration driving force; and Eq. (11–9), the transfer of the same number of molecules of solute through the gas film under a partial pressure driving force.

It is not possible to measure the partial pressure and concentration at the interface ($p_{A,I}$ or $y_{A,I}$ and $c_{A,I}$ or $x_{A,I}$, respectively). Therefore, it is common to employ *overall mass transfer coefficients* based on the overall driving force between $p_{A,G}$ (or $y_{A,G}$) and $c_{A,L}$ (or $x_{A,L}$). The overall coefficients may be defined on the basis of the gas film K_G or the liquid film K_L by the equations

$$N_A = K_G\left(p_{A,G} - p_A^*\right) = k_y'\left(y_{A,G} - y_A^*\right) \tag{11–10}$$

or

$$N_A = K_L\left(c_A^* - c_{A,L}\right) = k_x'\left(x_A^* - x_{A,L}\right) \tag{11–11}$$

where the starred terms represent equilibrium values.

11–3 Absorption

The removal of one or more selected components from a gas mixture by absorption is one of the most important operations in the field of mass transfer. The term *absorption* conventionally refers to the intimate contact of a mixture of gases with a liquid so that part of at least one of the gas constituents will dissolve in the liquid. The contact usually takes place in some type of packed column, although plate and spray towers are also used. The equilibrium of interest in gas absorption operations is that between a relatively nonvolatile absorbing liquid (solvent) and a soluble gas (solute). The solute is ordinarily removed from a relatively large amount of a carrier gas which does not dissolve in the absorbing liquid. Temperature, pressure, and solute concentration in one phase are independently variable. The equilibrium relationship of importance is a plot (or data) of x, the mole fraction of solute in the liquid, against y^*, the mole fraction in the vapor in equilibrium with x. For cases that follow Henry's law, Henry's law constant m can be defined by the equation

$$y^* = mx \tag{11–12}$$

The usual operating data to be determined or estimated for isothermal systems are the liquid rate(s) and the terminal concentrations or mole fractions. An *operating line*, which describes operating conditions in the column, is obtained by a mass balance around the column (as shown in Fig. 11–1). An overall balance is given by

$$\text{(total moles in)} = \text{(total moles out)}$$

$$G_{m1} + L_{m2} = G_{m2} + L_{m1} \tag{11–13}$$

For component A, the mass (or mole) balance becomes

$$(G_{m1})(y_{A1}) + (L_{m2})(x_{A2}) = (G_{m2})(y_{A2}) + (L_{m1})(x_{A1}) \tag{11–14}$$

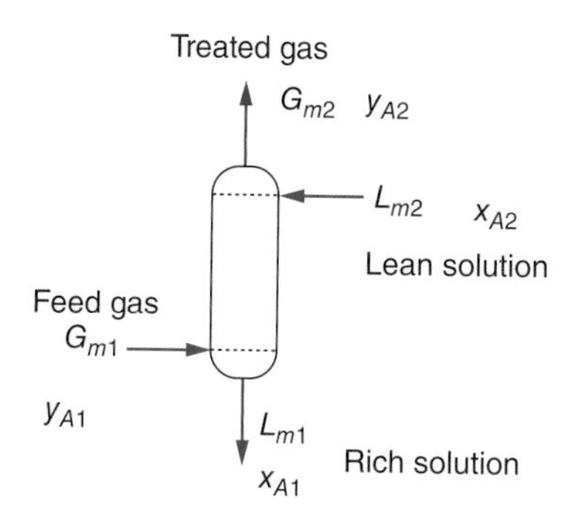

Figure 11–1

Absorption system.

Assuming $G_{m1} = G_{m2}$ and $L_{m1} = L_{m2}$ (reasonable for many applications where solute concentrations are usually extremely small), one obtains

$$(G_m)(y_{A1}) + (L_m)(x_{A2}) = (G_m)(y_{A2}) + (L_m)(x_{A1}) \tag{11–15}$$

Rearranging Eq. (11–15) gives

$$\frac{L_m}{G_m} = \frac{(y_{A1} - y_{A2})}{(x_{A1} - x_{A2})} \tag{11–16}$$

This is the equation of a straight line known as the *operating line*. On x,y coordinates it has a slope of L_m/G_m and passes through the points (x_{A1}, y_{A1}) and (x_{A2}, y_{A2}) as indicated in Fig. 11–2.

In the design of most absorption columns, the quantity of gas to be treated G_m, the terminal concentrations y_{A1} and y_{A2}, and the composition of the entering liquid x_{A2} are ordinarily fixed by process requirements. However, the quantity of liquid solvent to be used is subject to some choice. This can be resolved by setting a minimum liquid-gas ratio. With reference to Fig. 11–2, the operating line must pass through point A and must terminate at the ordinate y_{A1}. If such a quantity of liquid is used to give operating line AB, the exiting liquid will have the composition x_{A1}. If less liquid is used, the exit liquid composition will clearly be greater, as at point C, but the driving force for mass transfer is less since the displacement of operating from the equilibrium has been reduced, i.e., the absorption is more difficult. The duration of contact between gas and liquid must then be greater, and the absorber must be correspondingly taller.

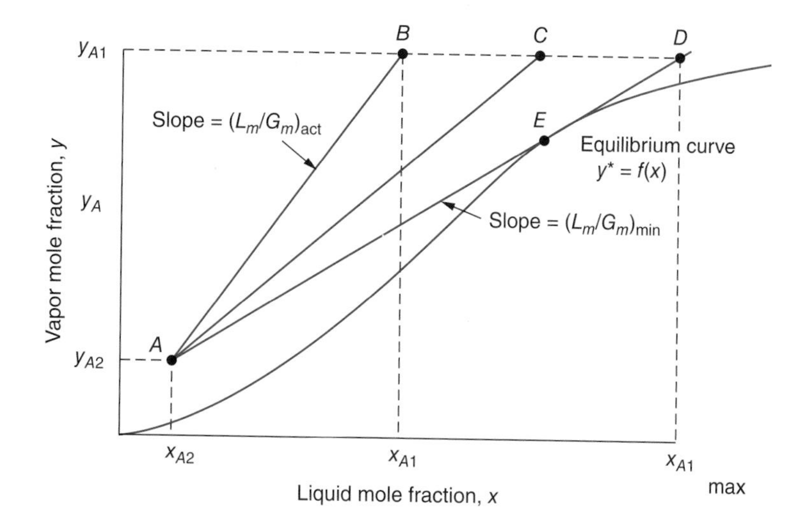

Figure 11–2

Operating and equilibrium lines.

The minimum liquid which may be used corresponds to the operating line *AD*, which has the slope of a line touching the equilibrium curve and is tangent to the curve at *E*. At point *E*, the concentration difference driving force is zero, the required contact time for the concentration change desired is infinite, and an infinitely tall column results. This then represents the minimum liquid to gas ratio. The importance of the minimum liquid to gas ratio lies in the fact that column operation is frequently specified as some factor of the minimum liquid to gas ratio. For example, a typical situation frequently encountered is that the actual operating line, $(L_m/G_m)_{act}$, is 1.5 times the minimum $(L_m/G_m)_{min}$.

Note that when the solute of a gas stream is fairly concentrated (usually >1 mol% solute), mole ratios should be used instead of mole fractions since a significant amount of solute may be absorbed which would change the gas and liquid flow rates and lead to a *curved* operating line. The mole concentration ratios for a concentrated solute would be

$$Y = \frac{\text{mole fraction of solute in gas}}{1 - \text{mole fraction of solute in gas}} \tag{11–17}$$

$$X = \frac{\text{mole fraction of solute in liquid}}{1 - \text{mole fraction of solute in liquid}} \tag{11–18}$$

The resulting mole ratios (X, Y) are treated in the same way as mole fractions.

In a countercurrent spray column, the gas enters at a point near the tower bottom and passes through the liquid which is introduced at the top of the tower by means of single or multiple nozzles. To provide a large liquid surface available for contact, relatively fine droplets are required. Hence, high-pressure drops across the spray nozzles are usually employed. There is also the danger of liquid entrainment at all except very low gas velocities. In addition to countercurrent spray columns, concurrent spray columns are used on rare occasions. In these columns, the gases enter and pass down with the liquid and are withdrawn from a side connection near the tower bottom. In general, the performance of these units is somewhat poor because the droplets tend to coalesce after they have fallen a short distance, and the interfacial area is thereby significantly reduced. Although there is considerable turbulence in the gas phase, there is little circulation of the liquid within the drops, and the resistance of the liquid phase tends to be high. Spray columns are therefore useful only where the main resistance to mass transfer lies within the gas phase (very soluble gases).

Packing Height

The height of a packed column is calculated by determining the required number of theoretical (gas transfer) separation units and multiplying this number by the height of a separation unit. In continuous-contact countercurrent operations, the theoretical separation unit is called the *transfer unit*, and the packing height producing one transfer unit is referred to as the *height of a transfer unit*. The transfer unit is essentially a measure of the degree of the separation (transfer of the solute out of the gas stream and into the liquid stream) and the average driving force producing this transfer. In actual gas absorption design practice, the number of overall gas transfer units N_{OG} can be estimated; one can also show that if Henry's law applies, the number of transfer units is given by Colburn's equation

$$N_{OG} = \frac{\ln[(y_1 - mx_2)/(y_2 - mx_2)][\{1-(1/A)\}+(1/A)]}{1-(1/A)} \tag{11–19}$$

where

$$A = \frac{L_m}{mG_m} \tag{11–20}$$

The term A is defined as the *absorption factor* and is related to the slope of the equilibrium curve. In cases where the gas is highly soluble in the liquid and/or reacts with the liquid, Theodore has shown that, $N_{OG} = \ln(y_1/y_2)$.[6]

The height of a transfer unit (H_{OG}) is usually determined experimentally for the system under consideration. Information on many different systems using various types of packing has been compiled by the manufacturers of gas absorption equipment and should be consulted prior to design. The data may be in the form of graphs depicting, for the specific system and packing, the H_{OG} versus the gas rate [in lb/(h · ft²)] with the liquid rate [in lb/(h · ft²)] as a parameter.

The packing height (Z) is then simply the product of the H_{OG} and the N_{OG}. Although there are many different approaches to determining the column height, the $H_{OG} - N_{OG}$ approach is the most commonly used at present for approximate calculations with the H_{OG} usually obtained from the manufacturer. Generalized correlations are also available for computing the height of an overall gas transfer unit. These use experimentally derived factors based on the type of packing and the gas and liquid flow rates.

A more detailed evaluation of the above design factors for the column height may be found in the literature.[1,2] Regardless of the method employed, a safety factor (typically 1.2) is frequently used as a multiplication factor for the calculated column packed height.

Tower Diameter

Regarding tower diameter, consider a packed column operating at a given liquid rate in which the gas rate is gradually increased. After a certain point, the gas rate is so high that the drag on the liquid is sufficient to keep the liquid from flowing freely down the column. Liquid begins to accumulate and tends to block the entire cross-section available for flow. This, of course, both increases the pressure drop and prevents the packing from mixing the gas and liquid effectively, and ultimately some liquid is even carried back up the column. This undesirable condition, known as *flooding*, occurs fairly abruptly, and the superficial gas velocity at which it occurs is called the *flooding velocity*. The calculation of column diameter is often based on flooding considerations; the usual operating range is 60 to 70% of the flooding rate.

The most commonly used correlation for pressure drop and diameter is the U.S. Stoneware's generalized correlation (see Fig. 11–3, where F is the packing factor and Ψ is the liquid-to-water density ratio). The procedure for determining the tower diameter is as follows:

1. Calculate the abscissa, $(L/G)(\rho/\rho_L)^{0.5}$; L,G = mass divided by time · area
2. Proceed to the flood line and read the ordinate.
3. Solve the ordinate equation for G at flooding.
4. Calculate the tower cross-sectional area S for the fraction f of flooding velocity chosen for operation using the equation:

$$S = m/fG = (\text{total lb/s})/(\text{lb/s} \cdot \text{ft}^2) = \text{ft}^2 \tag{11–21}$$

5. Then determine the tower diameter by:

$$D = 1.13S^{0.5} = \text{ft} \tag{11–22}$$

Liquid Distributors and Mist Eliminators

Liquid distribution plays an important role in the efficient operation of a packed column. From a process perspective, packing efficiency can be reduced by poor liquid distribution across the top of the column's upper surface. Poor distribution reduces the effect of wetted packing area and promotes liquid *channeling*. The final selection of the mechanism of

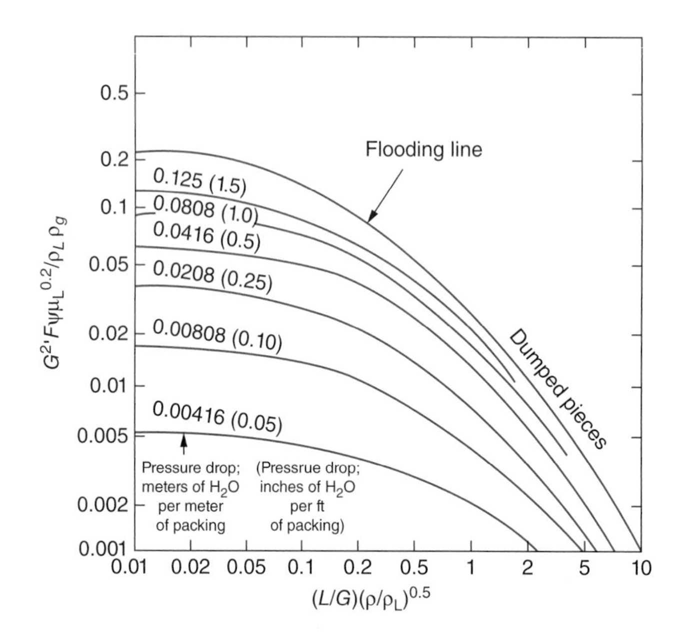

Figure 11-3

Generalized flooding and pressure drop correlation.

distributing the liquid across the packing depends on the size of the column, type of packing, tendency of packing to divert liquid to the column walls, and materials of construction for distribution.

Mist eliminators also play an important role in absorbers. They are used to remove liquid droplets entrained in the gas stream; the ease of separation depends on the size of the droplets. Droplets formed from liquids are usually large—up to hundreds of micrometers in diameter. Droplets formed during condensation or chemical reaction may be less than 1 μm in size. Entrainment removal (mist separation) is possible by a number of methods, including the following:[1]

1. Knitted wire or plastic mesh
2. Swirl vanes or zigzag vanes
3. Cyclones
4. Gravity settling chambers
5. Units in which the gas is forced to make 180° turn
6. Additional packing above the packed bed[2]

One of the simplest and most efficient means of mist separation is to use a porous blanket of knitted wire or plastic mesh. For most processes, pressure drops across these mist eliminators range from 0.1 to 1.0 in of water, depending on liquid loadings and the size of the eliminators. Efficiency of separation is generally high—usually at least 90%.

In many real-world applications, the need for detailed engineering calculations is rarely warranted or justified. This is particularly true for absorbers. For example, chemical engineers in industry are often confronted with the need for quickly either (1) providing a design of a packed tower absorber or (2) predicting the performance of a packed tower absorber. Certain assumptions can be made on the calculation procedures developed earlier. For this situation, the procedure reduces to relatively simple calculations that can be generated with little or no effort.

Details are also available to qualitatively determine how one can size (diameter, height) a packed tower to achieve a given degree of separation without any information on the physical and chemical properties of a gas to be processed.[6] To calculate the height,

Table 11–1

Packing Height

Packing diameter, in	Plastic packing H_{OG}, ft	Ceramic packing H_{OG}, ft
1.0	1.0	2.0
1.5	1.25	2.5
2.0	1.5	3.0
3.0	2.25	4.5
3.5	2.75	5.5

one needs both the aforementioned height of a gas unit H_{OG} and the number of gas transfer units N_{OG}. Since equilibrium data are not available, one can assume that m (slope of equilibrium curve) approaches zero. This is not an unreasonable assumption for most solvents that preferentially absorb (or react with) the solute. For this condition[3]

$$N_{OG} = \ln \frac{y_1}{y_2} \tag{11–23}$$

where y_1 and y_2 represent inlet and outlet concentrations, respectively. Since it is reasonable to assume the scrubbing medium to be water or solvent that effectively has the physical and chemical properties of water, H_{OG} can be assigned values usually encountered for water systems (listed in Table 11–1).

For ceramic and plastic packing, both the liquid and gas flow rates typically range from 500 to 1000 and from 1500 to 2000 lb/(h · ft^2) of cross-sectional area, respectively. For difficult to absorb gases, the gas flow rate is usually lower and the liquid flow rate higher. Superficial gas velocities (velocity of the gas if the column is empty) range from 3 to 6 ft/s.

The height is calculated from:

$$Z = (H_{OG})(N_{OG})(\text{SF}) \tag{11–24}$$

where SF is a safety factor whose value can range from 1.25 to 1.5. Pressure drops can vary from 0.15 to 0.40 in H_2O per foot of packing. Also note that packing size increases with increasing tower diameter.

In gas absorption operations the choice of a particular solvent is also important. Frequently, water is used as it is inexpensive and plentiful. However, the following properties must also be considered:

1. Gas solubility
2. Volatility
3. Corrosiveness
4. Cost
5. Viscosity
6. Chemical stability
7. Toxicity
8. Low freezing point

Plate columns are essentially vertical cylinders in which the liquid and gas are contacted in stepwise fashion (stage operation) on plates or trays.[1,2] The liquid enters at the top and flows downward via gravity. On the way down, it flows across each plate and through a downspout to the plate below. The gas passes upward through openings in the plate, e.g., a sieve tray[1,2] and then bubbles through the liquid to form a froth, disengages from the froth, and passes on to the next plate above. The overall effect is the multiple countercurrent contact between gas and liquid. Each plate of the column constitutes a *stage* since the fluids on the plate are brought into intimate contact, interphase diffusion occurs, and the fluids are separated. The number of theoretical plates (or stages) is dependent on the difficulty of the separation to be carried out and is determined primarily from material balances and equilibrium considerations.

The most important design considerations for plate columns include calculation of the column diameter, type and number of trays to be used (usually either bubble-cap or sieve trays), actual plate layout and physical design, and interplate spacing (which, in turn, determines the column height). To consider each of these factors to any great extent is beyond the scope of this section, but is treated to some extent in Sec. 11–5. Additional details are available in any standard chemical engineering unit operations or mass transfer text.[1,2]

The column diameter and consequently its cross-sectional area must be sufficiently large to accommodate the gas and liquid at velocities which will not cause flooding or excessive entrainment. The superficial gas velocity for a given type of flooding is expressed as

$$V_f = C_f \left(\frac{\rho_L - \rho_G}{\rho_G} \right)^{0.5} \tag{11–25}$$

where V_f = gas velocity through net column cross-sectional area for gas flow at flooding, $\text{ft}^3/\text{s} \cdot \text{ft}^2$
ρ_L = liquid density, lb/ft^3
ρ_G = gas density, lb/ft^3
C_f = empirical coefficient that depends on the type of plate and operation conditions

The net cross section is the difference between the column cross section and the area occupied by the downcomers. In actual design, 80 to 85% of V_f is often used for non-foaming liquids and 75% or less for foaming liquids.

The column height is determined from the product of the number of actual plates (theoretical plates divided by the overall plate efficiency) and the plate spacing chosen. The theoretical plate (or stage) is the aforementioned theoretical unit of separation in plate column calculations; it was previously defined as a plate in which two dissimilar phases are brought into intimate contact with each other and then are mechanically separated. The number of theoretical plates N may be calculated from the Kremser–Souders–Brown (KSB) equation:

$$N = \frac{\log[\{(y_{N+1} - mx_0)/(y_1 - mx_0)\}\left\{1 - \left(\frac{1}{A}\right)\right\} + \{1/A\}]}{\log A} \tag{11–26}$$

Here mx_2 is the gas composition in equilibrium with the entering liquid (m is Henry's law constant, i.e., slope of equilibrium curve). If the entering liquid contains no solute gas, then $x_0 = 0$. The solute concentrations in the gas stream y_1 and y_2 represent inlet and outlet gaseous mole fractions, respectively, and L and V (contained in the A term) are the total molar rates of liquid and gas flow per unit time per column cross-sectional area, respectively. An equivalent form of the KSB equation is

$$\frac{y_1 - y_2}{y_1 - mx_2} = \frac{A^{N+1} - A}{A^{N+1} - 1} \tag{11–27}$$

The general procedure to follow in sizing a plate tower is given below:

1. Calculate the number of theoretical stages, using Eq. (11–27).
2. Estimate the efficiency of separation E. This may be determined at the local (across–plate), plate (between–plate), or overall (across–column) level. The overall efficiency E_0 is generally employed to calculate the actual number of plates required for a given separation.
3. Calculate the actual number of plates $N_{act} = N/E$.
4. Obtain or select the height between plates h. This usually ranges from 12 to 36 in. Most towers use a 24-in plate spacing.
5. The tower height Z is then

$$Z = N_{act}\, h \tag{11–28}$$

6. The diameter may be calculated directly from the Eq. (11–28).
7. The plate or overall pressure drop is difficult to quantify; it usually ranges from 2 to 6 in H_2O per plate for most columns with the lower and upper values applying to small and large diameters, respectively.

The assumption of ideal solution and isothermal operation is certainly valid for many absorption operations. One generally also assumes no condensation, mixing (heat) effects, no chemical reaction, etc. Since the solute concentration is usually dilute, the liquid and gas rates essentially are constant. Thus, the absorption factor A is also a constant.

The two key factors to consider in multicomponent calculations is to assume (1) there are no interaction effects between the various solutes, and (2) absorption of each component occurs as if the other solutes are not present (i.e., treat each component separately). The KSB equation is again employed – only this time for each solute present in the gas mixture.[7]

A simplified equation[7] is available for calculating the total outlet concentration or loading for a given unit when more than two components are absorbed:

$$\sum_{1}^{n} y_{i,1} = \sum_{1}^{n} (1 - A_i)(y_{i,\text{in}})(1 - A_i^{N+1}) \tag{11–29}$$

where i = solute in question
N = number of theoretical plates
$y_{i,\text{in}}$ = inlet mole fraction of solute i

This method, assuming ideal conditions, can also be solved graphically. It can be set up to solve for either the outlet solute concentration or the required liquid flow to achieve a given separation in a particular application. Finally, this simplified technique can also be used to design or predict performance of packed towers:[7]

$$Z = (H_{OG})_A (N_{OG})_A = (H_{OG})_B (N_{OG})_B = \ldots \tag{11–30}$$

where $(H_{OG})_A = (H_{OG})_B = \ldots (H_{OG})_i$ = height of a gas transfer unit, ft

Stripping

Quite often, an absorption column is followed by a liquid absorption process in which the gas solute is removed from the absorbing medium by contact with an insoluble gas. This operation is called *stripping* and is utilized to regenerate the solute-*rich* solvent so that it can be recycled back to the absorption unit. The rich solution enters the stripping unit, and the volatile solute is stripped from solution by either reducing the pressure, increasing the temperature, using a stripping gas to remove the vapor solute dissolved in the solvent, or any combination of these process options. While the concept of stripping is opposite to that of absorption, it is treated in the same manner.[8]

The height of a stripping tray column can be calculated in a much the same way as outlined earlier. The describing equation is

$$N = \frac{\log\{(x_2 - y_1/m)/(x_1 - y_1)\}\{1-(1/R)+1/R\}}{R} \tag{11–31}$$

where the stripping factor R replaces the absorption factor A. The stripping factor $R = 1/A = mG_m/L_m$. The height H is then calculated from the equation

$$H = \frac{Nh}{E_o} \tag{11–32}$$

where h = intertray spacing (typically 24 in)
E_o = overall fractional trade efficiency (typically 0.4 to 0.7).

Summary of Key Equations[1,2]

The key equations for absorption and stripping calculations for tower height, including a summary of earlier material, are presented below:
For packed-tower absorption:

$$N_{OG} = \frac{\log[\{(y_1 - mx_2)/(y_2 - mx_2)\}\{1-(1/A)\}+\{1/A\}]}{1-(1/A)} \tag{11–33}$$

For stripping in packed towers:

$$N_{OG} = \frac{\log[\{(x_2 - y_1/m)/(x_1 - y_1/m)\}\{1-A\}+A]}{1-A} \tag{11–34}$$

where the subscripts 1 and 2 denote bottom and top conditions, respectively. In addition, $A = L_m/mG_m$ and $S = 1.0/A$.

For plate tower absorption:

$$N = \frac{\log[\{(y_{N+1} - mx_0)/(y_1 - mx_0)\}\{1-(1/A)\}+\{1/A\}]}{\log A} \tag{11–35}$$

NOTE The term *ln*, rather than *log*, may also be employed in both the numerator and denominator.

If A approaches unity, Eq. (11–35) becomes

$$N = \frac{y_{N+1} - y_1}{y_1 - mx_0} \tag{11–36}$$

or

$$\frac{y_{N+1} - y_1}{y_{N+1} - mx_0} = \frac{N}{N+1} \tag{11–37}$$

Note that the subscripts 1 and N denote the top and bottom of the column, respectively. For stripping in plate towers, one obtains

$$N = \frac{\log[\{(x_0 - y_{N+1}/m)/(x_N - y_{N+1}/m)\}\{1-(1/S)\} + S]}{\log S} \tag{11–38}$$

or

$$\frac{x_0 - x_N}{x_0 - (y_{N+1}/m)} = \frac{S^{N+1} - S}{S^{N+1} - 1} \tag{11–39}$$

If S is approximately 1.0, one may use either of the following equations:

$$N = \frac{x_0 - x_N}{x_0 - (y_{N+1}/m)} \tag{11–40}$$

or

$$\frac{x_0 - x_N}{x_0 - (y_{N+1}/m)} = \frac{N}{N+1} \tag{11–41}$$

11–4 Adsorption[3]

Adsorption is a mass transfer process in which a solute is removed from a fluid stream because it adheres to the surface of a solid. In a gas adsorption system, the gas stream is passed through a layer of solid particles referred to as the *adsorbent bed*. As the gas stream passes through the adsorbent bed, the solute absorbs or adheres to the surface of the solid adsorbent particles. Eventually, the adsorbent bed becomes filled or saturated with the solute. The adsorbent bed must then be desorbed before the adsorbent bed can be reused.

The process of adsorption is analogous to using a sponge to mop up water. Just as a sponge soaks up water, a porous solid (the adsorbent) is capable of capturing gaseous solute molecules. The gas stream carrying the solute must then diffuse into the pores of the adsorbent (internal surface) where they are adsorbed. Most gas molecules are adsorbed on the internal pore surfaces. The presentation to follow will primarily be on *gas* adsorption although much of the material can also be applied to liquid adsorption.

The relationship between the amount of substance adsorbed by the adsorbent at constant temperature and the equilibrium pressure or concentration is called the adsorbent *isotherm*. The adsorption isotherm is the most important and by far the most frequent use of the various equilibria relationships which can be employed. To represent the variation in the amount of adsorption per unit area or unit mass with pressure, Freundlich proposed the equation

$$Y = kp^{1/n} \tag{11–42}$$

where Y = weight or volume of gas (or vapor) absorbed per unit area or unit mass of adsorbent

p = equilibrium *partial* pressure

k,n = empirical constants dependent on the nature of solid and adsorbate, and on the temperature

Although the requirements of the equation are met satisfactorily at lower pressures, this equation does not have general applicability in reproducing adsorption of gases (or vapor) by solids at higher pressures. A much better equation was deduced by Langmuir from theoretical considerations. Langmuir postulated that gases (or vapors), following adsorption by a solid surface, cannot form a layer more than a single molecule in depth.

Further, he visualized the adsorption process as consisting of two opposing actions: (1) a condensation of molecules from the gas phase onto the surface and (2) an evaporation of molecules from the surface back into the body of the gas. If the amount of gas (or vapor) adsorbed per unit area or per unit mass of adsorbent is again defined as Y, the Langmuir equation takes the form

$$Y = \frac{ap}{1+bp} \tag{11–43}$$

The constants a and b are characteristic of the system under consideration and can be evaluated from experimental data. Their magnitude also depends on the temperature.

There are a host of other adsorption models. Two that are often used include the Brunauer–Emmet–Teller (BET) equation[9] and the Polyani equation.[10]

Fixed-bed adsorbers are the usual choice when adsorption is the desired method of recovery or control. Consider a binary solution containing a strongly adsorbed solute at concentration C_0 (see Fig. 11–4). The gas stream containing the solute (or adsorbent) is to be passed continuously down through a relatively deep bed of adsorbent which is initially free of adsorbate. The top layer of adsorbent, in contact with the inlet gas that is entering, at first adsorbs the adsorbate rapidly and effectively. The remaining adsorbate is removed in the lower part of the bed. At this point in time, the effluent concentration from the bottom of the bed is essentially zero. The bulk of the adsorption occurs over a relatively narrow adsorption zone [defined as the mass transfer zone (MTZ)] in which there is a rapid change in concentration. At some later time, roughly half of the bed is saturated with the adsorbate, but the effluent concentration C_2 is still essentially zero. Finally, at C_3, the lower portion of the adsorption zone has reached the bottom of the bed, and the concentration of adsorbate in the effluent has steadily increased to an appreciable value for the first time.

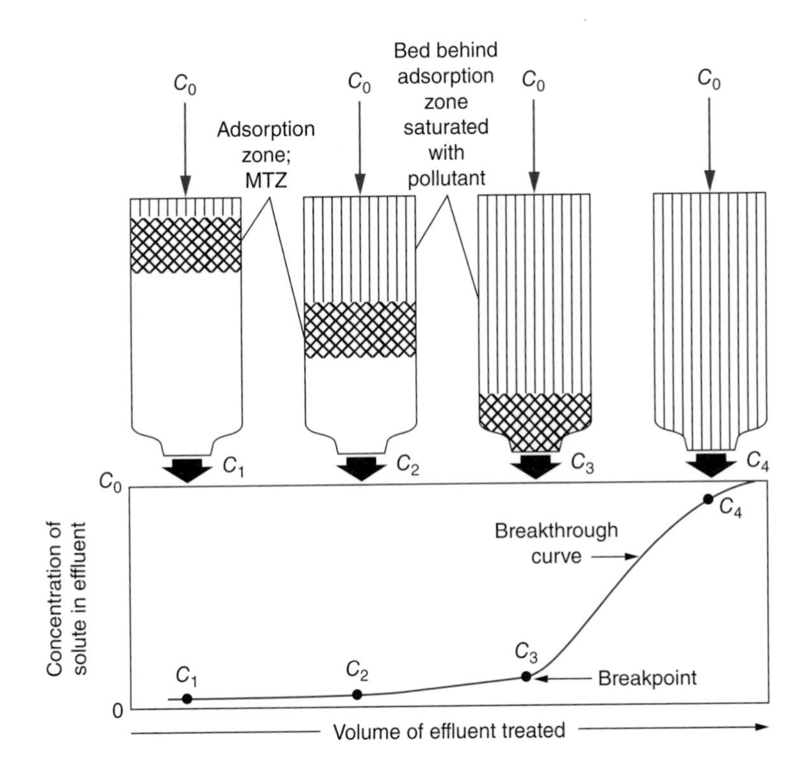

Figure 11–4
Adsorption system.

The system is said to have reached the *breakpoint*. The adsorbate concentration in the effluent gas stream now rises rapidly as the adsorption zone passes through the bottom of the bed and at C_4 has just about reached the initial value, C_0. At this point the bed is essentially fully saturated with adsorbate. The portion of the curve between C_3 and C_4 is termed the *breakthrough curve*.

The breakthrough capacity is the capacity when traces of vapor first begin to appear in the exit gas stream from the adsorption bed. It may be estimated from the following equation

$$\text{BC} = \frac{(0.5)(\text{CAP})(\text{MTZ}) + (\text{CAP})(\text{Z} - \text{MTZ})}{Z} \tag{11–44}$$

where BC = breakthrough capacity
CAP = saturation (equilibrium) capacity
MTZ = mass transfer zone height
Z = adsorption bed depth

The usual procedure in practice is to work with a term defined as the *working charge* (or *working capacity*). It provides a numerical value for the actual adsorbing capacity of the bed of height Z under operating conditions. If experimental data is available, the working charge (WC) may be estimated from

$$\text{WC} = \text{CAP}\frac{(\text{Z} - \text{MTZ})}{Z} + 0.5\frac{\text{MTZ}}{Z} - \text{HEEL} \tag{11–45}$$

where CAP is the equilibrium capacity and HEEL is the residual adsorbate present in the bed following regeneration. Since much of the data is rarely available, or simply ignored, the working charge may be taken to be some fraction f of the saturated (equilibrium) capacity of the adsorbent:

$$\text{WC} = (f)(\text{CAP}) \qquad (0 \leq f \leq 1.0) \tag{11–46}$$

The time at which the breakthrough curve appears and its shape greatly influence the method of operating a fixed bed absorber. The curves usually have an S shape, but they may be steep or relatively flat, and in some cases considerably distorted. The actual rate and mechanism of the adsorption process, the nature of the adsorption equilibrium, the fluid velocity, the solute concentration in the entering gas, and the adsorber bed length, all contribute to the shape of the curve produced for any system. The breakpoint is very sharply defined in some cases, while in others it is not. The breakpoint time generally decreases with decreases in height, increased adsorbent particle size, increased gas flow rate through the bed, and increased initial solute concentration in the entering gas stream. It is usual industrial practice to experimentally determine the breakpoint and breakthrough curve for a particular system under conditions to as close as possible to those expected in the process.

Several materials are used commercially as adsorbing agents. The most common adsorbents used industrially are activated carbon, silica gel, activated alumina (alumina oxide), and zeolites or molecular sieves. The important characteristics in determining the effectiveness of an adsorbent are its chemical structure, total surface area, pore size distribution (diameters of its pores), and particle size.

In physical adsorption, polar gas molecules prefer polar adsorbents, while nonpolar adsorbents are best for removing nonpolar gases. Since water vapor (polar) is present in most streams, polar adsorbents are not very effective in the recovery or control of nonpolar gases. Polar adsorbents quickly fill with water and have no room to adsorb the solute gases. Of the nonpolar adsorbents, activated carbon is the one primarily in use. Because of its nonpolar surface, activated carbon is used to recover organic gases such as solvents, odors, toxics, and gasoline vapors.

Adsorber systems have a number of different shapes and gas flow patterns. As described above, the most common industrial adsorption system is the fixed bed. The fixed-bed system consists of a vessel (horizontal or vertical) that contains a carbon bed. The depth of the carbon bed is usually between 0.3 and 1.2 m (1 and 4 ft) thick, depending on the solute concentration in the gas stream. The gas stream is often first passed through a filter to remove any dust that could blind the bed and decrease efficiency. The gas stream then passes *down* through the fixed bed of carbon or another adsorbent of choice. The gases are exhausted from the unit. Upward airflow-through the carbon bed is usually avoided to eliminate the risk of entraining carbon in the exhaust stream plus other factors.[10]

Fixed-bed adsorption systems use multiple beds. One or more of the beds treat the process exhaust while the others are regenerated or remain idle. *Regeneration*, a topic discussed later, is accomplished by reversing the conditions that promote adsorption. The most common way to desorb the vapors is by increasing the temperature or decreasing the pressure of the system.

Adsorption Design

A rather simplified overall design procedure for an activated carbon (AC) system adsorbing an organic that consists of two horizontal units (one on/one off) that are regenerated with steam is provided here:[11]

1. Select adsorbent type and size.
2. Select cycle time, estimate regeneration time, set adsorption time equal to regeneration time, set cycle time equal to twice the regeneration time, and generally try to minimize regeneration time.
3. Set gas throughput velocity; v is usually 80 ft/min but can increase to 100 ft/min.
4. Set the steam to solvent ratio.
5. Calculate (or obtain) WC for above.
6. Calculate the amount of solvent adsorbed (M_s) during half the cycle time (t_{ads}) using the equation

$$M_s = qC_i t_{\text{ads}} \qquad (C_i = \text{inlet solvent concentration}) \tag{11–47}$$

7. Calculate the adsorbent required M_{AC}:

$$M_{AC} = \frac{M_s}{\text{WC}} \tag{11–48}$$

8. Calculate the adsorbent volume requirement:

$$V_{AC} = \frac{M_{AC}}{\rho_B} \qquad (\rho_B = \text{carbon bulk density}) \tag{11–49}$$

9. Calculate the face area of the bed:

$$A_{AC} = \frac{q}{v} \tag{11–50}$$

10. Calculate the bed height Z.

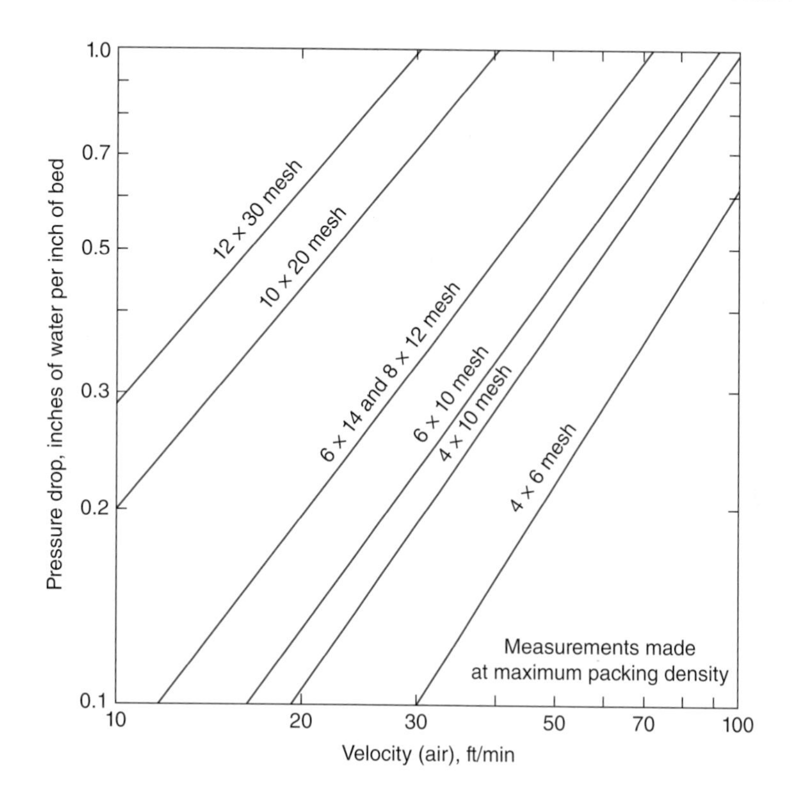

Figure 11–5 Activated-carbon pressure drop curves (USEPA chart).

11. Estimate the pressure drop from Fig. 11–5.
12. Set the L/D (length-to-diameter) ratio.
13. Calculate L and D, noting that

$$A = LD \tag{11–51}$$

Constraints: $L < 30$ ft, $D < 10$ ft; L/D of 3 to 4 is acceptable if $v < 30$ ft/min.

14. Design (structurally) to handle if filled with water.
15. Design vertically if $q < 2500$ actual cubic feet per minute (acfm).

The pressure drop (step 11) can be estimated from Fig. 11–5 or from a suitable equation. For example in lieu of any data, the pressure drop may be estimated from Eq. 11–52.

$$\Delta P = 0.37Z\left(\frac{v}{100}\right)^{1.56} \tag{11–52}$$

where Z is the bed depth in inches, v is the velocity in ft/min, and the pressure drop is in inches of water.

For multicomponent adsorption, the working charge WC may be calculated from the following equation for n number of components:[11]

$$\text{WC} = \frac{1.0}{\sum_{i=1}^{n}\left(\frac{w_i}{\text{CAP}}\right)} \tag{11–53}$$

where w_i = mass fraction of i in n components (*not* including the carrier gas)
CAP_i = equilibrium capacity of component i

For a two-component (A and B) system, Eq. (11–53) reduces to

$$WC = \frac{(CAP_A)(CAP_B)}{(w_A)(CAP_B) + (w_B)(CAP_A)} \tag{11–54}$$

Regeneration

Adsorption processes in practice employ various techniques to accomplish regeneration or desorption. The adsorption-desorption cycles are usually classified into four types, used separately or in combination:

1. *Thermal swing cycle* can be employed by either *direct transfer by* contacting the bed with a hot fluid or *indirect transfer* through a surface to reactivate the adsorbent by raising the temperature. Usually a temperature between 300°F and 600°F is reached; the bed is then flushed with a dry purge gas or is placed under reduced pressure, and then is returned to adsorption conditions. High design loadings on the absorbent can usually be obtained, but a follow-up cooling step is almost always needed.
2. *Pressure swing cycles* use either low pressure or a vacuum to desorb the bed. The cycle can be operated at nearly isothermal conditions with no heating or cooling steps. The advantages include fast cycling with a reduced adsorber size and adsorbent inventory, direct production of a high purity product, and the ability to utilize gas compression as the main source of energy.
3. *Purge gas stripping cycles* use an essentially nonadsorbate purge gas to desorb the bed by reducing the partial pressure of the adsorbed component. Such stripping becomes more efficient at higher operating temperatures and lower operating pressures. The use of a condenser purge gas has the advantages of reducing power requirements gained by using a liquid pump instead of a blower, and an effluent stream which may be condensed to separate the desorb material by simple distillation.
4. *Displacement cycles* use an adsorbable purge to displace the previously adsorbed material on the bed. The stronger the adsorption of the purge, the more completely the bed is desorbed using a reduced amount of purge, but the more difficult it becomes to remove the adsorbed purge itself from the bed.

To maintain the concentration of the discharge stream at its lowest value, the regeneration step is almost always conducted in a direction *opposite* to that for adsorption.[10] A residual concentration gradient exists in the bed after regeneration, with the minimum value located at the inlet side of the regeneration stream; in effect, the bottom of the bed contains a smaller HEEL, thus providing more efficient recovery.

Adsorbing down provides an additional advantage. Fine particles, impurities, polymeric materials, and high-molar-weight hydrocarbons will be deposited or captured at the inlet or top side of the bed. If the performance of the bed is reduced for this reason, one need to replace only a small portion of the bed (an easy skimming operation) rather than the entire bed.[10]

11–5 Distillation

Distillation is probably the most widely used separation process in the chemical and allied industries. Applications range from the rectification of alcohol to the fractionation of crude oil. The separation of liquid mixtures by distillation is based on differences in volatility between the components. The first component to vaporizing is considered to be more volatile than the other components in the system. The greater the *relative* volatilities, the easier the separation.

Batch Distillation

Batch distillation is typically chosen when it is not possible to run a continuous process because of limiting process constraints or the need to distill other process streams, or because the low frequency of distillation does not warrant a unit devoted solely to a specific product. A relatively efficient separation of two or more components can be accomplished through batch distillation employing a pot or tank. A feed is charged to a tank, the vapor generated by boiling the liquid is withdrawn from contact with the liquid and totally condenses as rapidly as it is formed. The condensed product is collected as distillate, D, with composition x_D, and the liquid remaining in the pot W has composition x_W. A material balance around a batch distillation unit yields the following:

$$F = W_{\text{final}} + D \tag{11–55}$$

$$Fx_F = x_{W,\text{final}} W_{\text{final}} + x_D D \tag{11–56}$$

The Rayleigh equation presents a convenient method for graphical analysis of a batch distillation system. It relates the composition and amount of material remaining in the pot to the initial feed charge F and composition x:

$$\ln\left(\frac{W_{\text{final}}}{F}\right) = -\int_{x_{W,\text{final}}}^{x_F} \left(\frac{dx}{x_D - x}\right) \tag{11–57}$$

When a total condenser is employed, the condensed product has the same composition as the vapor product. Thus, $y = x_D$, and Eq. (11–57) can be written as

$$W_{\text{final}} = (F)\exp\left(-\int_{x_{W,\text{final}}}^{x_F} \left(\frac{dx}{x_D - x}\right)\right) \tag{11–58}$$

Integration of the Rayleigh equation in closed analytical form previously required that relative volatility α (covered in the next section) be the assumed constant. Kunz[12] recently developed a new technique that produces a continuous analytical function for the Rayleigh equation integral regardless of whether α is constant. The newly developed equation reduces algebraically to the traditional function when α is absolutely constant. The new method can prove especially useful for any nonideal systems investigated in which α varies widely with composition. The approach requires expressing the factor $1/(\alpha - 1)$ as a function of mole fraction x. This can be fitted by a quadratic function of x in the form $1/(\alpha - 1) = ax^2 + bx + c$. With $1/(\alpha - 1)$ thus represented, the integrand of the Rayleigh equation may be analytically integrated to produce a solution in closed form. The only source of error is the goodness of fit of the above quadratic function.[12]

Flash Distillation

The separation of a volatile component from a liquid process stream can be achieved by *flash distillation*. The term *flash* is used since the more volatile component of a gas mixture rapidly vaporizes upon entering a tank or drum that is at a pressure lower and/or a temperature higher than that of the incoming feed. If the feed is considered to be cold, a pump and heater may be required to elevate the pressure and temperature, respectively, to achieve an effective flash (see Fig. 11–6). As the feed enters the tank or drum, it may impinge against the wall or an internal deflector plate, which would promote liquid-vapor separation of the feed mixture.

As a result of the flash, the vapor phase will contain most of the more volatile component. Typically, flash distillation is not an efficient means of separation. However, it can be a

Figure 11–6

Flash system.

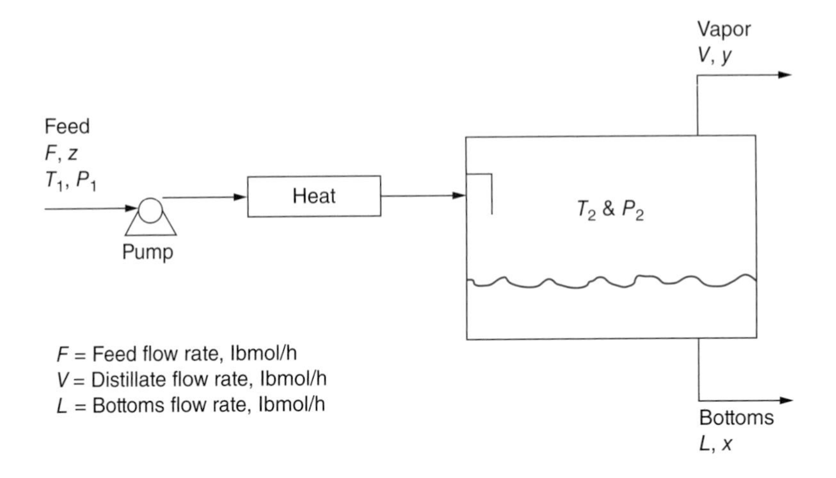

necessary and economical method of separating two or more components with discernible relative volatilities. The term *relative volatility* defines the lighter component as the one more likely to vaporize from a mixture when the temperature is raised or the pressure is lowered. For example, the relative volatility (α) of component A in mixture AB is defined as:

$$a_{AB} = \frac{K_A}{K_B} \tag{11–59}$$

where K is the phase equilibrium constant (see also Chaps. 8 and 9).

As with many process operations, an overall material balance can be written to describe the system illustrated in Fig. 11–6 as follows:

$$F = L + V \tag{11–60}$$

Based on the above, a material balance can be written from component i as follows:

$$z_i F = x_i L + y_i V \tag{11–61}$$

where z_i = feed composition of component i
x_i = liquid composition of component i
y_i = vapor composition of component i

Equation (11–61) can be rearranged in terms of the vapor composition to represent a straight line of the form $y = mx + b$ in the following manner:

$$y = -\frac{L}{V}x + \frac{F}{V}z \tag{11–62}$$

where $m = -L/V$ = slope of operating line
$b = F/V = y$ intercept
V/F = fraction of feed vaporized.

Equation (11–62) defines an operating line (see also Sec. 11–3) for this system. Since this operation is assumed to occur at equilibrium, it is defined as an *equilibrium stage*. The equation therefore relates the liquid and vapor compositions leaving this equilibrium stage. As shown in Fig. 11–7, a plot of the equilibrium data, specifically, the $y = x$

Figure 11–7

Graphical analysis of flash distillation.

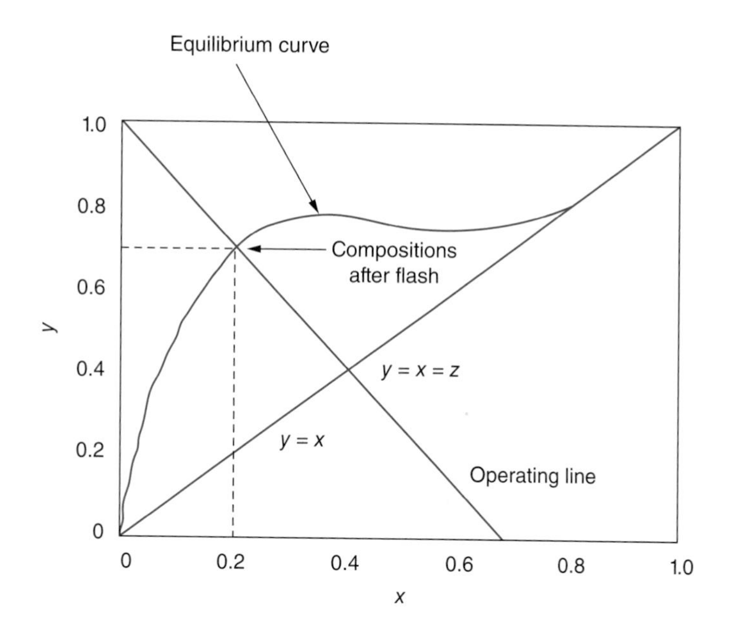

line, versus the operating line provides a procedure for calculating the unknown variables for a system. This is typically the outlet liquid and vapor compositions. The use of the $y = x$ line simplifies the graphical solution method and intersects the operating line and the equilibrium curve.

In some situations, the separation achieved in a single equilibrium stage, as is the case for simple batch distillation, is not large enough to obtain the desired distillate concentration. To improve the recovery of the desired product, a staged distillation column can be placed above a reboiler. The analysis of such a system is discussed next.

McCabe–Thiele Method[13]

In a distillation column (see Fig. 11–8), vapor flows up the column and liquid flows countercurrently down the column. The vapor and liquid are brought into contact on plates or packing as shown in Fig. 11–8. The vapor from the top of the column is sent to a condenser. Part of the condensate from the condenser is returned to the top of the column, as reflux, to descend countercurrently to the rising vapors. The remainder of the

Figure 11–8

Schematic of a multitray distillation column.

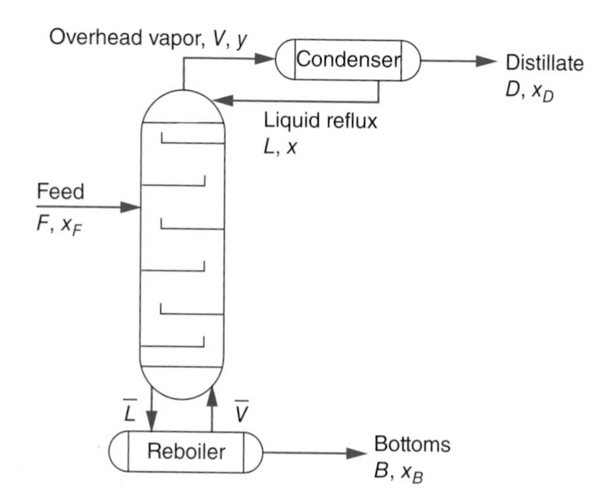

condensed liquid is drawn as product. The ratio of the amount of reflux returned to the column to the distillate product collected is known as the *reflux ratio*. As the liquid stream descends the column, it is progressively enriched with low-boiling constituents. The column internals serve as the mechanism for bringing these streams into intimate contact, so the vapor stream tends to vaporize the low-boiling constituent from the liquid and the liquid stream tends to condense the high-boiling constituent from the vapor. Thus, unlike the distillation approaches discussed above, columnwise distillation involves two sections: an enriching (top) section and a stripping (bottom) section. The *stripping* section lies below the feed, where the more volatile components are stripped from the liquid; above the feed, in the *enriching* (or *rectifying*) section, the concentration of the more volatile components is increased.

A column may consist of one or more feeds and may produce two or more product streams. The product recovered at the top of a column is referred to as the *tops*; product recovered at the bottom of the column, as the *bottoms*. Any products drawn at various stages between the top and bottom are referred to as *sidestreams*. Multiple feeds and product streams do not alter the basic operation of a column, but they do complicate the analysis of the process to some extent. If the process requirement is to strip a volatile component from a relatively nonvolatile solvent, the rectifying (bottom) section may be omitted, and the unit is referred to as a *stripping column*. Virtually pure top and bottom products can be achieved by using many stages or (occasionally) additional columns; however, this is seldom economically feasible.

The top of the column is cooler than the bottom, so that the liquid stream becomes progressively hotter as it descends and the vapor stream becomes progressively cooler as it rises. This heat transfer is accomplished by actual contact of liquid and vapor; for this purpose, effective contacting is desirable. Each plate in the column is assumed to approach equilibrium conditions. This type of plate is defined as a *theoretical plate*, i.e., is, a plate on which there is sufficient contact between the vapor and the liquid to ensure that the vapor leaving the plate has the same composition as the vapor in equilibrium with the liquid overflow from the plate. The vapor and liquid leaving the plate are related by the equilibrium curve described above. Distillation columns designed on this basis serve as a standard for comparison to actual columns. Such comparisons enable one to determine the number of actual plates that are equivalent to a theoretical plate and then to reapply this factor when designing other columns for similar service.

In some operations where the top product is required as a vapor, the liquid condensed is sufficient only to provide reflux to the column, and the condenser is referred to as a *partial condenser*. In a partial condenser, the reflux will be in equilibrium with the vapor leaving the condenser and is considered to be an equilibrium stage in the development of the operating line and for estimating the column height. When the liquid is totally condensed, the liquid returned to the column will have the same composition as the top product and will not be considered as an equilibrium stage. A *partial reboiler* is utilized to generate vapor to operate the column and to produce a liquid product if necessary. Since both liquid and vapor are in equilibrium, a partial reboiler is considered to be an equilibrium stage as well.

As discussed earlier, an operating line can be developed to describe the equilibrium relation between the liquid and vapor components. However, in staged column design, it is necessary to develop an operating line that relates the passing streams (liquid entering and vapor leaving) on each stage in the column. The following analysis will develop operating lines for the top, or enriching (rectifying) section and the bottom, or stripping section, of a column.

To perform this analysis, an overall material balance is applied for the condenser as $V = L + D$, which represents the vapor (V) leaving the top stage, the liquid reflux returning (L) to the column from the condenser (reflux), and the distillate (D) collected.

A material balance for component A is written as $Vy = Lx + Dx_D$. This can be rearranged in the form of an equation for a straight line, $y = mx + b$, as follows:

$$y = \frac{L}{V}x + \frac{D}{V}x_D \tag{11–63}$$

From the overall material balance, $D = V - L$, and Eq. (11–63) can be written as

$$y = \frac{L}{V}x + \left(1 - \frac{L}{V}\right)x_D \tag{11–64}$$

where L/V = slope. This is the internal reflux ratio (liquid reflux returned to the column or vapor from the top of the column). Also, the term $[1 - (L/V)]x_D$ represents the y intercept on Fig. 11–7. Although Eq. (11–64) has been developed around the condenser, it represents the equilibrium relationship of passing liquid-vapor streams and can be applied to the top of the column.

The corresponding operating line in the bottom or stripping section can be developed in a similar manner. The overall material balance around the reboiler is $\overline{L} = B + \overline{V}$ and the component material balance is given by $\overline{Lx} = Bx_B + \overline{Vy}$. Note that the terms with the bars over them represent the flow at the bottom of the column. Again, this material balance can be rearranged in the form of a straight line as

$$y = \frac{B}{\overline{V}}x_B + \frac{\overline{L}}{\overline{V}}x \tag{11–65}$$

Since $B = \overline{L} + \overline{V}$, however, Eq. (11–65) can be rewritten as

$$y = -\left(\frac{\overline{L}}{\overline{V}} - 1\right)x_B + \frac{\overline{L}}{\overline{V}}x \tag{11–66}$$

Equation (11–66) can then be rearranged into a form similar to that of Eq. (11–64):

$$y = \frac{\overline{L}}{\overline{V}}x - \left(\frac{\overline{L}}{\overline{V}} - 1\right)x_B \tag{11–67}$$

And, as before, the term $\overline{L}/\overline{V}$ is the slope and $[(\overline{L}/\overline{V}) - 1]x_B$ represents the y intercept. This slope can be calculated from the external reflux ratio as

$$\frac{\overline{L}}{\overline{V}} = \frac{(L/D)(z - x_B) + q(x_D - x_B)}{(L/D)(z - x_B) + q(x_D - x_B) - (x_D - z)} \tag{11–68}$$

where L/D = external reflux ratio, the ratio of liquid returned to the column to the distillate collected (not the flow rate)

q = feed quality, representing the fraction of feed remaining liquid below the feed stage

For further information on feed quality, the reader is referred to any chemical engineering text which addresses distillation.[1,2]

A procedure for designing a staged distillation column is provided here:

1. Plot the equilibrium data, the top and bottom operating lines, and the $y = x$ line on the same graph. This plot is called a *McCabe–Thiele diagram* as portrayed in Fig. 11–9.[13]

Figure 11–9

McCabe Thiele diagram.

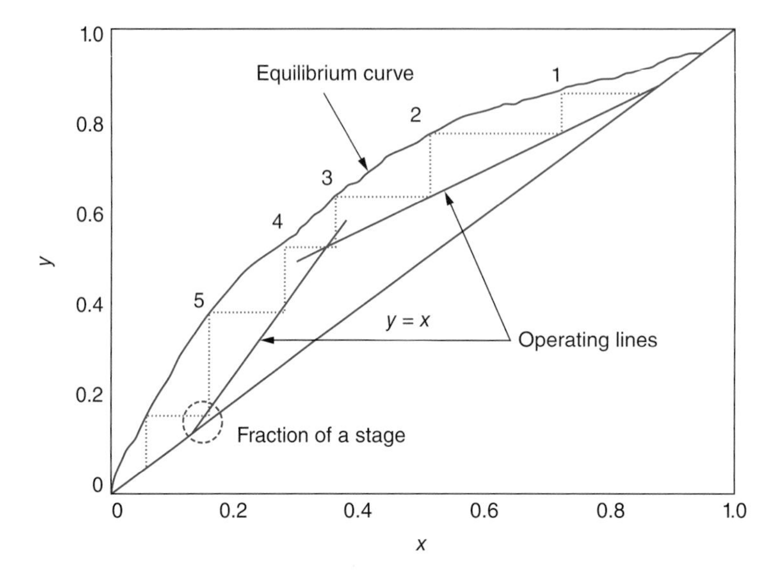

2. Step off the number of stages required beginning at the top of the column. The point at the top of the curve represents the composition of the liquid entering (reflux) and gas leaving (distillate) the top of the column. The point at the bottom of the curve represents the composition of liquid leaving (bottoms) and the gas entering (vapor from reboiler) the bottom of the column. All stages are stepped off by drawing alternate vertical and horizontal lines in a stepwise manner between the top and bottom operating lines and the equilibrium curve. The number of steps is the number of theoretical stages required. When a partial reboiler or partial condenser is employed, the number of equilibrium stages stepped off is $N + 1$, and the number of theoretical plates would be N. Thus, if both are to be included, the number of equilibrium stages stepped off would be $N + 2$.

3. To determine the actual number of plates required, divide the result in step 2 by the overall plate fractional efficiency, typically denoted by E_o. Values can range from 0.4 to 0.8. The actual number of plates can be calculated from

$$N_{act} = \frac{N}{E_o} \qquad (11\text{–}69)$$

4. The column height can be calculated by multiplying the result in step 3 by the tray spacing (9, 12, 15, 18, and 24 inches are typical tray spacings).

5. To determine the *optimum* feed plate location, draw a line from the feed composition on the $y = x$ line, through the intersection of the top and bottom operating lines, to the equilibrium curve. The step straddling the feed line is the correct feed plate location.

6. Because of vapor traffic variation throughout the column, the diameter will vary from top to bottom. Typical designs are *swaged* columns, where the diameter is larger at the bottom than at the top. A conelike vessel links the bottom of the column with the top of the column. The diameter at the top of a column can be roughly sized by assuming a superficial (column only contains vapor) vapor velocity of 5 ft/s and then using the following equation to calculate the diameter:

$$D = \left(\frac{4Q}{\pi v}\right)^{0.5} \qquad (11\text{–}70)$$

where D = diameter of the column, ft
Q = vapor volumetric flow rate, ft^3/s
V = gas velocity in column, ft/s

A somewhat similar procedure is employed to calculate the diameter at the bottom of a column. The interested reader is referred to any text on mass transfer which can provide a more detailed analysis.[1,2]

As indicated earlier, distillation can be carried out in both trayed and packed columns. The selection of which method to use depends on the specific application. Packing is frequently comparable in cost with trays and becomes more attractive where a low pressure drop and small liquid holdup are required. Packed towers should be used if the column diameter is less than 1.5 ft. Packed columns also provide continuous contact between the liquid and the gas. This intimate contact occurs on the packing, which also promotes the heat transfer required to separate the more volatile component from the process stream.

Fenske–Underwood–Gilliland Method

It is customary to design distillation columns for industry on a preliminary basis for economic estimates, for recycle calculations where distillation is only a small portion of the entire system, for calculations involving control systems, and as a first estimate for more detailed simulation calculations. Shortcut design methods for multicomponent distillation follow.

The Fenske equation calculates the minimum number of equilibrium stages required for multicomponent separation at total reflux (100% of the liquid reflux is returned to the column). On the other hand, Underwood developed an equation which estimates the maximum number of equilibrium stages required at minimum reflux, and the actual reflux ratio for a multicomponent system. But in order to estimate the number of actual stages, the Gilliland equation correlates the minimum number of stages at total reflux, the minimum reflux ratio, and the actual reflux rate for a multicomponent system. The application of all three of these equations in the design of a distillation column is referred to as the *Fenske–Underwood–Gilliland* (FUG) method. Details on each of these methods are provided below.

Fenske developed the following equation to calculate the minimum number of stages at total reflux for the separation of two key components (i.e., A and B in a mixture of A, B, and C):

$$N_{\min} = \frac{\ln\{[(\mathrm{FR}_{A,\mathrm{dist}})/(1-\mathrm{FR}_{A,\mathrm{dist}})]/[(1-\mathrm{FR}_{B,\mathrm{bot}})/(\mathrm{FR}_{B,\mathrm{bot}})]\}}{\ln(1/\alpha_{BA})} \qquad (11\text{–}71)$$

where $N_{\min}$ = minimum number of equilibrium stages
FR_i = fractional recovery in distillate (dist) or bottoms (bot) for component i
α_{BA} = relative volatility of component B versus that of component A

The following equation can be used to calculate the fractional recoveries of the nonkey components (e.g., of C) in the distillate and bottoms:

$$\mathrm{FR}_{C,\mathrm{dist}} = \frac{\alpha_{CB}^{N\min}}{[(\mathrm{FR}_{A,\mathrm{bot}})/(1-\mathrm{FR}_{A,\mathrm{bot}})] + \alpha_{CB}^{N\min}} \qquad (11\text{–}72)$$

(Theodore and Ricci provide additional details.[1])

The Underwood equation requires a trial-and-error solution and a subsequent material balance to estimate the minimum reflux ratio. First, the unknown α is determined by trial and error, such that both sides of Eq. (11–73) are equal. The unknown value of α should lie between the relative volatilities of the light and heavy key components. The key components are those whose fractional recoveries are specified. The most volatile component of the keys is the light key; the least volatile is the heavy key. All other components are referred to as *nonkey components*. If a nonkey component is lighter than the light key component, it is a light nonkey component; if it is heavier than the heavy key component, it is referred to a heavy nonkey component. The describing equation is

$$\Delta V_{\text{feed}} = F(1-q) = \sum_{i=1}^{n} \frac{\alpha_i F Z_i}{\alpha_i - \Phi} \tag{11–73}$$

where ΔV_{feed} = change in vapor flow rate at feed stage
F = feed rate
q = percentage of feed remaining liquid
Z_i = feed composition of component i
Φ = root of Underwood equation

Next, calculate the minimum vapor flow rate V_{min} from the following equation:

$$V_{\text{min}} = F(1-q) = \sum_{i=1}^{n} \frac{\alpha_i D_{xi,\text{dist}}}{\alpha_i - \Phi} \tag{11–74}$$

where $D_{xi,\text{dist}}$ = the amount of component i in the distillate
$= FZ_i(FR)_{i,\text{dist}}$

Once V_{min} is known, the minimum liquid flow rate is calculated from

$$L_{\text{min}} = V_{\text{min}} - D \tag{11–75}$$

The minimum reflux ratio is then calculated by dividing the minimum liquid flow rate by the distillate flow rate as follows:

$$R_{\text{min}} = \left(\frac{L}{D}\right)_{\text{min}} \tag{11–76}$$

Gilliland correlation used the results of the Fenske and Underwood equations to determine the actual number of equilibrium stages. The correlation has been fit to three equations:

$$\frac{N - N_{\text{min}}}{N+1} = 1.0 - 18.5715x \quad \text{for} \quad 0 \le x \le 0.01 \tag{11–77}$$

$$\frac{N - N_{\text{min}}}{N+1} = 0.545827 - 0.591422x + \frac{0.002743}{x} \quad \text{for} \quad 0.01 \le x \le 0.90 \tag{11–78}$$

$$\frac{N - N_{\text{min}}}{N+1} = 0.16595 - 0.16595x \quad \text{for} \quad 0.90 \le x \le 1.0 \tag{11–79}$$

where N = actual number of equilibrium stages

$$x = [(L/D)-(L/D)_{\text{min}} /[(L/D) + 1]$$

The values of N_{min} and $(L/D)_{min}$ were previously defined as the minimum number of equilibrium stages (Fenske equation) and minimum reflux ratio (Underwood equation).

To summarize, the FUG approach is used to

- Calculate the minimum number of equilibrium stages N_{min} from the Fenske equation.
- Calculate the minimum reflux ratio $(L/D)_{\text{min}}$ from the Underwood equation.
- Select an actual reflux ratio L/D, which usually ranges from 1 to 2.5 times the minimum ratio.
- Calculate x.
- Solve the Gilliland equation for the actual number of equilibrium stages.

Once the minimum number of equilibrium stages have been determined from the Fenske equation, a simple rule of thumb can be applied to estimate the actual number of stages as 2.5 times the minimum. Thus, this method requires only a calculation of N_{min} and is usually appropriate for preliminary design estimates.

Theodore and Ricci[1] provide details plus examples that can (1) locate the theoretical feed tray (Kirkbride equation) and (2) estimate the overall column efficiency (O'Connell correlation).

11-6 Liquid-Liquid Extraction and Leaching

Liquid-Liquid Extraction

Liquid-liquid extraction (or *liquid extraction*) is a process for separating a solute from a solution based on the concentration driving force between two immiscible (nondissolving) liquid phases. Thus, liquid extraction involves the transfer of solute from one liquid phase into a second immiscible liquid phase. The simplest example involves the transfer of one component from a binary mixture into a second immiscible liquid phase; such is the case with the extraction of an impurity from wastewater into an organic solvent. Liquid extraction is usually selected when distillation or stripping is impractical or too costly; this usually occurs when the relative volatility for two components falls between 1.0 and 1.2.

The feed to an extractor consists of a liquid component, referred to as the *feed solvent*, and the solute(s) to be removed from the stream. The extraction solvent is usually an immiscible liquid that removes the solute from the feed without any appreciable dissolution of either solvent into the other phase. The *extract* phase is the solute-rich extraction solvent leaving the unit. The treated liquid feed phase remaining after contact with the extraction solvent is referred to as the *raffinate*.

A *theoretical* or *equilibrium stage* provides a mechanism by which two immiscible phases intimately mix until an equilibrium concentration is reached and then physically separated into distinct layers. As shown in Fig. 11–10, crosscurrent extraction involves a series of stages in which the raffinate (R) from one extraction stage is contacted with additional fresh solvent (S) in a subsequent stage. Crosscurrent extraction is rarely economically attractive for large commercial processes because of the high solute usage and low solute concentration in the extract. Figure 11–11 illustrates countercurrent

Figure 11–10

Crosscurrent extraction.

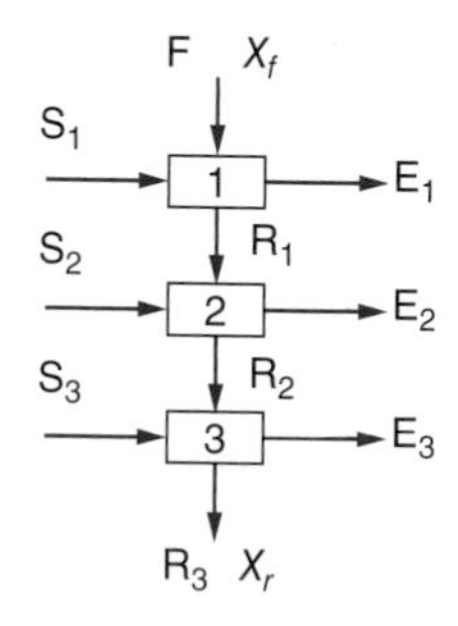

Figure 11–11

Countercurrent extraction.

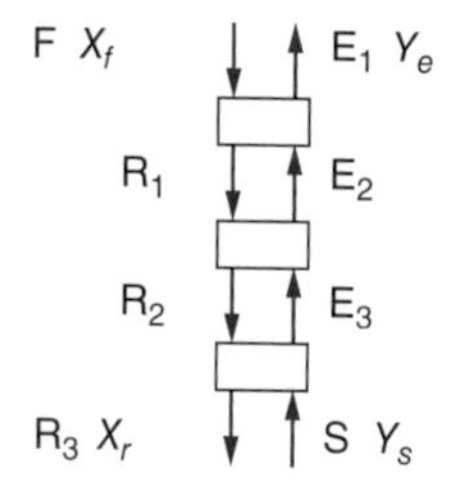

extraction in which the extraction solvent enters a stage at the opposite end from where the feed (F) enters and the two phases pass each other countercurrently. The purpose is to transfer one or more components from the feed solution to the extract (E).

The crosscurrent extraction process (see Fig. 11–10) is an ideal laboratory procedure since the extract and raffinate phases can be analyzed after each stage to generate equilibrium data as well as to achieve high solute removal. If the distribution coefficient K' and the ratio of extraction solvent to feed solvent (S'/F') are constant and the fresh extraction solvent is pure, then the number of crosscurrent stages (N) required to reach a specified raffinate composition can be estimated from the following equation:

$$N = \frac{\log \dfrac{X_f}{X_r}}{\log\left(\dfrac{k'S'}{F'} + 1\right)} \tag{11–80}$$

where X_f = mass ratio of solute in feed
X_r = mass ratio of solute in raffinate
S' = mass flow rate of solute-free extraction solvent
F' = mass flow rate of solute-free feed solvent

Note that the *distribution coefficient* k' is defined as the ratio of the mass fraction in the liquid phase to the mass fraction of solute in the raffinate phase.

Most liquid extraction systems can be treated as having either (1) immiscible (mutually nondissolving) solvents, (2) partially miscible solvents with a low solute concentration in the extract, or (3) partially miscible solvents with a high solute concentration in the extract. Only case 1 is addressed in this section. The reader is referred to *Perry's Handbook*[15] or any other chemical engineering book dealing with liquid extraction for further information on cases 2 and 3.

Since the solvents are immiscible for case 1, the rate of solvent in the feed stream (F') is the same as the rate of feed solvent in the raffinate stream (R'). Also, the rate of extraction solvent (S') entering the unit is the same as the extraction solvent leaving the unit in the extract phase (E'). Thus, the ratio of extraction-solvent to feed-solvent flow rates (S'/F') is equivalent to (E'/R'). However, the total flow rates entering and leaving the unit will be different since the extraction solvent is removing solute from the feed

By writing an overall material balance equation for around the unit illustrated in Fig. 11–11, the unit can be rearranged into a McCabe–Thiele type of operating line with a slope (F'/S'):

$$Y_e = \frac{F'X_f + S'Y_s - R'X_r}{E'} \tag{11–81}$$

where Y_e = mass ratio of solute leaving with the extraction solvent
Y_s = mass ratio of solute entering with the extraction solvent

If the equilibrium line is straight, its intercept is zero, and the operating line is straight. The number of theoretical stages can be calculated with one of the following equations, which are forms of the Kremser eqution.

When the intercept of the equilibrium line is greater than zero, Y_s/k'_s should be used instead of Y_s/m, where k'_s is the distribution coefficient at Y_s. Also, these equations contain an extraction factor ε, which is calculated by dividing the slope of the equilibrium line m, by the slope of the operating line, F'/S':

$$\varepsilon = \frac{mS'}{F'} \tag{11–82}$$

where m = slope of equilibrium line

If the equilibrium line is not straight, a geometric mean value of m should be used.[15] This quantity is determined by the following equation:

$$m = \sqrt{(m_1)(m_2)} \tag{11–83}$$

where m_1 = slope of equilibrium line at concentration leaving feed stage
m_2 = slope of equilibrium line at concentration leaving raffinate stage

When $\varepsilon \neq 1.0$, then

$$N = \frac{\ln\{[(X_f - Y_s/m)/(X_r - Y_s/m)][1-(1/\varepsilon)] + (1/\varepsilon)\}}{\ln\varepsilon} \tag{11–84}$$

when $\varepsilon = 1.0$, then

$$N = \frac{X_f - Y_s/m}{X_r - Y_s/m} - 1 \tag{11–85}$$

Equation (11–80) can be rearranged into the following form in order to estimate the amount of extraction solvent required to achieve a desired separation.

$$S' = \frac{F'}{k'}\left[\frac{X_f/X_r}{10^N} - 1\right] \tag{11–86}$$

For a crosscurrent extraction process, assume that the feed flow rate, as well as the compositions of the solute in the feed, raffinate at each stage, and extract are known. The above equation is also based on the assumption that the fresh extraction solvent contains no solute.

Leaching

Leaching is the preferential removal of one or more components from a solid by contact with a liquid solvent. The soluble constituent may be solid or liquid, and it may be chemically or mechanically held in the pore structure of the insoluble solids material. The insoluble solids material is often particulate in nature, porous, cellulate with selectively permeable cell walls, or surface-activated. In chemical engineering practice, leaching is also referred to by several other terms such as *extraction*, *liquid-solid extraction*, *lixiviation*, *percolation*, *infusion*, *washing*, and *decantation-settling*. A simple example of a leaching process is in the preparation of a cup of tea. Water is the solvent used to "extract," or leach, tannins and other solids from the tea leaf(s).

There are four categories of leaching operation: (1) operating cycle (batch, continuous, or multibatch intermittent), (2) direction of streams (concurrent, countercurrent, or hybrid flow), (3) staging (single-stage, multistage, or differential-stage), and (4) method of contact. Additionally, the following is a list of typical leaching systems:

1. Horizontal-basket design
2. Endless-belt percolation
3. Kennedy extraction
4. Dispersed-solids leaching
5. Batch-stirred tank leaching
6. Continuous dispersed-solids leaching
7. Screw conveyor extraction

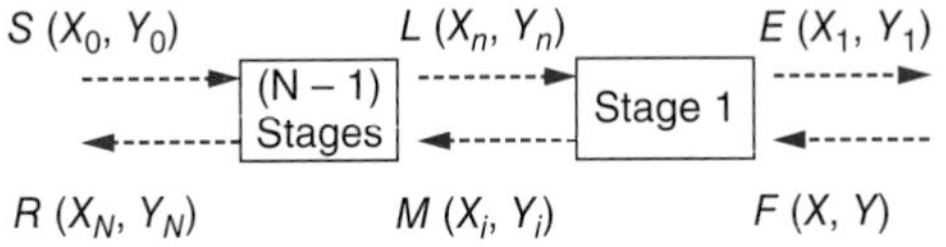

Figure 11–12

Flowchart of a countercurrent leaching system.

$S(X_0, Y_0)$ → (N – 1) Stages → $L(X_n, Y_n)$ → Stage 1 → $E(X_1, Y_1)$

$R(X_N, Y_N)$ ← (N – 1) Stages ← $M(X_i, Y_i)$ ← Stage 1 ← $F(X, Y)$

The selection of process and operating conditions, as well as the sizing of the extraction equipment, are important aspects in the design of the leaching system. Other parameters that must be addressed are the type of solvent to be used, the temperature, terminal stream compositions and quantities, operating cycle, and type of extractor.

Countercurrent leaching systems are more efficient, and thus more economical, than cross-flow continuous systems. (See Fig. 11–12.) The number of theoretical extraction stages can be calculated from the following equation:

$$N = \frac{\ln\left[(X_o - X_N)/(X_n - X_1)\right]}{\ln(R/S)} + 1 \qquad (11\text{–}87)$$

where S = extraction solvent flow rate (removes solute from feed)
R = raffinate leaving extraction unit (extracted feed stream)
L = extraction solvent flow rate into stage 1
M = raffinate flow leaving stage 1
E = extract flow leaving stage 1
F = feed stream to be extracted
X = weight ratio of solute to solvent
Y = weight ratio of solid containing solute to solid containing solvent

There are numerous applications of extraction and leaching in the production of consumer products. Coffee and tea are just two which are deeply rooted in this mass transfer application. The decaffeination of coffee can occur by several mechanisms: extraction with a solvent such as methylene chloride or naturally by a water decaffeination process. The methylene chloride solvent must be extracted to recover the solvent.

Humidification and Drying

In many unit operations, it is necessary to perform calculations involving the properties of mixtures of air and water vapor. Such calculations often require knowledge of

1. The amount of water vapor carried by air under various conditions
2. The thermal properties of such mixtures
3. The changes in enthalpy content and moisture content as air containing some moisture is brought into contact with water or wet solids, and other similar processes

Some mass transfer operations in the chemical process industry involve simultaneous heat and mass transfer. In the *humidification* process, water or another liquid is vaporized, and the required heat of vaporization must be transferred to the liquid. *Dehumidification* is the condensation of water vapor from air, or, in general, the condensation of any vapor from a gas.

Drying involves the *removal* of relatively small amounts of water from solids. In many applications, such as corn processing, drying equipment follows an evaporation step to provide an ultrahigh-solids-content product stream. Drying, in either a batch or continuous process, removes liquid as a vapor by passing warm gas (usually air) over, or indirectly heating, the solid phase. Three basic types of dryers are employed in the drying process: (1) *continuous-tunnel dryer* in which supporting trays with wet solids are moved through an enclosed system while warm air blows over the trays; (2) rotary dryer, which is similar in concept to the continuous-tunnel drier, consisting of an inclined rotating hollow cylinder—the wet solids are fed into one side and hot air is usually passed countercurrently over the wet solids, and the dried solids then pass out the opposite side of the dryer unit; and (3) *spray* dryer, in which a liquid or slurry is sprayed through a nozzle and fine droplets are dried by a hot gas passed either concurrently, countercurrently, or in both directions past the falling droplets; this unit has found wide application in air pollution control.[10]

Despite the differing methods of heat transfer, all three types of dryers can reduce the moisture content of solids to less than 0.01% when designed and operated properly. Additional information is available in the literature[1,2] on both humidification and drying.

References

1. L. Theodore and F. Ricci, *Mass Transfer Operations for the Practicing Engineer*, Wiley, Hoboken, N. J., 2010.
2. R. Treybal, *Mass Transfer Operations*, McGraw-Hill, New York, 1955.
3. L. Theodore and J. Barden, *Mass Transfer*, A Theodore Tutorial, Theodore Tutorials, East Williston, N.Y., originally published by USEPA/APTI, Research Triangle Park, N.C.,1990.
4. A. Fick, *Pogg. Ann.*, **XCIV,** 59, 1855.
5. W. Whitman, *Chem. Met. Eng.*, **29**, 147, 1923.
6. L. Theodore, "Engineering Calculations: Sizing Packed-Tower Absorbers without Data," *Chem. Eng. Prog.*, 18–19, N.Y., May 2005.
7. L. Theodore, personal notes, East Williston, N.Y., 1991.
8. L. Theodore, personal notes, East Williston, N.Y., 1996.
9. S. Brunauer, P. Emmett, and E. Teller, *J. Am. Chem. Soc.*, **60**, 309, 1938.
10. L. Theodore, *Air Pollution Control Equipment Calculations*, Wiley, Hoboken, N.Y., 2008.
11. L. Theodore, "Engineering Calculations: Adsorber Sizing Made Easy," *Chem. Eng. Prog.*, CEP, pp 9–20, N.Y., March 2005.
12. R. Kunz, "The Raleigh Equation Revisited: What to Do When Alpha (α) Isn't Constant," paper presented at AIChE spring meeting, Houston, TX., 2012.
13. W. McCabe and E. Thiele, *Ind. Eng. Chem.*, **17**, 605, N.Y., 1925.
14. E. Gilliland, *Ind. Eng. Chem.*, **32**, 1220, N.Y., 1940.
15. R. Treybal, *Liquid Extraction* McGraw-Hill, N.Y., 1951.
16. D. Green and R. Perry, eds., *Perry's Chemical Engineers' Handbook*, 8th ed., McGraw-Hill, N.Y., 2008.

12 Membrane Separation Processes

Chapter Outline

12–1 Introduction

Phase separation, as its name implies, simply involves the separation of one (or more) phase(s) from another phase. Most industrial equipment used for this class of processes involves the relative motion of two phases under the action of various external forces (gravitational, electrostatic, etc.). There are basically five phase separation processes:

1. Gas-solid (GS)
2. Gas-liquid (GL)
3. Liquid-solid (LS)
4. Liquid-liquid (LL)
5. Solid-solid (SS)

Phases 1 to 4 are immiscible processes. As one might suppose, the major phase separation process encountered in industry is GS. Traditional equipment for GS separation processes includes (1) gravity settlers, (2) centrifugal separators (cyclones), (3) electrostatic precipitators, (4) wet scrubbers, and (5) baghouses.

The overall collection-removal process for solid particles in a fluid consists of four steps:[1–3]

1. An external force (or forces) must be applied that enables the particle to develop a velocity that will displace or direct it to a collection or retrieval section or area or surface.
2. The particle should be retained at this area with sufficiently strong forces that it is not reentrained.
3. As collected-recovered particles accumulate, they are subsequently removed.
4. The ultimate disposition of the particles completes the process.

Obviously, the first is the most important step in the overall process. The particle collection mechanisms discussed below are generally applicable when the fluid is air; however, they may also apply if the fluid is another gas, water, or another liquid.

The following forces are basically the tools which may be used for particulate-recovery collection:[1,2] (1) gravity settling, (2) centrifugal action, (3) inertial impaction, (4) electrostatic attraction, (5) thermophoresis and diffusiophoresis, and (6) brownian motion. All of these collection mechanism forces are strongly dependent on particle size.[1]

Miscible phases are another matter. One popular separation technique involves the use of membranes. In operations with miscible phases separated by a membrane, the membrane is necessary to prevent intermingling of the phases.

Three different membrane separation phase combinations are briefly discussed below:

1. *Gas-Gas (GG)*. Operation in the gas-gas category is known as *gaseous diffusion, gas permeation*, or *effusion*. If a gas mixture whose components are of different molecular weights is brought into contact with a porous membrane or diaphragm, the various components of the gas will diffuse through the pore openings at different rates. This leads to different compositions on opposite sides of the membranes and, consequently, to some degree of separation of the gas mixture. The large-scale separation of the isotopes of uranium, in the form of uranium hexafluoride, can be carried out in this manner.

2. Liquid-Liquid (LL). The separation of a crystalline substance from a colloid by contact of the solution with a liquid solvent with a membrane permeable only to the solvent and the dissolved crystalline substance is known as *dialysis*. For example, aqueous sugar beet solutions, containing undesired colloidal material, are freed of the latter by contact with water with a semipermeable membrane; sugar and water diffuse through the membrane, but the larger colloidal particles cannot. *Fractional dialysis* for separating two crystalline substances in solution makes use of the difference in the membrane permeability of the substances. If an electronegative force is applied across the membrane to assist in the diffusional transport of charged particles, the operation is *electrodialysis.* If a solution is separated from the pure solvent by a membrane that is permeable only to the solvent, the solvent transported into the solution is termed *osmosis* by most chemical engineers in the field. This is not a separation operation, of course, but if the flow of solvent is reversed by superimposing a pressure to oppose the osmosis pressure, the process is then defined as *reversed osmosis*.

3. *Solid-Solid (SS)*. Operation in the solid-solid category has found little, if any, practical application in the chemical process industry.

Membrane processes are today state-of-the-art separation technologies that have shown promise for future technical growth and wide-scale industrial commercialization. They are used in many industries for process stream and product concentration, purification, separation, and fractionation. The need for membrane research and development (R&D) is important because of the increasing use of membrane technology in both traditional and emerging engineering fields. Membrane processes are increasingly finding their way into the growing chemical engineering areas of biotechnology, green engineering, specialty chemical manufacture, and biomedical engineering, as well as in the traditional chemical process industries. Membrane technology is also being considered as either a replacement for or supplement to traditional separations such as distillation or extraction (see Chap. 11). Membrane processes can be more efficient and effective since they can simultaneously concentrate and purify, and can also perform separations at ambient conditions.[4]

Membranes create a boundary between different bulk gas or liquid mixtures. Different solutes and solvents flow-through membranes at different rates; this enables the use of membranes in separation processes. Membrane processes can be operated at moderate temperatures for sensitive components (food, pharmaceuticals, etc.). Membrane processes also tend to have low relative capital and energy costs. Their modular format permits simple and reliable scale-up.

Key membrane properties include their size rating, selectivity, permeability, mechanical strength, chemical resistance, low fouling characteristics, high capacity, low cost, and consistency. Vendors characterize their filters with ratings indicating the approximate size (or corresponding molecular weight) of components retained by the membrane. Commercial membranes consist of polymers and some ceramics. Other membrane types include sintered metal glass and liquid film.

Topics addressed in this chapter include membrane separation principles, reverse osmosis (RO), ultrafiltration (UF), microfiltration (MF), and gas permeation (GP). Pevaporation (PER) and electrodialysis (ED) are briefly addressed at the conclusion of the chapter.

12-2 Membrane Separation Principles

Membrane unit operations are often characterized by the following parameters: (1) driving force utilized, (2) membrane type and structure, and (3) species being separated.

The following membrane unit operations utilize a pressure difference driving force to separate a liquid feed into a liquid *permeate* and *retentate:* reverse osmosis, nanofiltration,[5] ultrafiltration, and microfiltration. They are listed in *ascending* order in their ability to separate a liquid feed on the basis of solute size. *Reverse osmosis* (RO) uses nonporous membranes and can separate down to the ionic level, e.g., seawater in the rejection of dissolved salt. *Nanofiltration* performs separations at the nanometer range.[5] *Ultrafiltration* uses porous membranes and separates components of molecular weights ranging from the low thousand to several hundred thousand molecules; an example includes components in biomedical processing. *Microfiltration* uses much more porous membranes and is typically employed in the micro- or macromolecular range to remove particulate or larger biological matter from a feed stream (e.g., in the range of 0.05 to 2.5 μm).[4,6]

Gas separation processes can be divided into two categories: gas permeation through nonporous membranes and gas diffusion through porous membranes. Both of these processes utilize a concentration difference driving force. The gas permeation processes are used extensively in industry to separate air into purified nitrogen and enriched oxygen. Another commercial application is hydrogen recovery in petroleum refineries.[4] As noted in Sec. 12–1, dialysis membrane processes use a concentration difference as a driving force for separation of liquid feed across a semipermeable membrane, with the major application in the medical field of hemodialysis. *Electrodialysis* separates a liquid employing an electric driving force and is widely used in water purification and industrial processing.

Since membrane separation processes are one of the newer (relatively speaking) technologies being applied in practice, the subject matter is and has been introduced into the chemical engineering curriculum. There are four major membrane processes of interest to the chemical engineer:

1. Reverse osmosis (hyperfiltration)
2. Ultrafiltration
3. Microfiltration
4. Gas permeation

Each process is discussed in more detail in the sections to follow. As noted above, main difference between reverse osmosis (RO) and ultrafiltration (UF) is that the overall size and diameter of the particles or molecules in solution to be separated is smaller in RO. In microfiltration (MF), the particles to be separated or concentrated are generally solids

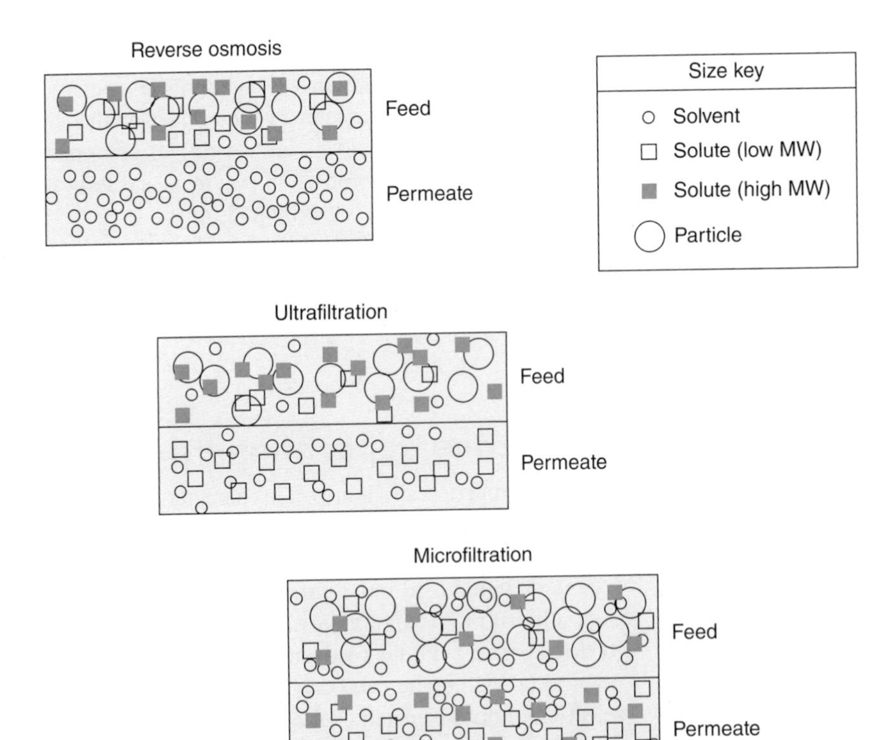

Figure 12–1

Three membrane separation processes.

or colloids rather than molecules in solution. Figure 12–1 illustrates the difference between the processes. Gas permeation (GP) is another membrane process that employs a nonporous semipermeable membrane to *fractionate* a gaseous stream.

Naturally, the heart of the membrane process is the membrane itself; it is an ultrathin semipermeable barrier separating two fluids that permits the transport of certain species through the membrane barrier from one fluid to the other. As noted above, membrane is typically produced from various *polymers* such as cellulose acetate or polysulfone, but ceramic and metallic membranes are also used in some applications. The membrane is referred to as *selective* since it permits the transport of certain species while rejecting others. The term *semipermeable* is frequently used to describe this selective action.[4]

12–3 Reverse Osmosis

The most widely commercialized membrane process by far is *reverse osmosis* (RO). It belongs to a family of pressure-driven separation operations for liquids that includes not only reverse osmosis but also *ultrafiltration* and *microfiltration*. Care should be exercised with these terms since some of them are used interchangeably. For example, RO is considered by some, and defined, as *hyperfiltration.*

Reverse osmosis is an advanced separation technique that may be used when low-molecular-weight (MW) solutes such as inorganic salts or small organic molecules (e.g., glucose) are to be separated from a solvent (usually but not always water). In normal (as opposed to reverse) osmosis, water flows from a less concentrated salt solution to a more concentrated salt solution as a result of driving forces. As a result of the migration of water, an *osmotic pressure* is created on the side of the membrane to which water flows. In reverse (as opposed to normal) osmosis, the membrane is permeable to the solvent or water and relatively impermeable to the solute or salt. To make water pass through an RO membrane in the desired direction (i.e., away from the concentrated salt solution), a pressure must be applied that is higher than that of the osmotic pressure.

Reverse osmosis is widely utilized today by a host of chemical process industries for a surprisingly large number of operations. Aside from the classic example of RO for seawater desalination, it has found a niche in the food industry for concentration of various fruit juices, in the galvanization industry for concentration of waste streams, and in the dairy industry for concentration of milk prior to cheese manufacturing.[7–9]

Reverse osmosis processes are classified into the following two basic categories:

1. *Purification* of the solvent such as in desalination where the *permeate* or purified water is the product
2. *Concentration* of the solute such as in concentration of fruit juices where the retentate is the product

The membranes used for RO processes are characterized by a high degree of semipermeability, high water flux, mechanical strength, chemical stability, and relatively low operating and high capital costs. Early RO membranes were composed of cellulose acetate, but restrictions on process stream pressure, temperature, and organic solute rejection spurred the development of noncellulosic and composite materials.

Reverse osmosis membranes may be configured or designed into certain geometries for system operation: plate and frame, tubular, spiral wound (composite), and hollow fiber.[9] In the *plate-and-frame* configuration, flat sheets of membrane are placed between spacers with heights of approximately 0.5 to 1.0 mm. These are, in turn, stacked in parallel groups. *Tubular* units are also commonly used for RO. This is a simpler design in which the feed flows inside of a tube whose walls contain the membrane. These types of membranes are usually produced with inside diameters ranging from 12.5 to 25 mm and are generally produced in lengths of 150 to 610 cm. There is also the *hollow fine-fiber* (HFF) arrangement; this geometry is used in 70% of worldwide desalination applications. Millions of hollow fibers are oriented in parallel and fixed in epoxy at both ends; the feed stream is sent through a central distributor where it is forced out radially through the fiber bundle. As the pressurized feed contacts the fibers, the permeate is forced into the center of each hollow fiber. The permeate then travels through the hollow bore until it exits the permeator. A *spiral-wound* cartridge is occasionally employed in this configuration; the solvent is forced inward towards the product tube while the concentrate remains in the space between the membranes. A flat film membrane is made into a "leaf." Each leaf consists of two sheets of membrane with a sheet of polyester tricot in between and serves a collection channel for the water product. Plastic netting is placed between each leaf to serve as a feed channel. Each leaf is then wrapped around the product tube in a spiral fashion.

It's no secret that water covers around 70% of Earth's surface, but 97.5% of it is unfit for human consumption. With the world facing a growing freshwater shortage from which the United States may not be spared, one method for producing freshwater is *desalination*. The major application of RO is water desalination. Some areas of the world that do not have a ready supply of freshwater have chosen to desalinate seawater or brackish water using RO to generate potable drinking water. Because no heating or phase change is required, the RO process is both a relatively low energy and economical water purification process. A typical saltwater RO system consists of an intake, a pretreatment component, a high-pressure pump, membrane apparatus, remineralization, and pH adjustment components, as well as a disinfection step. A pressure difference driving force of about 1.0 to 6.5 mPA is generally required to overcome the osmotic pressure of saltwater.

Another important application is *dialysis*. This technique is used in patients who suffer from kidney failure and can no longer filter waste products (urea) from the blood. In general, RO equipment used for dialysis can reduce ionic contaminants by up to 90%. In this process, the patient's blood flows in a tubular membrane while a dialysate flows countercurrently on the outside of the feed tube. The concentrations of undesirables (e.g., potassium, calcium, urea) are high in the blood (while low or absent in the dialysate). This treatment successfully mimics the filtration capabilities of the kidney.

Reverse osmosis membranes are designed for high salt retention, high permeability, mechanical robustness (to allow module fabrication and withstand operating conditions), chemical robustness (for fabrication materials, process fluids, cleaners, and sanitizers), low extractables, low fouling characteristics, high capacity, low cost, and consistency. The predominant RO membranes used in water applications include cellulose polymers, thin composites consisting of aromatic polyamides, and cross-linked polyetherurea.

The membrane operation for the purification of seawater that incorporates a selective barrier can be simply described using the line diagram provided in Fig. 12–2. This membrane operation typifies the case in which a feed stream (seawater) is separated by

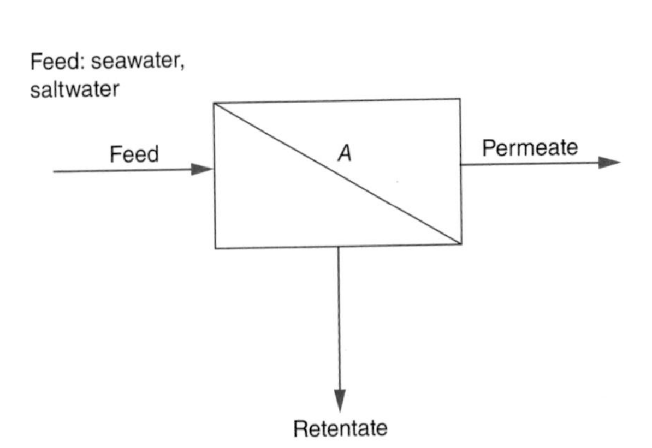

Figure 12–2

Desalination of seawater by RO.

Figure 12–3

Preosmosis equilibrium.

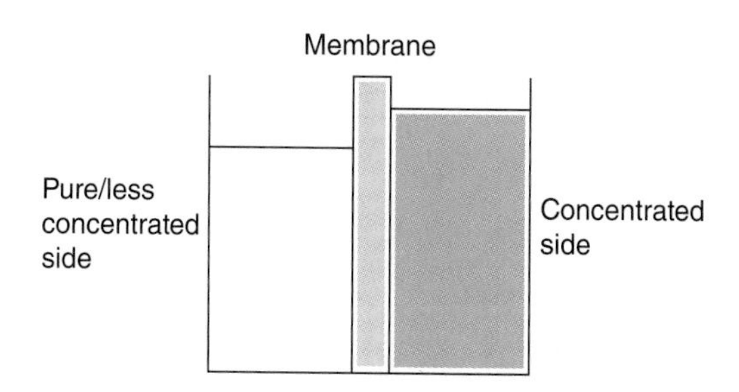

a semipermeable membrane that rejects salt but selectively transports water. A purified stream (the permeate) is therefore produced while (at the same time) a concentrated salt stream (the retentate) is discharged. With reference to Fig. 12–2, a simple material balance can be written on the overall process flows and for that of the solute:

$$q_f = q_r + q_p \tag{12–1}$$

$$C_f q_f = C_r q_r + C_p q_p \tag{12–2}$$

where q = volumetric flow rate
C = solute concentrate

Subscripts f, r, and p refer to the feed, retentate, and permeate, respectively.

Osmosis occurs when a concentrated solution is partitioned from a pure solute or relatively lower concentration solution by a semipermeable membrane. The semipermeable membrane allows only the solvent to flow-through it freely. Equilibrium is achieved when the solvent from the lower-concentration side ceases to flow-through the membrane to the higher-concentration side (thus reducing the concentration) because the mass transfer concentration difference driving force has been reduced. This is shown in Figs. 12–3 and 12–4. *Osmotic pressure* is the pressure needed to terminate the flux of solvent through the membrane or the force that pushes up on the concentrated side of the membrane (see Fig. 12–5). Applying a pressure on the concentrated side halts the solvent flux. Reverse osmosis (see Fig. 12–6) takes place when an applied force (pressure) overcomes the osmotic pressure and forces the solvent from the concentrated side through the membrane and leaves the solute on the concentrated side.

Osmotic pressure is related to both the solute concentration and the temperature of the solution as described in the Van't Hoff equation:

$$\pi = iC_s RT \tag{12–3}$$

where π = osmotic pressure, psi
i = Van't Hoff factor, dimensionless
C_s = solute concentration, mol/L
R = universal gas constant, (L·atm)/(mol·K)
T = absolute temperature, K

Figure 12–4

Osmosis of solvent.

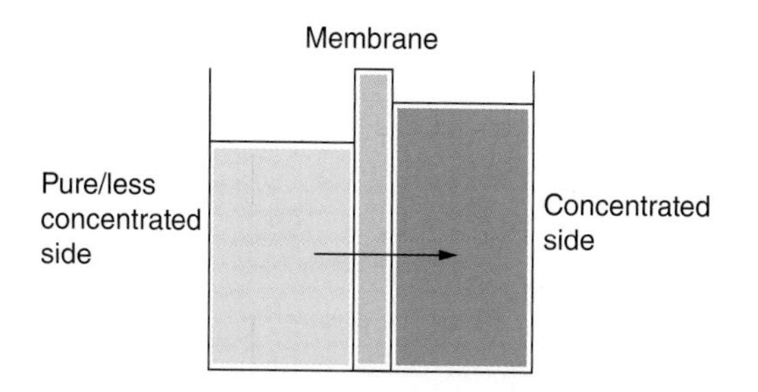

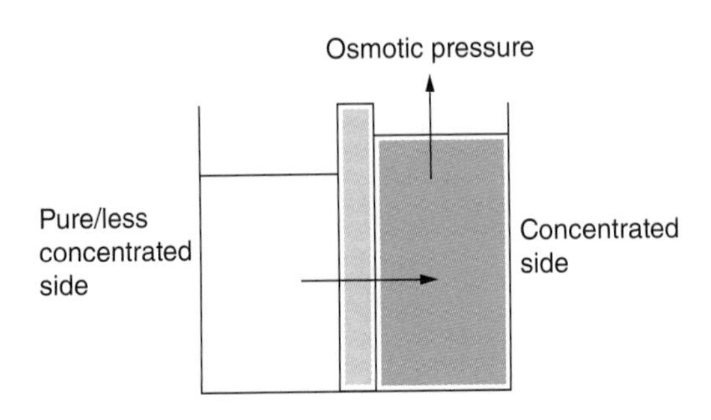

Figure 12–5

Osmotic pressure.

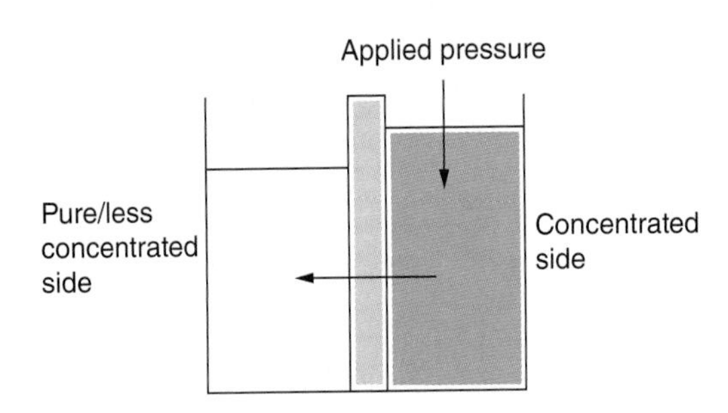

Figure 12–6

Reverse osmosis.

The Van't Hoff factor i in Eq. (12–3) factors in the number of ions in solution. For example, NaCl separates into two ions, Na^+ and Cl^-, therefore rendering a Van't Hoff factor equal to 2. Closer inspection of Eq. (12–3) reveals that the Van't Hoff equation is analogous to the ideal gas law.

The change in osmotic pressure across the membrane in this operation must be overcome in order to achieve RO. This is described in the following equation

$$\Delta\pi = \pi_f - \pi_p \qquad (12\text{–}4)$$

where $\Delta\pi$ = change in the osmotic pressure, psi
π_f = osmotic pressure in the feed, psi
π_p = osmotic pressure in the permeate, psi

This change in osmotic pressure can also be calculated using the concentrations of both the feed and the permeate as well as an empirical coefficient denoted as Ψ. This formula is

$$\Delta\pi = \Psi(C_f - C_p) \qquad (12\text{–}5)$$

where $\Delta\pi$ = change in the osmotic pressure, psi
Ψ = constant, (L · psi/g
C_f = feed concentration, g/L
C_p = permeate concentration, g/L

The *permeate flux* is an important characteristic of an RO operation. It is related to the permeate flow as well as the area of the membrane. This is represented in the following equation:

$$J_p = \frac{q_p}{A_m} \qquad (12\text{–}6)$$

where J_p = permeate flux, gal/(ft^2 · day)
q_p = permeate flow, gal/day
A_m = membrane surface area, ft^2

The flux can be determined experimentally by measuring each incremental volume of permeate ΔV collected in time period Δt, and dividing by the surface area of the membrane. In water-based processes such as desalination, the permeate naturally consists mostly of water. Therefore, the permeate flux can be considered to be equal to the *water flux*. Equation (12–7) defines the water flux:

$$J_p = J_w = A_w(\Delta P - \Delta\pi) = \frac{K_s}{t_m}(\Delta P - \Delta\pi) \tag{12–7}$$

where J_w = water flux, gal/(ft^2·day)
J_p = permeate flux, gal/(ft^2·day)
A_w = water permeability coefficient, gal/(ft^2·day·psi)
ΔP = pressure drop, psi
$\Delta\pi$ = osmotic pressure change, psi
K_s = permeability constant, gal/(ft·day·psi)
t_m = membrane thickness, ft

The water permeability coefficient can therefore be experimentally determined by obtaining data on the unit with pure water; this eliminates the change in osmotic pressure since both sides of the membrane contain pure water.

Another important factor is the *solute flux*, which can be determined through the utilization of

$$J_s = \frac{q_p C_p}{A_m} = \frac{\dot{m}_p}{A_m} \tag{12–8}$$

where J_s = solute flux, g/(ft^2·min)
q_p = permeate flow rate, L/day
C_p = permeate concentration, g/L
$\dot{m}_p$ = permeate flow rate, g/day
A_m = membrane surface area, ft^2

The solute flux can also be related to the solute concentration by utilizing the solute permeability factor. This relationship is given by

$$J_s = B_s(\Delta C_s) \tag{12–9}$$

where J_s = solute flux, g/(ft^2·min)
B_s = solute permeability coefficient, L/(ft^2·min)
ΔC_s = change in concentration, g/L

The *selectivity* of a membrane to filter out a solute can be expressed as the *percent rejection* (%*R*). Percent rejection represents a membrane's effectiveness and is a measure of the membrane's ability to selectively allow certain species to permeate and others to be rejected. This is an important characteristic or consideration when selecting a membrane for a separation process. The percent rejection represents the percentage of solute that is *not* allowed to pass into the permeate stream, and is given by

$$\%R = \left(\frac{C_f - C_p}{C_f}\right) \cdot 100\% \tag{12–10}$$

where %R = solute rejection, %
C_p = permeate concentration, g/L
C_f = feed concentration, g/L

Finally, the *solvent recovery* Y is a measure of the quantity of solvent that is allowed to pass through the membrane. This is defined as the quotient of the permeate flow divided by the feed flow, as shown by

$$Y = \frac{q_p}{q_f} \tag{12–11}$$

where Y = solvent recovery, dimensionless-fractional basis
q_p = permeate flow, L/min
q_f = feed flow, L/min

12–4 Ultrafiltration

Ultrafiltration (UF) is a membrane separation process that can be used to concentrate single solutes or mixtures of solutes. Transmembrane pressure (the membrane pressure drop) is the main driving force in UF operations, and separation is achieved via a *sieving* mechanism. The UF process can be used for the treatment or concentration of oily wastewater, for pretreatment of water *prior* to RO, and for the removal of bacterial contamination (pyrogens). In the food industry, UF is used to separate lactose and salt from cheese whey proteins, to clarify apple juice, and to concentrate milk for ice cream and cheese production.[10–12] The most energy-intensive step in ice cream production is concerned with the concentration of skimmed milk, where membrane processes are more economical for this step than vacuum evaporation.[11] UF processes are also used for concentrating or dewatering fermentation products as well as purifying blood fractions and vaccines.

Ultrafiltration may be regarded as a membrane separation technique where a solution is introduced on one side of a membrane barrier while water, salts, and/or other low-molecular-weight materials pass through the unit under an applied pressure. As noted, these membranes separation processes can be used to concentrate single solutes or mixtures of solutes. The variety in the different membrane materials allows for a wide temperature-pH processing range.

The main economic advantage of UF is a reduction in both design complexity and energy usage since UF processes can simultaneously concentrate and purify process streams. The fact that no phase change is required leads to highly desirable energy savings. A major disadvantage is the high capital cost that might be required if low flux rates for purification demand a large system. However, UF processes are usually economically sound in comparison to other traditional separation techniques.

In addition to the applications described above, UF membrane processes are utilized in various commercial applications. They are found in the treatment of industrial effluents and process water; concentration, purification, and separation of macromolecular solutions in the chemical, food, and drug industries; sterilization, clarification, and prefiltering of biological solutions and beverages; and, production of ultrapure water and preheating of seawater in RO processes. The most promising area for the expansion of UF process applications is the biochemical industry. Some of its usage in this area includes purifying vaccines and blood fractions; concentrating gelatins, albumin, and egg solids; and, recovering proteins and starches.

The rejection of a solute is a function of the size, size distribution, shape, and surface binding characteristics of the hydrated molecule. It is also a function of the pore size distribution of the membrane; therefore, molecular weight (MW) cutoff values can be used only as a rough guide for membrane selection. The retention efficiency of the solutes depends to a large extent on the proper selection and condition of the membrane.

Replacement of highly used membranes and regular inspections of the separation units averts many problems that might otherwise occur because of clogging and gel formation.

Ultrafiltration processes use driving forces of 0.2 to 1.0 MPa to drive liquid solvent (usually water) and small solutes through membranes while retaining solutes of 10 to 1000 Å diameter. The membranes consist primarily of polymeric structures, such as polyethersulfone, regenerated cellulose, polysulfone, polyamide, polyacrylonitrile, or various fluoropolymers. Hydrophobic polymers are surface modified to render them hydrophilic.

The general design factors for any membrane system (including UF), as reported by Wankat, are[13]

1. Thin active layer of membrane
2. High permeability for species A and low permeability for species B
3. Stable membrane with long service life
4. Mechanical strength
5. Large surface area of membrane in a small volume
6. Concentration polarization elimination or control
7. Ease in cleaning, if necessary
8. Low construction costs
9. Low operating costs

System performance is usually defined in terms of the permeate *flux* J_p with dimensions of volume/(area·time). The typical units are L/(m²·h). As with RO, J_p can be obtained experimentally by measuring the incremental volume of the permeate ΔV collected in a time period Δt. Thus, the permeate flux describing equation for the operation is

$$J_p = \frac{\Delta V/\Delta t}{A_{surf}} \tag{12–12}$$

where J_p = permeate flux, L/(m²·h)
ΔV = incremental volume of permeate, m³
Δt = collection time period, h
A_{surf} = surface area of membrane, m²

Other consistent SI units for the flux may be employed, for example, cm³/(cm²·s).

Transmembrane pressure is the main driving force in UF operations, and separation is achieved through the sieving mechanism mentioned above. Since UF is a pressure-driven separation process, it is appropriate to examine the effects of pressure on flux. Equation (12–13) illustrates how the flux varies with pressure. It is seen that the flux of a pure solvent through a porous membrane is directly proportional to the applied pressure gradient across the membrane ΔP and inversely proportional to the membrane thickness t_m:

$$J_s = \frac{K_s}{t_m}(\Delta P - \Delta\pi) \tag{12–13}$$

where K_s = permeate constant, cm²/psi·s
ΔP = pressure drop across membrane (transmembrane pressure), psi
$\Delta\pi$ = osmotic pressure difference across membrane, psi
t_m = membrane thickness, cm

Such factors as the membrane porosity, pore size distribution, and viscosity of the solvent are accounted for by the permeability constant K_s. When t_m is not available or is not known, the water permeability coefficient A_w may be used in place of K_s in Eq. (12–13). The water permeability coefficient is a function of the distribution coefficient (solubility), diffusion coefficient, membrane thickness, and temperature. The value of A_w can be determined by conducting ultrapure water flux experiments at varying operating pressures while the permeate collection occurs at or near atmospheric pressure.

Since the pressure is relatively low for macromolecular solutions, which are typically the ones recommended for UF processes, the $\Delta\pi$ term can be neglected in Eq. (12–13). This occurs because the molar concentration of the high-MW molecules separated by UF is low, even when the mass concentrations are high. When the $\Delta\pi$ term is neglected and K_s/t_m is replaced by A_w, the following equation is obtained:

$$J_s = A_w \, \Delta P \tag{12–14}$$

where A_w = water permeability coefficient, $cm^2/psi \cdot s$

When a solute such as milk solids dissolved in water flows through a typical UF process, some of the solute usually passes through the membrane since real membranes are *partially* permeable. The *apparent rejection* on a fraction basis is then once again calculated as follows [see also Eq. (12–10)]:

$$R_{app} = \frac{C_r - C_p}{C_r} \tag{12–15}$$

where R_{app} = apparent rejection; dimensionless, fractional basis
C_p = permeate concentration, g/cm^3
C_r = retentate concentration, g/cm^3

Three additional factors that need to be considered in UF separations: concentration polarization, gel formation, and fouling. A concentration gradient or boundary layer typically forms during a UF process. This concentration difference driving force appears near the membrane surface and is referred to as *concentration polarization*. It results from the counteracting effects of the convective flow of solute toward the membrane and diffusion of the solute toward the bulk fluid in the reverse direction. While concentration polarization is regarded as a reversible boundary-layer phenomenon that causes a rapid initial drop in flux to a steady-state value, fouling is considered as an irreversible process that leads to a flux decline over the long term. However, the process of *gel formation* may be reversible or irreversible. When the gel is difficult to remove, the membrane is said to be *fouled* and thus the gel formation is irreversible. Concentration polarization may occur with or without gel formation.

Concentration polarization occurs in many separations and for large solutes where the osmotic pressure can be safely neglected. Concentration polarization without gelling is predicted to have no effect on the flux. Therefore, if a flux decline is observed, it is usually attributed to the formation of a gel layer with a concentration C_g. The gel layer, once formed, usually controls mass transfer so that Eq. (12–13) is no longer applicable.[14]

12–5 Microfiltration

Microfiltration (MF) can separate particles from liquid or gas phase solutions. It is alone among the membrane processes, since microfiltration may be accomplished without the use of a membrane. The usual materials retained by a microfiltration membrane range in size from several micrometers down to 0.2 μm. Very large soluble molecules are retained by a microfilter at the low end of this spectrum.

Microfiltration (MF) is employed in modern industrial biochemical and biological separation processes. For example, MF can be used instead of centrifugation or precoat rotary vacuum filtration to remove yeast, bacteria, or mycelia organisms from fermentation broth in cell harvesting. Both MF and UF are used for cell harvesting. Microfiltration is used to retain cells and colloids, while allowing passage of macromolecules into the permeate stream. UF is also used to concentrate macromolecules, cells, and colloidal material, while allowing small organic molecules and inorganic salts to pass into the permeate stream. As stated above, pore sizes in microfiltration are around 0.10 to 10 μm in diameter as compared with 0.001 to 0.02 μm for ultrafiltration (ranges vary slightly depending on the source).[15,16]

Similar types of equipment are used for MF and UF; however, membranes with larger pore sizes are generally employed in MF applications. MF and UF belong to a group of separation processes that depend on pressure as the driving force for separation. MF processes operate at lower pressure than UF but at a *higher pressure difference driving force* (PDDF) than does conventional particulate filtration.[14] Ideal membranes possess high porosity, a narrow pore size distribution, and a low binding capacity.

When separating microorganisms and cell debris from fermentation broth, a biological cake is formed. Principles of cake filtration[1,14] apply to MF systems, except that the small size of the yeast particles produces a cake with a relatively high resistance to flow and a relatively low filtration rate.

In *dead-end filtration*, feed flow is perpendicular to the membrane surface, and the thickness of the cake layer on the membrane surface increases with filtration time; consequently, the permeation rate decreases. *Cross-flow filtration*, on the other hand, features feed flowing parallel to the membrane surface, which is designed to decrease formation of cake by *sweeping* previously deposited solids from the membrane surface and returning them to the bulk feed stream. Cross-flow filtration is far superior to dead-end filtration for cell harvesting because the biological cake is highly compressible, which causes the accumulated layer of biomass to rapidly blind the filter surface in dead-end operation. Therefore, MF experiments are often conducted using cross-flow filtration because of the advantages that this mode offers.[17,18]

Separation principles and governing equations for MF are similar to those developed in the two previous sections for RO and UF, respectively. This representation is primarily directed toward the theoretical aspects of MF. System performance is usually defined in terms of permeate flux J_p with dimensions (once again) of (volume/area · time); typical units are L/(m^2 · h). As noted earlier, the flux can be determined experimentally by measuring each incremental volume of permeate ΔV collected in time period Δt and dividing by surface area of the membrane as follows:

$$J_p = \frac{\Delta V / \Delta t}{A_s} \tag{12–16}$$

Since MF is a pressure-driven separation process, it is appropriate to comment on the effects of PDDF on the flux. The flux of a liquid solution through a porous membrane is directly proportional to the applied pressure gradient across the membrane ΔP and inversely proportional to the solution viscosity μ and membrane thickness t_m [see also Eq. (12–12)]:

$$J_s = \frac{K_s \, \Delta P}{t_m} = \frac{\Delta P}{\mu R_m} \tag{12–17}$$

The *hydrodynamic resistance* of the membrane R_m is inversely related to the solvent permeability constant K_s. The permeability constant accounts for factors such as membrane porosity, pore size distribution, and viscosity of the liquid. Permeate is normally collected at atmospheric pressure.

In membrane separation processes with pure solvents, temperature effects on the flux generally follow the Arrhenius relationship provided in Eq. (12–18), where J_0 is the flux at 25°C, E_a is the activation energy, R is the universal gas constant, and T is absolute temperature (all in consistent units):

$$J_s = J_0 e^{E_a/RT} \tag{12–18}$$

Equation (12–18) may also be linearized and written as follows:

$$\ln(J_s) = \ln(J_0) + \left(\frac{E_a}{R}\right)\frac{1}{T} \tag{12–19}$$

Changes in flux with temperature result from changes in solution viscosity. As stated earlier, the viscosity of liquids decrease as the temperature increases; thus, water permeability through the membrane subsequently increases. This relationship can be shown to hold for a Newtonian fluid (see also Chap. 9) such as distilled water. Fermentation broth containing suspended microorganisms is a non-Newtonian fluid; therefore, increased temperatures tend to improve the flux but not to the same magnitude as observed with dilute aqueous solutions.

It can be seen from Eq. (12–17) that the products ($J_s\mu$) should be a constant value in temperature studies on water at constant ΔP. Substituting Eq. (12–17) into Eq. (12–19) and taking the logarithm of both sides of the resultant equation leads to the Arrhenius-type relationship similar to Eq. (12–19). Thus, by employing ultrapure water at varying temperatures and constant transmembrane pressure, one can determine E_a from the slope of a graph (linear-linear coordinates) of $\ln(1/\mu)$ versus $1/T$ as follows:

$$\ln\left(\frac{1}{\mu}\right) = \ln\left(\frac{1}{\mu_0}\right) + \left(\frac{E_a}{R}\right)\frac{1}{T} \tag{12–20}$$

The primary factor limiting flux in MF processes is *cake buildup* and *fouling*; both are caused by factors such as the absorption of proteins on the membrane surface. The increased cake resistance arises because of cell debris, antifoam, precipitates, and other substances that fill the void space in the biological cake and contribute to the flux decline.

Cross-flow filtration is designed to sweep the membrane surface so as to remove deposited solids from the membrane surface. The cross-membrane flow rate can be varied and its effect on flux determined. While the cross-flow mode is a significant improvement over dead-end filtration, the permeate flux still decreases to some steady-state value of *limiting flux* J_∞. The limiting flux can be modeled in terms of the resistances to permeation through the membrane R_g as follows:

$$J_\infty = \frac{\Delta P}{\mu(R_m + R_c + R_g)} \tag{12–21}$$

The resistances in Eq. (12–21) can be measured experimentally. For example, the value of R_m can be found by initial clean-water flux measurements [see also Eq. 12–17)]. The total resistance (R_m+R_c+R_g) is measured from the final steady-state flux through the system after cake buildup, e.g., with yeast slurry as feed.

After completing yeast runs, the MF systems may be cleaned in two steps:

1. Cleaning with water to remove yeast cake and other reversible deposits
2. Chemical cleaning with a solution (e.g., hypochlorite) to remove fouling deposits

After cleaning with water, the value for (R_m+R_g) is measured by clean-water fluxes. As R_g is usually negligible in microfiltration processes, thorough cleaning with water should result in flux values that are very close to the original water values. Therefore, chemical cleaning should not be required under normal operating conditions but may be required if membranes are to be reused after high-pressure or low-cross-flow studies.

The final concentration of the retentate C_r can be used as an absolute measure of system performance. A relative measurement of performance is the *concentration factor* Ψ, defined as the ratio of the initial feed volume V_0 to the final retentate volume V_r; specifically, $\Psi = V_0/V_r$. Initial and final volumes and concentrations can also be used to calculate the recovery Y, where C_0 is the initial cell concentration in the feed:

$$\text{Recovery } Y = \frac{(C_r)(V_r)}{(C_0)(V_0)} \times 100\% \tag{12–22}$$

Solute rejection $R°$ is another parameter that can be used to measure performance in these systems[19] where C_p is the concentration of yeast cells in the permeate. This parameter is given by

$$R° = 1 - \frac{C_p}{C_0} \tag{12–23}$$

12–6 Gas Permeation

Several different types of membrane separation processes are used in the chemical process industries, including systems for gas separation. These processes are generally considered as relatively new and emerging technologies because they are not included in the curriculum of many traditional chemical engineering programs.[18–23]

Gas permeation systems have gained popularity in both traditional and emerging engineering areas. These systems were originally developed primarily for hydrogen recovery. There are presently numerous applications of gas permeation in industry, and other potential uses of this technology are in various stages of development. Applications today include gas recovery from waste gas streams, landfill gases, and ammonia and petrochemical products. Gas permeation membrane systems are also employed in gas generation and purification, including the production of nitrogen and enriched oxygen gases.[20]

Gas permeation (GP) is the term used to describe a membrane separation process using a nonporous, semipermeable membrane. In this process, a gaseous feed stream is fractionated into permeate and nonpermeate streams. The nonpermeating stream is typically referred to as the *nonpermeate* in gas separation terminology, although it is defined as the retentate in liquid separation. Transport separation occurs by a solution diffusion mechanism. Membrane selectivity is based on the relative permeation rates of the components through the membrane. Each gaseous component transported through the membrane has a characteristic individual permeation rate that depends on its ability to dissolve and diffuse through the membrane material. The mechanism for transport is based on solubilization and diffusion; the two describing relationships on which the transport are based are Fick's law (diffusion) and Henry's law (solubility),[20] as defined in Chap. 11.

As noted above, gas separation membranes separate gases from other gases. Some gas filters, which remove liquids or solids from gases, are microfiltration membranes. Gas membranes generally work because individual gases differ in their solubility and diffusivity through nonporous polymers.

There must be a driving force for the rate process of permeation to occur. For gas separations, that force is *partial pressure*. Since the ratio of the component fluxes determines the separation, the partial pressure of each component at each point is important. There are two ways of driving the process: employing a high partial pressure on the feed or provide a low partial pressure on the permeate side which may be achieved by either vacuum or inert-gas flushing.

Diffusive flux through the membrane can be expressed by Fick's law, as related to the membrane system, and given by[20]

$$J_i = \frac{D_i}{t_m}(C_{im2} - C_{im1}) \tag{12–24}$$

where J_i = flux of component i, mol/(m$^2 \cdot$ s)
D_i = diffusivity of component i, m^2/s
t_m = membrane thickness, m
C_{im2} = concentration of component i inside membrane wall on feed side, mol/m^3
C_{im1} = concentration of component i inside membrane wall on permeate side, mol/m^3

A modified form of Henry's law (see also Chap. 11) may be written in the following form:[20]

$$C_{im} = S_i p_i \tag{12–25}$$

where S_i = solubility constant for component i in membrane
p_i = partial pressure of component i in gas phase

Substituting Eq. (12–25) into Eq. (12–24) yields[20]

$$J_i = \frac{D_i}{t_m}(S_i p_{i2} - S_i p_{i1}) \tag{12–26}$$

The terms p_{i2} and p_{i1} are the *partial pressure* of gas i on the feed and permeate side of the membrane, respectively. Permeation through the membrane is given by:

$$P_i = D_i S_i \tag{12–27}$$

Substitution of Eq. (12–27) into Eq. (12–26) provides the relationship for local flux through the membrane:[20]

$$J_i = \frac{P_i}{t_m}(p_{i2} - p_{i1}) \tag{12–28}$$

The *separation efficiency* a_{ij} is based on the different rates of permeation of the gas components:

$$a_{ij} = \frac{P_i}{P_j} \tag{12–29}$$

These data are available for some commonly separated gases and the polymer(s) used.[16,24]

An experimental *separation factor* a_{ij} is frequently used to quantify the separation of a binary system of components i (oxygen, O_2) and j (nitrogen, N_2), where C_p and C_r represent molar concentrations in the permeate and retentate (nonpermeate) streams, respectively.[24] The separation factor can also be defined in terms of the concentrations in the permeate and feed streams.[20,23] These relationships can be written in terms of mole fraction terms y_p, y_r, and y_f, which is often more convenient since (the oxygen) analyzers measure concentrations in mol%:[19]

$$a'_{ij} = \frac{C_{ip}/C_{jp}}{C_{ir}/C_{jr}} = \frac{y_{ip}/y_{jp}}{y_{ir}/y_{jr}} \tag{12–30}$$

$$a''_{ij} = \frac{C_{ip}/C_{jp}}{C_{if}/C_{jf}} = \frac{y_{ip}/y_{jp}}{y_{if}/y_{jf}} \tag{12–31}$$

Recovery is defined by the equations below, where q_p, q_r, and q_f represent the volumetric flow rates (m^3/s) of the permeate, retentate (or nonpermeate), and feed streams, respectively. Volumetric flow rates of the permeate and nonpermeate are measured as the difference between final and initial cumulative gas, volumes for the permeate and nonpermeate ΔV (m^3) measured during time period Δt; specifically, $q = \Delta V/\Delta t$.[20] For air,

$$\text{Recovery of } O_2 = \frac{q_p C_{O_2,p}}{q_f C_{O_2,f}} \tag{12–32}$$

$$\text{Recovery of } N_2 = \frac{q_r C_{N_2,r}}{q_f C_{N_2,f}} \tag{12–33}$$

The term *stage cut* is used to define the ratio of permeate flow rate to total flow rate as shown in Eq. (12–34). The concentrations and volumetric flow rates are usually measured at atmospheric pressure for both the permeate and the nonpermeate streams. If this were not the case, stage cut would be defined as the ratio of molar flows rather than volumetric flows:[20]

$$\text{Stage cut} = \frac{q_p}{q_p + q_r} \tag{12–34}$$

The *total flux* of a component J_i may be calculated from Eq. (12–35):

$$J_i = \frac{q_{ip}\rho}{nA} \tag{12–35}$$

where q_{ip} = volumetric flow rate of species i in the permeate, m^3/s
ρ = density of permeate gmol/m^3
A = area of membrane, m^2 per module
n = number of modules used

If the P_i and t values cannot be determined independently from an experiment or the literature, an *intrinsic permeability* P_i^* may be used where P_i^* has units of lb/ft^2·psi·h, and given by

$$P_i^* = \frac{P_i}{t_m} = \frac{J_i}{p_{i2} - p_{i1}} \tag{12–36}$$

Note that permeate pressure is assumed to be atmospheric (0 psi gauge or 14.7 psi absolute) in these equations. The operating pressure should be expressed as a pressure difference (usually psi), although some researchers prefer employing absolute pressure on the feed side of the membrane.[24]

12-7 Pervaporation

Pervaporation (PER) is a separation process in which a liquid mixture contacts a nonporous permselective membrane. One component is transported through the membrane preferentially. It evaporates on the downstream side of the membrane, leaving as a vapor. The process is induced by lowering the partial pressure of the permeating component, usually by a vacuum or occasionally with an inert gas. The transferred component is then condensed or recovered.

12-8 Electrodialysis

Electrodialysis (ED) is a membrane separation process in which ionic species can be separated from water macrosolutes and all charged solutes. Ions are induced to move by an electrical potential, and separation is facilitated by ion-exchange membranes. The membranes are highly selective, passing primarily either anions or cations. In the ED process, the feed solution containing ions enters a compartment whose walls contain either a cation exchange or an anion exchange membrane.[24]

References

1. L. Theodore, *Air Polution Control Equipment Calculations*, Wiley, Hoboken, N.J., 2008.
2. J. Reynolds, J. Jeris, and L. Theodore, *Handbook of Chemical and Environmental Engineering Calculations*, Wiley, Hoboken, N.J., 2004.
3. L. Theodore, personal notes, East Williston, N.Y., 1975.
4. S. Slater, "Membrane Technology," NSF Workshop Notes, Manhattan College, Bronx, N.Y., 1991.
5. L. Theodore, *Nanotechnology: Basic Calculations for Engineers and Scientists*, Wiley, Hoboken, N.J., 2006.
6. P. Schweitzer, *Handbook of Separation Techniques for Chemical Engineers*, McGraw-Hill, N.Y., 1979.
7. L. Applegate, Membrane Separation Processes," *Chem. Eng.*, 64–89, N.Y., June 11, 1984.
8. G. Parkinson, "Reverse Osmosis: Trying for Wider Applications," *Chem. Eng.*, 26–31, N.Y., May 30, 1983.
9. K. Brooks, "Membranes Push into Separations," *Chem. Week*, pp. 21–24, Washington DC, Jan. 16, 1985.
10. F. V. Kosikowski, "Membrane Separations in Food Processing," in W. C. McGregor, ed., *Membrane Separations in Biotechnology*, Marcel Dekker, N.Y., 1986, Chap. 9.
11. A. Garcia, B. Medina, N. Verhoek, and P. Moore, "Ice Cream and Components Prepared with Ultrafiltration and Reverse Osmosis Membranes," *Biotechnol. Prog.*, **5**, 46–50, 1989.
12. J. Maubois, "Recent Developments of Membrane Ultrafiltration in the Dairy Industry," in A. R. Cooper, ed., *Ultrafiltration Membranes and Applications*, Plenum Press, N.Y., 1980, pp. 305–318.
13. L. Theodore and F. Ricci, *Mass Transfer Operations for the Practicing Engineer*, Wiley, Hoboken, N.J., 2010.
14. M. C. Porter, *Handbook of Industrial Membrane Technology*, Noyes Publications, Park Ridge, N.J., 1990, Chap. 2.
15. M. Mulder, *Basic Principles of Membrane Technology*, 2nd ed., Kluwer Academic, Boston, 1996, Chaps. VI–VII.
16. H. Hollein, C. Slater, R. D'Aquino, and A. Witt, "Bioseparation Via Cross Flow Membrane Filtration," *Chem. Eng. Educ.*, **29**, 86–93, 1995.
17. C. S. Slater and H. Hollein, "Educational Initiatives in Teaching Membrane Technology," *Desalination, Chem. Eng. Educ.,* **90**, 625–634, 1993.
18. C. S. Slater, H. Hollein, P. P. Antonechia, L. S. Mazzella, and J. Paccione, "Laboratory Experiences in Membrane Separation Processes," *Int. J. Eng. Educ.*, **5**, 369–378, 1989.
19. C. S. Slater, C. Vega, and M. Boegel, "Experiments in Gas Permeation Membrane Processes," *Int. J. Eng. Educ.*, **8**, 1–7, 1992.
20. C. S. Slater, M. Boegel, and C. Vega, "Membrane Gas Separation Experiments for a Chemical Engineering Laboratory," *American Society for Engineering Education (ASEE) Annual Conference Proceedings*, pp. 648–650, Washington, D.C., 1990.
21. *Prism Separators*, Bulletin PERM-6-008, Permea Inc., St. Louis, MO, 1986.
22. R. Davis and O. Sandall, "A Membrane Gas Separation Experiment for the Undergraduate Laboratory," *Chem. Eng. Educ.*, 10–21, 1990.
23. L. Clements, M. Otten, and P. Bhat, "Laboratory Membrane Gas Separator—a New Teaching Tool," Paper 53b presented at the AIChE Annual Meeting, Miami Beach, FL, 1986.
24. P. C. Wankat, *Rate-Controlled Separations*, Chapman & Hall, Boston, 1990, Chap. 13.

13 Chemical Reactors

Chapter Outline

13–1 Introduction

Almost every chemical process is designed to produce an economically desirable product from a variety of starting materials through a succession of treatment steps. The raw materials may first undergo a number of physical treatment steps to ensure their viability in chemical reactions. They then pass through a reactor. The discharge (i.e., the reaction products) then usually undergoes additional physical treatment processes, such as separations or purifications, to obtain the final desired product. The physical treatment processes were discussed in earlier chapters. This chapter is concerned with the chemical treatment processes that involve chemical reactors.

The design of a chemical reactor requires information, knowledge, and experience from various areas, including thermodynamics, stoichiometry, fluid flow, heat transfer, a host of mass transfer processes, and process control (see also Chap. 14). The successful design or prediction of the performance of a chemical reactor requires an understanding of chemical kinetics, a topic that is addressed in the next section.

A major objective of this chapter is to prepare the reader to solve real-world engineering and design problems involving chemical reactors. There are several classes of reactors. The three most commonly encountered in practice—and hence the most heavily discussed in this chapter—are batch (B), continuous stirred-tank (CST), and tubular flow (TF) reactors. Another reactor reviewed is the semibatch unit.

This chapter addresses seven topics that are either directly or indirectly concerned with chemical reactors: chemical kinetics, batch reactors, continuous stirred-tank reactors (CSTRs), tubular flow reactors (TFs), catalytic reactors, thermal effects, and interpretation of kinetic data.

NOTE The bulk of the material presented in this chapter is drawn from Theodore.[1] Also, in an attempt to ensure consistency with the chemical reactor literature, the author has denoted the volumetric flow rate by Q (not q, as in most of the other chapters).

13–2 Chemical Kinetics

Chemical kinetics involves the study of reaction rates and the variables that affect these rates. It is a topic that is critical for the analysis of reacting systems. The rate of a chemical reaction can be described in any of several different ways. The most commonly used definition involves the time rate of change in the amount of one of the components participating in the reaction; this rate is also based on some arbitrary factor related to the reacting system size or geometry, such as volume, mass, or interfacial area.

A consistent and correct definition of the rate of a reaction is essential before meaningful kinetic and reactor applications can be discussed. As described above, the *rate* of a chemical reaction is defined as the time rate of change in the quantity of a particular species (say, A) participating in a reaction divided by a factor that characterizes the reacting system's geometry. The choice of this factor is also a matter of convenience. For homogenous media, the factor almost always employed is the volume of the reacting system. One may therefore write

$$r_A = \left(\frac{1}{V}\right)\left(\frac{dN_A}{dt}\right) \tag{13–1}$$

where r_A = rate of reaction for species A
N_A = moles of A at time t
t = time
V = volume of reacting system

If the volume term is constant, one may rewrite Eq. (13–1) as

$$r_A = d\left(\frac{N_A}{V}\right)\left(\frac{1}{dt}\right) = \frac{dC_A}{dt} \tag{13–2}$$

where C_A is the molar concentration of A. This equation states that the reaction rate is equal to the rate of change of concentration with respect to time, all in consistent units. Rates expressed using concentration changes require the assumption of constant volume. The units of the rate become (lbmol/(h·ft³) and (gmol/(s·L) in the engineering and metric systems, respectively. For fluid-solid reaction systems (a topic discussed in a later section), the factor is often the mass of the solid, e.g., in gas phase catalytic reactions, the factor is the mass of catalyst (cat). The units of the rate are then lbmol/(h·lbcat. In line with this definition for the rate of reaction, the rate is positive if species A is being formed or produced since C_A increases with time. The rate is negative if A is reacting or disappearing due to the reaction because C_A decreases with time. The rate is zero if the system is at *equilibrium*.

Experimental evidence reveals that the rate of reaction is a function of

1. The concentration of components existing in the reaction mixture (this includes reacting and inert species)
2. Temperature
3. Pressure
4. Catalyst variables

This may be expressed in equation form as

$$r_A = r_A(C, P, T, \text{catalyst variables}) \tag{13–3}$$

or simply

$$r_A = \pm k_A f(C) \tag{13–4}$$

where k_A incorporates all the variables other than concentration. The plus/minus (±) notation is included to account for the reaction or formation of species A. One may think of k_A as a constant of proportionality. It is defined as the *specific reaction rate* or more commonly the *reaction velocity constant*. It is a "constant" which is *independent* of concentration but *dependent* on the other variables. This approach has, in a sense, isolated one of the variables. The reaction velocity constant *k*, like the rate of reaction, *must* refer to one of the species in the reacting system. However *k* almost always is based on the same species as the rate of reaction.

Consider, for example, the reaction

$$a(\text{A}) + b(\text{B}) \rightarrow c(\text{C}) + d(\text{D}) \tag{13–5}$$

where the notation → represents an irreversible reaction; in other words, if stoichiometric amounts of A and B are initially present, the reaction will proceed to the right until all the A and B have reacted (disappeared) and C and D have been formed. If the reaction is *elementary*, the rate of reaction (13–5) is given by

$$r_A = -k_A C_A^a C_B^b \tag{13–6}$$

where the negative sign is introduced to account for the disappearance of A and the product concentrations do not affect the rate. For elementary reactions, the reaction mechanism for r_A is simply obtained by multiplying the molar concentrations of the reactants raised to powers of their respective stoichiometric coefficients (power-law kinetics). For *nonelementary* reactions, the mechanism can take any form.

The order of reaction for Eq. (13–5) with respect to a particular species is given by the exponent of that concentration term appearing in the rate of expression. The above reaction is, therefore, of order a with respect to species A and of order b with respect to species B. The *overall order n*, usually referred to simply as the *order*, is the sum of the individual orders:

$$n = a + b \tag{13–7}$$

All real and naturally occurring reactions are reversible. A *reversible* reaction is one in which products react to form reactants. Unlike *irreversible* reactions, which proceed to the right until completion, reversible reactions achieve an *equilibrium state* after an infinite period of time. Reactants and products are still present in the system. At this (equilibrium) state, the reaction rate is zero. For example, consider the following reversible reaction:

$$a(\text{A}) + b(\text{B}) \rightleftarrows c(\text{C}) + d(\text{D}) \tag{13–8}$$

where the notation $\rightleftarrows$ is a reminder that the reaction is reversible; $\rightarrow$ represents the forward reaction contribution to the total or net rate, while $\leftarrow$ represents the contribution of the reverse reaction. The notation = is employed if the reaction system is at equilibrium. The rate of this reaction is [see Eq. (13–8)] then given by

$$r_A = \underbrace{-k_A C_A^a C_B^b}_{\text{forward reaction}} + \underbrace{k'_A C_C^c C_D^d}_{\text{reverse reaction}} \tag{13–9}$$

and

$$K_A = \frac{k_A}{k'_A} \tag{13–10}$$

This topic is covered later in this section.

With regard to chemical reactions, two important issues are of concern to the chemical engineer: (1) how far the reaction will go and (2) how rapidly the reaction will proceed? *Chemical thermodynamics* (reaction equilibrium principles) resolves the first issue; however, it provides no information regarding the second issue, which is concerned with kinetics. Reaction rates fall within the domain of chemical kinetics.

Conversion Variables

The two most common reaction conversion variables are α and X. The term α is employed to represent the change in the number of moles of a particular species due to chemical reaction. However, the most commonly used conversion variable is X, which is used to represent the change in the number of moles of a particular species (say, A) relative to the number of moles of A initially present or initially introduced (to a flow reactor). Thus

$$X_A = \frac{\text{moles of A reacted}}{\text{initial moles of A}} = \frac{N_A}{N_{A,0}} \tag{13–11}$$

Other conversion (-related) variables include N_A, the number of moles of species A at some later time (or position); C_A, the concentration of A at some later time (or position); and X_A^*, the moles of A reacted per total moles initially present. Also note that all of the conversion variables presented above can also be based on mass, but this is rarely employed in practice.

Volume Correction Factor

Once again, consider the reaction

$$a(\text{A}) + b(\text{B}) \rightarrow c(\text{C}) + d(\text{D}) \tag{13–5}$$

The volume of an *ideal* gas in a reactor at any time can be related to the initial conditions by the following equation

$$V = V_0\left(\frac{P_0}{P}\right)\left(\frac{T}{T_0}\right)\left(\frac{N_T}{N_{T,0}}\right) \tag{13–12}$$

where P = absolute pressure
T = absolute temperature
V = volume
N_T = total number of moles
0 = subscript denoting initial conditions

If δ is defined as the increase in the total number of moles per moles of A reacted; then

$$\delta = \left(\frac{d}{a}\right) + \left(\frac{c}{a}\right) - \left(\frac{b}{a}\right) - 1 \tag{13–13}$$

The term ε denotes the change in the total number of moles when the reaction is complete versus the total number of moles initially present in (or fed to) the reactor. This term may be expressed as

$$\varepsilon = y_{A,0}\,\delta \tag{13–14}$$

where $y_{A,0}$ is the initial mole fraction of A. The gas volume can now be expressed as follows:

$$V = V_0\left(\frac{P_0}{P}\right)\left(\frac{T}{T_0}\right)(1 + \varepsilon X) \tag{13–15}$$

If the pressure does not change significantly, the equation becomes

$$V = V_0\left(\frac{T}{T_0}\right)(1 + \varepsilon X) \tag{13–16}$$

For isothermal operation, one obtains

$$V = V_0(1 + \varepsilon X) \tag{13–17}$$

Note that both the temperature, and pressure, as well as the extent of reaction (X), affect the concentration of a reacting species. For non-ideal-gas conditions, the equation for the volume is written as

$$V = V_0\left(\frac{P_0}{P}\right)\left(\frac{T}{T_0}\right)(1 + \varepsilon X)\left(\frac{z}{z_0}\right) \tag{13–18}$$

where z = compressibility factor (see also Chap. 8)

For ideal gases, $z = 1$. For most chemical engineering applications, one may assume ideal conditions; deviations from ideality only come into play at very high pressures (see also Chap. 8).

13–3 Arrhenius Equation

Equations describing the rate of reaction at the macroscopic level have been developed in terms of meaningful and measurable quantities. The reaction rate is affected not only by the concentration of species in the reacting system but also by the temperature. An increase in temperature will almost always result in an increase in the rate of reaction; in fact, the literature states that as a general rule, a 10°C increase in reaction temperature will double the reaction velocity constant. However, this theory is generally no longer regarded as valid, particularly at elevated temperatures.

The Arrehenius equation relates the reaction velocity constant with temperature:

$$k = A \exp\left(\frac{-E_a}{RT}\right) \tag{13–19}$$

where A = frequency factor constant, which is usually assumed to be independent of temperature

R = universal gas constant

E_a = activation energy and is also usually assumed independent of temperature

Accordingly, the Arrhenius equation should yield a straight line of slope $(-E_a/R)$ and intercept A if $\ln(k)$ is plotted against $(1/T)$. Implicit in this statement is the assumption that E_a is constant over the temperature range in question. Although E_a generally varies significantly with temperature, the Arrhenius equation has wide applicability in industry. This method of analysis can be used to test the law, describe the variation of k with T, and/or evaluate E_a. The numerical value of E_a will depend on the choice and units of the reaction velocity constant.

As noted earlier, equations describing the rate of reaction at the macroscopic level have been presented in terms of meaningful and measurable quantities. Reaction rate theory *attempts* to provide some theoretical foundation for these equations. It has, in a few isolated cases, provided meaningful information on the controlling mechanism for the rate of reaction. But keep in mind that because the chemical engineer's concern is not with a detailed description of the reaction process at the molecular level, this approach has rarely been used in industry. A satisfactory rigorous approach to the evaluation of reaction velocity constants from basic principles has yet to be developed. At this time, industry still relies on the procedures presented in Sec. 13–9 to provide information on reactions for which data (in the form of rate equations) are not available.

It should also be noted that a meaningful presentation of the theory of absolute reaction rates requires much background work in quantum and statistical mechanics and is beyond the scope of this text.[2]

13–4 Batch Reactors

Batch reactors are commonly used in experimental studies. Their industrial applications are somewhat limited. They are seldom used for gas phase (e.g., combustion) reactions since small quantities (mass) of product are produced with even a very large reactor. Batch reactors are used for liquid phase reactions when small quantities of reactants are to be processed. They find major application in the pharmaceutical industry. As a rule, batch reactors are less expensive to purchase but more expensive to operate than are either continuous stirred-tank or tubular flow reactors.

The extent of a chemical reaction and/or the amount of product produced can be affected by the relative quantities of reactants introduced to the reactor. For two reactants, each is usually introduced through separate feed lines normally located at or near the top of the reactor. Both are usually fed simultaneously over a short period of time. Mixing is accomplished with the aid of a turning-spinning impeller. (See also Fig. 13–1.) The reaction is assumed to begin after both reactants are in the reactor and no spatial variations in concentration, temperature, or other properties are present.

In most liquid phase batch systems, the reactor volume is considered to be constant. However, in many systems, these parameters may vary, depending on the phase and reaction stoichiometry. Such variance must be included in the analysis of gas phase reaction systems. [See also Eqs. (13–15) to (13–18).]

As noted above, a batch reactor is normally used for small-scale operation, for testing new processes that have not been fully developed, for the manufacture of expensive products, and for processes that are difficult to convert to continuous operation. The batch reactor has the advantage of high conversion rates that can be obtained by leaving the reactant in the reactor for long periods of time, but it also has the disadvantage of high labor costs per unit production, and large-scale production is usually difficult.

A batch reactor is a solid vessel or container. It may be open or closed. As noted, reactants are usually added to the reactor simultaneously. The contents are then mixed (if necessary) to ensure that there are no spatial variations in the concentration of the species present. The reaction then proceeds. There is no transfer of mass into or out of the reactor during this period. The concentration of reactants and products change with time; thus, this is a *transient* or *unsteady-state* operation. The reaction is terminated when the desired chemical change has been achieved. The contents are then discharged and sent elsewhere, usually for further processing.

The describing equation for chemical reaction mass transfer is obtained by applying the conservation law for either mass or moles on a time-rate basis to the contents of a batch reactor. It is best to work with moles rather than mass since the rate of reaction is most conveniently

Figure 13–1

Batch reactor.

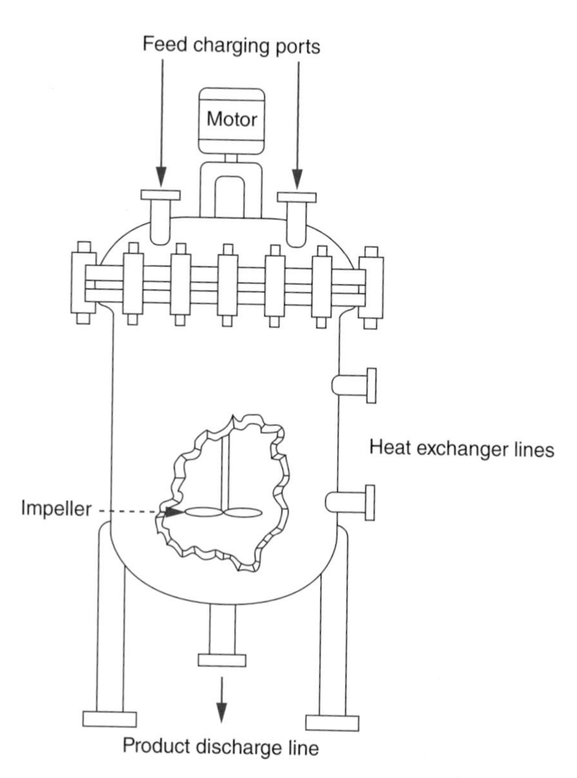

described in terms of molar concentrations. The describing equation for species A in a batch reactor takes the form

$$\frac{dN_A}{dt} = r_A V \qquad (13\text{–}20)$$

where N_A = moles of A at time t
r_A = rate of reaction of A; change in moles of A per unit time per unit volume
V = reactor volume *contents*

Equation (13–20) may also be written in terms of the conversion variable X since

$$N_A = N_{A,0} - N_{A,0}X_A; \qquad (13\text{–}21)$$

where $N_{A,0}$ = initial moles of A
X_A = moles of A reacted/moles of A initially present

Thus (setting $X = X_A$), one obtains

$$N_{A,0}\left(\frac{dX}{dt}\right) = -r_A V \qquad (13\text{–}22)$$

The integral form of this equation is

$$t = N_{A,0}\int_0^X \left(\frac{-1}{r_A V}\right) dX \qquad (13\text{–}23)$$

If V is constant (as is the case with most liquid phase reactions), then

$$t = \frac{N_{A,0}}{V}\int_0^X \left(\frac{-1}{r_A}\right) dX \qquad (13\text{–}24)$$

$$= C_{A,0}\int_0^X \left(\frac{-1}{r_A}\right) dX$$

The reaction time t employed above represents the time required to reduce a reactant concentration (or moles) from some initial value to some final value in a batch reactor. The total cycle time in any batch operation is considerably longer, as one must account for the time necessary to fill (t_f) and empty or dump (t_e) the reactor contents, the time necessary to clean the reactor between batches (t_c), and any possible standby (t_s) and/or heating or cooling ($t_{h,c}$) time. Thus, in some cases, the reaction time t may be only a small fraction of the total cycle time t_t. Typical downtimes (t_d) are listed in Table 13–1.

The production rate (PR) for reactant A is given by

$$\text{PR}_A = \frac{N_{A,0}X_A}{(t + t_d)} \qquad (13\text{–}25)$$

Table 13–1 Reactor Downtime

Operation	Time, min
Fill	5–30
Empty	5–30
Clean	5–120
Standby	5–15
Heat/cool	15–90

Thus PR_A is a function of both X_A and t. Normally, the production rate refers to one of the products formed. This production rate, using the same conversion variable, is then [for product C—see Eq. (13–5)]

$$PR_C = \left(\frac{c}{a}\right)\frac{N_{A,0}X_A}{(t+t_d)} \tag{13–26}$$

where c and a are the stoichiometric coefficients in the reaction equation for species C (product) and A (reactant), respectively.

To maximize production, one needs to express PR in terms of X (replacing t by the design equation that contains X) and solve the equation

$$\frac{d(\mathrm{PR})}{dX} = 0 \tag{13–27}$$

to yield the value of X that gives the maximum production. If PR is expressed in terms of t (with X replaced by the design equation in terms of t), one may solve the equation

$$\frac{d(\mathrm{PR})}{dt} = 0 \tag{13–28}$$

for t. The derivative may be solved analytically or numerically, or one may simply graph PR versus either t or X and locate the maximum on the resulting curve.

13–5 Continuous Stirred-Tank Reactors (CSTRs)

Another reactor where mixing is important is the tank flow or *continuous stirred-tank reactor* (CSTR). This type of reactor, like the batch reactor, also consists of a tank or kettle equipped with an agitator. It may be operated under steady or transient conditions. Reactants are fed continuously, and the products are withdrawn continuously (see Fig. 13–2). The reactants and products may be liquid, gas, or solid or any combination of these. If the contents are perfectly mixed, the reactor design problem is greatly simplified for steady-state conditions because the mixing results in uniform concentrations, temperature, etc., throughout the reactor. This means that the rate of reaction is constant and the describing equations are not differential and, therefore, do not require integration. In addition, since the reactor contents are perfectly mixed,

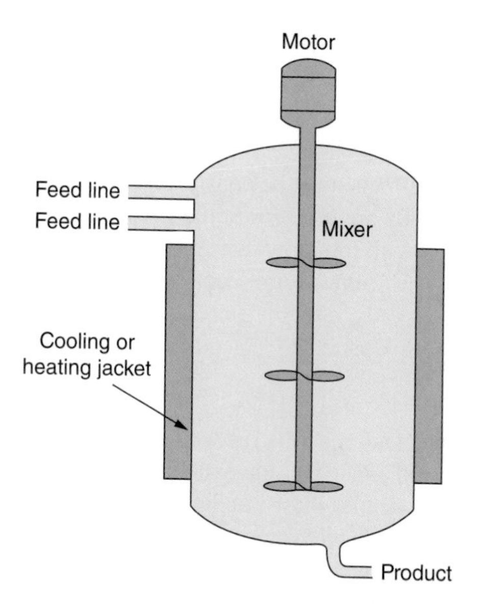

Figure 13–2

Schematic representation of a CSTR.

the concentration or conversion in the CSTR is exactly equal to the concentration or conversion *leaving* the reactor. The describing equation for the CSTR can then be shown to be

$$V = \frac{F_{A,0}X_A}{(-r_A)} \tag{13–29}$$

where V = volume of reacting mixture
$F_{A,0}$ = inlet molar feed rate of A
X_A = conversion of A
$-r_A$ = rate of reaction of A

If the volumetric flow rate entering and leaving the CSTR are constant (this is equivalent to a constant density system), Eq. (13–29) becomes

$$\frac{V}{Q} = \frac{C_{A,1} - C_{A,0}}{(r_A)} = \frac{C_{A,0} - C_{A,1}}{(-r_A)} \tag{13–30}$$

where Q = total volumetric flow rate through CSTR
$C_{A,0}$ = inlet molar concentration of A
$C_{A,1}$ = exit molar concentration of A

An important equation that is employed in many flow reactors, including both CSTRs and tubular flow reactors (see next subsection and Sec. 13–6) is the equation

$$F_i = C_i Q \tag{13–31}$$

where F_i = molar flow of i
C_i = molar concentration of i
Q = total volumetric flow rate

This equation applies to any species and for both liquid and gas phase reactions. However, for gas phase reactions, one must use the general form of the equation presented earlier since the density is not constant:

$$V = \frac{F_{A,0}X_A}{(-r_A)} \tag{13–29}$$

However, volume effects must be considered in most gas phase reactions.

In general, CSTRs are used for liquid phase reactions. High reactant concentrations can be maintained with low flow rates so that conversions approaching 100% can often be achieved. However, the overall economics of the system is reduced because of the low throughput.

CSTRs in Series

Continuous stirred-tank reactors (as well as tubular flow reactors) are often connected in series in such a manner that the exit stream of one reactor serves as the feed stream for another reactor. Under these conditions, it is convenient to define the conversion at any point downstream in the battery of CSTR reactors in terms of *inlet* conditions, rather than with respect to any one of the reactors in the series. The conversion X is then the total moles of A that have reacted up to that point per mole of A fed to the *first* reactor. However, this definition should be employed only if there are no sidestream withdrawals and the feed stream only enters the first reactor in the series. The conversion from reactors 1, 2, 3, and so on in the series are usually defined as X_1, X_2, X_3, and so forth, respectively, and effectively represent the overall conversion for that reactor relative to the feed stream to the *first* reactor.

CSTRs in series are usually designed so that the volumes of the individual reactors are equal. For almost all reactions, the total volume requirement for achieving a given conversion decreases as the number of reactors in series increases. This can significantly impact on the economics, particularly the capital cost. However, the total volume requirement to achieve a particular conversion can be further reduced, particularly for nonelementary reactions, if the constraint of equal reactor volumes is removed; in other words, the volumes of each reactor need not be the same. Although these systems can be designed to lower the volume requirements, the impact on the overall economics can be negative.[2]

The procedure described above can obviously be extended to more than two reactors. The solutions to the describing equations for constant density first-order reactions in a series of CSTRs are quite simple. Derived solutions are also available in the literature for various rate expressions, including reversible, consecutive, and simultaneous reactions.[2, 4] Graphical procedures, which have the advantage of pictorially representing the concentration (or conversion) in each reactor, are also available.[2, 4]

As indicated earlier in Eq. 13–30, the design equation for liquid phase reactions can be written as

$$\theta = \frac{V}{Q} = \frac{C_{A,1} - C_{A,0}}{(r_A)} \tag{13–32}$$

The term on the left-hand side (LHS) of this equation has the units of time and represents the average holdup (residence) time in the reactor. It is usually denoted by the symbol θ. The reciprocal of θ is defined as the space velocity (SV) and finds wide application with tubular flow reactors. However, there is a distribution around this average, and it is often important that this distribution effect be included in the analysis of certain types of systems. For example, a high or variable residence time for certain organic reactions in the pharmaceutical industry can lead to undesirable side reactions. For polymeric reactions, this distribution effect can lead to a product of variable chain length.

13–6 Tubular Flow Reactors

The last traditional reactor to be examined is the tubular flow reactor. The most common type is the single-pass cylindrical tube. Another type is one consisting of a number of tubes in parallel. The reactor(s) may be vertical or horizontal. The feed is charged continuously at the inlet of the tube, and the products are continuously removed at the outlet. If heat exchange with surroundings is required, the reactor setup includes a jacketed tube (see Fig. 13–3). If the reactor is empty, a *homogenous* reaction (where only one phase is

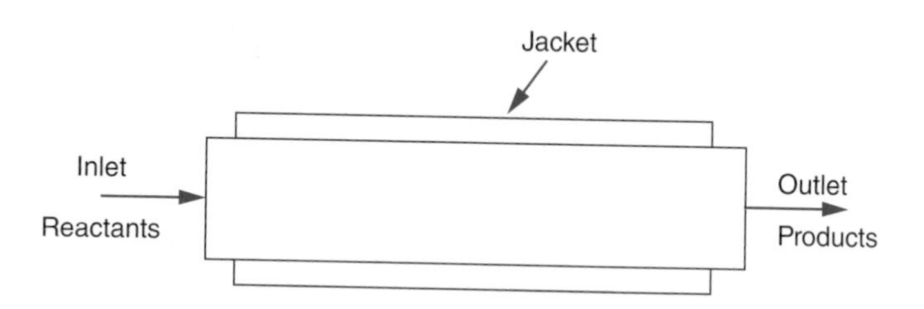

Figure 13–3

Tubular flow reactor; single pass.

present) usually occurs. If the reactor contains catalyst particles, the reaction is said to be *heterogeneous*. This type will be considered in the next section.

Tubular flow reactors are usually operated under steady-state conditions, so that physical and chemical properties do not vary with time. Unlike, the batch and tank flow reactors, there is no mechanical mixing. Thus, the state of the reacting fluid will vary spatially from point to point in the system, and this variation may be in both the radial and axial directions. The describing equations are then differential, with position as the independent variable.

For the describing equations presented below, the reacting system is assumed to move through the reactor in plug flow (no velocity variation through the cross section of the reactor). It is further assumed that there is no mixing in the radial direction so that the concentration, temperature, and other parameters do not vary through the cross section of the tube. Thus, the reacting fluid flows through the reactor in an undisturbed plug of mass. For these conditions, the describing equation for a tubular flow reactor is

$$V = F_{A,0}\int\left(\frac{-1}{r_A}\right)dX_A \tag{13–33}$$

Since $F_{A,0} = C_{A,0}\,Q_0$, it follows that

$$\frac{V}{Q_0} = C_{A,0}\int\left(\frac{-1}{r_A}\right)dX_A \tag{13–34}$$

The RHS of Eq. (13–34) represents the residence time in the reactor based on inlet conditions. If Q does not vary through the reactor, then

$$\theta = \frac{V}{Q_0} \tag{13–35}$$

where θ is once again the residence time in the reactor.

One can show that for a zeroth (0)-order reaction

$$X = \frac{kt}{C_{A,0}} \tag{13–36}$$

and for a half (½)-order reaction

$$X = 1 - \left[\pm\left\{\frac{kt}{2(C_{A,0})^{1/2}}\right\} + 1\right]^2 \tag{13–37}$$

In actual practice, tubular flow reactors deviate from a plug flow model because of velocity variations in the radial direction. For this condition, the residence time for annular elements of fluid within the reactor will vary from some minimal value at a point where the velocity approaches zero. Thus, the concentration and temperature profiles, as well as the velocity profile, are not constant across the reactor, and the describing equations based on the plug flow assumption are not applicable. This situation can be further complicated if the reaction occurs in the gas phase. Volume effect changes that impact the concentration term(s) in the rate equation also need to be taken into account.

From a qualitative perspective, as the length of the reactor approaches infinity, the concentration of a (single) reactant approaches zero for irreversible reactions (except zeroth order) and the equilibrium concentrations for reversible reactions. Since infinite time is required to achieve equilibrium conversion, this value is approached as the reactor length approaches infinity. For reactors of finite length, where the reaction is reversible, some fraction of the equilibrium conversion is achieved.

Note that the time for a hypothetical plug of material to flow-through a tubular flow reactor is the same as the contact or reaction time in a batch reactor. Under these conditions, the form of the describing equations for batch reactors will also apply to tubular flow reactors.

Another design variable is *pressure drop*. This effect is usually small for most liquid and gas phase reactions conducted in short or small reactors. This effect may be estimated using Fanning's equation (see also Chap. 9)

$$\Delta P = \frac{4fLV^2}{2g_c D} \tag{13–38}$$

where ΔP = pressure drop
f = Fanning friction factor
L = reactor length
V = average flow velocity
g_c = gravity conversion constant
D = reactor diameter

For some gas phase reactions, the pressure drop can be significant because of changes in temperature and the number of moles. In the literature for additional details on this subject the reader is referred to any fluid flow text[5].

Some inorganic reactions follow the reaction rate law for elementary reactions. However, most reactions (particularly organic ones) do not follow elementary reaction-law kinetics. The reaction rate equation(s) is(are) rare and usually complex, and this can lead to analytical or numerical problems in integrating the tubular flow design equation. Equations in this form can sometimes be solved analytically. If this is not possible, a numerical solution can often be accomplished. One very common numerical technique that is employed in practice to solve nonlinear differential equations is the Runge–Kutta method (see Chap. 1 for additional details).

Combinations of Reactors

There are certain inherent advantages and disadvantages to using each of the three reactors discussed above. These reactors are sometimes used in combination for any one of a variety of reasons. The sequence in which the reactors are placed can impact on the design or the final conversion. An example is provided in Fig. 13–4.

The reader should note that for any order of reaction other than first order, the sequence in which the reactors are placed may affect performance. The reader is referred to the literature for details.[2]

The two reactors shown in Fig. 13–4 have both advantages and disadvantages. For example, the CSTR has certain advantages because of the near-uniform temperature,

Figure 13–4

Combination of CSTR and tubular flow (TF) reactors.

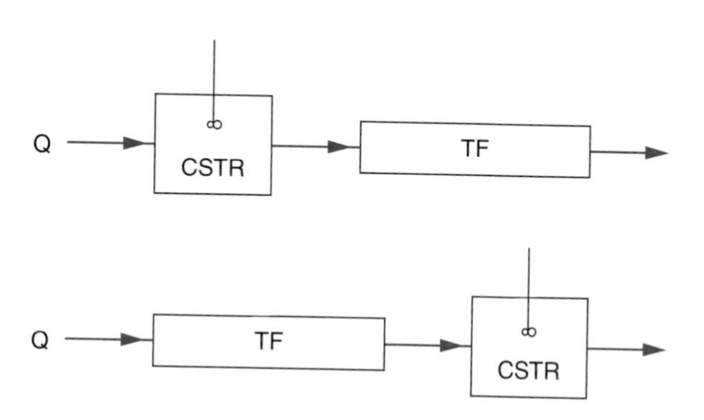

concentration, etc., that results because of mixing. However, the reaction occurs at a rate determined by the concentration of the discharge (or exit) stream from the reactor. Since the rate usually decreases with the extent of reaction (conversion), the CSTR operates in the range between the high reaction rate corresponding to the concentration in the feed stream and the (normally) lower rate corresponding to the concentration leaving the reactor. On the other hand, the tubular flow reactor takes maximum advantage of the high reaction rate corresponding to the high concentration(s) near the entrance to the reactor. The CSTR therefore requires a larger volume than a tubular reactor to accomplish a given degree of conversion. However, the preceding analysis occasionally does not apply, particularly if the reaction mechanism is complex and/or nonelementary.[2]

13-7 Catalytic Reactors

Metals in the platinum family are recognized for their ability to promote reactions at low temperatures. Other catalysts include various oxides of copper, chromium, vanadium, nickel, and cobalt. These catalysts are subject to poisoning, particularly from halogens, sulfur compounds, zinc, arsenic, lead, mercury, and particulates. High temperatures can also reduce catalytic activity. It is therefore important that catalyst surfaces be clean and active to ensure optimum performance. Catalysts can usually be regenerated with superheated steam.

Catalyst may be porous pellets, usually cylindrical or spherical in shape, ranging from ⅟16 to ½ inch in diameter. Small sizes are recommended, but the pressure drop through the reactor increases. Other shapes include honeycombs, ribbons, and wire mesh. Since catalysis is a surface phenomenon, an important physical property of these materials is that the internal pore surface be many magnitudes greater than the outside surface. The reader is referred to the literature for more information on catalyst preparation, properties, comparisons, costs, and impurities.[2,4]

Some of the advantages of catalytic reactors are

1. Low fuel requirements
2. Lower operating temperatures
3. Little or no insulation requirements
4. Reduced fire hazards
5. Reduced flashback problems

The disadvantages include

1. High initial cost
2. Catalyst poisoning
3. (Large) particles must first be removed
4. Some liquid droplets must first be removed
5. Catalyst regeneration problems

Catalytic or heterogeneous reactors are an alternative to homogenous reactors. If a solid catalyst is added to the reactor, the reaction is said to be *heterogeneous*. For simple reactions, the effect of the presence of a catalyst is to

1. Increase the rate of reaction
2. Permit the reaction to occur at a lower temperature
3. Permit the reaction (usually) to occur at a more favorable pressure
4. Reduce the reactor volume
5. Increase the yield of reactants

The basic problem in the design of a heterogeneous reactor is to determine the quantity of catalyst and/or reactor size required for a given conversion and flow rate. In order to obtain this, information on the rate equations and their parameters must be made available. A rigorous approach to the evaluation of reaction velocity and other constants for many industrial applications has yet to be accomplished for catalytic reactions. At this time, industry still relies on the procedures described earlier and, in particular, in Sec. 13–9.

In many catalytic reactions, the rate equation is extremely complex and cannot be obtained either analytically or numerically. A number of equations may result, and some simplification is warranted. As mentioned earlier, in many cases it is safe to assume that the rate expression may be satisfactorily expressed by the rate equation of a single step.

It is common practice to write the describing equations for mass and energy transfer for homogeneous and heterogeneous flow reactors in the same way. However, the (units of the) rate of reaction may be expressed as either

$$\frac{\text{moles reacted}}{\text{time}} \times \text{volume of reactor} \tag{13–39}$$

or

$$\frac{\text{moles reacted}}{\text{time}} \times \text{mass of catalyst} \tag{13–40}$$

Equation (13–40) is normally the preferred method employed in industry since it is the mass of catalyst present in the reactor that significantly impacts the reactor design. The rate expression is often more complex for a catalytic reaction than for a noncatalytic (homogeneous) one, and this can make the design equation of the reactor difficult to solve analytically. Numerical solution of the reactor design equation is usually required when designing tubular flow reactors for catalytic reactions.

As indicated above, the principal difference between reactor design calculations involving homogeneous reactions and those involving catalytic (fluid-solid) heterogeneous reactions is that for the latter the reaction rate is based on the mass of solid W, rather than on reactor volume V. The rate of reaction of a substance A for a fluid-solid heterogeneous system is then defined as

$$-r_A' = \frac{\text{moles of A reacted}}{\text{mass catalyst} \times \text{time}} \tag{13–41}$$

Fluidized Bed Reactors

One type of catalytic reactor in common use is the *fluidized-bed reactor,* which is analogous to the CSTR in that its contents, although heterogeneous, are well mixed, resulting in a uniform concentration and temperature distribution throughout the bed. The fluidized-bed reactor can therefore be modeled, as a first approximation, as a CSTR. For the ideal CSTR, the reactor design equation based on volume was previously shown to take the form:

$$\mathrm{V} = \frac{F_{A,0} X_A}{-r_A} \tag{13–29}$$

The companion equation for catalytic or fluid-solid reactor, with the rate based on the mass of solid, is

$$\mathrm{W} = \frac{F_{A,0} X_A}{-r_A'} \tag{13–42}$$

The reactor volume is simply the catalyst weight W divided by the fluidized-bed density ρ_{fb} of the catalyst in the reactor:

$$V = \frac{W}{\rho_{fb}} \qquad (13\text{–}43)$$

The fluidized-bed catalyst density is normally expressed as some fraction of the catalyst bulk density ρ_B.

Fixed-Bed Reactors

A *fixed-bed (packed-bed) reactor* is essentially a tubular flow reactor that is packed with solid catalyst particles. This type of heterogeneous reaction system is most frequently used to catalyze gaseous reactions. The design equation for tubular flow reactor was previously shown to be

$$V = F_{A,0}\int_0^X \left(\frac{-1}{r_A}\right) dX; \quad X = X_A \qquad (13\text{–}33)$$

The companion equation based on the mass of catalyst for a fixed-bed reactor is

$$W = F_{A,0}\int_0^X \left(\frac{-1}{r'_A}\right) dX \qquad (13\text{–}44)$$

The volume of the reactor V is then

$$V = \frac{W}{\rho_B} \qquad (13\text{–}45)$$

where ρ_B = bulk density of the catalyst

The Ergun equation is normally employed to estimate the pressure drop for fixed-bed units.

13–8 Thermal Effects

It was shown in Sec. 13–2 that the rate of reaction r_A is a function of temperature and concentration. The application of r_A to the reactor equations is simplest for isothermal conditions since r_A is generally solely a function of concentration. If nonisothermal conditions exist, another equation must be written to describe temperature variations with time or position within the reactor.

Batch Reactors

The equation describing temperature variations in batch reactors is obtained by applying the conservation law for energy on a time-rate basis to the reactor contents. Since batch reactors are stationary (fixed in space), kinetic and potential effects can be neglected. The equation describing the temperature variation in reactors due to energy transfer, subject to the assumptions in its development, is[2]

$$mc_p\left(\frac{dT}{dt}\right) = -UA(T - T_a) + V(-\Delta H_A)\,|-r_A| \qquad (13\text{–}46)$$

where m = mass of the reactor contents
c_p = heat capacity
V = reactor volume
$-\Delta H_A$ = enthalpy of reaction of species A
$|-r_A|$ = absolute value of the rate of reaction of A
T = reactor temperature

and

$$\dot{Q} = UA(T - T_a) \tag{13–47}$$

where $\dot{Q}$ = heat transfer rate across reactor walls
U = overall heat transfer coefficient
A = area available for heat transfer
T_a = temperature surrounding reactor walls

For nonisothermal reactors, one of the reactor design equations—the energy transfer equation (see above), and an expression for the rate in terms of concentration and temperature—must be solved simultaneously to give the conversion as a function of time. Note that the equations may be interdependent, e.g., each can contain terms that depend on the other equation(s). These equations, except for simple systems, are usually too complex for analytical treatment.

Regarding the energy equation, note that for an adiabatic system, Eq. (13–46) reduces to

$$mc_p\left(\frac{dT}{dt}\right) = V(-\Delta H_A)|-r_A| \tag{13–48}$$

Since, by definition

$$-r_A = \left(\frac{1}{V}\right)\left(\frac{dN_A}{dt}\right) \qquad [N_A = N_{A,0}(1 - X)] \tag{13–49}$$

one obtains

$$-r_A = \left(\frac{1}{V}\right)\frac{d[N_{A,0}(1 - X)]}{dt}$$

Thus,

$$V|-r_A| = N_{A,0}\left(\frac{dX}{dt}\right) \tag{13–50}$$

and Eq. (13–48) becomes

$$mc_p\,dT = (-\Delta H_A)N_{A,0}\,dX \tag{13–51}$$

This equation may be integrated directly to give the temperature as a function of conversion:

$$mc_p(T - T_0) = (-\Delta H_A)N_{A,0}(X - X_0) \tag{13–52}$$

where the zero (0) subscript refers to initial conditions. This equation directly correlates conversion with temperature. The need for the simultaneous solution of the mass and energy equations is, therefore, removed for this case. Note that all the terms in this equation must be dimensionally consistent. The reader should also note that both the heat capacity and the enthalpy of the reaction have been assumed to be constant.

Fogler[4] has derived a somewhat similar equation relating the conversion to temperature

$$T = T_0 - \frac{[\Delta H^o(T_0)]X}{[\theta_i C_{pi} + X\,\Delta C_p]} \tag{13–53}$$

where $\Delta H^\circ(T_0)$ = standard enthalpy of reaction at inlet reference temperature T_0
ΔC_p = heat capacity difference between products and reactants
$\theta_i = y_{io}/y_{AO}$

and $\theta_i C_{pi}$ applies only to the initial reaction feed mixture (including inerts).

For the reaction $a(\mathrm{A}) + b(\mathrm{B}) \rightarrow c(\mathrm{C}) + d(\mathrm{D})$

$$\Delta C_p = cC_{pC} + dC_{pD} - aC_{pA} - bC_{pB} \tag{13–54}$$

Note that this term accounts for enthalpy of reaction variation with temperature.[8]

CSTRs

If the conservation law for energy is applied to a CSTR, information on temperature changes and variations within the reactor can be obtained. For constant heat capacity C_p and enthalpy of reaction ΔH_A, the describing equation becomes

$$\dot{Q} = FC_p(T_0 - T_1) + (-r_A)V(-\Delta H_A) \tag{13–55}$$

where F = flow rate through the reactor, units consistent with C_p
T_0 = inlet temperature
T_1 = outlet temperature

For adiabatic conditions, and noting that

$$(-r_A)V = F_{A,0}X_A \tag{13–56}$$

where $F_{A,0}$ is the feed rate of A. Equation (13–55) becomes (with $T_1 = T$)

$$FC_p(T_0 - T_1) = F_{A,0}X_A(-\Delta H_A) = 0 \tag{13–57}$$

and

$$T_i = T_0 + \left[\frac{F_{A,0}(-\Delta H_A)}{FC_p}\right]X_A \tag{13–58}$$

Thus, the need for the simultaneous solution of the mass and energy transfer equations is again removed.

Fogler[4] has accounted for enthalpy of reaction variation with temperature through the equation

$$\dot{Q} - [\Delta H^\circ(T_R) + \Delta C_p(T - T_R)]F_{A,0}X_A = F_{A,0}\theta_i C_{pi}(T - T_{i,0}) \tag{13–59}$$

where $\Delta H^o(T_R)$ = enthalpy of reaction at temperature T_R
ΔC_p = heat capacity difference between products and reactants

and $\theta_i C_{pi}$ applies only to the species in the feed stream.

Note again that for the reaction $a(\mathrm{A}) + b(\mathrm{B}) \rightarrow c(\mathrm{C}) + d(\mathrm{D})$, one obtains

$$\Delta C_p = cC_{pC} + dC_{pD} - aC_{pA} - bC_{pB} \tag{13–54}$$

Note that for a nonadiabatic reactor, the term $\dot{Q}$ must be retained in the energy equation.

Tubular Flow Reactors (TFRs)

The temperature in a tubular flow reactor (TFR) can vary with position (volume) as a result of enthalpy of reaction effects or transfer of energy in the form of heat across the walls of the reactor. The reactor design equation must then include temperature variations before being solved. To obtain information on the temperature at every point in the reactor, the conservation law for energy is applied to the system. For adiabatic operation, heat transfer with the surroundings is zero. The energy equation reduces to[2]

$$FC_p\,dT = |-r_A|(-\Delta H_A)\,dV \tag{13–60}$$

Since

$$|-r_A|\,dV = F_{A,0}\,dX_A \tag{13–61}$$

for a tubular flow reaction, Eq. (13–60) becomes

$$FC_p\,dT = F_{A,0}(-\Delta H_A)\,dX_A \tag{13–62}$$

which may be integrated to give the previously developed equation for CSTRs:

$$T = T_0 + \left[\frac{F_{A,0}(-\Delta H_A)}{FC_p}\right]X_A \tag{13–58}$$

The term in brackets is a constant if the enthalpy of reaction and the average heat capacity are assumed independent of temperature. The temperature is then a linear function of conversion.[2] Fogler,[4] also provides an equation describing temperature variation within a tubular flow reactor. This equation also takes the same form as that provided for CSTRs.

The energy equation developed above for tubular flow reactors is based on the following assumptions:

1. Heat capacity is not a function of temperature.
2. Conditions are steady state.
3. Plug flow exists.
4. Radial (perpendicular to the axial direction of flow) temperature effects are negligible so that the reacting mass and products at any r (radial) position for a given z (axial) position are at an average temperature.

In adiabatic operation, the enthalpy (heat) effect accompanying the reaction will be completely absorbed by the system and result in temperature changes in the reactor. In an exothermic reaction, the temperature increases; this, in turn, increases the rate of reaction, which, in turn, increases the conversion for a given interval of time. The conversion, therefore, is higher than that obtained under isothermal conditions. When the reaction is endothermic, the decrease in temperature of the system results in a lower conversion than that associated with the isothermal case. If the endothermic enthalpy of reaction is large, the reaction may essentially stop as a result of the sharp decrease in temperature. Finally, temperature effects in a reactor become more complicated if the reaction is reversible. However, the Arrhenius equation can provide information on both the forward and reverse velocity constants.[2, 6, 7]

13–9 Interpretation of Kinetic Data

The application of chemical kinetic principles to the design of reactors for the chemical process industry is hindered in many cases by the lack of information on the rate of reaction. Therefore, one of the more important aspects of chemical kinetics is the prediction of the form of the reaction rate equation from experimental data. This is a very difficult task and is, for all intents and purposes, still state-of-the-art. Yet, it is worthwhile to present and discuss some of the elementary methods employed in determining reaction mechanisms. Information on the reaction mechanism can provide a complete description of the behavior of reacting systems.

Kinetic data can be obtained experimentally in any one of several different ways. The choice is usually dictated by

1. Convenience
2. Reaction systems (species)
3. Availability of equipment
4. Simplicity in experimental arrangement
5. Cost of experimentation
6. Time of experimentation

and not necessarily in that order. Some of the more common experimental measurement methods of analysis include

1. Refractive index
2. Colorimetric indicators (dyes, tracers, etc.)
3. Thermal conductivity (gas chromatograph)
4. Dielectric constant
5. Electric conductance
6. Total pressure

All of these and other methods, excluding method 6, require calibration charts or the equivalent, e.g., a plot of refractive index versus concentration is needed with method 1.

Batch Reactors

There are two experimental methods of operation, and both are applicable to CSTRs and tubular flow reactors as well as batch reactors. In one, the data are obtained over a very *short* interval of reaction time—so small, in fact, that it is defined as a *differential* reaction system. Changes in concentration are usually so small that they may be assumed constant during the short period of reaction time. This permits a direct calculation of the rate. The rate calculated may be interpreted as an average rate since it is based on average concentration(s) and an average temperature. This represents an acceptable method of obtaining information on the rate equation. However, experimental problems can arise in the measurement of small concentration changes. This technique is also used for systems in which it is difficult to maintain isothermal conditions. In the second method, concentration (or pressure) data are gathered over a *long* interval of time. This mode of operation is defined as an *integral* reaction system. The analytical methods of analyses to follow will focus on integral reaction systems.

Assume in the development below that concentration (or the equivalent) data are available as a function of time. Two analytical techniques can be used to analyze the experimental data: (1) differentiation and (2) integration. In the *differentiation* method, not to be confused with a differential reaction system, solutions have traditionally been obtained using graphical means. Consider the elementary irreversible reaction

$$nA \rightarrow \text{products}; \; n = \text{reaction order} \tag{13–63}$$

The rate equation for $-r_A$ is (for constant volume)

$$\frac{-dC_A}{dt} = k_A C_A^n \tag{13–64}$$

If C_A is given as a function of time, one can plot C_A versus t. The rate of change of C_A with respect to t is the LHS of this equation and may be obtained by drawing a tangent to the curve or analytically calculating the slope at that point in time. The rate dC_A/dt will then be available as a function of time. Taking the logarithm of both sides of the above equation gives

$$\log\left(\frac{-dC_A}{dt}\right) = \log k_A + n \log C_A \tag{13–65}$$

This equation indicates that a plot of log $(-dC_A/dt)$ versus log C_A will yield a straight line (if the assumed mechanism above is correct) of slope n and intercept log k_A. Thus, one sees that this method, with the help of two graphs (or the equivalent), provides information on both the order of the reaction and the numerical value of the reaction velocity constant.

The integration method requires that the rate equation be integrated analytically for an *assumed* reaction mechanism or order. The integrated rate equation that compares most favorably with the experimental data then corresponds to the correct equation. Once again, consider the equation

$$nA \rightarrow \text{products} \tag{13–63}$$

The rate equation is written in the form

$$\frac{dC_A}{C_A^n} = k_A\, dt \tag{13–66}$$

This equation may be integrated to give

$$k_A = \left[\frac{1}{(t)(n-1)}\right]\left[\left(\frac{1}{C_A^{n-1}}\right) - \left(\frac{1}{C_{A,0}^{n-1}}\right)\right] \quad (n \neq 1) \tag{13–67}$$

If $n = 2$, then

$$k_A = \left(\frac{1}{t}\right)\left[\left(\frac{1}{C_A}\right) - \left(\frac{1}{C_{A,0}}\right)\right] \tag{13–68}$$

Values of k_A can be calculated from this equation for various values of C_A and t. The assumed order is correct if relatively *constant* values of k_A are generated from the data. This procedure can be repeated for different orders until the above criterion is met. Thus, it can be seen that the integration method requires a trial-and-error calculation of n.[2]

CSTR Reactors

Tank flow reactors are occasionally used to obtain information on reaction velocity constants and orders of reaction. These reactors are operated under conditions approaching perfect mixing and steady state. Two basic methods of analysis are available. These are again defined as the integration and differentiation methods, although there is no need to integrate or differentiate the describing equations.

The integration method is first presented. Consider once again the reaction

$$nA \rightarrow \text{products} \tag{13–63}$$

where

$$-r_A = k_A C_A^n \tag{13–69}$$

For a CSTR, the describing equation becomes

$$\frac{V}{Q} = \theta = \frac{C_{A,0} - C_{A,1}}{k_A C_{A,1}^n} \tag{13–70}$$

which may also be written as

$$k_A = \frac{C_{A,0} - C_{A,1}}{\theta C_{A,1}^n} \tag{13–71}$$

If θ (or Q and V), $C_{A,0}$ and $C_{A,1}$ experimental data are available for the system, one can assume an order n until the calculated values for k_A become constant and do not exhibit a *trend* in any direction.

In the differentiation method, Eq. (13–71) is written in the following form:

$$\frac{C_{A,0} - C_{A,1}}{\theta} = k_A C_{A,1}^n \tag{13–72}$$

Taking the log of both sides of this equation gives

$$\log\left(\frac{C_{A,0} - C_{A,1}}{\theta}\right) = \log k_A + n \log C_{A,1} \tag{13–73}$$

This is a linear equation of the form

$$Y = A + BX \tag{13–74}$$

Therefore, a plot of $[(C_{A,0} - C_{A,1})/\theta]$ versus $C_{A,1}$ on log-log coordinates will provide direct information on both k_A and n.

Tubular Flow Reactors

Tubular flow reactors are usually employed to obtain information on the reaction mechanism, particularly for gas reactions. For tubular flow reactors, experimental data can be obtained in the some way as for batch reactors, using a differential or integral reactor, and the analysis performed using the differentiation or integration method.

In a differential reactor, inlet and outlet concentrations (or the equivalent) and flow rate data are obtained for a small reactor volume. A number of runs are made varying the inlet concentration. Changes in concentration must be extremely small so that both the concentration and rate may be assumed constant in the reactor. This permits a direct calculation of the rate:

$$-r_A = \frac{F_{A,0} X_A}{V} \tag{13–75}$$

with

$$C_A = \frac{C_{A,0} + C_A}{2} \tag{13–76}$$

If the differentiation method of analysis is employed for an assumed mechanism, e.g.,

$$-r_A = k_A C_A^n \tag{13–77}$$

a plot of $-r_A$ versus C_A on log-log coordinates will then once again yield the reaction velocity constant k_A and order n. In the integration scheme, one assumes n and checks to see whether calculated values for k_A are constant.

If the small differences in concentration between the inlet ($C_{A,0}$) and outlet (C_A) from the reactor cannot be experimentally determined accurately, the above method is not satisfactory and an integral reactor system is employed. Here, concentration and flow rate data are gathered for a larger reactor. The flow and/or inlet concentration may be varied. For the differentiation method of analysis, Eq. (13–33) is written in the form

$$\frac{dX_A}{d(V/F_{A,0})} = -r_A \tag{13–78}$$

and a plot of X_A versus $V/F_{A,0}$ yields information on $-r_A$. A log-log plot of $-r_A$ versus C_A will provide information on k_A and n. In the integration method of analysis, one integrates the design equation for an assumed n and checks to see whether the calculated values for k_A are constant.

Complex Systems

The application of any interpretation of kinetic data method to reaction rate equations containing more than one concentration term can be somewhat complex. Consider the "mixed" elementary reaction

$$iA + jB \rightarrow \text{products} \tag{13–79}$$

where

$$\frac{dC_A}{dt} = -k_A C_A^i C_B^j \tag{13–80}$$

For the differentiation technique, this equation is written

$$\log\left(\frac{-dC_A}{dt}\right) = \log k_A + i \log C_A + j \log C_B \tag{13–81}$$

Three unknowns appear in this equation and can be solved by regressing the data.[8]

Using the integration technique is also complex, and a trial-and-error graphical calculation is required. If one assumes $i = 1$ and $j = 1$, then

$$\frac{dC_A}{dt} = -k_A C_A C_B \tag{13–82}$$

which may be integrated to give

$$k_A = \left[\frac{1}{t(C_{A,0} - C_{B,0})}\right] \ln\left(\frac{C_{B,0} C_A}{C_{A,0} C_B}\right) \tag{13–83}$$

A constant value of k_A for the data would indicate that the assumed order ($i = 1, j = 1$) is correct.

The isolation method has often been applied to the mixed-reaction system described above. The procedure employed here is to set the initial concentration of one of the reactants, say, B so large that the change in concentration of B during the reaction is vanishingly small. The rate of equation may then be approximated by

$$\frac{dC_A}{dt} \approx k_A^* C_A^i \tag{13–84}$$

where

$$k_A^* = k_A C_A^j \approx \text{constant} \tag{13–85}$$

Either the differentiation or integration method may then be used. For example, in the differentiation method, a log-log plot of $(-dC_A/dt)$ versus C_A will once again give a straight line of slope i. A similar procedure is employed to obtain j.

For a reversible reaction

$$iA \rightleftarrows jB \tag{13–86}$$

the rate is given by

$$\frac{dC_A}{dt} = -k_A C_A^i + k_A' C_B^j \tag{13–87}$$

and

$$K = \frac{k_A}{k_A'} \tag{13–88}$$

The equilibrium constant K is usually available or can be easily obtained (see also Chap. 8). If K is large, one may experimentally set the concentration of C_A equal to its maximum value and C_B to zero. The reaction is observed for a period of time during which $C_B \approx 0$. The rate equation may then be approximated by

$$\frac{dC_A}{dt} = -k_A C_A^i \tag{13–89}$$

If a reaction has a complex reaction mechanism, the rate equation can become unwieldy and difficult to handle. These equations may contain fractional or even negative reaction order terms. It is sometimes nearly impossible or inconvenient to treat these systems in the manner outlined above. The application of statistical principles can provide quantitative solutions to the equation(s).[8]

A chemical engineer who finds it necessary to perform statistical analyses of experimental data probably has little to no knowledge of the reaction mechanism and is therefore often more interested in obtaining information on (1) how well the proposed reaction mechanism is in predicting system behavior, rather than (2) how well the data fit the proposed model (reaction mechanism) or in obtaining the best fit of the data to the model by regressing the data. A least-squares analysis, or the equivalent, provides *no* information on issue 1; an analysis of variance (ANOVA) must be performed to address this issue. The reader is referred to the statistics literature for details on ANOVA.[8]

References

1. L. Theodore, *Chemical Reactor Kinetics*, A Theodore Tutorial, Theodore Tutorials, East Williston, N.Y., originally published by USEPA/APTI, Research Triangle Park, N.C., 1995.
2. L. Theodore, *Chemical Reactor Design and Applications for the Practicing Engineer*, Wiley, Hoboken, N.J., 2012.

3. L. Theodore, personal notes, East Williston, N.Y., 1968.
4. S. Fogler, *Elements of Chemical Reaction Engineering*, 4th ed., Prentice-Hall, Upper Saddle River, N.J., 2006.
5. P. Abulencia and L. Theodore, *Fluid Flow for the Practicing Chemical Engineer*, Wiley, Hoboken, N.J., 2009.
6. S. Ergun, "Fluid Flow-through Packed Columns," *Chem. Eng. Prog.*, **48**, 89–94, N.Y., 1952.
7. L. Theodore, "Heat Transfer Applications for the Practicing Engineer," Wiley, Hoboken, N.J., 2011.
8. L. Theodore, F. Ricci, and T. VanVliet, *Thermodynamics for the Practicing Engineer*, Wiley, Hoboken, N.J., 2005.
9. S. Shaefer and L. Theodore, *Probability and Statistics Applications in Environmental Science*, CRC Press/Taylor & Francis Group, Boca Raton, Fla., 2007.

14 Process Control

Chapter Outline

14–1 Introduction

Chemical processes can be controlled to yield not only more products but also more uniform and higher-quality products, usually resulting in a profit increase. In addition, some processes respond so rapidly to changes in the system that they cannot be properly controlled by a plant operator; these systems are amenable to some form of automatic control. However, the decision to apply automatic control(s) should be based on an applicable and appropriate cost analysis that includes process objectives. As discussed in Part IV, Chap. 28, the economics should be based on a cost-benefit analysis.

NOTE Much of the material on process control has been drawn from the work of P. T. Vasudevan.[1] Although various topics (e.g., Routh criteria for stability, root locus analysis, Bode plots) are not treated in this chapter, numerous excellent illustrative examples are available in Vasudevan's tutorial.[1] In addition, Vasudevan's notation has been retained in the presentation.

Topics addressed in this chapter include process control fundamentals, calibration of flow and temperature transducers, sizing concerns, feedback and feedforward control, and cascade control.

14–2 Process Control Fundamentals

Automatic control can perhaps be best described via a continuous stirred-tank reactor[2] (CSTR), as shown in Fig. 14–1. The contents in the reactor are heated to a design temperature by the steam flowing through a heat exchanger, e.g., heating coils.[3] The temperature of the product flow (the variable controlled) and the CSTR mixture are affected by the flow rate and inlet temperature of the reactant(s), the temperature, pressure and flow ratio of the steam, the degree of mixing, and (any) heat losses to the surrounding environment. Certain process control terms may now be introduced and defined.

Figure 14–1 represents an *open-loop* system since the output temperature is not employed to adjust or change any of the reactor variables; i.e., the system cannot compensate or correct any of the reactor valuables. In a *closed-loop* system, the measured value of the temperature (the system variable to be controlled) is used to compensate or change one or more reactor variables, e.g., the system temperature.

In *feedback control* (see Fig. 14–2), the temperature is compared to a particular value—often referred to as a *set-point* or *design value*. The degree of displacement of the temperature from the set point provides the correction to one of the reactor variables in a manner to reduce the displacement (often referred to as the *error*). *Feedforward control* (see Fig. 14–3) allows compensation for any reactor disturbance prior to a change to the controlled variable, i.e., the product temperature. This type of control has an obvious advantage if the controlled variable cannot be measured. Feedforward control and feedback

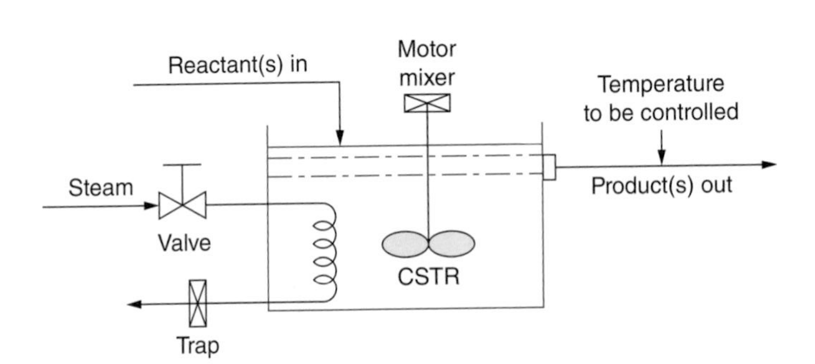

Figure 14–1

Continuous stirred-tank reactor (CSTR).

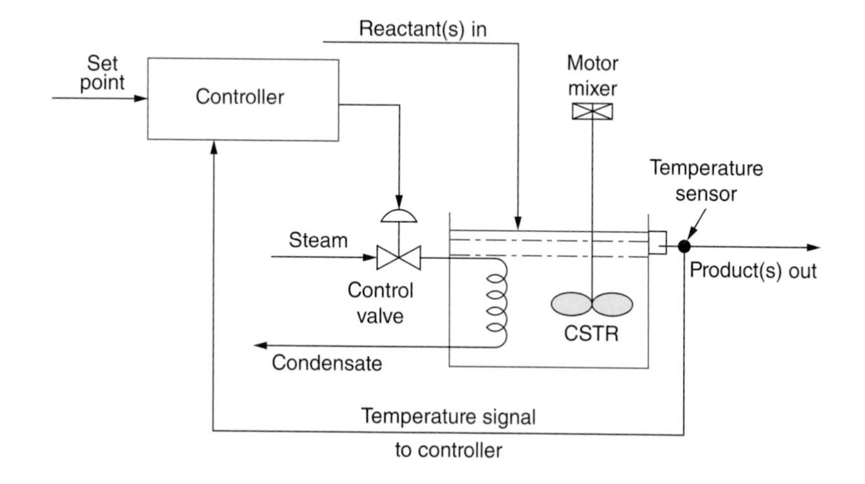

Figure 14–2

Feedback control of a CSTR process.

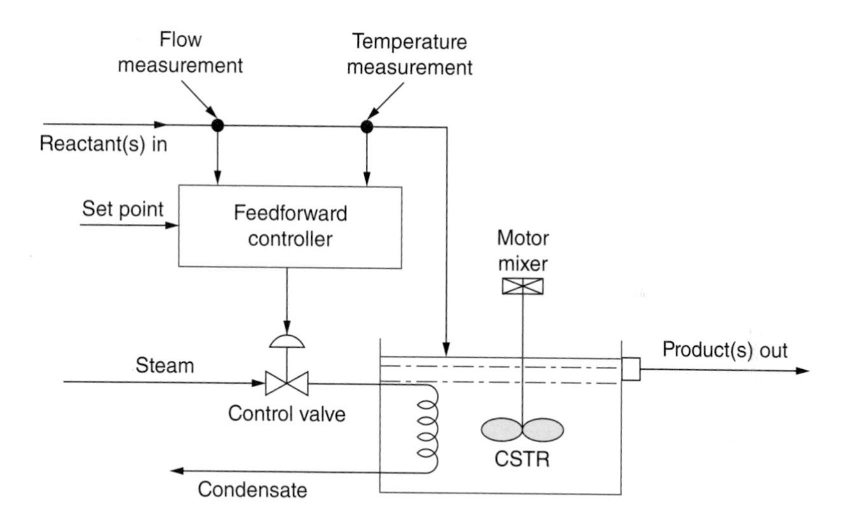

Figure 14–3

Feedforward control of a CSTR process.

control are often combined (see Fig. 14–4) in certain systems. (Both topics are discussed later in this chapter.) The employment of *block diagrams* (see Fig. 14–5) is the standard method of depicting the controlled system with its variables. The block diagram is obtained from the physical system by dividing it into functional, noninteracting sections whose inputs and outputs are readily identifiable.

Four parameters that are employed in process control are

1. Transfer function
2. Time constraint
3. Steady-state gain
4. Dead-time or lag

These four terms are briefly discussed below:

1. The *transfer function* relates two variables in a process: (a) the forcing function or input variable and (b) the response or output variable. The input and output variables are usually expressed in the Laplace[3] domain, and are in derivative form. The transfer function describes the dynamic characteristics of a system.
2. The *process time constant* provides an indication of the speed of the response of the process, i.e., the speed of change of the output variable to either force function

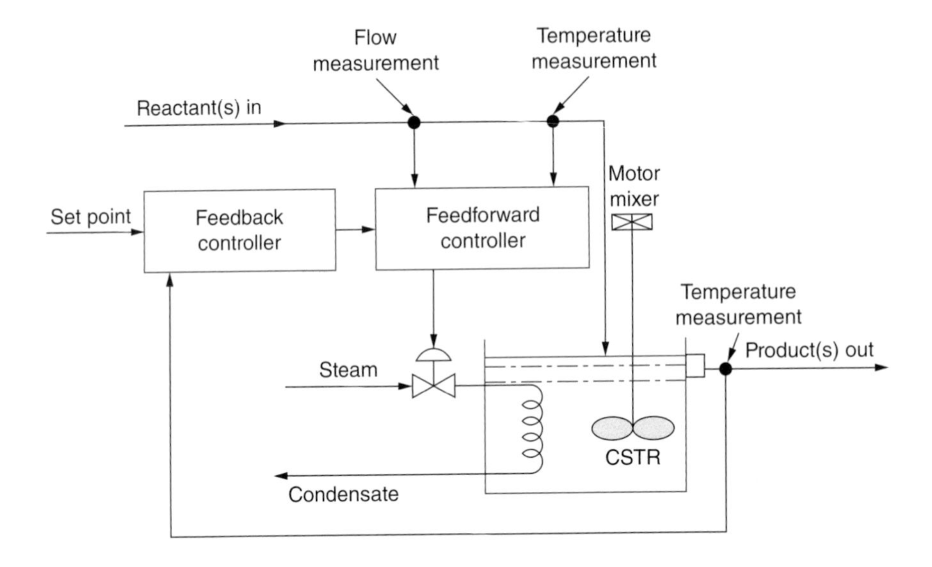

Figure 14–4 Feedforward control and feedback control of a CSTR exchange.

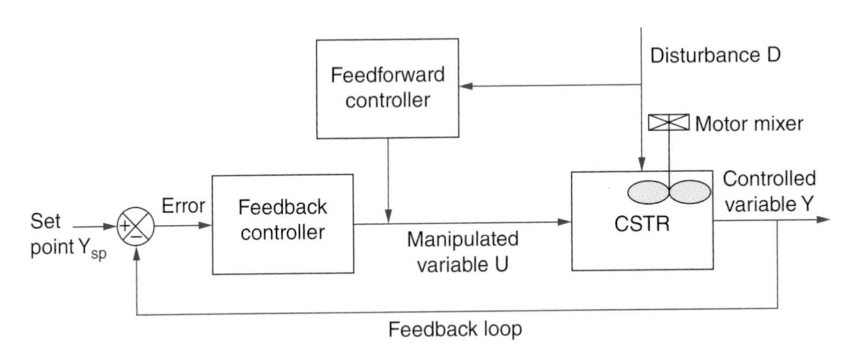

Figure 14–5 Block diagram for feedforward control and feedback control.

or change in the input variable, or both. The slower the response, the larger the time constant is, and vice versa. The time constant usually depends on the different physical properties and operating parameters of the process. The time constant has units of time.

3. The *steady-state gain* indicates the extent of change of the output variable (in the time domain) for a given change in the input variable (also in the time domain). The gain is also dependent on the physical properties and operating parameters of the process. The term *steady-state* is applied since a step change in the input variable results in a change in the output variable, which reaches a new steady state that can be predicted by application of the final-value theorem.[1] Gain may or may not be dimensionless.
4. *Process dead-time* refers to the delay in time before the process starts responding to a disturbance in an input variable. It is sometimes referred to as *transportation lag* or *time delay*. Dead-time or delays can also be encountered in measurement sensors such as thermocouples, pressure transducers, and in transmission of information from one point to another. In these cases, it is referred to as measurement *lag*.

14–3 Calibration of Flow and Temperature Transducers

Controller calibration and tuning are very important in the chemical and petrochemical industries, or in any industry that employs controllers. The selection of optimum values for controller variation and parameters is crucial for the smooth running of processes.

Sensor transducers are very important elements in a control system since they measure the value of the aforementioned controlled variable, and relay the information to the controller. Sensors may be used for measuring pressure, flow rates, temperature, level, humidity, and concentration.[1, 4]

The most common pressure sensors are the Bourdon gauge and the bellows and diaphragm sensors. Flow sensors include the time-tested orifice meter, turbine meter, venturi meter, rotameter, pilot tube, and magnetic flowmeter. The most popular temperature sensor in the chemical industry is the thermocouple. Other temperature sensors include expansion thermometers, resistance-sensitive devices, and optical and radiation pyrometers. Level sensors include differential pressure, air bubbler, and float. Sensors are also available for measuring viscosity, density, pH, and composition.

The purpose of a *transmitter* is to convert the output from a sensor to a signal strong enough to be transmitted to a controller. Pneumatic and electrical transmitters are very popular in the process industry. The process industry employs three types of signals to provide communication between sensors and controllers, and also between various elements in a control system. The pneumatic signal ranges between 3 and 15 psig. The electronic signal ranges between 4 and 20 mA (sometimes 1 to 5 V or 0 to 10 V). Discrete digital signals are popular, and these are based on microprocessor control. Sometimes, it is necessary to convert one type of signal to another, and this is achieved by a transducer. An *I*/*P* transducer converts a current (*I*) signal to a pneumatic (*P*) signal, whereas an *E*/*P* transducer converts a voltage (*E*) signal to a pneumatic (*P*) signal.

Consider a temperature sensor-transmitter that has been calibrated to measure a process temperature between the values of 100°C and 400°C, has a range of 100°C to 400°C, and a span of 300°C. The gain K_T of the sensor-transmitter (assuming that the transmission is electric) for the signal range given above can be calculated as follows:

$$K_T = \frac{20\text{ mA} - 4\text{ mA}}{400°\text{C} - 100°\text{C}} = 0.053\frac{\text{mA}}{°\text{C}} \qquad (14\text{–}1)$$

For flow transmitters, the differential pressure (e.g., across an orifice), is proportional to the square of the flow rate q. The equation that describes the output signal from a pneumatic differential pressure transmitter (DP cell) can be written as

$$P = 3 + \left(\frac{12}{q_{max}^2}\right)q^2 \qquad (14\text{–}2)$$

where P = output signal, lb_f/in^2 gauge (psig)
q = volumetric flow rate
q_{max} = maximum flow rate (or span)

The gain of the transmitter can be calculated as follows:

$$K_T = \frac{dP}{dq} = \frac{2(12)}{q_{max}^2}q = \frac{24\,q}{q_{max}^2} \qquad (14\text{–}3)$$

It is clear from Eq. (14–3) that the gain is not a constant, but varies with flow. One should therefore note that the gain is a function of flow rate, and increases linearly with an increase in flow rate. Most manufacturers of DP cells offer transmitters with built-in square-root extractors, thereby yielding a linear transmitter. Generally, the response time of sensors-transmitters is much faster than the response time of the process. The transfer function can therefore be assumed to be a simple gain as the time constant and dead-time can be considered negligible.

14-4 Sizing Concerns

Control valves represent the most common final control elements used in the chemical industry. The sizing of control valves requires a procedure to calculate the valve coefficient C_v. Once the C_v value is determined, it is possible to determine the valve size from a standard manufacturer's catalog. The development to follow addresses both liquid service and gas service.

Liquid Service

For liquid service, the following equation is used for sizing control valves:

$$q = C_v f(l)\sqrt{\frac{\Delta P_{cv}}{SG}} \tag{14-4}$$

where q = flow rate through control valve, gal/m
C_v = valve coefficient, (gal·min)/psia
$f(l)$ = flow characteristic, dimensionless
ΔP_{cv} = pressure drop across control valve, psi
SG = specific gravity of liquid at flowing temperature, dimensionless

The flow characteristics $f(l)$ depends on the type of valve being used. The three most popular valves are linear valves, equal-percentage valves, and quick-opening valves. The flow characteristic for a linear valve is given by

$$f(l) = 1 \qquad l = \text{valve lift} \tag{14-5}$$

The range of values for l is 0.0 to 10, where 0 represents a fully shut valve and 1 represents a fully open valve. For an equal-percentage valve, the flow characteristic is given by

$$f(l) = R^{l-1} \tag{14-6}$$

where R is a design value/rangeability parameter (whose value usually lies between 20 and 50). The *rangeability* of a valve is defined as the maximum controllable flow to the minimum controllable flow. For a quick-opening valve, the flow characteristic is given by

$$f(l) = \sqrt{l} \tag{14-7}$$

Some valve manufacturers combine the C_v and flow characteristic $f(l)$ terms in Eq. (14-4) and define the combined term as the valve coefficient. In this case, it is important to remember that this combined term, also denoted as C_v, is given by

$$C_v = C_v\big|_{l=1} f(l) \tag{14-8}$$

The first step in sizing a control valve is to determine the pressure drop across the valve. The total system pressure drop (ΔP_s) should be specified. Usually, the *beginning* of the system will consist of a pump, or compressor, or a high-pressure header, and the *end* of the system will be at a constant pressure. In between these two, there are usually several pieces of equipment and connected piping. The total system pressure drop is the sum of the dynamic pressure drop (including valve, pipe, fittings, equipment, etc.) and the static pressure drop. The static pressure drop is constant at all flow rates, and the dynamic pressure drop is a function of the flow rate.[5] The total system pressure drop can be described by the following equation

$$\Delta P_s = \Delta P_{stat} + \Delta P_{eq} + \Delta P_p + \Delta P_{cv} \quad (14\text{–}9)$$

where ΔP_{stat} = static pressure drop, psi
ΔP_{eq} = pressure drop across any equipment, psi
ΔP_p = pressure drop as a result of friction loss through piping, fittings, etc., psi
ΔP_{cv} = pressure drop across control valve, psi

The available dynamic pressure drop at low flow rates is higher than the available dynamic pressure drop at higher flow rates. Since the total system pressure drop is a constant, and the dynamic pressure drop is a function of flow rate, the control valve functions by absorbing the rest of the pressure drop. In general, a good rule of thumb is to specify that the pressure drop across the control valve be 25 to 50 percent (0.25 to 0.50 on a fractional basis) of the total dynamic pressure drop across the system at the design flow rate. Once the pressure drop across the control valve is calculated from Eq. (14–9), the C_v value can easily be determined. A good rule of thumb is to assume that the dynamic pressure drop, excluding the control valve, is proportional to the square of the flow rate. Accordingly, the dynamic pressure drop at any flow rate may be calculated according to the following equation:

$$\Delta P_{dyn,2} = \Delta P_{dyn,1}\left(\frac{q_2}{q_1}\right)^2 \quad (14\text{–}10)$$

Under normal conditions, when a fluid passes through a valve, there is a point in the stream just beyond the narrowest passage where the stream converges to a minimum area and then diverges again. This point is called the *vena contracta*, at which point the static pressure also drops to a minimum point. The fluid pressure partially recovers further downstream, and the pressure increases again. "Butterfly" valves and the ball valves have a high-pressure recovery characteristic, whereas most reciprocating-steam valves show a low-pressure recovery characteristic.[5]

If the static pressure of the liquid falls below the vapor pressure at the temperature at which it is flowing, some of the liquid will start to change phase, resulting in a phenomenon known as *flashing*. Flashing can result in serious damage to the valve plug and seat. As the liquid continues to change phase, the flow capacity of the valve is lowered, resulting in a condition known as *choking*. Any increase in pressure drop will not result in an increased flow.

It is important for the design engineer to determine the maximum pressure drop that is effective in producing flow. Valve manufacturers specify the maximum allowable pressure drop to avoid the aforementioned flashing. For example, Masoneilan proposes the following equation:

$$\Delta P_{allowable} = C_f^2\,\Delta P_s \quad (14\text{–}11)$$

where C_f = critical flow factor
ΔP_s = upstream pressure, psia

ΔP_s is given by the following equation:

$$\Delta P_s = P_1 - P_v\left(0.96 - 0.28\sqrt{\frac{P_v}{P_c}}\right) \quad (14\text{–}12)$$

where P_v = vapor pressure of liquid, psia
P_c = critical pressure, psia

When $P_v < 0.5\ P_1$, then

$$\Delta P_{\text{allowable}} = P_1 - P_v \tag{14–13}$$

Other valve manufacturers propose different correlations for checking for flashing. If the liquid experiences sufficient pressure recovery to raise its pressure above the vapor pressure of the liquid, then the vapor bubbles will implode or collapse, a phenomenon known as *cavitation*.[5] Cavitation almost always results in severe damage to the valve material. For low-pressure recovery valves, choked flow and cavitation occur at nearly the same pressure drop. Equation (14–11) can be used to calculate the pressure drop at which cavitation occurs. For higher-pressure recovery valves, Masoneilan recommends the following equation:

$$\Delta P_{\text{cavitation}} = K_c(P_1 - P_2) \tag{14–14}$$

where K_c = coefficient of incipient cavitation.

In addition to checking for flashing and cavitation, it is important to consider the effect of liquid viscosity in calculating the C_v coefficient. If the flow-through the control valve falls in the laminar region as a result of high viscosity or low velocity (or both), a correction for viscosity effects is required.[5] The correction for viscosity is a complex function involving the Reynolds number, and valve manufacturers normally provide correction charts or nomographs for high-viscosity fluids. It is important to remember that valve manufacturers have their own set of equations and nomographs for flashing and cavitation.

Gas Service

The sizing of control valves in liquid service was discussed above. The sizing of control valves in gas, vapor, or steam service is another important topic. Critical flow is an important consideration in the flow of compressible fluids. This condition arises when the fluid reaches sonic velocity at the vena contracta, and the flow rate is no longer a function of the square root of the pressure drop across the valve, but only of the upstream pressure. Under critical flow, any decrease in the downstream pressure does not affect the flow rate.[6] The following equations are available for compressible fluids. For gases, the equation is

$$C_v = \frac{q\sqrt{(\text{SG}_f)}}{836 C_f P_1 (y - 0.148 y^3)} \tag{14–15}$$

where q = gas flow rate, standard ft³/h
SG = specific gravity at 14.7 psia and 60°F (with respect to air)
SG_f = specific gravity of gas at flowing temperature, (SG)(520/T)
T = flowing temperature, °R
C_f = critical flow factor (usually within 0.6 to 0.95; actual value can be found in valve manufacturer's catalog)
P_1 = upstream pressure, psia

The equation for steam is

$$C_v = \frac{W(1 + 0.0007 T_{\text{sh}})}{1.83 C_f P_1 (y - 0.148 y^3)} \tag{14–16}$$

where, in addition to the terms described earlier, T_{sh} is the degree of superheat in °F and W is the steam flow rate. The term y provides a measure of the critical or subcritical flow condition and is defined as

$$y = \frac{1.63}{C_f}\sqrt{\frac{\Delta P_{cv}}{P_1}} \quad (14\text{–}17)$$

When $y = 1.5$, the term $y - 0.148y^3$ equals 1, and this indicates critical flow conditions; the flow rate is then, as noted, a function of the upstream pressure only.

Another equation used for compressible fluids is the *universal sizing equation*. This equation was developed from experimental data on airflow-through valves. The flow rate that the control valve can pass under a set of operating conditions is given by

$$q = K_f \frac{C_g P_1}{\sqrt{(\mathrm{MW})T_1}} \sin\left(\frac{3417}{C_1}\sqrt{\frac{\Delta P}{P_1}}\right)_{\text{degrees}} \quad (14\text{–}18)$$

where C_1 = ratio of gas sizing coefficient to control valve coefficient
C_g = gas sizing coefficient
MW = molecular weight of flowing fluid
T_1 = inlet temperature of fluid

The value of K_f = constant is as follows. When q is given in scf/h_1, P_1 is in psia and T_1 is in °R, K_f = 122.72; when q is given in m³/h, P_1 is in kg_f (absolute)/cm², T_1 is in kelvin, and K_f = 34.778.

Since different control valves, but having the same C_v, deliver different gas flow rates under the same operating conditions, a separate coefficient called the *gas sizing coefficient* C_g is used in the universal sizing equation:

$$C_g = C_1 C_v \quad (14\text{–}19)$$

Values of C_1 for different types of valves are established by air tests under critical conditions.

For vapors, Eq. (14–18) can be modified as

$$\dot{m} = K_g \sqrt{P_1 \rho_1 C_g} \sin\left(\frac{3417}{C_1}\sqrt{\frac{\Delta P}{P_1}}\right)_{\text{degrees}} \quad (14\text{–}20)$$

where $\dot{m}$ is in lb/h, P_1 is in psia and ρ_1 is in lb/ft³, with K_g = 1.06. When $\dot{m}$ is in kg/h, then P_1 is given in kg_f (absolute)/cm², and ρ is in kg/m³, with K_g = 0.453. It should also be clear that the equations presented above are empirical. The reader is therefore advised to refer to valve manufacturer catalogs for more information.

Finally, safety control valve or control valve action relates to the *tail position* of the valve. In case of instrument air failure, it becomes exceedingly important to decide whether the valve should close or remain fully open, or retain the status quo (be fail-safe). Most of the control vales in the chemical industry are operated by pneumatic air pressure, and hence it is important to decide whether a certain valve in a certain application should be fail-closed or fail-open. A *fail-closed valve* requires air pressure to open it, so it is also referred to as an *air-to-open* (AO) valve. A *fail-open* valve requires air pressure to close it, so it is also referred to as an *air-to-close* (AC) valve. In designing control systems, safety should always be kept in mind. This includes normal operation as well as start-up and shutdown. Safety interlocks are also part of the overall control system, and sometimes override the control action.[5]

14–5 Feedback Control

Feedback control is a very important aspect of process control. Its role is best described in terms of an example (see also Fig. 14–2). Assume that one desires to maintain the temperature of a polymer reactor at 70°C. Temperature is thus the controlled variable, and the desired temperature level 70°C, is called the *set point*. In feedback control, the temperature is measured using a sensor (such as a thermocouple device). This information is then continuously relayed to a controller, and a device known as a *comparator* is used to compare the set point with the measured signal (or variable). The difference between the set point and the measured variable is the previously defined error. With respect to the magnitude of error, the controller element in the feedback loop takes corrective action by adjusting the value of a process parameter, known as the *manipulated variable*.

The controller logic (how it handles the error) is an important process control criterion. Generally, feedback controllers are either proportional *P* (send signals to the final control element proportional to the error), proportional-integral *PI* (send signals to the final control element that is proportional to the magnitude of the error at any instant and to the sum of the error), and proportional-integral-derivative *PID* (send signals that are also based on the slope of the error).

In the example above, the manipulated variable may be cooling water flow-through the reactor jacket. This adjustment or manipulation of the flow rate is achieved by a *final control element*. In most chemical processes, the final control element is usually a pneumatic control valve. However, depending on the process parameter being controlled, the final control element could very well be a motor whose speed is regulated. Thus, the signal from the controller is sent to a final control valve that manipulates the manipulated variable in the process.

In addition to the controlled variable, other variables may disturb or affect the process. In the reactor example above, a change in the inlet temperature of the feed or inlet flow rate are considered to be *load* disturbances. A *servo* problem is one in which the response of the system to a change in set point is recorded, whereas a load or regulator problem is one in which the response of a system to a disturbance or load variable is measured.

Before selecting a controller, it is very important to determine its action. Consider another example involving the heat exchanger shown in Fig. 14–6. Steam is used to heat the process fluid. If the inlet temperature of the process fluid increases, this will result in an increase in the outlet temperature. Since the outlet temperature moves above the set point (or desired temperature), the controller must close the steam valve (the control valve is air-to-open or fail-closed). This is achieved by the controller sending a lower output (pneumatic or current) signal to the control valve, i.e., an increase in the input signal from the controller to the valve. The action of the controller is considered to be *reverse*. If the input signal to the controller and the output signal from it act in the same direction, the controller is *direct acting*. To determine the action of the controller, it is important to consider the process

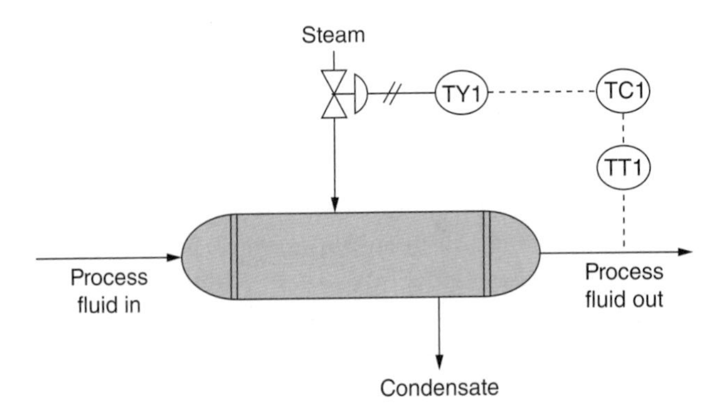

Figure 14–6

Controller action.

requirements for control, and the action of the final control element. The controller action is usually set by a switch on electronic and pneumatic controllers. On microprocessor-based controllers, the setting can be made by changing the sign of the scale factor in the software (which then changes the sign of the proportional gain of the controller).

The functions of a feedback controller in a process control loop are twofold: (1) to compare the process signal from the transmitter (the controlled or measured variable) with the set point, and (2) to send a signal to the final control element with the sole purpose of maintaining the controlled variable at its set point. As noted above, the most common feedback controllers are proportional controllers (P control), proportional-integral controllers (PI control), and proportional-integral-derivative (PID) controllers. The proportional controller can be described by the following equation:

$$p(t) = \overline{p} + K_c[r(t) - c(t)] \tag{14–21}$$

where $p(t)$ = output from controller, psig or mA
$r(t)$ = set point, psig or mA
$c(t)$ = controlled variable, psig or mA
K_c = controller gain, dimensionless
$\overline{p}$ = bias value

The *bias value* is the output from the controller when the error is zero; this value is usually set at midscale. Equation (14–21) can be written in terms of the error as follows:

$$p(t) = \overline{p} + K_c\varepsilon(t) \tag{14–22}$$

where $\varepsilon(t)$ = error or difference between the set point and controlled variable.

Equation (14–22) applies for a reverse-acting controller. In the case of a direct-acting controller, K would be negative. The controller gain is the only parameter that needs to be adjusted in proportional controllers. Unfortunately, this leads to the problem of *offset*, which is the difference between the set point and the steady-state value of the controlled variable. Offset can be minimized by increasing the value of the controller gain. However, above a certain value (known as the *ultimate gain*), the process becomes unstable. Some controller manufacturers use the term *proportional band,* which is $100/K_c$. Proportional controllers provide satisfactory control in processes that do not require the controlled variable to be exactly at the set point. Such applications include liquid-level control or maintenance of gas pressure in a tank.

In some process applications such as reactor temperature or pressure or flow control, it may be necessary to maintain the controlled variable at the desired set point. Offset has to be eliminated at all costs. A proportional-integral (PI) controller can be used, and the equation for a PI controller is given by

$$p(t) = \overline{p} + K_c[r(t) - c(t)] + \frac{K_c}{\tau_1}\int_0^{t^*}[r(t) - c(t)]dt \tag{14–23}$$

where τ_1 = integral or reset time, minutes per repeat.

A PI controller thus has two parameters, K_c and τ_1. The integral time τ_1 is the time it takes to repeat the proportional action, and a small value for this parameter results in a steeper and quicker response. The integral term [see Eq. (14–23)] is a summation of the errors. Consequently, even though the error at any instant may be zero (i.e., the offset may be zero), the integral itself does not have to be zero.

The third mode of control is the proportional-integral derivative (PID) action. The derivative mode is also known as *anticipatory control* since the controller anticipates where the process is headed by measuring the rate of change of the error. The describing equation is

$$p(t) = \bar{p} + K_c[r(t) - c(t)] + \frac{K_c}{\tau_I}\int_0^{t^*} \varepsilon(t)\,dt + K_c\tau_D\frac{d\varepsilon(t)}{dt} \tag{14–24}$$

where τ_D is the derivative or rate parameter in units of time. PID controllers thus have three tuning parameters.

A comment on transform functions (TFs) is in order before proceeding to transform functions for controllers. Transform functions convert linear differential equations to algebraic expressions, which include a new variable s, referred to as the Laplace variable. These TFs are used to represent the process dynamics of linear systems. A replacement of Laplace transform is beyond the scope of this text, but is available in the literature.[4] The transfer functions for a proportional (P), proportional-integral (PI), and proportional-integral-derivative (PID) controller are as follows:

$$\frac{P(s)}{\varepsilon(s)} = K_c \tag{14–25}$$

and

$$\frac{P(s)}{\varepsilon(s)} = K_c\left(1 + \frac{1}{\tau_I s}\right) \tag{14–26}$$

$$\frac{P(s)}{\varepsilon(s)} = K_c\left(1 + \frac{1}{\tau_I s} + \tau_D s\right) \tag{14–27}$$

respectively. For digital controllers, the following equations are applicable:

$$p_n = \bar{p} + K_c\varepsilon_n + \frac{K_c}{\tau_I}\Delta t\sum_{k=0}^{n}\varepsilon_k + K_c\tau_D\frac{\varepsilon_n - \varepsilon_{n-1}}{\Delta t} \tag{14–28}$$

where p_n = output from controller at nth sampling instant
$\bar{p}$ = bias value
ε_n = error at nth sampling instant
ε_{n-1} = error at $(n-1)$th sampling instant
Δt = sampling interval, time units

Equation (14–28) is known as the *position form* of the controller. The *velocity form* of the controller is obtained by writing an equation at the $(n-1)$th sampling interval and taking the difference between the nth and $(n-1)$th sampling interval; this equation is

$$p_n - p_{n-1} = K_c(\varepsilon_n - \varepsilon_{n-1}) + \frac{K_c}{\tau_I}\Delta t\,\varepsilon_n + K_c\tau_D\frac{\varepsilon_n - 2\varepsilon_{n-1} + \varepsilon_{n-2}}{\Delta t} \tag{14–29}$$

Some controller manufacturers use the term *proportional band* instead of *gain*, which is $100/K_c$. A small controller gain indicates a large value for the proportional band. One problem with the derivative mode of control [the way Eq. (14–24) is written)], is that whenever the process requires a change in the set point, an unnecessary and sudden change in the controller output is introduced because the term $d\varepsilon(t)/dt$ is large. This can be avoided by substituting $-dc(t)/dt$ instead of $d\varepsilon(t)/dt$ in the equation since the two terms are equal when the set point is constant. This option is known as the *derivative on controlled variable*. Another problem with the integral mode is essentially a summation of the error (difference between set point and controlled variable). When an error is introduced (either because of a change in the set point or a disturbance in the process, known as a *load disturbance*), the summation of the error term can lead to a fairly substantial value for the controller output, and the control valve can be fully open or shut (depending on its action). In other words, the controller output becomes saturated because of the integration or reset action of the controller. Even after the process reaches the new set point (in case of a set point change), the control valve still remains fully open or shut because the controller must integrate in the reverse direction before the valve can start closing or opening (depending on the reverse position). Consequently, this results in the controlled variable overshooting the set point, and it takes a while before the process can "settle down." The advantage of the velocity form of the PID equation for digital controllers is that it eliminates the problem of reset-windup.

The different types of feedback controllers were discussed earlier. Two important problems in feedback control systems are the choice of a feedback controller for a given application (P, PI, or PID), and the adjustment or selection of the optimum controller parameters (controller gain K_c, integral time or reset time τ_I, and rate parameter alternative τ_D). Once a controller is selected for a given application, the selection of the best values for the controller parameters is known as *controller tuning*.

The selection of a feedback controller may be based on purely quantitative considerations. However, one can evaluate the effects of proportional, integral, and derivative action on the response of a feedback control system. Advantages and disadvantages are outlined below:

1. Proportional action
   a. Accelerates the response of feedback control systems
   b. Produces an offset (difference between set point and steady-state value of the controlled variable) which may be reduced by increasing the value of the controller gain K_c
2. Integral action
   a. Eliminates offset
   b. Produces sluggish, oscillating responses (due to an overall increase in the order of the system)
   c. Becomes more oscillatory and unstable if K_c is increased to produce faster response
3. Derivative action
   a. Does not eliminate offset
   b. Anticipates future errors (hence also known as *anticipatory control*)
   c. Permits the selection of large values of gain by introducing a stabilizing effect (hence is more *robust*)

From the discussion above, it appears that the PID controller should be used in all applications since it offers the highest flexibility. Also, in the case of digital controllers (or computer-based control), selection of the PID mode is achieved without any increase in capital cost. However, the PID mode introduces a more complex tuning problem since there are three adjustable parameters.

The following rules of thumb may be used in the selection of controllers:

1. In process control applications, where offset is not a major problem, proportional controllers can be used. Also, if the process itself exhibits an integrating action, then a simple proportional controller will suffice. In general, P-only controllers are used in liquid-level and gas pressure (surge tanks) control applications.
2. PI controllers are used in applications in which offset has to be completely eliminated. In applications where the response of the process is very fast (small time constants),

a PI controller can be used. A typical application is flow control, where the sluggish response of the integral action does not hamper the overall performance of the feedback control loop.

3. Processes with large time constants usually require the addition of the derivative mode of action to the control system. Typical examples are multicapacity processes, or temperature and composition control. The addition of the derivative mode makes the control action more robust by stabilizing its response.

The selection of a controller is based on qualitative, rather than quantitative considerations. It is important to note that with the development of digital or PC-based control, the choice of a P, PI, or PID controller is based on the reasons presented above and not on economic considerations.

The selection and tuning processes are based on the choice of a performance criterion which may be simple (based on steady-state performance) of the controlled variable, or more complex (based on the dynamic performance criteria) of the controlled variable. Simple performance criteria may include control objectives such as keeping the maximum deviation from the set point as small as possible, or returning to the set point as soon as possible in the event of a disturbance. Simple performance criteria are therefore based on some feature of the feedback control behavior such as decay ratio, overshoot, or rise time. In general, steady state performance criteria call for zero error at steady-state. It should be clear that the error or offset can be zero if integral action is selected. It is not possible to design a controller response that is based on multiple criterion as they may conflict with one another. The most popular steady-state criterion is the decay ratio, and a one-quarter (1/4) decay ratio provides a good compromise between a rapid rise time and a fairly good settling time.

14-6 Feedforward Control

Feedforward control has several advantages. Unlike feedback control, a feedforward control measures the disturbance directly, and takes preemptive action before the disturbance can affect the process. Consider the heated-tank example discussed earlier. A conventional feedback controller would entail measuring the temperature in the tank, the controlled variable, and maintaining it at the desired set point by regulating the heat input, the manipulated variable. In feedforward control, the load disturbances would first be identified, i.e., the inlet flow rate and the temperature of the inlet fluid. Any change in the inlet temperature, for example, would be monitored in a feedforward control system, with corrective action taken by adjusting the heat input (again the manipulated variable), before the process is affected. Thus, unlike in the case of feedback control, the error is not allowed to propagate through the system.

It is clear that a feedforward controller is not a PID controller, but a special computing or digital machine. Good feedforward control relies to a large extent on sound knowledge of the process, which is the major drawback. Finally, the stability of a feedforward/feedback system is determined by the roots of the characteristic equation of the feedback loop (feedforward control does not affect the stability of the system). Vasudevan provides additional developmental material and illustrates examples.[1]

14-7 Cascade Control

The simple feedback control loop considered earlier is an example of a single-input single-output (SISO) system. In some instances, it is possible to have more than one measurement but one manipulated variable, or one measurement and more than one manipulated variable. In *cascade control*, for example, there is one manipulated variable but more than

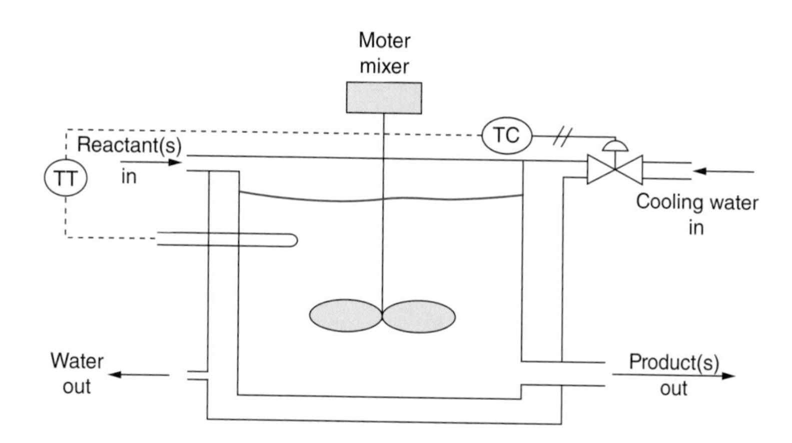

Figure 14–7

Simple feedback temperature control in a CSTR.

one measurement. Consider a slight modification of the continuous stirred-tank reactor (CSTR) shown in Fig. 14–7. The control objective is to maintain the reactor temperature at a set value by regulating the cooling-water flow to the exchanger. The load disturbances to the reactor include changes in the feed inlet temperature or in the cooling-water temperature. In simple feedback control, any change in the cooling-water temperature will affect the reactor temperature, and the disturbance will propagate through the system before it is corrected. In other words, the control loop will respond more rapidly to changes in the feed inlet temperature than to changes in the cooling-water temperature.

Now consider a hypothetical scenario in which any change in the cooling-water temperature is corrected by a new controller added to the process control loop before its effect propagates through the reaction system. There are two different measurements, reaction temperature and cooling-water temperature, but only one manipulated variable (cooling-water flow rate). The loop that measures the reaction temperature is known as the *primary* control loop and the loop that measures the cooling-water temperature, the *secondary* control loop. The secondary control loop uses the output of the primary controller as its set point, whereas the set point to the primary controller is supplied by the operator. Cascade control has wide applications in chemical processes. Usually, flow rate control loops are cascaded with other control loops.

References

1. P. T. Vasudevan, *Process Dynamics and Control*, A Theodore Tutorial, Theodore Tutorials, East Williston, N.Y., originally published by USEPA/APTI, Research Triangle Park, N.C., 1996.
2. L. Theodore, *Chemical Reactor Analysis and Applications for the Practicing Engineer*, Wiley, Hoboken, N.J., 2012.
3. L. Theodore, *Heat Transfer Application for the Practicing Engineer*, Wiley, Hoboken, N.J., 2011.
4. M. Spiegel, *Laplace Transforms*, Schaum's Outline Series, McGraw-Hill, N.Y., 1965.
5. D. Considine, *Instruments and Control Handbook*, McGraw-Hill, N.Y., 1985.
6. P. Abulencia and L. Theodore, *Fluid Flow for the Practicing Chemical Engineer*, Wiley, Hoboken, N.J., 2009.
7. L. Theodore and R. Dupont, *Environmental Health and Hazard Risk Assessment: Principles and Calculations*, CRE Press/Taylor & Francis Group, Boca Raton, Fla., 2012.

15 Process and Plant Design

Chapter Outline

15–1 Introduction

Current process and plant design practices can be categorized as state-of-the-art and pure empiricism. Past experience with similar applications is commonly used as the sole basis for the design procedure. The vendor (seller) maintains proprietary files on past installations, and these files are periodically revised and expanded as new installations are evaluated. During the design of a new unit, the files are consulted for similar applications and old designs are heavily relied on.

By contrast, the engineering profession in general, and the chemical engineering profession in particular, have developed fairly well-defined procedures for the design, construction, and operation of chemical plants. These techniques are routinely used by today's chemical engineers. These same procedures may also be used in the design of other facilities.

The purpose of this chapter is to introduce the reader to some of the process design fundamentals. Such an introduction to design principles, however sketchy, can provide the reader with a better understanding of the major engineering aspects of new or modified facilities, including some of the operational and economic factors, controls, instrumentation for safety and regulatory requirements, and environmental factors associated with the process. The author[1] has simplified the process by keying in five topics that are summarized in the acronym *SCORE*: safety, costs, operability, reliability, and environment.

It should also be noted that process plant location and layout considerations are not reviewed. However, in a general sense, physical plant considerations should include process flow, construction, maintenance, operator access, site conditions, space limitations, future expansion, special piping requirements, structural supports, storage space, utility requirements, and (if applicable) energy conservation. Finally, the reader should note that no attempt is made in the sections that follow to provide extensive coverage of this topic; only general procedures and concepts are presented and discussed.

The topics covered in this chapter include preliminary studies, process schematics, material and energy balances, equipment design, instrumentation and controls, design approach, and the design report.

Application of the somewhat crude approach to a plant design presented in this chapter can produce valuable information. Perhaps its greatest value is that it can provide early signals as to when the proposed process is technically unfeasible or economically prohibitive.

NOTE The bulk of the material presented in this chapter has been drawn from the author's personal notes[1] and from Shen et al.[2] and Theodore.[3]

15–2 Preliminary Studies

A *process* chemical engineer is usually involved in one of two principal activities: building a plant or deciding whether to do so. The skills required in both cases are quite similar, but the money, time, and detail involved are not as great in the latter situation. It has been estimated that only one out of 15 proposed new processes ever achieves the construction stage.

In general design practice, there are usually five levels of sophistication for evaluating projects and estimating costs. Each is discussed in the following list, with particular emphasis on cost:

1. The first level of analysis requires little more than identification of products, raw materials, and utilities. This is what is known as an *order-of-magnitude estimate* and is often made by extrapolating or interpolating from data on similar

existing processes. The evaluation can be done quickly and at minimum cost, but with a probable error exceeding ±50%.

2. The next level of sophistication is called a *study estimate* and requires a preliminary process flowchart (to be discussed in Sec. 15–3) and a first attempt at identification of equipment, utilities, materials of construction, and other processing units. Estimation accuracy improves to within ±30% probable error, but more time is required and the cost of the evaluation can escalate to over $30,000 for a $5 million plant.
3. A *scope* or *budget authorization* is the next level of economic evaluation. It requires a more defined process definition, detailed process flowcharts, and prefinal equipment design (discussed in a later section). The information required is usually obtained from pilot plant, marketing, and other studies. The scope authorization estimate could cost upward of $80,000 for a $5 million facility with a probable error potentially exceeding ±20%.
4. If the evaluation is positive at this stage, a *project control estimate* is then prepared. Final flowcharts, site analyses, equipment specifications, and architectural and engineering sketches are employed to prepare this estimate. The accuracy of this estimate is about ±10% probable error. Because of the increased intricacy and precision, the cost of preparing such an estimate for the process can approach $150,000.
5. The fifth and final economic analysis is called a *firm* or *contractor's estimate*. It is based on detailed specifications and actual equipment bids. It is employed by the contractor to establish a project cost and has the highest level of accuracy: ±5% probable error. The cost of preparation results from engineering, drafting, support, and management-labor expenses.

Because of unforeseen contingencies, inflation, and changing political and economic trends, it is impossible to ensure actual costs for even the most precise estimates.

15–3 Process Schematics

To the practicing engineer, particularly the chemical engineer, the process flowchart is the key instrument for defining, refining, and documenting a chemical process. The process flow diagram is the authorized process blueprint, the framework for specifications used in equipment designation and design; it is the single, authoritative document employed to define, construct, and operate the chemical process.[4]

Beyond equipment symbols and process stream flow lines, there are several essential constituents contributing to a detailed process flowchart. These include equipment identification numbers and names; temperature and pressure designations; utility designations; mass, molar, and volumetric flow rates for each process stream; and, a material balance table pertaining to process flow lines. The process flow diagram may also contain additional information such as energy requirements, major instrumentation, environmental equipment (and concerns), and physical properties of the process streams. When properly assembled and employed, a process schematic provides a coherent picture of the overall process; it can pinpoint some deficiencies in the process that may have been overlooked earlier in the study, e.g., instrumentation overkill, by-products (undesirable or otherwise), and recycle needs. Basically, the flowchart symbolically and pictorially represents the interrelation between the various flow streams and equipment, and permits easy calculation of material and energy balances. These two topics are considered in the next section. Controls and instrumentation must also be considered in the overall requirements of the system; these concerns are covered later in this chapter as well as in Chap. 14.

As one might expect, a process flow diagram for a chemical or petroleum plant is usually significantly more complex than that for a simple facility. For the latter case, the flow sequence and determinations reduce to an approach that employs a "railroad" or sequential type of calculation that does not require iterative calculations.

Various symbols are universally employed to represent equipment, equipment parts, valves, piping, etc. Some of these are depicted in the schematic in Fig. 15–1. Although a significant number of these symbols are used to describe some of the chemical and petrochemical processes, only a few are needed for simpler facilities. These symbols obviously reduce, and in some instances replace, detailed written descriptions of the process. Note that many of the symbols are pictorial, which helps in better describing process components, units, and equipment.

The degree of sophistication and details of a flowchart usually vary with both the preparer and time. It may initially consist of a simple freehand block diagram with limited information that includes only the equipment; later versions may include line drawings with pertinent process data such as overall and componential flow rates, utility and energy requirements, environmental equipment, and instrumentation. During the later stages of the design project, the flowchart will consist of a highly detailed

Figure 15–1

Flowchart symbols.

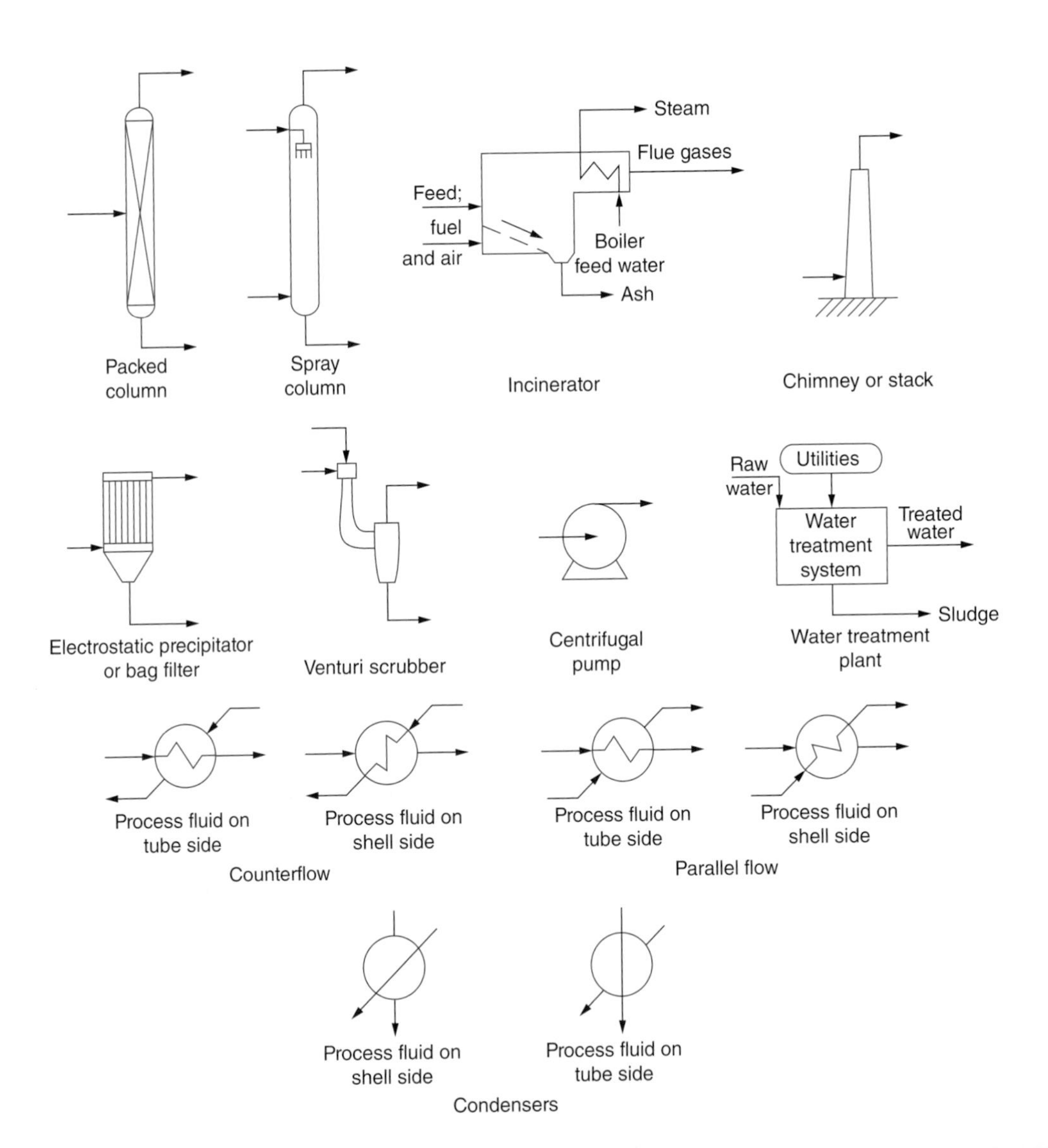

P&I (piping and instrumentation) diagram; this aspect of the design procedure is beyond the scope of this text; for information on P&I diagrams, the reader is referred to the literature.[2]

In a sense, flowcharts are the international language of the engineer, particularly the chemical engineer. Chemical engineers conceptually view a (chemical) plant as consisting of a series of interrelated building blocks that are defined as *units* or *unit operations*. The plant essentially ties together the various pieces of equipment that make up the process. Flow schematics follow the successive steps of a process by indicating where the pieces of equipment are located, and the material streams entering and leaving each unit.[5–7]

15-4 Material Balances

Overall and componential material balances have already been described in some detail in Chap. 7. Material balances may be based on mass, moles, or volume, usually on a rate (time-rate of change) basis. Care should be exercised here since moles and volumes are *not* conserved; i.e., the quantities may change during the course of a reaction. Thus, the initial material balance calculation should be based on *mass*. Mole balances and molar information are important in not only stoichiometric calculations but also chemical reaction and phase equilibria calculations. Volume rates play an important role in equipment sizing calculations.

The units (pound, kilogram, etc.) employed may also create a problem. Despite earlier efforts by the Environmental Protection Agency (EPA) and other government agencies, industry still primarily employs the British and/or American engineering units in practice; there has been a reluctance to accept standard international units, commonly referred to as SI that employ a modified set of metric units. As indicated earlier, this text uses primarily engineering units. However, for those individuals who are more comfortable with the SI system, a short write-up on conversion of units has been prepared by the author and included in Chap. 1.

Some design calculations in the chemical industry today include transient effects that can account for process upsets, start-ups, shutdowns, etc. The describing equations for these time-varying (unsteady-state) systems are *differential*. The equations usually take the form of a first-order derivative with respect to time, where time is the independent variable. However, design calculations for most facilities assume steady-state conditions, with the ultimate design based on worst-case or maximum flow conditions. This greatly simplifies the calculations, since the describing equations provide an accounting or inventory of all mass entering and leaving one or more pieces of equipment, or the entire process.

The heart of any material balance analysis is the *basis* selected for the calculation. The usual basis is a unit of time, such as, minutes, hours, days, or years. For more complex calculations, and this may include multicomponent systems and recycle streams, one may choose a convenient amount of a *key* component or element as a basis. Note also that the calculation may be based on either a feed stream, an intermediate stream, or a product stream. Selecting a feed stream as a basis is preferred since it often allows one to follow calculations through the process in a railroad manner, i.e., in sequential order.[2,5–7]

The number of material balance equations can be significant, depending on the number of components in the system, process chemistry, and pieces of equipment. These are critical calculations since (as noted above) the size of the equipment is often linearly related to the quantity of material being processed. This can then significantly impact—often linearly or even exponentially—capital and operating costs. In addition, componential rates can impact on (other) equipment needs, energy considerations, materials of construction, and so on.

Once the material balance is completed, one may then proceed directly to energy calculations, some of which play a significant role in the design of a facility. As indicated earlier, energy calculations are also usually based on steady-state conditions. An extensive treatment of this subject has already been presented earlier in this text, and need not be repeated here. However, a thorough understanding of thermodynamic principles—particularly the enthalpy calculations—is required for most of the energy (balance) calculations. Entropy calculations are employed in meaningful energy conservation analysis; this subject is beyond the scope of this presentation but is a topic that may be given more serious attention by the engineering community in the not-so-distant future.

Industry has always recognized that wasting energy reduces profits. The cost of energy was often a negligible part of the overall process cost, and immense operational inefficiencies were tolerated before the Arab oil embargo. The sudden decrease in the availability of natural gas and oil resulting from the embargo raised the cost of energy and encouraged the elimination of unnecessary energy consumption. These energy-saving measures have continued to be employed, even when energy costs are low, and will continue in the near future.

One of the principal jobs of a chemical engineer involved in the design of a facility is to account for the energy that flows into and out of each process unit and to determine the overall energy requirement for the process. This is accomplished by performing energy balances on each process unit and on the overall process. These balances play an important role as material balances in facility design. They find particular application in determining fuel requirements, heat exchanger design, heat recovery systems, specifying materials of construction, and calculating fan and pump power requirements.

15–5 Equipment Design

As noted previously, chemical engineers describe the application of any piece of equipment that operates on the basis of mass, energy, and/or momentum transfer as a *unit operation*. A combination of two or more of these operations is defined as a *unit process*. A *whole chemical process* can be described as a coordinated set of unit operations and unit processes. This subject matter has received much attention in recent years and, as a result, is adequately covered in the literature. From details on these unit operations and processes, it is therefore possible to design new plants more efficiently by coordinating a series of *unit actions*, each of which operates according to certain laws of physics regardless of the other operations being performed along with it.

The unit operation of combustion, for example, is used in many different types of industries; many of the critical design parameters for the combustion processes, however, are common to all combustion systems and independent of the particular industry. For example, in a hazardous waste incineration plant, the major pieces of equipment that must be considered are all of the following:[8]

1. Storage and handling facilities (feed and residuals)
2. Incinerator (rotary kiln or liquid injection)
3. Waste-heat boiler (primary quench system or energy recovery, if economically practical)
4. Quench system
5. Wet scrubber – venturi scrubber (particulate removal scrubber)
6. Absorber (packed tower for acid gas absorption)
7. Spray dryer (quench and acid gas absorption)

8. Baghouse or electrostatic precipitator (ESP)
9. Peripheral equipment (cyclone)
10. Fan(s) and blower(s)
11. Stack
12. Pumps (feed, recycle, and scrubber)

Calculation procedures for some of these facilities have been presented in Chaps. 7 to 14. Since design calculations are generally based on the maximum throughput capacity for the proposed process or for each piece of equipment, these calculations are never completely accurate. It is usually necessary to apply reasonable *safety factors* when setting the final design. Safety factors vary widely and are a strong function of the accuracy of the data involved, calculation procedures, and past experience. Attempting to justify these by a process engineer is a difficult task.

Unlike many of the problems encountered and solved by the chemical engineer, there is absolutely no correct solution to a design problem; however, there is usually a *better* solution. Many alternative designs when properly implemented will function satisfactorily, but one alternative will usually prove to be economically more efficient and/or attractive than the others. This leads to the general subject of optimization, a topic briefly addressed in Chap. 1.

15-6 Instrumentation and Controls

The control of a system or process requires careful consideration of all operational and regulatory requirements. The system is usually designed to process materials. Safety should be a primary concern of all individuals involved with the handling, treatment, or movement of the materials. The safe operation of any unit requires that the controls keep the system operating within a safe operating envelope. The envelope is based on many of the design, process, and regulatory constraints. These are placed on the unit to ensure proper operation. Additional controls may be installed to operate additional equipment needed for energy recovery, neutralization, or other peripheral operations.

The control system should also be designed to vary one or more of the process variables to maintain the appropriate conditions with the unit. These variations are programmed into the system on the basis of past experience of the process in question. The operational parameters that may vary include the temperatures and system pressure.

The proper control system should also be subjected to extensive analysis on operational problems and items that could go wrong. A hazard-and-operation (HAZOP) analysis is conducted on the control system to examine and identify all possible failure mechanisms.[9] (See also Chap. 26.) It is important that all of the failure mechanisms have appropriate responses by the control system. Several of the failure mechanisms that must be addressed within the appropriate control system response are excess or minimal temperature, excessive or subnormal flow rate, equipment failure, sticking or inoperable components, and broken circuits. All are usually examined, including the response time of the control system to the problem as well as the appropriate response of the system to the problem. The system must identify the problem and integrate necessary actions and alarms into the system.

A complete review of the maximum and minimum process (variable) time rate of change and the equipment time rate of change must be identified before the control system can be defined. The system must also be reviewed to define the primary control parameters for operation and safety. The regulatory limitations imposed on the unit must be identified and monitored to ensure that they are not exceeded.

The purpose of the control system is to ensure that the system is operating in a reliable and safe manner, and within the guidelines of the design. The control system is responsible for all of the variables that occur during operation of the system. The reader is referred to Chap. 14 for additional details on instrumentation and control.

15-7 Design Approach

Although all chemical engineers approach design problems somewhat differently, six major steps are generally required. These six steps are discussed below and may also be applied to the design of most (other) facilities:

1. The first step is to conceptualize and define the process. A designer must know the bases and assumptions that apply, the plant capacity, and the process time involved. Some of the answers to a host of questions pertaining to the process operation will be known from past experience.
2. After a problem has been defined, a method of solution must be sought. Although a method is seldom obvious, a good starting point is the preparation of a process flowchart. This effort usually produces valuable results. For example, it may suggest to the designer ways of reducing the complexity of the problem; it can allow for easier execution of material and energy balances, which, in turn, can point up the most important process variables. It is an efficient way to become familiar with the process and information that is initially lacking.
3. The third step is the actual design of the process equipment that involves the numerous calculations needed to arrive at specifications of operating conditions, equipment geometry, size, materials of construction, controls, instrumentation, monitors, safety equipment (automatic feed cutoff), etc. As part of this step, equipment costs must be established. Cost-estimating precision is dependent on the desired accuracy of the estimate. As noted earlier in the chapter, if the decision based on an estimate is positive, a detailed project control or contractor's estimate will follow.
4. An overall economic analysis must also be performed in order to determine the process feasibility. The main purpose of this step is to answer the question of whether a process will ultimately be profitable. To answer this, raw material, labor, equipment, and other processing costs are estimated to provide an accurate economic forecast for the proposed operation.
5. In a case where alternate design possibilities exist, economics analysis and engineering optimization is necessary. Since this is often the case, optimization calculations are usually applied several times during most design projects.
6. The final step of this design scheme is the compilation of a design report. A design report may represent the only relevant product of months or even years of effort. This is discussed in the last section. At a minimum, the report should utilize the latest computer-generated layouts and graphics to provide an impressive final product.

These six activities are prominent steps in the traditional development of all modern chemical processes. Today, safety and regulatory (if applicable) concerns have also been integrated into the approach.[9]

As noted earlier, the safe operation of equipment requires that some of the operational parameters be constrained within specific bounds. Each system has parameters that must remain within the appropriate bounds to assure that the system is stable. All systems will have safety equipment to prevent the system from being operated at a condition outside of the safe limits. The equipment may be electronic or mechanical.

The insurance companies such as Industrial Risk Insurers (IRI) and Factory Mutual (FM) and national groups such as the National Fire Protection Agency (NFPA) have recommended specific training requirements; the most important is to assume that all personnel are properly trained on operational limitations of the equipment.

Any environmental regulation requires that each operational limit be monitored to ensure that the system has not been operated when the parameters have been exceeded. However, any of the permit parameters often cannot be monitored on a continuous basis. The design of a system must include both standard and nonstandard operational conditions. Most design conditions are based on the best guess at a maximum flow or operational condition. Variations in the steady-state conditions should be incorporated into the design. As part of any environmental regulatory review, the flow rate, composition, system size, and other physical and chemical characteristics of all streams must be reviewed to assess any potential problems. (See also Chaps. 25 and 26.)

15–8 The Design Report

As pointed out in step 6 in Sec. 15–7, a comprehensive plant design project report is often required. This material should be written in a clear and concise fashion. In addition, the project leader might be requested to make informal or formal presentations to management on the study. The report and presentation should explain what has been accomplished and how it has been carried out. There are many different formats for design reports; the format will vary with the organization and the project. One possible format for a project report is given in the following list:

1. Title page
2. Table of contents
3. Executive summary (abstract)
4. Prefactory comments (optional)
5. Introduction
6. Discussion
7. Laboratory studies (if applicable)
8. Pilot plant studies (if applicable)
9. Comprehensive process design, process flow diagrams, including mass and energy balances and annualized costs
10. Calculation and design limitations
11. Optimization studies (if applicable)
12. Illustrative calculations
13. Recommendations
14. Acknowledgement(s)
15. Appendix

Great care should be exercised with the preparation of the *executive summary*. As discussed in Part I, it is recommended that the executive summary be no longer than one single-spaced typewritten page. It should contain a short introduction, important results, and pertinent recommendations and conclusions. It should *not* refer to the body of the report. In many instances, the executive summary is the only portion of the report that upper-level management will initially review; this one section is unquestionably the most important part of the report.

References

1. L. Theodore, personal notes, East Williston, N.Y., 1967.
2. T. Shen, Y. Choi, and L. Theodore, *Hazardous Waste Incineration Manual*, USEPA/APTI, Research Triangle Park, N.C., 1985.
3. L. Theodore, *EPA Hazardous Waste Incineration*, USEPA Instructor's Guide, Research Triangle Park, N.C., 1986.
4. G. Ulrich, *A Guide to Chemical Engineering Process Design and Economics*, Wiley, N.Y., 1984.
5. R. Felder and R. Rousseau, *Elementary Principles of Chemical Processes*, 2nd ed., Wiley, N.Y., 1986.
6. R. Perry and D. Green, *Perry's Chemical Engineers' Handbook*, 7th ed., McGraw-Hill, N.Y., 1997.
7. W. McCabe, J. Smith, and P. Harriott, *Unit Operations of Chemical Engineering*, 5th ed., McGraw-Hill, N.Y., 1993.
8. J. Santoleri, J. Reynolds, and L. Theodore, *Introduction to Hazardous Waste Incineration*, 2nd ed., Wiley, N.Y., 2000.
9. L. Theodore and R. Dupont, *Environmental Health and Hazard Risk Assessment: Principles and Calculations*, CRC Press/Taylor & Francis Group, Boca Raton, Fla., 2012.

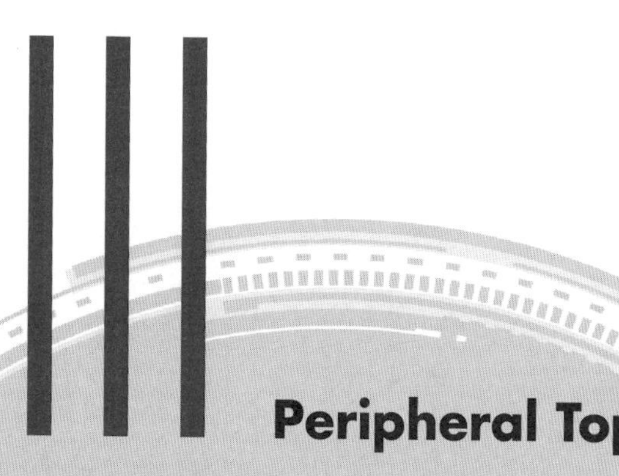

PART III Peripheral Topics

Part Outline

16 Biomedical Engineering

Chapter Outline

16–1 Introduction

Biomedical engineering (BME) is a relatively new discipline in the chemical engineering profession, and, as one might expect, it has come to mean different things to different people. Terms such as *biomedical engineering, biochemical engineering, bioengineering, biotechnology, biological engineering*, and *genetic engineering* have been used interchangeably by many in the technical community. To date, standard definitions have not been created to distinguish between these terms. Consequently, the author has lumped them all together using the acronym BME for the sake of simplicity. What one may conclude here is that BME involves applying the concepts, knowledge, basic fundamentals, and approaches of virtually all engineering disciplines (not only chemical engineering) to solve specific health and health-care related problems; the opportunities for interaction between chemical engineers and health-care professionals are therefore many and varied.

Because of the broad nature of this subject, this chapter can serve only as an introduction. The reader is referred to three excellent references in the literature for an extensive comprehensive treatment of this new discipline.[1–3] However, the bulk of the material in this chapter has been drawn from P. T. Vasudevan.[4]

This subject has served as the title for numerous books. Condensing this subject matter into one chapter was a particularly difficult task. In the end, the author decided to provide a superficial treatment that would introduce the reader to the subject. As noted, Vasudevan[4] provides an excellent and detailed review of key-related topics.

Following this introductory section, the chapter consists of three subsections on basic operations, biological oxygen demand (BOD), design considerations, and design procedures. Additional information is available in Vasudevan's tutorial for the interested reader.[4] Also note that the notation adopted by Vasudevan is employed throughout this chapter.

16–2 Basic Operations

Three basic operations are reviewed in this section. The material has been drawn directly from the work of Vasudevan,[4] where numerous illustrative examples are also provided.

Enzyme and Microbial Kinetics

Enzyme and microbial kinetics involve the study of reaction rates and the variables that affect these rates. It is a topic that is critical for the analysis of enzyme and microbial reacting systems. The rate of a biochemical reaction can be described in many different ways. The most commonly used definition is similar to that employed for traditional reactors. (See also Chap. 13.) It involves the time rate of change in the amount of one of the components participating in the reaction or in one of its products; this rate is also based on some arbitrary factor related to the system size or geometry, such as volume, mass, or interfacial area. In the case of immobilized enzyme-catalyzed reactions, it is common to express the rate per unit mass or per unit volume of the catalyst.

The Michaelis–Menten rate equation is as follows:

$$v = \frac{v_{\max}[\mathrm{S}]}{K_m + [\mathrm{S}]} \tag{16–1}$$

where v = reaction time = $-(d[\mathrm{S}]/dt)$
$v_{\max}$ = constant defined as the maximum reaction rate
K_m = Michaelis constant (dissociation constant)
$[\mathrm{S}]$ = substrate or reactant concentration
t = time

Note that when $K_m >> [S]$, the equation reduces to the following first-order rate of reaction

$$v = \frac{v_{max}}{K_m}[S] \tag{16–2}$$

The rate equation for microbial cell growth can be written as

$$\frac{d[X]}{dt} = \mu[X] \tag{16–3}$$

where [X] = cell growth concentration or mass
t = time
μ = specific growth rate

An enzyme reaction in which $[S] << K_m$ yields

$$\ln\left(\frac{[S]}{[S]_0}\right) = -\frac{v_{max}}{K_m}t \tag{16–4}$$

For the microbial reaction, one obtains

$$\ln\left(\frac{[X]}{[X]_0}\right) = \mu t \tag{16–5}$$

where $[S]_0$ and $[X]_0$ represent the initial substrate and cell concentrations at time = 0, respectively.

Unlike chemical reactions, both enzyme and microbial reactions are generally complex. The mechanism of enzyme-catalyzed reactions is discussed in the next subsection.

Enzyme Reaction Mechanisms

The rate of a chemical or biochemical reaction is similar to that defined earlier, i.e., the time rate of change in the quantity of a particular species participating in a reaction divided by a factor that characterizes the reacting system's geometry. As noted in the previous section, the choice of this factor is also a matter of convenience. For homogeneous media, the factor is almost always the volume of the reacting system. For most fluid-solid reaction systems, the factor is often the mass of the solid. For example, in immobilized enzyme reactions, the factor is the mass of the immobilized or supported enzyme catalyst.

Consider the following scenario for a simple enzyme-catalyzed reaction; the reaction scheme is as follows:

$$E + S \underset{k_{-1}}{\overset{k_1}{\rightleftarrows}} ES$$

$$ES \xrightarrow{k} E + P \tag{16–6}$$

where E = free enzyme
S = substrate or reactant
ES = primary enzyme-substrate complex
P = product

The decomposition of the primary complex ES to the free enzyme E and the product P is assumed to be the rate-determining (slow) step. The expression below is valid for both homogeneous (where the enzyme is used in the native or soluble form) and for immobilized enzyme reactions. The reaction rate v is given by

$$v = -\frac{d[\mathrm{S}]}{dt} = \frac{d[\mathrm{P}]}{dt} \tag{16–7}$$

where [S] and [P] are once again the concentrations of substrate and product, respectively.

There are two approaches in deriving an expression for the reaction rate. In the *Michaelis–Menten approach,* the first reaction is assumed to be in equilibrium. Decomposition of the enzyme-substrate complex ES to form E and P is the rate-determining step. In the second approach, it is assumed that after an initial period, the rate of change of the concentration of the enzyme-substrate complex is essentially zero. Mathematically, this can be expressed as

$$\frac{d([\mathrm{ES}])}{dt} \approx 0 \tag{16–8}$$

This is known as the *quasi-steady-state approximation,* and is valid for enzyme-catalyzed reactions if the initial total enzyme concentration is much less than the initial substrate concentration ($[\mathrm{E}]_0 \ll [\mathrm{S}]_0$).

The maximum reaction rate v_{max} is equal to $k[\mathrm{E}]_0$. When $K_m = [\mathrm{S}]$, then $v = v_{max}/2$. Assuming that Michaelis–Menten kinetics is valid, one can derive an expression for the reaction rate. The total enzyme balance can be written as

$$[\mathrm{E}]_0 = [\mathrm{E}] + [\mathrm{ES}] \tag{16–9}$$

where $[\mathrm{E}]_0$ = the total enzyme concentration
$[\mathrm{E}]$ = concentration of free enzyme
$[\mathrm{ES}]$ = concentration of enzyme-substrate complex

Since Michaelis–Menten kinetics is valid, the reaction between the free enzyme and substrate to form ES [Eq. (16–6)] may be assumed to be in equilibrium:

$$K_m = \frac{k_{-1}}{k_1} = \frac{[\mathrm{E}][\mathrm{S}]}{[\mathrm{ES}]} \tag{16–10}$$

where K_m is the Michaelis–Menten constant. This equation can be combined with the total enzyme balance to provide a relationship between [ES] and the total enzyme concentration $[\mathrm{E}]_0$:

$$[\mathrm{ES}] = \frac{[\mathrm{E}]_0[\mathrm{S}]}{K_m + [\mathrm{S}]} \tag{16–11}$$

The reaction rate v is equal to

$$v = k[\mathrm{ES}] = \frac{k[\mathrm{E}]_0[\mathrm{S}]}{K_m + [\mathrm{S}]} \tag{16–12}$$

In the quasi-steady-state approximation, the constant K_m is known as the *dissociation constant*. Assuming quasi steady-state, the rate of disappearance of the enzyme-substrate complex [ES] is

$$\frac{d[\mathrm{ES}]}{dt} = k_1[\mathrm{E}][\mathrm{S}] - k_1[\mathrm{ES}] - k[\mathrm{ES}] = 0 \tag{16–13}$$

Eliminating [E] by combining it with the equation for the total enzyme balance, and solving for [ES], yields

$$[\mathrm{ES}] = \frac{[\mathrm{E}]_0[\mathrm{S}]}{[(k + k_{-1})/k_1] + [\mathrm{S}]} \quad (16\text{–}14)$$

The reaction rate v then equals

$$v = k[\mathrm{ES}] = \frac{k[\mathrm{E}]_0[\mathrm{S}]}{K_m + [\mathrm{S}]} \quad (16\text{–}15)$$

which is identical to the expression obtained earlier; the only difference lies in the definition of K_m, which is now equal to $(k + k_{-1})/k_1$ instead of k_{-1}/k_1 equilibrium assumption).

Effectiveness Factor

The *effectiveness factor* (η) is the ratio of the reaction rate in the presence of internal or pore diffusion to the reaction rate in the absence of pore diffusion. The value of the effectiveness factor is thus a measure of the extent of diffusional limitations. For isothermal reactions (most biomedical reactions are isothermal), diffusional limitations are negligible when the effectiveness factor (η) is close to unity. If $\eta < 1$, the reaction is diffusion limited.

The problem of pore diffusion is limited only to immobilized enzyme catalysts, and not enzyme-catalyzed reactions in which the enzyme is used in the native or soluble form. Immobilized enzymes are supported catalysts in which the enzyme is supported or immobilized on a suitable inert substance such as alumina, kieselguhr (diatomaceous earth) or silica, or is microencapsulated in a suitable polymer matrix. The shape of the immobilized enzyme pellet may be spherical, cylindrical, or rectangular (as in a slab). If the reaction follows the Michaelis–Menten kinetics discussed previously, then a shell balance around a spherical enzyme pellet can be shown to result in the following second-order differential equation:

$$D_e\left(\frac{d^2[\mathrm{S}]}{dr^2} + \frac{2}{r}\frac{d[\mathrm{S}]}{dr}\right) = \frac{v_{\max}[\mathrm{S}]}{K_m + [\mathrm{S}]} \quad (16\text{–}16)$$

where D_e = effective diffusivity, cm^2/s

The boundary conditions are

1. $[\mathrm{S}] = [\mathrm{S}]_0$ at $r = R$, where R is the radius of the spherical catalyst pellet and $[\mathrm{S}]_0$ is the substrate concentration in the bulk liquid
2. $D[\mathrm{S}]/dr = 0$ at $r = 0$

Because of the symmetry of the pellet, the concentration gradient is zero at the center in condition 2. The preceding differential equation can be solved numerically to determine the concentration profile inside the pellet.

From the definition of the effectiveness factor (η), the actual or observed reaction rate v_s (in the presence of pore diffusion) is equal to

$$v_s = \eta v = \eta\frac{v_{\max}[\mathrm{S}]_0}{K_m + [\mathrm{S}]_0} \quad (16\text{–}17)$$

When the substrate concentration is low, that is, when $[\mathrm{S}] \ll K_m$, the reaction rate becomes first order. In this case, the preceding differential equation can be solved analytically to obtain the concentration profile inside the catalyst pellet. By defining the following dimensionless parameters, one can express the differential equation in a dimensionless form as follows:

$$\bar{S} = \frac{[S]}{[S]_0} \qquad \bar{r} = \frac{r}{R} \tag{16–18}$$

and

$$\left(\frac{d^2 S}{d\bar{r}^2} + \frac{2}{\bar{r}}\frac{d\bar{S}}{d\bar{r}}\right) = 9\phi^2 \bar{S}$$

where

$$\phi = \frac{R}{3}\sqrt{\frac{v_{max}/K_m}{D_e}} = \phi = \text{Thiele modulus} \tag{16–19}$$

Once the concentration profile is known, the effectiveness factor can be expressed as a function of the Thiele modulus by the following relationship:

$$\eta = \frac{1}{\phi}\left(\frac{1}{\tanh 3\phi} - \frac{1}{3\phi}\right) \tag{16–20}$$

The observed reaction rate can then be determined from Eq. (16–17).

For a rectangular catalyst slab or a rectangular membrane in which both sides are exposed to the substrate, the effectiveness factor is related to the Thiele modulus as follows:

$$\eta = \frac{\tanh \phi}{\phi} \tag{16–21}$$

where

$$\phi = \frac{L}{2}\sqrt{\frac{v_{max}/K_m}{D_e}}$$

where L is the thickness of the membrane or catalyst slab.

16–3 Biological Oxygen Demand (BOD)

One of the most widely used parameters of organic pollution in water systems is the 5-day biological oxygen demand (BOD), BOD_5. This involves the measurement of the dissolved oxygen used by microorganisms in the biochemical oxidation of organic matter. The carbon strength of wastewater can be expressed as BOD, chemical oxygen demand (COD), or total organic carbon (TOC). The BOD_5 is the amount of dissolved oxygen consumed when a wastewater sample is seeded with active bacteria and is incubated at 20°C for 5 days.

Biological oxygen demand tests are used to determine the approximate quantity of oxygen that is needed to biologically stabilize the organic matter present, measure the efficiency of treatment processes, determine compliance with wastewater discharge standards, and assist in determining the size of wastewater treatment plants. When the dilution water is seeded, one obtains

$$\text{BOD} = \frac{(D_1 - D_2) - (B_1 - B_2)R}{f} \tag{16–22}$$

where D_1 = dissolved oxygen of fresh diluted sample, mg/L
D_2 = dissolved oxygen after 5-day incubation at 20°C, mg/L
B_1 = dissolved oxygen of seed control before incubation, mg/L
B_2 = dissolved oxygen of seed control after incubation, mg/L
f = volume fraction of sample used
R = ratio of seed in sample to seed in control

The kinetics of the BOD reaction may be assumed to be first-order and can be expressed as

$$\frac{dW_t}{dt} = -kW_t \tag{16–23}$$

where W_t = amount of BOD remaining in water at time t
k = reaction velocity constant

Integrating Eq. (16–23), one obtains

$$\frac{W_t}{W_0} = \exp(-kt) = e^{-kt} \tag{16–24}$$

where W_0 = BOD at time $t = 0$ (total or ultimate BOD initially present)

The amount of BOD that has been exerted (spent) at any time t equals

$$\text{BOD} = y_t = W_0 - W_t = W_0[1 - \exp(-kt)] \tag{16–25}$$

The 5-day BOD equals

$$\text{BOD}_5 = y_5 = W_0 - W_5 = W_0[1 - \exp(-5k)] \tag{16–26}$$

The reader is reminded that BOD is an indication of the dissolved oxygen consumed by the microorganism. For wastewater, a typical value of k is 0.23 day^{-1}. This, however, varies significantly with the type of waste. The oxygen consumption may vary with time and with different k values, even if the ultimate BOD is the same.

16–4 Design Considerations

Numerous factors come into play when designing biochemical systems. These include the power number, key dimensionless numbers, and time-temperature profiles. Each of these factors is briefly discussed below.

Power Number

The power required to drive the impeller is an important consideration in the design of agitated tanks. The power number N_p is analogous to the friction factor or drag coefficient in fluid flow studies.[5] It is proportional to the ratio of the drag force acting on a unit area of the impeller and the inertial stress. To estimate the power requirements for an impeller, the empirical correlations relating power or the power number with other variables of the system such as tank and impeller dimensions, liquid viscosity, liquid density, liquid depth, distance of the impeller from the bottom of the vessel, impeller speed, and baffle dimensions are available. The number and arrangement of baffles and the number of impeller blades also (naturally) affect the power consumption. By dimensional analysis, one can relate the power number to the Reynolds number, the Froude number, and various shape factors.[6] These shape factors are obtained by converting the various linear measurements to dimensionless ratios (usually obtained by dividing these measurements by the impeller diameter). The Froude number provides a measure of the ratio of the inertial stress to the gravitational force per unit area on the fluid.

The power number is not usually affected by the shape of the tank; the power number is affected if the tank is baffled. However, for Reynolds numbers below 300, the power number for baffled and unbaffled tanks is almost identical. Finally, power number calculations are useful in determining the agitator requirements in fermenters.

Key Dimensionless Numbers: Air Sterilization:

Particulate carryover is an important problem in air sterilization. Particulate material (in this case, microbes) is removed by flowing air through a fibrous filter or collector. The particles are removed by a combination of three mechanisms: collision or impaction, interception, and diffusion.[7] Which of these three mechanisms is predominant depends on the particle size and mass. In the case of impaction, the efficiency of collection is a function of the Reynolds and Stokes numbers. The Reynolds number provides information on flow behavior.

The region of particle Reynolds number between 10^{-4} and 0.5 is known as the *Stokes, streamline* or *creeping flow region*. When a particle settles by gravity, it reaches a constant settling velocity, known as the *terminal velocity*. In terms of a force balance, the gravity force on the particle is equal to the sum of the drag force plus the buoyancy force. The terminal velocity for spherical particles in the Stokes region can be easily derived.[6] The velocity is equal to

$$v_t = \frac{g d_p^2 \rho_p}{18\mu} \tag{16–27}$$

where d_p = diameter of particle
ρ_p = density of particle
v_t = terminal velocity
g = acceleration due to gravity
μ = viscosity of medium

Equation (16–27), referred to as the *Stokes equation*, is accurate for spherical particles with diameters of <50 μm, and is frequently used with little error for particle sizes of ≤100 μm. The range of roughly 1 to 100 μm is an important size range for industrial dusts. For particle sizes less than ~2 μm, the particle diameter approaches the mean free path of the gas molecules, and the settling velocity becomes greater than that predicted by Stokes law. Below this lower limit it is necessary to apply Cunningham's[8] correction to the equation for terminal velocity.

The Cunningham correction factor C_f is given by

$$C_f = 1 + \frac{2\lambda}{d_p}\left[1.257 + 0.4\exp\left(\frac{-1.10 d_p}{2\lambda}\right)\right] \tag{16–28}$$

where λ is the mean free path of the molecules in the gas phase. This quantity is given by

$$\lambda = \frac{\mu}{0.499 \rho u_m} \tag{16–29}$$

where u_m = mean molecular speed
μ = given viscosity
ρ = gas density

From the kinetic theory of gases, u_m is given by

$$u_m = \left[\frac{8RT}{\pi(\text{MW})}\right] \qquad (16\text{–}30)$$

where MW = molecular weight of gas
T = temperature in absolute units

The Stokes number is given by

$$\text{St} = \frac{C_f \rho_p d_p^2 u}{18\mu D} \qquad (16\text{–}31)$$

where D = diameter of collection device
u = fluid velocity *within* void space

and

$$u = \frac{u_s}{1-\varepsilon} \qquad (16\text{–}32)$$

where u_s = upstream fluid velocity
$1-\varepsilon$ = volume fraction of void space

The *Peclet number* is an important dimensionless parameter in convective diffusion, and is the product of the Reynolds number and Schmidt number. It finds applications in both fluid flow and mass transfer studies. The Peclet number (Pe = 0) corresponds to complete back mixing, while Pe = ∞ corresponds to plug flow. Large values of the Peclet number result when the mixing is not extensive.[6,7]

The *Schmidt number* is the ratio of the momentum diffusivity (kinematic viscosity) to the molecular diffusivity. In this case, the diffusivity is due to the brownian motion of submicrometer particles. The Schmidt number is given by[7]

$$\text{Sc} = \frac{\mu}{\rho D'} \qquad (16\text{–}33)$$

where D' = diffusivity due to brownian motion, given by

$$D = \frac{C_f kT}{3\pi\mu d_p} \qquad (16\text{–}34)$$

where k is the Boltzmann constant (1.38054×10^{-23} J/K). The Peclet number is the product of the Reynolds number ($Du\rho/\mu$) and the Schmidt number ($\mu/\rho D'$), and is equal to uD/D'. A large value of the Peclet number suggests that there is very little back mixing.

These dimensionless groups are very useful in solving air sterilization problems.

Batch Sterilization: Time-Temperature Profiles

In batch operation, the medium in question is heated to the sterilization temperature, held at the temperature for a definite time, and then cooled to the fermentation temperature. The time-temperature profile depends on the type of heating and cooling. The equations for different types of heating and cooling are available in the literature.[9]

Depending on the type of heating, the appropriate time-temperature equation(s) can be selected for determining the time needed to reach the sterilization temperature when the medium is heated from ambient temperature. If steam sparging is used, it must be remembered that this steam will condense. This amount must be added to the mass of the medium when calculating the time taken to cool the medium from the sterilization temperature to the fermentation temperature. Obviously, the time taken to cool the medium will be larger in this case. The describing equation is[4]

$$T = T_{co}[1 + b\exp(-\alpha t)] \tag{16–35}$$

with

$$\alpha = \frac{wc}{M_p C_\rho}\left[1 - \exp\left(\frac{-UA}{wc}\right)\right] \tag{16–36}$$

where w = covalent flow rate
c = covalent heat capacity
M = initial sterilizer contents
ρ = density of catalyst
T_c = initial constant temperature
T_{co} = muted constant temperature

and

$$b = \frac{T_0 - T_{co}}{T_{co}} \tag{16–37}$$

16–5 Design Procedures

There are a host of design topics associated with biochemical reactors.[10] Discussing these in any detail is beyond the scope of this book. Topics of interest to the practicing chemical engineer are primarily concerned with sterilization. Four specific areas include

1. Design of a batch sterilization unit
2. Design of a continuous sterilization unit
3. Design of an air sterilizer
4. Scale-up of a fermentation unit

Each of these briefly receives qualitative treatment below. Vasudevan[4] provides quantitative details and analyses plus illustrative examples.

Design of a Batch Sterilization Unit

Sterilization is the process of inactivation or removal of viable organisms. Sterilization can be accomplished by the steaming of equipment and/or medium, and by steaming the additives; it is an important operation in the fermentation industry. The main objective of media sterilization is to kill all living organisms present before inoculation and to eliminate any possible competition or interference with the growth and metabolism of the desired organism. This objective should be accomplished with minimal damage to the media[11] ingredients. The thermal inactivation procedure is designed on a probability basis. A typical design value might be a 1-in-1000 chance that at least one organism will survive the sterilization process. One drawback of batch sterilization is that the Del factors[4] for heating and cooling are scale dependent. Large fermenters require longer heating and cooling times. The consequences of longer times can be severe if the medium is thermolabile since the destruction due to heat is dependent on the value of the thermal rate constant. A better alternative for heat sensitive material is to use continuous sterilization.

Design of a Continuous Sterilization Unit

As noted above, the consequences of long heatup and cooldown times in batch sterilization can be severe if the medium components are heat sensitive. The destruction due to heat is dependent on the value of the *thermal death constant*.[4] It can be shown that the spore *Bacillus stearothermophilus* is significantly inactivated only above 110°C because of its high activation energy of 67.7 kcal/mol. On the other hand, many organic nutrients, which also follow the Arrhenius relationship for thermal degradation, have a much lower activation energy of approximately 25 to 30 kcal/mol. This analysis implies that longer exposure to lower sterilization temperatures, due to slow heatup or cooldown, can cause more damage to nutrients. The best alternative for heat-sensitive materials is to use *continuous sterilization*, in which the raw medium is mixed with water, and then continuously pumped through the sterilizer to a sterilized fermenter. In the sterilizer, the medium is instantaneously heated by either direct or indirect contact with steam,[9] and held at a very high temperature (~140°C) for a relatively short time. The residence time or holding time of the medium is fixed by adjusting the flow rate and length of the insulated holding pipe. The hot steam from the sterilizer is rapidly cooled by a heat exchanger (with or without heat recovery) and/or by flash cooling before it enters the fermenter.

The design of continuous sterilizers must also allow some flexibility in operating conditions to adapt the system to a different medium. The system must incorporate automatic recycle to recirculate the medium if the temperature falls below the design value. The design should include:

1. The ability to fill the fermenter within 2 to 3 h (this might be considered downtime)
2. The recovery of 60 to 70 percent of the heat
3. Plug flow in the holding section
4. The option for either direct or indirect heating

The continuous sterilization process has several advantages:

1. The temperature profile of the medium is almost one of instant heating and cooling, allowing an easy estimate of the Del factor required.[4]
2. Scale-up is very simple since the medium is exposed to high temperatures for very short times, thereby minimizing nutrient degradation.
3. The energy requirements of the sterilization process can be dramatically reduced by using the incoming raw medium to cool the hot sterile medium.

The difficulties with continuous sterilizers are typically due to exchanger fouling and control instability. In general, any medium containing starches requires special attention.

Design of an Air Sterilizer

In aerobic fermentations, it is necessary to sterilize air. Since the volume of air required in aerobic fermentations is usually large, conventional techniques of heat sterilization are uneconomical. Effective and viable alternatives include the use of membranes or fibrous filters. An important consideration of the filter medium is that it should not be wetted, since this can lead to deposition contamination. Materials such as glass fibers are used to avoid this problem.

The mechanisms by which particles suspended in a flowing stream of air are removed include the aforementioned impaction, diffusion, and interception. Impaction occurs when particles in the air collide with the fibrous filter as a result of their higher momentum as compared to air. Smaller particles, on the other hand, travel toward the fiber as a result of molecular diffusion caused by brownian motion (see also Sec. 16–4). Particles under 1 μm are collected by this mechanism. Particles that are not small or heavy but are large in size are intercepted by the fiber. The efficiency of air filtration is therefore a combination of the three mechanisms.

The mathematical equations for designing air sterilizers may be developed by considering the effect of each of the above mechanisms separately, and then developing a combined expression. In the case of collision or impaction, the efficiency of the process is a function of the aforementioned Reynolds (Re) and Stokes (St) numbers with the Reynolds number based on the diameter of the filter or collection device.

The effect of the air velocity on efficiency is considerable. Removal of particles by collision and interception is enhanced as the air velocity is increased, whereas the efficiency of particulate removal by diffusion is lowered. This effect is, of course, dependent on the particle size.[5] In general, it should be remembered that the equations for the removal efficiency are empirical.

Scale-up of a Fermentation Unit

Scale-up is a fundamental problem in the fermentation industry because of the need to perform microbial operations in different-size equipment. The scale-up problem arises from the difference in transport phenomena when the scale and geometry of the equipment are changed. Describing transport and kinetic phenomena in fermenters requires knowledge of both the kinetics and flow patterns. Since the balances are very complex and the flow patterns largely unknown, the use of fundamental principles for design is limited. Instead, various empirical procedures are usually employed.

The first step in scaling up a fermentation process is to use the production requirements to determine the size and number of fermenters that will be required. In sizing a new unit, or evaluating an existing fermenter for a biological process, it is necessary to establish the desired product rate R on an annual or daily basis. Then the *productivity*—defined as the weight of product that the fermenter can produce per unit volume per unit time—of the individual fermenter may be determined.

In addition to the fermentation phase, the overall productivity must account for the time spent in activities such as cleaning, filling and sterilizing; this is normally referred to as *downtime*. This is important if the process is to be operated batchwise. The total installed fermenter capacity V required can then be calculated by dividing the desired product rate by the productivity. The number of fermenters can then be easily calculated. The next step is to size the number of inoculum stages. Fermenter precultures must be made to ensure sufficient inoculum for a large fermenter. The production fermenter volume and the optimal inoculum levels will determine how many preculture stages are necessary. Typical inoculum concentrations are bacteria (0.1–3%), actinomycetes (5–10%), and fungi (5–10%).

In practice, the first stage in inoculum production, usually a shaker flask and the final product fermented, are fixed. The number of intermediate levels is usually a compromise between the optimal inoculum levels and available equipment. Because of the requirement for many preculture stages in scaling up to a large volume, the culture may undergo a large number of generations. It is therefore critical to establish whether the culture is genetically stable for these numerous generations. Assuming exponential growth, the number of generations required to achieve a certain biomass concentration can be easily calculated. The number of generations is proportional to the logarithm of the fermenter volume.

References

1. J. Enderle, S. Balnchard, and J. Bronzing, *Introduction to Biomedical Engineering*, 2nd ed, Elsevier/Academic Press, N.Y., 2000.
2. J. Bronzing, ed., *Biomedical Engineering Fundamentals*, 3rd ed, CRC/Taylor & Francis Group, Boca Raton, Fla., 2000.
3. S. Vogel, *Life in Moving Fluids*, 2nd ed, Princeton University Press, Princeton, N.J., 1994.
4. P. Vasudevan, *Biomedical Engineering*, A Theodore Tutorial, Theodore Tutorials, East Williston, N.Y., originally published by USEPA/APTI, Research Triangle Park, N.C., 1994.
5. L. Theodore, *Air Pollution Control Equipment Calculations*, Wiley, Hoboken, N.J., 2008.
6. P. Abulencia and L. Theodore, *Fluid Flow for the Practicing Chemical Engineer*, Wiley, Hoboken, N.J., 2009.
7. L. Theodore and F. Ricci, *Mass Transfer Operations for the Practicing Engineer*, Wiley, Hoboken, N.J., 2010.
8. C. E. Cunningham, *Proc. Roy. Soc. Lond., Ser. A*, **83**, 357, 1910.
9. L. Theodore, *Heat Transfer Applications the Practicing Engineer*, Wiley, Hoboken, N.J., 2011.
10. L. Theodore, *Chemical Reactor Design and Applications for the Practicing Engineer,* Wiley, Hoboken, N.J., 2012.
11. S. Shaefer and L. Theodore, *Probability and Statistics Applications for Environmental Science*, CRC Press/Taylor & Francis Group, Boca Raton, Fla., 2007.

17 Medical Applications

Chapter Outline

17–1 Introduction

The application of engineering principles to the solution of medical problems is a relatively new discipline in both the engineering and medical professions. As one might expect, and as stated in the previous chapter, this "new" discipline has come to mean different things to different people. Terms such as *biomedical engineering*, *biochemical engineering, bioengineering*, *biotechnology*, *biological engineering*, and *genetic engineering* are often used interchangeably in the technical community. As standard definitions have not yet been created to distinguish between these terms, the author has referred to them collectively as *engimed*. Similar to the acronym *BME* (biomedical engineering) introduced in Sec. 16–1, the term *engimed* implies the application of the basic principles and technologies of most engineering disciplines (in addition to chemical engineering) to treatment of human disease and disabilities. There is, therefore, significant potential for successful collaboration between chemical engineers and health care professionals.[1]

On a personal note, the author views engimed as the application of engineering mathematics, and physical sciences to principles in medicine and, to a lesser degree, biology. Furthermore, the terms *biophysics* and *bioengineering* can also imply the interactions of physics or engineering with either biology or medicine. Because of the broad nature of this subject, this chapter (which some may justifiably call *introductory*) addresses the application of engimed to the autonomy of humans and attempts to relate this topic to engineering.[1]

This chapter addresses the following issues: early history, definitions, key anatomic parts, current medical problems, health risk factors, chemical engineering applications to treat cardiac (heart) conditions and sleep apnea, and future applications of chemical engineering.

17–2 Early History

Serious diseases were of primary interest to early humans, although they did not function effectively in treating them. Numerous diseases were attributed to the influence of malevolent demons, which were believed to project an alien spirit, a stone, or a worm into the body of the unsuspecting patient. Such disabilities were to be warded off by incantation, dancing, magic effects, charms and talismans, and various other measures.

Operative procedures practiced in ancient societies included cleaning and treating wounds by cautery, poultices, and sutures, resetting dislocations and fractures, and using splints. Additional therapy included the use of purges, diuretics, laxatives, emetics, and enemas. Perhaps the greatest success was achieved by the use of plant extracts which featured narcotic and stimulating properties. Several prescientific systems of medicine, based primarily on magic, folk remedies, and elementary surgery, existed in various diverse societies before the advent of the more advanced Greek medicine that appeared about the sixth century BC.

No abrupt change in medical thought occurred until the Renaissance. Renaissance artists undertook the study of human anatomy, particularly muscles, in order to better portray the human body. Leonardo da Vinci made remarkably accurate drawings based on the dissection of human corpses. Unfortunately, his work, the bulk of which was lost for centuries, exerted little effect at the time.

Many discoveries made in the nineteenth century led to advances in diagnosis and treatment of disease and in surgical methods. Diagnostic procedures for chest disorders were advanced to an extent in the eighteenth century by the method of percussion first described by the Austrian physician Leopold Auenbrugger von Auenbrug. His work was ignored, however, until 1808, when it was publicized in a French translation by the personal physician to Napoleon. About 1819 the French physician René Hyacinthe Laënnec invented the stethoscope, a tool that remains the most useful single tool of the physician. A number of British clinicians assimilated the new methods of diagnosing diseases, with the result that their names have become familiar through their identification with commonly recognized diseases. The physician Thomas Addison discovered the disorder of the adrenal glands now known as *Addison's disease*. Richard Bright diagnosed nephritis or *Bright's disease*. Thomas Hodgkin described a malignant disease of lymphatic tissue now known as *Hodgkin's disease*. James Parkinson described the chronic nervous disease called *Parkinson's disease*. Robert James Graves diagnosed exophthalmia, sometimes referred to as *Grave's disease*.

In the twentieth century, many infectious diseases were conquered through vaccines, antibiotics, and improved living conditions. Cancer has become a more common illness, but treatments have been developed that effectively combat many forms of the disease. Basic research into life systems also began in the twentieth century. Important discoveries were made in many areas, especially concerning the basis for the transmission of hereditary traits, and the chemical and physical mechanisms for brain function.

17–3 Definitions[2]

There are numerous terms in this topic area that the chemical engineer may not be familiar with. The definition of a host of terms related to health and medicine is provided below. A one-sentence (in most instances) description or explanation of each word or phrase is provided; as one might suppose, the decision of what to include (as well as what to omit) was somewhat difficult.

1. *Anatomy* the structure of an organism or body.
2. *Aorta* the key artery of the body that carries oxygen-rich blood from the left ventricle of the heart to other organs.
3. *Artery* any one of the thick-walled tubes that carry blood from the heart to the other parts of the body.
4. *Atrium* either the left or right upper chamber of the heart.
5. *Autonomic nervous system* the functional division of the nervous system that innervates most glands, the heart, and smooth muscle tissue in order to maintain the internal environment of the body.
6. *Capillary* any of the extremely small blood vessels connecting the arteries with veins.
7. *Cardiac muscle* involuntary muscle possessing much of the anatomic attributes of skeletal voluntary muscle and some of the physiologic attributes of involuntary smooth muscle tissue; it allows the heart to expel blood during systole.
8. *Cell* an extremely small complex unit of protoplasm, usually with a nucleus, cytoplasm, and an enclosing membrane; the semielastic, selectively permeable cell membrane controls the transport of molecules into and out of a cell.
9. *Chronotropic effects* affecting the periodicity of a recurring action such as the slowing (bradycardia) or accelerating (tachycardia) of the heartbeat that results from extrinsic control of the sinoatrial node.
10. *Circulatory system* the course taken by the blood through the arteries, capillaries, and veins and returning to the heart.
11. *Clot* consists primarily of red corpuscles enmeshed in a network of fine fibrils or threads, composed of a substance called fibrin.
12. *Corpuscle* an extremely small particle, especially any of the erythrocytes or leukocytes that are carried and/or float in the blood.

13. *Cytoplasm* the protoplasm outside the nucleus of a cell.
14. *Endocrine system* the system of ductless glands and organs secreting substances directly into the blood to produce a specific response from another organ or body part.
15. *Endothelium* flat cells that line the innermost surfaces of the blood, lymphatic vessels, and the heart.
16. *Erythrocytes* red corpuscles.
17. *Gland* any organ or group of cells that separates certain elements from the blood and secretes them in a form for the body to either use or discard.
18. *Heart* the organ that receives blood from the veins and pumps it through the arteries by alternate dilation and contraction.
19. *Homeostasis* a tendency to uniformity or stability in an organism by maintaining, within narrow limits, certain variables that are critical to life.
20. *Inotropic* affecting the contractility of muscular tissue such as the increase in cardiac power that results from extrinsic control of the myocardial musculature.
21. *Leukocytes* white corpuscles.
22. *Nucleoplasm* the protoplasm that composes the nucleus of a cell.
23. *Plasma* the fluid (primarily water) part of the blood, as distinguished from the corpuscles.
24. *Precapillary sphincters* rings of smooth muscle surrounding the entrance to capillaries where they branch off from upstream metarterioles.
25. *Protoplasm* a semifluid viscous fluid that is the living matter of humans and is differentiated into nucleoplasm and cytoplasm.
26. *Pulse* the expansion and contraction of the arterial walls that can be felt or sensed in all the arteries near the surface of the skin.
27. *Stem cells* a generalized parent cell spawning descendants that become individually specialized.
28. *Vein* any blood vessel that carries oxygen-deficit blood from one part of the body back to the heart; in effect, any blood vessel.
29. *Ventricle* either of the two lower chambers of the heart that receive blood from the atria and pump it into the arteries.

17–4 Key Anatomic Parts

The three key anatomic components discussed in this section are the blood, blood vessels, and plasma and cell flow. The relation of these topics to the general subject of health and medicine should be obvious.

Blood

Blood has been justifiably described as the "river of life." It is the fluid transport medium that serves as a dispenser and collector of nutrients, gases, and waste that allows life to be sustained. In terms of composition, blood is primarily composed of water (H_2O). A human contains about 5 L (5000 mL) of blood. This 5 L volume of blood contains approximately 5×10^{16} red corpuscles (erythrocytes), 1×10^{14} white corpuscles (leukocytes), and 2.5×10^{15} platelets (thrombocytes). Details on these three blood constituents are as follows:

1. The red blood cells are typically round disks, concave on two sides, and approximately 7.5 μm in diameter. They are formed in the bone marrow and have a relatively robust

lifespan of 120 days. These cells take up oxygen as blood passes through the lungs, and subsequently release the oxygen in the capillaries of tissues. From a mechanical perspective, red blood cells have the ability to deform. This is important since erythrocytes often have to navigate through irregular shapes within the vasculature, as well as squeeze through small diameters such as those encountered in capillaries.

2. In contrast, white blood cells serve a wider variety of functions. These cells can be classified into two categories based on the type of granule within their cytoplasm and the shape of the nucleus: granulocytes or polymorphonuclear (PMN) leukocytes, and mononuclear leukocytes. Examples of PMNs include neutrophils, eosinophils, and basophils, while those of mononuclear leukocytes include lymphocytes and monocytes.
3. Platelets are small, round, nonnucleated disks with a diameter about one-third that of red blood cells. They are formed from megakaryocytes in the bone marrow, and have a 7- to-10-day lifespan within the vasculature. The main function of platelets is to stop bleeding.

The final component of blood is *plasma*, the fluid in which the cells remain in suspension. It is composed primarily of water (90%), and the rest plasma protein (7%), inorganic salts (1%), and organic molecules such as amino acids, hormones, and lipoproteins (2%). The primary functions of plasma are to allow both the exchange of chemical "messages" between distant parts of the body and to maintain body temperature and osmotic balance.[3]

In terms of physical properties, the density of blood (as one might suppose) is approximately that of water: is 1.0 g cm^3 or 62.4 lb/ft^3. The other key property is viscosity, which is approximately 50 percent greater than that for water. Although plasma (the main constituent of blood) is a newtonian fluid, blood (with the added blood cells) is a nonnewtonian fluid.[2]

Blood Vessels

The study of the motion of blood is defined as *hemodynamics*. Part of the cardiovascular system involves the flow of blood through a complex network of blood vessels. Blood flows through organs and tissues either to nourish, cleanse, and sanitize them or to be processed in some sense, such as to be oxygenated (pulmonary circulation) or filtered of dilapidated red blood cells (splenic circulation).[4] The aforementioned river of life flows through a "piping" network, which is made up of blood vessels, by the action of two "pump stations" arranged in series. This complex network also consists of thousands of miles of blood vessels. The network also consists of various complex branching configurations. The analogy to some fluid flow systems may be obvious to the reader.

Regarding branching, the aorta divides the discharge from the heart into a number of main branches, which, in turn, divide into smaller ones until the entire body is supplied by an elaborately branching series of blood vessels. The smallest arteries divide into a fine network of still more minute vessels, defined as *capillaries,* which have extremely thin walls; thus, the blood comes into contact with the fluids and tissues of the body. In the capillaries, the blood performs three key functions:

1. It furnishes oxygen to the tissues.
2. It furnishes the nutrients and other essential substances that it carries to the body cells.
3. It takes up waste products from the tissues.

The capillaries then unite to form small *veins*. The veins, in turn, unite with each other to form larger veins until the blood is finally collected into the venal cava, from where it goes to the heart, thus completing the blood vessel circuit (see later section for additional details). This complex fluid flow network is designed to bring blood to within a capillary size of each and every one of more than 10^{14} cells of the body. Which cells receive blood at any given time, how much blood they receive, the composition of the fluid flowing by them, and related physiologic considerations are all matters that are not left up to chance.[5]

The blood vessel branching discussed above may be viewed as flow-through a number of pipes or conduits, an application that often arises in engineering practice. If flow originates from the same source and exits at the same location, the pressure drop across each conduit must be the same. Additional details can be found in the literature.[2]

Plasma and Cell Flow

As described earlier, blood is composed of fluid called *plasma* and suspended cells that primarily include erythrocytes (red blood cells), leukocytes (white blood cells), and platelets. From a fluid dynamics perspective, blood motion can be viewed as a fluid-particle application involving two-phase flow.[2] The blood vessel through which the blood flows has dimensions that are small enough that the effects of the particulate nature of blood cannot be ignored since blood consists of a suspension of the aforementioned red blood cells, white blood cells and platelets in plasma.

17–5 Current Medical Problem Areas

The major medical problem areas currently being faced by society are listed below

Infectious diseases
Immunities
Brain function
Mental illness
Heart disease
Cancer
Pregnancy
Childbirth
Geriatrics (care of the elderly)

Specific categories associated with the medical profession include[6]

Acupuncture
Anesthesiology
Cardiology
Dermatology
Dentistry
Endocrinology
Gastroenterology
Geriatrics
Hematology
Nephrology
Neurology
Oncology
Optometry
Pediatrics
Podiatry
Pulmonary medicine
Rheumatology
Urology
Veterinary medicine

The medical profession has become more focused on expensive medicines and complicated medical devices. However, a countermovement arose in the United States in the past half century. Individuals attempted to take control of their own health largely through preventive medicinal practices, basically employing a system of health care which assists individuals in improving mind, body, and spirit. Among the most important therapies are good nutrition, physical exercise, and *self-regulation* techniques such as relaxation; individuals utilize these so-called unconventional methods in various ways. Included in this category are *herbs*, which are leaves of certain herbaceous plants that grow in temperate climates. Some of the regions include the Far East and Mediterranean areas. These seed plants and the stems wither away to the ground after each season's growth. Ultimately, the resultant herb may be taken either whole, ground, or in the form of an extract which consists of its essential oil(s) dissolved in a solvent alcohol.

Herbs have long been used for both preserving agents and medicinal purposes. Much of the recorded information was accumulated by monks who planted, grew, and studied some of the herbs in use today. Many of the herbs were believed to have mystical properties. For example, thyme was reputed to be a source of courage and valor, while sesame was associated with immortality. Notwithstanding some of the above comments, herbs were recognized for their curative properties. They have withstood the test of time, and their use as a natural medicine is on the rise in today's high-tech society.

17-6 Health Risk Factors

Thomas and Hrudey[7] have selected some risk factors relating to health for individuals, including biological, lifestyle-behavioral, societal, and environmental factors. They are listed in Table 17–1 and the contributors to each risk factor is provided.

For example, age is clearly one of the strongest health risk factors. Although a complete range of health problems occur at all age levels, on average, there is a consistent and exponential increase in health effects starting in the teens. However, the *Weibull distribution*—often referred to as the "bathclub curve"[8]—has been employed to describe the death rate for humans over their *entire* lifespan. (This distribution is especially attractive in describing failure time distributions in industrial and plant applications.[9]) On the other hand, the sex of an individual is not as strongly related to health problems and mortality as is age, while smoking is the single greatest lifestyle-behavioral risk factor. Extensive details are provided in the informative work of Thomas and Hrudey.[7]

Table 17–1

Some selected health risk factors

Biological Factors	Lifestyle-Behavioral Factors	Societal Factors	Environmental Factors
Age	Smoking	Education level	Air pollution
Sex	Alcohol	Economic status	Water pollution
Genetics (heredity)	Drugs	Employment status	Food additives and contaminants
Blood pressure	Diet	Marital status	Chemicals
Cholesterol levels	Exercise	Occupation	Radiation
Diabetes	Weight	Geographic area or region	Natural disasters
	Sexual and reproductive factors	Medical	
	Risk avoidance	Transportation	
	Personality	Recreational	
	Stress	Other	

Source: Ref. 7.

17–7 Chemical Engineering Medical Applications

The Heart

The river of life flows through the cardiovascular circulatory system by the action of the heart, which essentially provides two pumps that are arranged in series. In simple terms, the heart is an organ that receives blood from the veins and propels it into and through the arteries. It is held in place mainly by its attachment to the arteries and veins, and by its confinement, via a double-walled sac with one layer enveloping the heart and the other attached to the breastbone. Furthermore, the heart consists of two parallel independent systems, each containing an *atrium* and a *ventricle* that have been referred to as the *right heart* and *left heart*, respectively.

The following is a description of the cardiovascular circulatory system. The heart is divided into four chambers through which blood flows. These chambers are separated by valves that help keep the blood moving in the correct direction. The chambers on the right side of the heart receive blood from the body and pump it through the lungs to pick up oxygen. The left side takes blood from the body and pumps it out through the *aorta*. The blood squeezes through a system of arteries, capillaries, and veins that then reach every part of the body. The cycle is completed when the blood is returned to the heart by the *superior* and *inferior vena cava* and enters the atrium (or chambers) on the upper right side of the heart. Once again, the analogy to some fluid systems should be obvious to the chemical engineer.

Regarding details of the action of the heart, blood is drawn into the right ventricle by a partial vacuum when the lower chamber relaxes after a beat. On the next contraction of the heart muscle, the blood is squeezed into the pulmonary arteries that carry it to the left and right lungs. The blood receives a fresh supply of oxygen in the lungs and is pumped into the heart by way of the pulmonary veins, entering at the left atrium. The blood is first sucked (drawn) into the left ventricle and then pumped out again through the aorta (the major artery) which connects with smaller arteries and capillaries reaching all parts of the body. Thus, the cardiovascular circulatory system consists of the heart (a pump), the arteries (pipes) that transport the blood from the heart, and the capillaries and veins (pipes) that transport the blood back to the heart. These form a complete *recycle* process.

A line diagram of the recycle circulatory system is provided in Fig. 17–1. The cycle begins at point 1. Oxygenated blood is pumped from the left lower ventricle (LLV) at an elevated pressure through the aorta and discharges oxygen to various parts of the body. Deoxygenated blood then enters the right upper atrium (RUA), passes through the right lower ventricle (RLV), and then enters the lungs, where its oxygen supply is replenished. The oxygenated discharge from the lungs then enters the left upper atrium (LUA) and returns to the lower left ventricle, completing the cycle.

The concepts regarding recycle, bypass, and purge presented in Chap. 7 readily apply to the discussion above. Blood is not only being recycled through the circulatory system but is also being reused. *Bypass* occurs when part of the blood is bypassed to the liver for cleansing purposes. The purging process may be viewed as occurring during the oxygen transfer process as well as the cleansing process.

Disorders of the heart kill more Americans than does any other disease. They can arise from congenital defects, infection, narrowing of the coronary arteries, high blood pressure, or disturbances of heart rhythm. The immediate cause of death in many heart attacks, whether atherosclerosis is present or not, is ventricular fibrillation, also called *cardiac arrest*. (The author has a defibrillator implanted adjacent to his heart.) This is a rapid ineffective beating of the ventricles. As many now know, a human heart from one person can be transplanted into the body of another. In addition, artificial hearts have been under development since the 1950s.

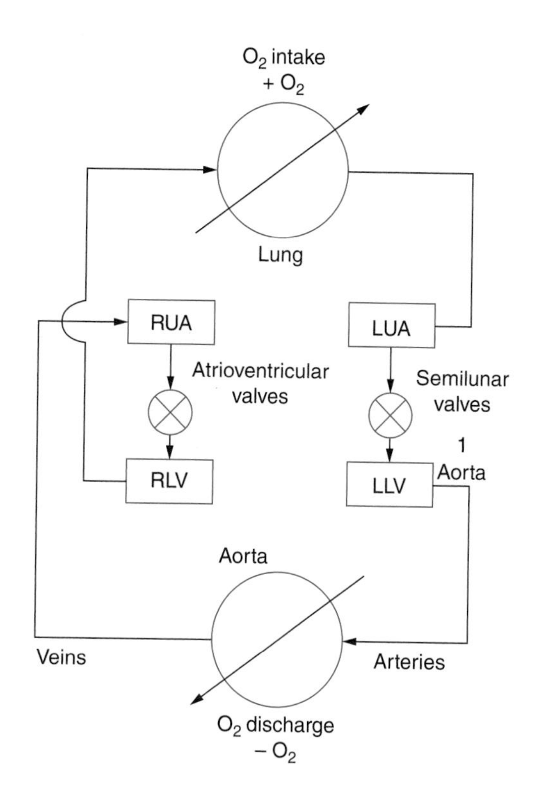

Figure 17–1

Cardiovascular circulatory system.

To summarize, from a fluid flow-engineering perspective, the systemic circulation carries blood to the neighborhood of each cell in the body and then returns it to the right side of the heart low in oxygen and rich in carbon dioxide. The pulmonary circulation carries the blood to the lungs, where its oxygen supply is replenished and its carbon dioxide content is purged before it returns to the left side of the heart to repeat the cycle. The driving force for flow arises from the pressure difference between the high-pressure left side of the heart (systemic) to the lower-pressure right side (pulmonary). This pressure difference provides the impetus for the river of life to flow. Thus, the heart may be viewed as an engine pump that has many of the characteristics of a centrifugal pump; it functions as a double pump—one side of the heart pumps blood to the lung (pulmonary) circuit while the other squeezes blood through a system of arteries, capillaries, and veins that reach every part of the body.

Sleep Apnea[10]

As noted in the previous subsection, the delivery and deposition of oxygen to the heart is a requisite to sustaining life. The breathing process provides a regular and continuous supply of air to the lungs which later delivers the oxygen content of the air to various locations within the body.

One of the oxygen passageways to the lungs is the larynx (often referred to as the *windpipe*); its opening is approximately ½ in in diameter (see Fig. 17–2). However, the passageway can be partially blocked by muscle tissue at the entrance to the windpipe. This tissue normally hangs loosely in the pharynx during most hours of the day. During sleep, particularly when one is dozing face up, the tissue can settle downward, partially blocking the opening. When blockage occurs, the resistance to the flow of air increases; in turn, this reduces the flow of air to the lungs. This sleeping disorder has come to be defined as *sleep apnea*, a disease that has had a direct impact on the author.

Figure 17–2

Air passageway.

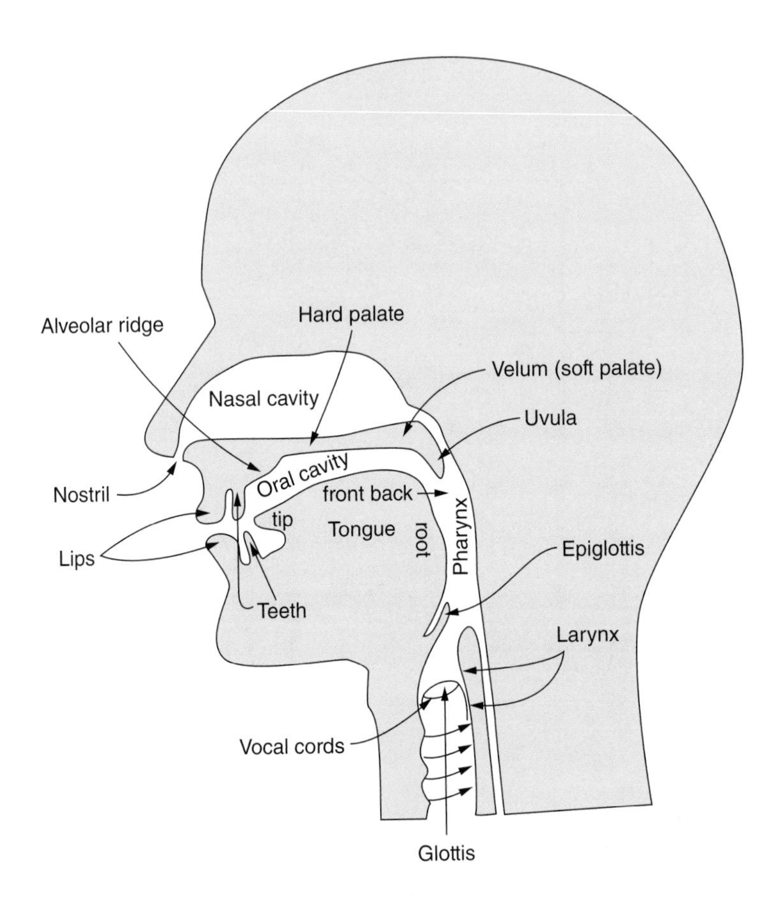

The behavior of the muscle tissue during apnea is similar to the action of a particular type of valve used in the chemical process industry.[2] The valve type that most closely approaches the closing action in the windpipe is the *check valve*, which is activated by the flowing material in the pipeline. The pressure of the fluid passing through the system causes the valve to open. Closure is accomplished by the weight that the check mechanism experiences due to gravity.[2]

Proposed is a simple, inexpensive, painless, and easy-to-implement method that can solve the sleep apnea problem for many; it involves the individual sleeping with a simple device that makes it difficult to turn from sleeping face down.[10] There are several possible solutions to sleep apnea as proposed by the author.[11] One is to sleep on one's stomach with a prop pillow to support one side of the face. This prop pillow can be similar to a doughnut used by pregnant women to relieve pressure from one side of her body. By sleeping on the stomach, the gravitational effect on the trachea and the blockage of the airway by the tongue will be minimized. (Theodore[10] provides details on the solutions.)

In lieu of treatment, other options are available that may reduce or even eliminate snoring and apnea:[12]

1. Lose weight.
2. Don't drink alcohol before going to sleep.
3. Use adhesive strips to hold your nostrils open.
4. Raise the head off the bed while sleeping.
5. Stop smoking.
6. Sleep on your side.
7. Avoid heavy snacks before bed.

17–8 Chemical Engineering Opportunities

Present major activities in the medical field include:[2,4,13]

1. Development of improved species of plants and animals for food production
2. Invention of new medical diagnostic tests for diseases
3. Production of synthetic vaccines from clone cells
4. Bioenvironmental engineering applications to protect human, animal, and plant life from toxicants and pollutants
5. Study of protein-surface interactions
6. Modeling of flow dynamics
7. Modeling of mass transfer through membranes
8. Modeling of the growth of kinetics of yeast and hybridoma cells
9. Research in immobilized enzyme technology
10. Development of therapeutic proteins and monoclonal antibodies
11. Development of artificial hearts

Applications that have emerged since the early 1960s include the following:[2,4,13]

1. Application of engineering system analysis (physiologic modeling, simulation, and control of biological problems)
2. Detection, measurement, and monitoring of the physiologic signals (e.g., biosensors and biomedical instrumentation)
3. Diagnostic interpretation via signal-processing techniques of bioelectric data
4. Therapeutic and rehabilitation procedures and devices (rehabilitation engineering)
5. Devices for replacement of body parts and functions (artificial organs)
6. Computer analysis of patient-related data and clinical decision-making (medical information and artificial intelligence)
7. Medical imaging (the graphical display of anatomic detail or physiologic functions)
8. The creation of new biologic products (biotechnology and tissue engineering)

Present and future job-related activities for chemical engineers may include the following:[4,13]

1. Research into new materials for implanted artificial organs
2. Development of new diagnostic instruments for blood analysis
3. Developing software for analysis of medical research data
4. Analysis of medical device hazards for both safety and efficacy
5. Development of new diagnostic imaging systems
6. Design of telemetry systems for patient monitoring
7. Design of biomedical sensors
8. Development of expert systems for diagnosis and treatment of diseases
9. Design of closed-loop control systems for drug administration
10. Modeling of the physiologic systems of the human body
11. Design of instrumentation for sports medicine
12. Development of new dental materials
13. Design communication aids for individuals with disabilities
14. Study of pulmonary fluid dynamics

15. Study of biomechanics of the human body
16. Development of material to be used as replacement for human skin
17. Applications of nanotechnology to many of the above activates

Obviously, these three lists are not intended to be all-inclusive. Many other applications are evolving that use the talents and skills of the chemical engineer. This is a field where there is continual change and creation of new areas due to the rapid advancement in technology.

In conclusion, this new area of study is now an important, vital interdisciplinary field. The ultimate role of the chemical engineer and the chemical engineering profession is to serve society. The great potential, challenge, and promise in this relatively new endeavor offers both technological and humanitarian benefits. The possibilities appear to be unlimited.

References

1. L. Theodore, personal notes, East Williston, N.Y., 2010.
2. P. Abulencia and L. Theodore, *Fluid Flow for the Practicing Chemical Engineer*, Wiley, Hoboken, N.J., 2010.
3. K. Konstantopoulos, class notes.
4. S. Vogel, *Life in Moving Fluids*, 2nd ed., Princeton University Press, Princeton, N.J., 1994.
5. J. Enderele, S. Blanchard, and J. Bronzino, *Introduction to Biomedical Engineering*, 2nd ed., Elsevier/Academic Press, N.Y., 2010.
6. L. Theodore, personal notes, East Williston, N.Y., 2011.
7. S. Thomas and S. E. Hrudey, *Risk of Death in Canada: What We Know and How Well We Know It*, University of Alberta Press, Alberta, Canada, 1997.
8. S. Shaefer and L. Theodore, *Probability and Statistics Applications for Environmental Science*, CRC Press/Taylor & Francis Group, Boca Raton, Fla., 2007.
9. L. Theodore and R. Dupont, *Environmental Health Risk and Hazard Risk Assessment; Principles and Calculations*, CRC Press/Taylor & Francis Group, Boca Raton, Fla., 2012.
10. L. Theodore, unpublished work, East Williston, N.Y., 2010.
11. ©, L. Theodore, sleep apnea unit, 2008.
12. L. Theodore, "On Sleep Apnea," *The Williston Times*, East Williston, N.Y., Sept. 26, 2003.
13. J. Bronzino, ed., *Biomedical Engineering Fundamentals*, 3rd ed., CRC press/Taylor & Francis Group, Boca Raton, Fla., 2000.

18 Legal Considerations

Chapter Outline

18–1 Introduction

Technology affects almost every area of human activity in one way or another. Therefore, the legal aspects of human interaction should be taken into account. Even though unique developments in technology will continue to be in the realm of reality, those involved with any field of engineering and science, as well as those involved in the legal profession, should consider how technology and law might interact.

It is incumbent upon those engaged in any area of technological development to acquire a basic understanding of patent law, and legal considerations in general, because the patent portfolio of a company, particularly one focused on research and development, may represent its most valuable asset(s). Certain activities, such as premature sale or public disclosure, can jeopardize one's right to obtain a patent; the reader should note that patents are creatures of the national law of the issuing country and are enforceable only in that country. Thus, a U.S. patent is enforceable only in the United States. To protect one's invention in foreign countries, one must apply in the countries in which protection is sought. In addition one can obtain a general idea of the development/progress of a new technological field by monitoring the number of patents issued in that field.

As with any invention, the qualities that make technology-related inventions patentable are *novelty*, *nonobvious*, and *utility*. While many unique aspects of technology are already known, more are being discovered, as are new ways of exploiting known properties, all of which can lead to patentable inventions. The growth of patents in the chemical engineering industry is a clear indication of the recognition of its importance in this field.

Another area of concern is government regulations. Environmental regulations, in particular, have become a concern for new technological development. In fact, for many technologies, it is the increasingly stringent clean air and water regulations that drive new technological development. This is particularly true, for example, in the automotive industry and in heavy industries where emissions are released into the atmosphere or water. What effects will new technology have on the environment? This remains unknown because the developments are still, relatively speaking, in their infancy. But suppose, for example, that new products are developed, products that can be ingested or inhaled, and that act on the interior organs of the human body. Suppose also that such products are released into the atmosphere. They might be carried by air within the state or across state lines. The state and federal environmental regulatory agencies would understandably be very interested in the environmental impact of such a release.

Although a host of topics are examined in this chapter, patents receive the bulk of the treatment. The following topics are also discussed: differences between laws and regulations, intellectual property law; contract law; patents and copyright infringement and interference; proprietary trademarks; and the engineering professional licensing process.

NOTE Some of the material presented in this chapter was drawn from the work of A. Calderone on legal considerations.[1]

18–2 Laws and Regulations: The Differences

The following are some of the major differences between a federal *law* and a federal *regulation*:[2, 3]

1. A law (or Act) is passed by both houses of Congress and signed by the President. A regulation is issued by a government agency such as the USEPA or the Occupational Safety and Health Administration (OSHA).
2. Congress can pass a law on any subject it chooses. It is limited only by the restrictions in the Constitution. A law can be challenged in court if it is unwise, unreasonable, or even ridiculous. If, for example, a law was passed that placed

a tax on coughing, spitting, or sneezing, it could not be challenged in court just because it was unenforceable. A regulation can be issued by an agency only if the agency is authorized to do so by a law passed by Congress. When Congress passes a law, it usually assigns an administrative agency to implement that law. A law regarding radio stations, for example, may be assigned to the Federal Communications Commission (FCC). Sometimes a new agency is created to implement a law. This was the case with the Consumer Product Safety Commission (CPSC). The OSHA is authorized by the Occupational Safety and Health Act to issue regulations that protect workers from not only exposure to the hazardous chemicals that they use in manufacturing processes but also accidents.

3. Laws can include a Congressional mandate directing an agency to develop a comprehensive set of regulations. Regulations, or rule makings, are issued by an agency that translates the general mandate of a statute into a set of requirements for the agency or regulatory body and the regulated community.
4. Regulations are developed in an open and public manner according to an established process. When a regulation is formally proposed, it is published in an official government document called the *Federal Register* to notify the public of the intent to create new regulations or modify existing ones. The public, which includes the potentially regulated community affected, is provided an opportunity to submit comments. Following an established comment period, the proposed rule may be revised on the basis of both an internal review process and public documents.
5. Once promulgated, the final regulation is published in the *Federal Register*. Included with the regulation is a discussion of the agency's rationale for the regulatory approach, known as *preamble language*. Final regulations are compiled annually and incorporated in the Code of Federal Regulations (CFR) according to a structured format based on the topic(s) of the regulation. This latter process is called *codification*, and each CFR title corresponds to a different regulatory authority.
6. A regulation may be challenged in court because the issuing agency exceeded the mandate given to it by Congress. If the law requires the agency to consider costs versus benefits of the regulation, the regulation could be challenged in court if the cost-benefit analysis was not correctly or adequately performed.
7. Laws are usually brief and general while regulations are usually lengthy and detailed. The Hazardous Materials Transportation Act (HMTA), for example, is only approximately 20 pages long. It speaks in general terms about the need to protect the public from dangers associated with transporting hazardous chemicals and identifies the Department of Transportation (DOT) as the regulatory agency responsible for issuing permits and implementing the law. The regulations issued by the DOT are several thousand pages long and are very detailed—details that include the exact size, shape, design, and color of the warning placards that must be used on trucks carrying any of the thousands of regulated chemicals.
8. Generally, laws are passed *infrequently*. Often, many years pass between amendments to an existing law. A completely new law on a given subject already addressed by an existing law is unusual. Laws are published as a "Public Law…" and are eventually codified into the United States Code.
9. Regulations are issued and amended *frequently*. Proposed and final new regulations and amendments to existing regulations are published daily in the *Federal Register*. Final regulations have the force of law when published.

Regulations are codified annually, in the Code of Federal Regulations (CFR), (see item 5 above). The CFR is divided into 50 volumes called *Titles*. Each Title is devoted to a subject or agency. For example, labor regulations are in Title 29, while environmental regulations are in Title 40.

18–3 Intellectual Property Law

One area of law with which the chemical engineer must be concerned is *intellectual property law*. Technological development is all about ideas. Ideas have commercial value only if they can be protected by excluding others from exploiting those ideas. Typically, ideas are protected through intellectual property rights such as *patents*, *trademarks*, *copyrights*, *trade secrets*, and *financial secrets*.

Patents can be used to protect useful inventions, ornamental designs, and even botanical plants. The patent allows the owner of the patent the right to prevent anyone else from making, using, or selling the "invention" covered by the claims of the patent. Trademark symbols are distinctive marks associated with a product or service (these are usually referred to as *service marks*), which the owners of the mark can use exclusively to identify themselves as the source of the product or service. Copyrights protect the expression of an idea, rather than the idea itself, and are typically used to protect literary works (also as this book) plus visual and performing arts, such as photographs, diagrams and figures (such as in this book), paintings, drawings, sculptures, movies, and songs, Trade-secret laws protect technical or business information that a company uses to gain a competitive business advantage by virtue of the secret being unknown to others. Customer or client lists, secret formulations, or methods of manufacture are typical business secrets.

18–4 Contract Law

Another legal area that is relevant to chemical engineers is *contract law*. Whenever two or more parties agree on something, the principles of contract law come into play. The essential components of a contract are:

1. Parties competent to enter into a contractual agreement on subject matter (what the contract is about)
2. Legal considerations [the inducement to contract such as the promise(s) or payment exchanged, or some other benefit or loss or responsibility incurred by the parties]
3. Mutuality of agreement
4. Mutuality of contract

While oral contracts can be legally binding, in the event of a dispute, it may be difficult to establish in court who said what. It is far better to memorialize the agreement in the form of a written contract. Who are the entities engaged in the contract? Typically, these are business entities. One must also consider whether there may be some peripheral issues of corporation or partnership law.

One of the basic principles of contract law is that the parties should have a meeting of the minds, i.e., they should have a common understanding of what the terms of the contract mean. Sometimes it is not clear what particular terms mean or represent, or the meaning or its implications may change in time. What, for example, qualifies as *technology*? Not only is technology not well defined today; it may encompass future entities or concepts that which are not even imagined today.

Generally, contracts are employed with the sale and licensing of exclusive rights to a technology. There are also agreements to fund technological research and development.

Finally, it should be noted that *most* chemical engineers leave *most* legal activities and decisions to the legal department. However, the chemical engineer *should* recognize that detailed contractual relation(s) should be maintained. Unfortunately, this is seldom the case.

18-5 Tort Law

There is a branch of law that can retrospectively address certain situations in which property or people are harmed. That is *tort law*, a topic that continues to receive significant attention by politicians. A *tort* is a civil wrong, other than a breach of contract, for which the law provides a remedy. One can recover damages under tort law if a legal duty has been breached that causes foreseeable harm. These duties are created by law other than duties created by criminal law, governmental regulations, or those agreed to under a contract. Tort law can be very encompassing.

Chemical engineers have to consider the possibilities of reasonably foreseeable harm arising from their developments and activities, and take prudent precautions to avert such harm. In the event that a technology is inherently dangerous, chemical engineers and/or their company may be held to a standard of strict liability for any harm caused by the technology regardless of whether an accident or problem was foreseeable.

Most chemical engineers are not at all interested in deliberately causing harm. But some have and are doing so presently. Technology can include military applications, where governments may be interested in developing new weapons. Suppose that new weapons are developed that can invade the human body and do harm. Is a cloud of toxic gas considered a poisonous gas? Or is that cloud a collection of antipersonnel objects such as shrapnel? And suppose such a cloud drifts over, or is released over, a civilian population? How will new weapons be treated under the Geneva Convention? The devastating effect of land mines, which remain lethal long after hostilities are ended and which wreak havoc on unsuspecting civilians wandering into minefields, has been amply documented. Will new weapons remain harmful years after their deployment? What responsibilities do government(s) have morally, and under international law?[4]

But suppose that new products are not designed to cause harm but simply to obtain and transmit information. For example, suppose that such new products, if ingested, provide information about bodily functions. Or suppose that they enable groups to be tracked wherever they go. Devices are already known that enable a person to be tracked by global positioning satellites. These devices are worn voluntarily. These are also implantable devices (the author of this text has one). But such smaller products would be undetectable and could very well be implanted in someone without that person's knowledge or consent. Under what circumstances should such invasions of privacy be allowed or forbidden? One could easily envision other questions or concerns.

Then there is the matter of the criminal use of new technology. Since the beginning of this new millennium there has been growth of a new area of crime: computer crime. For example, in addition to conventional theft, law enforcement agencies must now become technologically proficient to handle computer fraud identity theft, theft of information, embezzlement, copyright violation, computer vandalism, and similar activities, all accomplished over the computer network under conditions such that not only is the criminal difficult to trace but even the crime may go undetected. The computer criminals are technologically savvy and willing to exploit the potentials of any new technology. About the only thing one can expect is that if technology provides great new potentials, someone may use these new inventions for criminal purpose, and the laws will again be forced to play catch-up in response to the crimes after the harm has occurred.

18-6 Patents

In order to encourage new discoveries for the benefit of society and humankind, the U.S. Constitution provides for *patents*. These are *limited monopolies* provided in exchange for the public disclosure of new products and inventions. Patents are an integral part of a free enterprise system. The process discourages secret behavior by rewarding the monopoly for prompt and adequate disclosure. The U.S. patent system is responsible for much of the growth in the chemical engineering industry because it encourages research on which the growth is based. Attaining a patent is a procedure requiring skilled and experienced guidance. The patent must be fully disclosed, and the essentials must be covered by the claims.

Of all the intellectual property rights, the most pertinent for products and inventions developed by chemical engineers are patents. A patent can protect, for example, a composition of matter, an article of manufacture, or a method of doing something. Patent rights are private property rights. *Infringement* (see Sec. 18–7) of a patent is a civil offense, not criminal. Patent owners must come to their own defense through litigation, if necessary. Also, this is a very expensive undertaking. Lawsuits costing more than a million dollars are not unusual. But at stake can be exclusive rights to a technology worth several orders of magnitude more.

Patents pertaining to any new product, process, equipment, use, or application should be reviewed by a patent attorney. The product or invention of concern may be involved in other patents. Patents available for purchase and lease as well as participation in patent pools should be reviewed by a legal expert.

As with any invention, the qualities that make technology-related inventions patentable are *novelty*, *nonobvious*, and *utility*. While many processes of technology are already known, more are being discovered as are new ways of exploiting these, all of which can lead to patentable inventions. As noted above, the growth of patents is a clear indication of the industry's recognition of the potential in this field.

It is fair to say that a patent is essentially a contract between an inventor and the public. By full disclosure of the invention to the public, the patentee is granted exclusive rights to control the use and practice of the product or invention. A patent gives the holder the power to prevent others from using or practicing the invention for a period of years from the date of granting. In contrast, *trade secrets* (see Sec. 18–9) receive protection as long as the information is not public knowledge.

A patent may be obtained on any new and useful process, method of manufacture, or composition of matter, provided it has not been known or used by others before the patentee made the invention or discovery. The invention must not have been described in a printed publication or been in public use or on sale for more than 1 year prior to the patent application. A patentable item must result from the use of creative ability above and beyond that which would be expected of a person working in the particular field. A patentable item cannot be something requiring merely mechanical skill. Furthermore, a patent will not be granted for a change in a previously known item or process unless the change involves something entirely *new*.

A patent application consists of

1. A petition, directed to the Commissioner of Patents and requesting the grant of a patent
2. An oath, sworn to before a notary public or other designated officer
3. Specifications and claims, in which the claims to be patented are indicated along with detailed specifications including drawings and other pertinent information
4. The application filing fee

When the application is examined and after a period of time, official action on the claim is taken.[5]

The publication of patent applications by the U.S. Patent and Trademark Office (USPTO) has given researchers in both academia and industry a means of following new developments in the particular fields of interest. This is particularly true in chemical engineering. The U.S. Patent Classification System (USPC) is a system for organizing all U.S. patent documents and many other technical documents into relatively small collections based on common subject matter. Each subject matter division in the USPC includes a major component called a *class* and a minor component called a *subclass*. A class generally delineates one technology from another. Subclasses delineate processes, structural features, and functional features of the subject matter encompassed within the scope of a class. Every class has a unique alphanumeric identifier, as do most subclasses. A class-subclass pair identifies a subclass within a class (for example, the identifier "2/456" represents Class 2; Apparel, Subclass 456, Body Cover). This unique identifier is called a *classification symbol*, or simply a *classification*, or USPC classification, to distinguish it from classifications of other patent classification schemes. A subclass represents the smallest division of subject matter in the USPC under which documents may be collected. A *collection of documents* is defined as a set of documents sharing a common classification. A classification assigned to a document associates the document to the class and subclass identified by the classification. Documents are "classified in a subclass" if a classification corresponding to the unique subclass has been assigned to it. A document may be a member of more than one collection; i.e., it may have more than one classification assigned to it. Classifications are assigned to documents according to disclosure(s) in the document. The USPC serves both to facilitate the efficient retrieval of related technical documents and to route patent applications within the USPTO for examination.

18-7 Copyright Infringement and Interference

Two topics directly related to patents are infringement and interference. Infringement is of greater concern to the chemical engineer. However, both subject areas are briefly discussed below.

Infringement

The *infringement* of a patent may consists of making, using, or selling an invention covered by the patent without permission of the patentee. A *contributory infringement* involves the assistance or cooperation with another in the unauthorized making, using, or selling of a patented invention. If the infringement is deliberate, the court may award the patentee as much as three times the actual damages caused *plus* three times the earned profits. In the case of unintentional infringement the award to the plaintiff is no more than the actual loss.

The infringement process is conducted by a search in the U.S. Patent Office (USPO). Every feature of the patent is studied. The determination of the scope and the validity of a patent with respect to infringement is a question of law. As such, it should be undertaken by a patent attorney before any infringement action is taken. The USPO offers the following related to infringement:

1. Except as otherwise provided, whoever without authority makes, uses, offers to sell, or sells any patented invention, within the United States, or imports into the United States any patented invention during the term of the patent therefore, infringed the patent.
2. Whoever actively induces infringement of a patent shall be liable as an infringer.
3. Whoever offers to sell or sells within the United States or imports into the United States a component of a patented machine, manufacture, combination, or composition, or a material or apparatus for use in practicing a patented process, constituting a material part of the invention, knowing the same to be especially made or especially adapted for use in an infringement of such a patent, and not a staple article or commodity of commerce suitable for substantial noninfringing use, shall be liable as a contributory infringer.

4. Whoever without authority supplies or causes to be supplied in or from the United States all or a substantial portion of the components of a patented invention, where such components are uncombined in whole or in part, in such manner as to actively induce the combination of such components outside of the United States in a manner that would infringe the patent if such a combination occurred within the United States, shall be liable as an infringer.

Interference

A situation can arise in which two or more independent patent applications covering essentially the same invention is on file in the USPO. Although it rarely occurs, a procedure called *interference* is instituted to determine who is entitled to the patent. For example, an interference may also be instituted between a pending application and a granted patent. Generally, interferences are decided on the basis of priority. The patent is granted to the applicant who was first to conceive the idea for the invention.

Because of the role of priority in any type of interference process, it is very important for an inventor to maintain complete records. A written description and sketches should be prepared by an inventor as soon as possible after the conception of an idea that might eventually be patented. This material should preferably be disclosed to one or more witnesses who should indicate in writing that they understand the purpose, method, and structure of the invention. The disclosure should be signed and dated by the inventor and the witnesses. Additional details are available in the USPO.[7]

18-8 Copyrights

Copyright is a form of protection provided by U.S. law to the author of "original works of authorship" fixed in any tangible medium of expression. The manner and medium of fixation are virtually unlimited. Creative expression may be captured in words, numbers, notes, pictures, or any other graphic or symbolic medium. The subject matter of copyright is extremely broad. Although protection is available to both published and unpublished works, the author believes that this rarely arises in the publication of science and engineering works since the law of gravity is the law of gravity, the heat transfer equation is the heat transfer equation, and the multiplication and/or log tables are just that; as the author once said: "if you've read one thermodynamic textbook, you've read them all."[8]

Under the 1976 Copyright Act, the copyright owner has the exclusive right to reproduce, adapt, distribute, publicly reform, and publicly display the work. These exclusive rights are transferable and may be licensed, sold, donated to charity, or bequeathed to one's heirs. It is illegal for anyone to violate any of the exclusive rights of the copyright owner. If the copyright owner prevails in an infringement claim, the available remedies include preliminary and permanent injunctions (court orders to stop current or prevent future infringements), impounding, and destroying the infringing articles.

The exclusive rights of the copyright owner, however, are limited in a number of important ways. Under the *fair-use* doctrine, which has long been a part of U.S. copyright law and was expressly incorporated in the 1976 Copyright Act, a judge may excuse unauthorized uses that may otherwise be infringing. Section 107 of the Copyright Act lists criticism, comments, news reporting, teaching, scholarship, and research as examples of uses that may be eligible or the fair-use defense. In other instances, the limitation takes the form of a *compulsory license* under which certain limited uses of copyrighted works are permitted on payment of specified royalties and compliance with statutory conditions. The Copyright Act also contains a number of statutory limitations covering specific uses for educational purposes.

A copyright is secured automatically when the work is created, and a work is "created" when it is fixed in a "copy or a phonorecord for the first time." For example, a song can be fixed in sheet music or on a CD (compact disk), or both. Although registration with the Copyright Office is not required to secure protection, it is highly recommended for the following reasons:

1. Registration establishes a public record of the copyright claim.
2. Registration is necessary before an infringement suit may be filed in court (for works of U.S. origin).
3. If made before or within 5 years of publication, registration establishes prima facie evidence in court of the validity of the copyright and of the facts stated in the certificate.
4. If registration is made within 3 months after publication of the work or prior to an infringement of the work, statutory damages and attorneys' fees will be available to the copyright owner in any court actions. Otherwise, only an award of actual damages and profits is available to the copyright owner.
5. Registration allows the owner of the copyright to record the registration with the U.S. Customs Service for protection against the importation of infringing copies.

There is no such thing as an "intentional copyright" that will automatically protect an author's works in countries around the world. Instead, copyright protection is *territorial* in nature, which means that copyright protection depends on the national laws where protection is sought. However, most countries are members of the Berne Convention on the Protection of Literary and Artistic Works and/or the Universal Copyright Convention, the two leading international copyright agreements, which provide important protections for foreign authors. The U.S. Patent Office provides additional details.[9]

18–9 Trademarks

A *trademark* is a brand name. A trademark includes any word, name, symbol, device, or any combination, used, or intended for use, in commerce to identify and distinguish the goods of one manufacturer or seller from goods manufactured or sold by others, and to indicate the source of the goods. A *service mark* is any word, name, symbol, device, or any combination, used, or intended for use in commerce to identify and distinguish the services of one provider from services provided by others, and to indicate the source of the services. Not all trademarks need to be registered. But federal registration has several advantages, including a notice to the public of the registrant's claim of ownership of the mark, a legal presumption of ownership nationwide, and the exclusive right to use the mark on or in connection with the goods or services set forth in the registration.

The trademark process initially consists of the following seven steps:

1. Determines whether protection is required
2. Determines whether to hire a trademark attorney
3. Identifies the mark format
4. Clearly identifies the precise goods or services to which the mark will apply
5. Determines whether anyone is already claiming trademark rights on a particular mark through a federal registration
6. Identifies the proper basis for filing a trademark application
7. Files the application online through the Trademark Electronic Application System

Filing is a relatively simple process. Additional details are provided by the U.S. Patent Office.[10]

18–10 The Engineering Professional Licensing Process

Becoming a licensed professional engineer (PE) was really not that important in the "old days," particularly for non–civil engineers. In fact, the author of this text is not a PE. Interestingly, the author has claimed in all honesty, that it has not affected his professional development activities. However, this situation has changed. Chemical engineers who are not licensed will eventually impose significant constraints on their future professional development. In effect, licensing has become a necessity for the chemical engineer of the future.

The following four requirements must be satisfied for one to become a licensed professional engineer:

1. Obtaining a degree from a 4-year engineering program accredited by the Accreditation Board for Engineering and Technology, Inc. (ABET)
2. Passing the Fundamentals of Engineering (FE) examination
3. Completion of 4 years of acceptable engineering experience
4. Passing the Principles and Practice of Engineering (PE) examination

The two main obstacles to licensing for chemical engineers are passing the two examinations. Unfortunately, the author of this text feels that the administration of the FE exam is conducted in a shoddy, unprofessional, and perhaps unethical manner. Despite numerous complaints over the last half century, the current exam (as well as its predecessors) continues to be geared and directed toward the civil engineering profession. Another major negative criticism is that the FE exam administrators have refused to distribute all old exams. However, these exams are reportedly available to a select few. This, too, places an unfair burden on the majority of engineers who do not have access to these exams.

It appears that the engineering profession is moving toward a goal of requiring a license for all individuals who desire to practice engineering. This statement also applies to chemical engineers.

References

1. A. Calderone (contributing author), L. Theodore, and R. Kunz, *Nanotechnology: Environmental Implications and Solutions*, Wiley, Hoboken, N.J., 2005.
2. W. Matystik, personal notes communicated to L. Theodore, East Williston, N.Y., 1998.
3. W. Matystik, L. Theodore, and R. Diaz, *State Environmental Agencies on the Internet*, Theodore Tutorials (originally published by Government Institute), East Williston, N.Y., 1999.
4. L. Theodore, *Nanotechnology: Basic Calculations for Engineers and Scientists*, Wiley, Hoboken, N.J., 2006.
5. M. Peters, *Plant Design and Economics for Chemical Engineers*, McGraw-Hill, N.Y., 1958.
6. http://www.uspto.gov/web/offices/pac/mpep/documents/appxl_3
7. http://www.uspto.gov/ip/boards/bpai/index.jsp
8. L. Theodore, personal notes, East Williston, N.Y., 1974
9. http://www.uspto.gov/web/offices/dcom/olia/copyright/copyright
10. http://www.uspto.gov/trademarks/process/index.jsp
11. L. Theodore, personal notes, East Williston, N.Y., 1994

19

Purchasing Equipment

Chapter Outline

19–1 Introduction

All purchasing equipment processes have beginnings, middles, and ends. Since processing equipment projects arise out of needs, the whole process begins when someone or something has a need to be fulfilled that involves equipment. Choices have to be made regarding equipment selection. Some equipment options are selected, while others are rejected. Decisions are made on the basis of available resources, the needs which must be addressed, the cost of fulfilling those needs, and the relative importance of satisfying one set of needs and ignoring others.

The potential sources of financing for equipment are varied and depend on a host of factors. It is important to adhere to the equipment project budget. The success or failure of the purchasing process will often be judged according to whether the project comes in under, on, or over budget. Exceeding the budget can have serious consequences for a chemical engineer or organization. For example, if a project is funded through a contract, a cost overrun may lead to litigation, penalties, and other financial losses. If the equipment is funded internally, an overrun may lead to a drain of organizational resources.

Project reports provide one of the best means of tracking the progress of a project. Progress reports are used to monitor the purchasing activities and compare them to a schedule and budget. These reports should contain information on the status, schedule, budget, goals achieved, goals not achieved, goals due, important meetings, correspondence, etc., as they relate to the purchase of equipment.

A successful equipment purchasing project is dependent on a number of factors, including:

1. Updating the purchasing plan
2. Staying within the scope of work specified for the equipment
3. Obtaining authorization for any changes
4. Conducting project review meetings, if applicable
5. Checking any technical calculations associated with the equipment

Ten relatively short sections on the following topics complement the presentation for this chapter: factors in equipment selection, preliminary studies, equipment design and specifications, materials selection, instrumentation and controls, equipment selection and fabrication, installation procedures, and purchasing equipment guidelines.

NOTE A significant portion of the material presented in this chapter was drawn from a 1982 reference text by Theodore and Buonicore.[1] There is significant overlap between the material in this chapter and that in Chaps. 15, 24, and 28.

19–2 Factors in Equipment Selection

Numerous factors must be considered before selecting a particular piece of hardware.[1] In general, they can be grouped into three categories: engineering, economic, and environmental:

1. Engineering
   a. System characteristics (i.e., physical and chemical properties, concentration, chemical reactivity, corrosivity, abrasiveness, toxicity, etc.)
   b. Fluid stream characteristics (i.e., volume flow rate, temperature, pressure, humidity, composition, viscosity, density, reactivity, combustibility, corrosivity, toxicity, etc.)
   c. Design and performance characteristics of the particular equipment (i.e., size and weight, pressure drop, reliability and dependability, turndown capability, power requirements, utility requirements, temperature and pressure limitations, maintenance requirements, etc.)

2. Economic (see also Chap. 28)
   a. Capital cost (equipment, installation, engineering, etc.)
   b. Operating cost (utilities, maintenance, etc.)
   c. Expected equipment lifetime and salvage value
3. Environmental (see also Chaps. 25 and 26)
   a. Equipment location
   b. Available space
   c. Ambient conditions
   d. Availability of adequate utilities (e.g., power, water) and ancillary system facilities (e.g., wastewater treatment and disposal)
   e. Maximum allowable emissions (environmental regulations)
   f. Aesthetic considerations (e.g., visible steam or water vapor plume)

19-3 Preliminary Studies

Knowledge at the preliminary stage of a purchasing project is vital to prevent financial loss on one hand and provide opportunity for success on the other. In well-managed process organizations, the engineer's evaluation is a critical activity that usually involves preliminary research on the proposed equipment. Successful process development consists of a series of actions and decisions, the most significant of which takes place well before purchasing and (if applicable) construction.

It is important to determine whether an equipment project has promise as early in the purchasing stage as possible. There may be some preparatory work required before the equipment purchasing process can be formally initiated. In the chemical process industry, there may be an extended period of preparatory work required if the proposed equipment is a unique or first-time application. This can involve bench-scale work to develop and better understand the unit. This is often followed by pilot experimentation to obtain scale-up and equipment performance information. However, these two steps are seldom required in the purchase of an established unit. This is presently the situation with most equipment purchases, although some bench-scale or pilot work may be necessary and deemed appropriate by management

In some cases, equipment data on similar and/or existing units are normally available and economic estimates or process feasibility are determined from these data. It should be pointed out again that most equipment purchases in real practice are based on duplicating or *mimicking* similar existing equipment. Simple algebraic correlations that are based on past experience are the rule rather than the exception. This stark reality is often disappointing and occasionally depressing to students and novice engineers involved in design. The only preparatory work normally required is the gathering of all existing physical and chemical property data of the system and auxiliary materials. Process chemistry (including physical chemistry) information may also be needed, but this, too, may be obtained directly from the literature, company files, or a similar type of application.

19-4 Equipment Design

Equipment design calculations are often based on maximum throughput capacities. The reason for this is to enable the unit to perform satisfactorily under the most extreme conditions, e.g., maximum flow. Material and energy balances based on these conditions are required before the individual equipment design. Each piece of equipment is designed individually and independently of all others. Equipment design is also the first major step toward assessing the capital costs or purchase price (see also Chaps. 15 and 28).

Six conceptual steps are usually considered with the design of equipment:

1. Selecting the equipment in question
2. Identification of the parameters that must be specified
3. Application of the fundamentals underlying theoretical equations or concepts
4. Enumeration, explanation, and application of simplifying assumptions
5. Possible use of correction factors for nonideal behavior
6. Identification of other factors that must be considered for adequate equipment specification

Although all chemical engineers approach equipment design somewhat differently, six sequential steps are generally required; these six steps are discussed below and may be applied to the design of most equipment:

1. Conceptualize and define the equipment.
2. A method of solution must be sought after the need has been defined.
3. The actual physical design of the process equipment that involves the numerous calculations needed to arrive at specifications for operating conditions, such as equipment geometry, size, materials of construction, controls, monitors, instrumentation, and safety equipment (e.g., automatic waste feed cutoff) must be performed.
4. An economic analysis must also be performed.
5. In a case where alternate equipment possibilities exist, economics and engineering optimization may be necessary.
6. The final step of this design scheme is the preparation of a design report of activities 1 to 5.

19–5 Equipment Specifications

It is safe to say that standard equipment should be selected whenever possible. If the equipment is standard, the manufacturer may have the desired unit and size in stock. In addition, the manufacturer can usually quote a lower price and provide better guarantees for standard equipment than for special equipment. The chemical engineer should evaluate the needs associated with the equipment and prepare a preliminary specifications sheet for the equipment. These specifications can be used by the chemical engineer as a basis for the preparation of the final specifications for the vendor. They can also be used by the vendor to provide suggestions to the purchasers.

Preliminary specifications for equipment should include some or all of the following:[2]

1. Identification
2. Function
3. Operating constraints
4. Materials handled
5. Basic design data
6. Essential controls and instrumentation
7. Insulation requirements, if applicable
8. Allowable tolerances, if applicable
9. Special information
10. Special design details

11. Materials of construction
12. Delivery detail(s)

The chemical engineer should allow the vendor to make suggestions before final specifications are submitted. The final decision on the unit can therefore include any changes that reduce the cost with little to no decrease in the performance of the equipment.

19-6 Materials Selection

The effects of corrosion and erosion must be considered by the chemical engineer in the design and selection of equipment. Chemical resistance and the physical properties of constructional materials are also important factors to consider in the selection and analysis process. The materials of construction must be resistant to the corrosive action of any chemicals that may contact the exposed surfaces. Structural strength, resistance to physical or thermal shock, cost, ease of fabrication, necessary maintenance, and general type of service required, including operating temperatures and pressures, are additional factors that influence the final choice of construction materials.[2]

Materials in most engineering applications can be classified in the following five categories:

1. Ferrous metals and alloys, including steel, stainless steel, low-alloy steel, cast irons, alloy cast irons, medium alloys, and high alloys
2. Nonferrous metals and alloys, including nickel and nickel alloys, aluminum and its alloys, copper and its alloys, lead and its alloys, titanium, and zirconium
3. Inorganic nonmetallics, including glass and glassed steel, porcelain and stoneware, brick construction, cement and concrete, and soil
4. Organic nonmetallics, including plastic materials
5. Thermoplastics, including thermosetting plastics, rubber and elastomer, asphalt, carbon and graphite, and wood

19-7 Instrumentation and Controls

The applicable controls require careful consideration of all operational and maintenance requirements. The control system should also be designed to vary one or more of the process variables to maintain the appropriate conditions with the equipment in question. These variations are usually programmed into the system on the basis of past experience with the unit. The operational parameters that may vary include the system temperature and system pressure. The control system should be subjected to extensive analysis on operational problems and items that might become problematic. The system must identify the problem and integrate necessary actions and alarms or trips into the system. Any environmental regulatory limitations that are imposed on the unit must also be identified and monitored to ensure that they are not exceeded. In the end, the purpose of the control system is to ensure that the equipment will be operating in a safe reliable manner within the guidelines of the equipment design.

Process controls, together with their associated instrumentation, can be considered as the mechanical brain of the equipment of concern. Variables in the process must be measured and then controlled and integrated for optimum processing conditions. These components and systems have been designed by instrumentation and control engineers to reduce labor costs and improve performance capabilities of the equipment. The economic advantage of process control is well documented, thus accounting for its widespread use in the chemical process industry. (See also Chap. 14 for additional details.)

Chemical engineers not specializing in process control should have a working knowledge of this field so that they can appreciate its relative merits.

19–8 Equipment Selection

The final selection of equipment for a process requires some experience, particularly if the process is partially or completely new. If the process is a traditional one or in operation elsewhere, then the task consists chiefly of comparative calculations, while incorporating pertinent innovations and improvements that past experience suggests. Any new equipment requires a study of the process involved before a selection of the type and size of equipment required for guaranteed performance. Literature reviews and background reading plus applicable equipment books will assist equipment selection.

A specification form for the equipment should be prepared after the chemical engineer has made the necessary equipment calculations. If outside bids are acquired, detailed specification sheets may be required. Standard forms are available from both vendors and equipment manufacturers' associations.

The value of using standard equipment was discussed earlier. The experience of others should be used whenever possible; significant information is usually available from vendors who usually see possibilities of additional orders for other equipment. Thus, in many instances, they are not anxious to enter into a bid where their equipment will not provide satisfactory service.

Chemical engineers are often confronted with situations for which the equipment requires a special design and the use of special materials. In such cases, they must draw upon their training and expertise to design the requisite equipment.

Notwithstanding the above, the reader should note that traditional standard equipment has withstood the test of time and the rigorous test of use and service. These units have undergone long periods of experimentation and many modifications of its original design. *Standardization* means not only a minimum cost in manufacturing but also that standard units in standard sizes have usually been given the best of thought in its ultimate state of design.

19–9 Equipment Fabrication

The fabrication process varies with the equipment, equipment size, and the vendor. Thus, the process of equipment fabrication is *site specific*, and no set procedure can be provided. One can say that fabrication expenses often account for a large fraction of the purchase cost for equipment. A chemical engineer should therefore be acquainted with the methods for fabricating equipment, and the problems involved in the fabrication should be considered when selected equipment specifications are prepared.

Many fabrication details are governed by various codes, such as the American Society of Mechanical Engineers (ASME) codes. These codes can vary depending on the complexity and type of equipment being reviewed. The following steps are often involved in the fabrication process:[2]

1. Material selection
2. Cutting to correct dimensions
3. Forming into desired shapes
4. Fastening, if applicable
5. Testing

6. Heat treating, if applicable
7. Final preparatory steps

19–10 Installation Procedures

Most equipment is shipped by truck or rail and should be inspected immediately on receipt for any signs of damage to the exterior and interior of the equipment. Specific attention should be given to checking for any protruding parts, points that come in contact with the shipping media, or areas where lifting lugs, tie-down lugs, ladder clips, or other exterior components are attached, and around all internal components. Any damage observed should be reported to the carrier and the vendor promptly, and should be repaired before placing the equipment into service. When unloading large units and setting them on their foundations, proper rigging and handling procedures, as recommended by the vendor, should be employed. Some helpful suggestions in handling are:

1. Exercise care not only in lifting and moving but also to prevent dropping, abrasion, and impact.
2. Always use padding where necessary to prevent abrasion and impact.
3. Always provide sufficient support to prevent undue deflection, and distribute pressure over a large surface area.
4. Use lifting lugs, keeping the pull on the lug's centerline in a radial direction only.
5. Always use recommended guidelines when moving equipment to prevent striking objects.
6. Never place concentrated stresses at any one point.
7. Do not roll or drag the equipment over anything that will cause damage.
8. Never roll over a fitting, lug, or other component.
9. Do not lift or pull the equipment by fittings or other components except lifting lugs.
10. Use experienced workers.

After the installation is completed, each component of the unit should be checked prior to actual start-up of the system. Once the prestart-up check has been completed, a routine start-up procedure for the equipment can be initiated.

The installation becomes an important factor in the future activities of operating and maintenance procedures on the system. Improper installation may contribute to excessive energy consumption and other unnecessary losses. Everything possible should be done to ensure correct operation and ease of maintenance. Depending on the type of unit chosen, the installation of the equipment can take a few days to install or several months to erect. In either case, proper planning and installation procedures will save time and money during both the installation and future operation/maintenance periods. The reader should be aware that installation errors and oversights can lead to potential immediate and long-term consequences.

19–11 Equipment Purchasing Guidelines

Purchasing guidelines obviously vary with the equipment in question. This section provides information as it applies to a heat exchanger. However, the recommendations can be adapted or modified for other equipment.

Prior to the purchase of a heat exchanger (for example), experience has shown that the following points should be emphasized:

1. Refrain from purchasing any heat exchanger without reviewing *certified independent test data* on its performance under a similar application. Request that the manufacturer provide performance information and design specifications.
2. In the event that sufficient performance data are unavailable, request that the equipment supplier provide a small pilot model for evaluation under existing conditions.
3. Prepare a good set of specifications. Include a strong performance guarantee from the manufacturer to ensure that the heat exchanger will meet all design criteria and specific process conditions.
4. Closely review the overall process, other equipment, and economic fundamentals.
5. Make a careful energy balance study.
6. Refrain from purchasing any heat exchanger until firm installation cost estimates have been added to the total cost. Escalating installation costs are the rule rather than the exception.
7. Give operation and maintenance costs high priority on the list of exchanger selection factors.
8. Refrain from purchasing any heat exchanger until a solid commitment from the vendor(s) is obtained. Make every effort to ensure that the exchanger is compatible with the (plant) process.
9. The specification should include written assurance of prompt technical assistance from the supplier. This, together with a completely understandable operating manual (with parts list, full schematics, *consistent* units and notations, etc.) is essential and is too often forgotten in the rush to get the heat exchanger operating.
10. Schedules can be critical. In such cases, delivery guarantees should be obtained from the manufacturers and penalties identified.
11. The heat exchanger should be of fail-safe design with built-in indicators to show when performance is deteriorating.
12. Perhaps most importantly, withhold 10 to 15% of the purchase price until satisfactory operation is clearly demonstrated.

These 12 points have, in a real sense, become the "bible" to those involved with and/or responsible for the purchase of equipment.

The usual design procurement, installation, and/or start-up problems can be further compounded by any one or a combination of the following:

1. Unfamiliarity of chemical engineers with heat exchangers
2. New suppliers, frequently with unproven heat exchanger equipment
3. Lack of industry standards
4. Compliance schedules that are too tight
5. Vague specifications
6. Weak guarantees
7. Unreliable delivery schedules
8. Process reliability problems

These eight problems represent the standard that the purchaser should also be aware of.

Elan[4] offers the following four purchasing guidelines from a cash flow perspective. Chemical engineers representing *buying* organizations can contribute to responsible and effective cash flow management by observing the following practices:

1. Clearly define the project scope and delivery schedule.
2. Ensure that the suppliers are qualified to support the project.
3. Help the suppliers understand the payment terms and process.
4. Approve or deny invoices promptly.

Elan[4] also suggests the following for project managers representing a *selling organization*:

1. Fully understand the project scope and deliverable schedule.
2. Order a credit report before writing the proposal.
3. Understand the payment terms.
4. Agree on the invoice schedule, format, and delivery method at project outset.
5. Close payments when they are not received by the due date *and* stop work until payments are brought up to current status.

Thus, although project managers of procuring and selling organizations may sit across the table from one another, they all face similar pressures and are best served by similar approaches to cash flow management. The general subject of purchasing equipment is again discussed in Chap. 24.

Proper selection of a particular unit for a specific application can be extremely difficult and complicated. The final choice in equipment selection is usually dictated by that unit capable of achieving the aforementioned design performance criteria and required process conditions at the lowest uniform annual cost (amortized capital investment plus operation and maintenance costs; see also Chap. 28).

As noted earlier, in order to compare specific equipment alternatives, knowledge of the particular application and site is also essential. A preliminary screening, however, may be performed by reviewing the advantages and disadvantages of each type or class of unit. However, in many other situations, knowledge of the capabilities of the various options combined with common sense, will simplify the selection process.

References

1. L. Theodore and A. J. Buonicore, *Air Pollution Control Equipment: Selection, Design, Operation and Maintenance*, Theodore Tutorials (originally published by Prentice-Hall), East Williston, N.Y., 1982.
2. M. Peters, *Plant Design and Economics for Chemical Engineers*, McGraw-Hill, N.Y., 1958.
3. F. Vilbrandt and C. Dryden, *Chemical Engineering Plant Design*, McGraw-Hill, N.Y., 1934.
4. D. L. Elan, Jr., *Managing Cash Flow*, EM (Environmental Management), Air & Waste Management Association (AWMA), Pittsburg, Pa., May, 2012.

20 Operation, Maintenance, and Inspection

Chapter Outline

20–1 Introduction

This chapter is concerned primarily with operation, maintenance, and inspection (OMI) issues as they apply to equipment, processes, and plants. However, the presentation to follow will address OMI from a heat exchanger perspective. These issues obviously vary with the type of heat exchanger under consideration. For the purposes of this chapter, the material will primarily address condensers since most of the heat exchangers in industrial use can be employed for condensation operations. However, this material can be applied to virtually all heat exchangers which have been used in process operations for decades: shell-and-tube, double-pipe, finned, air-cooled, flat-plate, spiral-plate, barometric jet, spray, and other heat exchangers. Because of the generic nature of the material to follow the reader should note that this can also be applied to most equipment. This is demonstrated in the last section, which examines absorbers and adsorbers from an OMI perspective.

Chapter contents include installation procedures; operation, maintenance, and inspection; testing; improving operation and performance; and, other equipment.

NOTE The bulk of the material for this chapter has been drawn from the original work of Connery[1] and subsequent publicactions based on his work.[2–4]

20–2 Installation Procedures

The preparation of a condenser or heat exchanger for installation begins on delivery of the unit from the manufacturer. Condensers are shipped domestically using skids for complete units, and boxes or crates for bare tube bundles. Units are normally removed from trucks using a crane or forklift. Lifting devices should be attached to lugs provided for that purpose (i.e., for lifting of the complete unit as opposed to individual parts), or used with slings wrapped around the main shell. Shell supports are acceptable lugs for lifting, provided that the complete sets of supports are used together; any nozzles should not be used for attachment of lifting cables. On delivery of the unit, the general condition should be noted to determine any damage sustained during transit and prior to receipt. Any dents or cracks should be reported to the manufacturer prior to attempting to install the unit. Flanged connections are usually blanked with suitable pipe plugs. These closures are necessary to avoid entry of debris into the unit during shipping and handling, and should remain in place until actual piping connections are made.

Clearance Provisions

Sufficient clearance is required for at least inspection of the unit or in-place maintenance. The inspection of heat exchangers requires minimal clearances for the following: access to inspection parts (if provided), removal of channel or bonnet covers, and inspection of tube sheets and tube-to-tube sheet joints. If the removal of tubes or tube bundles already in place is anticipated, provision should be provided in the equipment layout. The actual clearance requirements can be determined from the condenser setting plan.

Foundations

Heat exchangers must be supported on structures of sufficient rigidity and strength to avoid imposing excessive strains due to settling. Horizontal units with saddle-type shell supports are normally supplied with slotted holes in one support to allow for expansion. Foundation bolts in these supports should be loose enough to allow movement.

Leveling

Heat exchangers should be carefully leveled and squared to ensure proper drainage, venting, and alignment with attached piping. On occasions, these units are purposely angled to facilitate venting and drainage, and the alignment with piping then becomes the prime concern.

Piping Considerations

The following guidelines for piping are necessary to avoid excessive strains, mechanical vibration, and access for regular inspection:

1. Sufficient support devices are required to prevent the weight of piping and fittings that may be imposed on the unit.
2. Piping should have sufficient expansion joints or bends to minimize expansion stresses arising with temperature excursions.
3. Surge drums or sufficient length of piping to the condenser should be provided to minimize pulsations and potential mechanical vibrations.
4. Valves and bypasses should be provided to permit inspection or maintenance in order to isolate the condenser during periods other than complete system shutdown (outage).
5. Plugged drains and vents are normally provided and located at low and high points of shell-tube sides not otherwise drained or vented. These connections are functional during start-up, operation, and shutdown, and should be piped up for either continuous or periodic use and never left plugged.
6. Instrument and controls connections should be provided either on condenser nozzles (if applicable) or in the piping close to the condenser. Pressure and temperature indicators should be installed to validate the initial performance of the unit as well as to demonstrate the need for inspection or maintenance.

20–3 Operation

The maximum allowable working pressures and temperatures are normally indicated on the heat exchanger's nameplate. These excursion values should not be exceeded. Special precautions should be taken if any individual part of the unit is designed for a maximum temperature lower than the unit as a whole (the most common example is some copper-alloy tubing with a maximum allowable temperature lower than the actual inlet gas temperature). This is necessary to compensate for the low strength levels of some brasses or other copper alloys at elevated temperatures. In addition, maintaining an adequate flow of the cooling medium may be required at all times.

Equipment such as condensers are designed for a particular fluid throughput. Generally, a reasonable overload can be tolerated without causing damage. If the equipment is operated at excessive flow rates, erosion or destructive vibration could result. Erosion could occur at normally acceptable flow rates if other conditions, such as entrained liquids or particulates in a gas stream or abrasive solids in a liquid stream, are present. Evidence of erosion should be investigated to determine the cause. Vibration can be propagated by problems other than flow overloads, e.g., improper design, fluid misdistribution, or corrosion-erosion of internal flow-directing devices such as baffles. Considerable study and research have been conducted to develop a reliable vibration analysis procedure to predict or correct damaging vibration.

Start-up

Most equipment, and exchangers in particular, should be warmed up slowly and uniformly; the higher the temperature ranges, the slower the warm-up should be. This is generally accomplished by introducing the coolant or heated fluid and increasing the flow rate to the design level and gradually adding the other stream. For fixed-tube-sheet units with different shell-and-tube material, expansion of shell and tubes should be considered. The respective areas should be vented to ensure complete distribution. It is recommended that gasketed joints be inspected after continuous full-flow operation has been established. Handling, temperature fluctuations, and yielding of gaskets or bolting may necessitate retightening of the bolting.

Shutdown

Exchangers are usually cooled down by shutting off the vapor stream first and then the cooling stream. Again, fixed-tube-sheet units require consideration of differential expansion of the shell and tubes. Condensers containing flammable, corrosive, or high-freezing-point fluids should be thoroughly drained for any prolonged outages.

20–4 Maintenance and Inspection

Recommended maintenance of all equipment, including exchangers, requires regular inspection to ensure the mechanical integrity of the unit and a level of performance consistent with the original design criteria. A brief general inspection should be performed on a regular basis while the unit is operating. Vibratory disturbance, leaking gasketed joints, excessive pressure drop, decreased heat thermal efficiency indicated by higher gas outlet temperatures or lower outlet coolant temperature, lower condensate rates, and intermixing of fluids, are all signs that a thorough inspection and maintenance procedure are required.

Complete inspection requires a shutdown of the unit for access to internals plus pressure testing and cleaning. Scheduling can be determined only from experience and general inspections. For exchangers, tube internals and exteriors, where accessible, should be visually inspected for fouling, corrosion, and/or damage. The nature of any metal deterioration should be investigated to properly determine the anticipated life of the equipment or possible corrective action. Potential causes of deterioration include general corrosion, intergranular corrosion, stress cracking, galvanic corrosion, impingement, erosion attack, and the lack of a formal maintenance-and-inspection program.

Cleaning

Fouling of exchangers occurs because of the deposition of foreign material on the interior or exterior of the tubes. Evidence of fouling during operation is increased pressure drop and a general decrease in performance. Fouling can be so severe that the tubes become completely plugged, resulting in thermal stresses, leading to mechanical damage of the equipment.

The nature of the deposited fouling determines the method of cleaning that should be employed. Soft deposits can be removed by steam, hot water, various chemical solvents, or brushing. Cooling water is sometimes treated with four parts of chlorine per million parts water to prevent algae growth and the consequent reduction in the overall heat transfer coefficient of the exchanger. Past experience usually determines the method to be used. Chemical cleaning should be performed by contractors and workers specialized in the field who will consider the deposit to be removed and the materials of construction. If the cleaning method involves elevated temperatures, consideration should be given to thermal stresses induced in the tubes, e.g., steaming out individual tubes can loosen the tube-to-tube sheet joints.

Mechanical methods of cleaning are useful for both soft and hard deposits. There are numerous tools for cleaning tube interiors: brushes, scrapers, and various rotating cutter-type devices. The exchanger manufacturer or suppliers of tube tools should be consulted in the selection of the correct tool for a particular deposit. When cutting or scraping deposits, care should naturally be exercised to avoid damaging tubes.

Cleaning of tube exteriors is generally performed using chemicals, steam, or other suitable fluids. Mechanical cleaning is performed but requires that the tubes be exposed, as in a typical air-cooled condenser, or capable of being exposed, as in a removable bundle shell-and-tube condenser. The layout pattern of the tubes must provide sufficient intersecting empty lanes between the tubes, e.g., as in a square pitch. Mechanical cleaning of tube bundles, if necessary and as noted, requires the utmost care to avoid damaging the tubes (or fins, if present).

20-5 Testing

Proper maintenance requires testing of the unit to check the integrity of the following: tubes, tube-to-tube sheet joints, welds, and gasketed joints. The normal procedure consists in pressuring the shell with water or air at the nameplate-specified test pressure and viewing the shell welds and the face of the tube sheet for leaks in the tube sheet joints or tubes. Any water employed should be at or near ambient temperature to avoid false indications due to condensation. Pneumatic testing requires extra care because of the destructive nature of a rupture or explosion, or fire hazards when residual flammable materials are present.[5] Condensers of the straight-tube floating head construction require a test gland to perform the test. Tube bundles without shells are tested by pressuring the tubes and viewing the length of the tubes and back face of the tube sheets.

Corrective action for leaking tube-to-tube sheet joint requires expanding the tube end with a suitable roller-type tube expander. Good practice calls for an approximate 8% reduction in wall thickness after metal-to-metal contact between the tube and the tube hole. Tube expansion should not extend beyond 1/8 inch of the inner tube sheet face to avoid cutting the tube.

Defective tubes can be either replaced or plugged. Replacing tubes requires special tools and equipment. The user should contact the manufacturer or a qualified repair contractor. Plugging of tubes, although a temporary solution, is acceptable provided that the percentage of the total number of tubes per tube pass to be plugged is not excessive. The author has addressed this problem for several types of equipment from a risk perspective.[5] The type of plug to be used is a tapered one-piece or two-piece metal plug suitable for the tube material and inside diameter. Care should be exercised in seating plugs to avoid damaging the tube sheets.

20-6 Improving Operation and Performance

Within constraints of the existing system, *improving operation and performance* generally refers to maintaining or improving operation and original (or consistent) performance. Several factors previously mentioned are critical to the design and performance of a condenser: operating pressure, amount of noncondensable gases in the vapor stream, coolant temperature and flow rate, fouling resistance, and mechanical soundness. Any pressure drop in the vapor line upstream of the condenser should be minimized. Deaerators or similar devices should be operational where necessary to remove gases in solution with liquids. Proper and regular venting of equipment and leakproof gasketed joints in vacuum systems are all necessary to prevent gas binding and alteration of any condensation equilibrium. Coolant flow rate and temperatures should be checked regularly to ensure that they are in accordance with the original design and performance criteria. The importance of this can be illustrated simply by comparing the winter and summer performance of a condenser using cooling tower or river water. Decreased performance due to fouling will generally be exhibited by a gradual decrease in thermal efficiency and should be corrected as soon as possible. Mechanical malfunctions can also be gradual, but will eventually lead to a near-total or total inability for the unit to perform.

Fouling and mechanical soundness can be controlled only by regular and complete maintenance. In some cases, fouling is much worse than originally predicted and requires frequent cleaning regardless of the precautions taken in the original design. These cases require special action to alleviate the problems associated with fouling.

Most condenser manufacturers will provide designs for alternate conditions as a guide to estimating the cost of improving efficiency via other coolant flow rates and temperatures as well as alternate configurations (i.e., vertical, horizontal, shell side, or tube side).

20–7 Other Equipment

This last section provides *operation, maintenance, and inspection* (OMI) details on two other equipment: absorbers and adsorbers.

Absorbers

Since packed towers are used primarily for absorbers, this device will be emphasized in the presentation, although much of this material will also be applicable to other absorption towers.

Normal preventive maintenance requires only periodic checks of the fan, pumps, chemical feed system, pipings, duct, and liquid distributor. The normally irrigated packing may never have to be cleaned during the life of the absorber as long as the absorber has been properly designed and operated, However, should the absorber run dry or the entering air stream contain unexpectedly high solids loading, a heavy formation of solids, crystallized salts, or other foreign matter may accumulate on the packing, and this must be removed. In most cases, removal can be accomplished by recirculating (for a short period of time) a chemical solution into which the solids will dissolve or react. Chemical treatment will not always remove the buildup and it may be necessary to use high-pressure water or hot water, or atmospheric steam, or any combination of these. The absorber manufacturer should be consulted to verify the resistance of the internals and shell to the cleaning process. Prior to using any chemical, hot water, or steam it may be necessary, in a few instances, to remove the packing medium from the absorber for cleaning.

The normally irrigated entrainment separator must be periodically flushed with sprays or the equivalent to prevent buildup and eventual plugging. The intervals between routine washings must be determined by experience since the collection of solid materials is a function of specific operating conditions, i.e., it is a site-specific problem.

A maintenance checklist is suggested to ensure proper operation and unexpected problems with the absorber system. The periodic time intervals indicated for maintenance will vary with the specific system's operating conditions and the equipment manufacturer's recommendation(s).

Certain options (see previous chapters) should be considered when purchasing a packed absorber (or any absorber) in order to reduce maintenance costs and allow the equipment to be more operationally reliable between scheduled maintenance shutdowns. Any of these options generally increase the initial cost of the equipment, and for this reason are seldom considered. However, some of these items can pay for themselves in a short time by reducing maintenance costs.

Another source of periodic maintenance relates to the strainers or filters that may be located in the recycle piping and the recycle pumps. To prevent a total system shutdown during cleaning or maintenance of these items, dual strainers or pumps may be considered, and valved separately, so that the system can be maintained in a fully operational mode during strainer cleaning or pump maintenance checks.

Proper instrumentation and controls used to provide operational data is useful in determining areas where problems of maintenance may be required. The type of instrumentation will depend on the system requirements. Areas of specific interest would be pressure drop across the absorber, both the wetted bed and entrainment separator individually, low liquid recycle flow, low liquid pressure, liquid makeup rates to the system, and liquid temperatures.

The absorber can be designed to provide future additional packing height requirements should greater future gas absorption capability be required. This can be accomplished in one of two ways:

1. The shell of the absorber can be flanged to allow a future subsections to be inserted
2. The tower can be initially designed with a void space above the packing between the liquid distributors so that future packing heights can be added without any shell modification.

Although the initial cost of this may appear high, it could preclude the need to field-modify the absorber or to purchase a new absorber for greater efficiency at a later date. Additional details are available in the literature.[6,7]

Adsorbers

Utilizing adsorption for recovery and control purposes has proved to be extremely effective when a proper design procedure is applied and rigid operating procedures are established and followed. In terms of operation and maintenances, it is always desirable to carefully review the process to ensure that normal operation is being experienced. Initial start-ups will probably require a system pressure balance of the exhaust ducts. If multiple processes are being exhausted into the same system, adjustments of individual slide dampers may be required to obtain the correct airflow in each branch duct. Airflow and switches are inexpensive and should be placed in each branch to sense airflow and allow the process to operate when the airflow is adequate.

Air velocities below 100 ft/min through the (carbon) beds should provide adequate retention time for solvents in an airstream to be adsorbed on the adsorbants. For activated carbon systems, excessive flows will reduce carbon efficiencies and allow volatiles to escape into the atmosphere. Excessive velocities are also detrimental to process operations where unnecessary solvents are evaporated and lost from process tanks and delivered to the carbon beds, posing additional loads on the system. Additional details are available in the literature.[6,8]

Optimizing performance of equipment such as carbon adsorbers involves consideration and monitoring of the following factors:

1. The type of adsorber
2. Operation of process controls to minimize solvent emissions
3. Quality of solvent or air inlet stream
4. Characteristics of the inlet stream, such as concentration, temperature, and flow rate
5. Duration of the adsorption cycle
6. Quality and quantity of available steam for regeneration, if applicable
7. Duration of steam-stripping cycle, if applicable
8. Saturation and retentivity of the carbon
9. Quantity and quality of any cooling water
10. Effectiveness of the water-solvent separator

11. Quality of reclaimed solvent
12. Quality of any wastewater
13. Quality of the exhaust stream from the adsorber bed

References

1. W. Connery et al., "Energy and the Environment," *Proceedings of the 3rd National Conference*, AIChE, N.Y., 1975, pp. 276–282.
2. W. Connery, "Condensers," in L. Theodore and A. J. Buonicore, eds., *Air Pollution Control Equipment*, Chap. 6, Theodore Tutorials Application (originally published by Prentice Hall) East Williston, N.Y., 1992.
3. L. Theodore, *Heat Transfer Applications for the Practicing Engineer*, Wiley, Hoboken, N.J., 2011.
4. L. Theodore, *Chemical Reactor Analysis and Applications for the Engineer*, Wiley, Hoboken, N.J., 2012.
5. L. Theodore and R. Dupont, *Environmental Health and Hazard Risk Assessment: Principles and Calculations*, CRC Press/Taylor & Francis Group, Boca Raton, Fla., 2012.
6. L. Theodore, *Air Pollution Control Equipment Calculations*, Wiley, Hoboken, N.J., 2010.
7. L. Theodore and F. Ricci, *Mass Transfer Operations for the Practicing Engineer*, Wiley, Hoboken, N.J., 2010.
8. L. Theodore, "Engineering Calculations: Adsorber Sizing Made Easy," *Chem. Eng. Prog.*, March, 2005, N.Y., pp. 19–20.

21 Energy Management

Chapter Outline

21-1 Introduction

Since the early 1970s, an acute awareness of energy as a problem of impending critical magnitude on the national scene has arisen among informed leaders of industry, government, and the environmental, movement. The energy *crisis*, or *problem*, or *shortage*, or *dilemma*, as it has been called, is created by the continually increasing demand for energy and lack of a management policy. This situation has resulted in two issues that are rapidly becoming pervasive concerns of this nation: (1) the adequate, reliable supply of all forms of energy and (2) the growing public concern with the environment and social consequences of producing this much energy and further, of the environmental and social ramifications of its expenditure.

The solution to the energy problem amazingly *may* simply involve the application of meaningful conservation measures and the development of new, less destructive energy forms. Energy conservation may sharply reduce the terrible waste of resources that has also been central to the energy problem. Moreover, an extensive conservation program can be implemented in a very short period of time. Such an effort can play a major role in slowing the growth in the demand for energy and leading to more effective use of energy. However, at the same time, new sources of energy must be developed to ensure the availability of adequate, long-term energy supplies. The feasibility of wind, tidal, geothermal, and other so-called unconventional sources of energy should continue to be investigated and developed further.

In the final analysis, one can either accept or reject the grim projections for the future obtained by extending the energy consumption patterns and trends of the past, which establish the basis for defining *energy demand*. Once it has been determined that the demand exists, the choice among the various sources of energy and the means of energy conversion systems, either available at present or in some stage of development, can be made. This requires an evaluation for each means of power generation of the available fuel resources, including the environmental implications and their relation to relevant economic, political, and social issues. However, all of these considerations are themselves influenced by assumptions regarding future demands for power; these, too, must be reexamined. For example, through analysis of the various components that presently constitute energy demand, resources and transmission or transportation options, various alternatives can be devised to *maximize* the long-term social return per unit of energy consumed. In turn, such alternatives may have important implications for the economic system, social processes, and lifestyles. Topics (where applicable) such as resource quantity and availability, economics, energy quality, conservation requirements, transportation requirements, delivery requirements, operation and manufacturing, regulatory issues, political issues, environmental concerns, cost consequences, advantages and disadvantages, and public acceptance need to be reviewed. A quantitative detailed review and practical evaluation of all viable energy options, categories, availability, cost, etc., has yet to be accomplished. These considerations define the energy issues and provide a means of solving and managing energy problems that exist today, and for defining the optimal course for future generations.

Regarding the future, are conservation and fossil fuels the answer? The author believes that fossil fuels will dominate the energy landscape for at least the next generation (25 years), if not longer; it is for this reason that much of the material in this chapter deals with fossil fuels (in addition to conservation).

Topics to be reviewed in this chapter include early history, energy resources, energy quantity and availability, general conservation practices in industry, general domestic and commercial property conservation applications, architecture and the role of urban planning, the U.S. energy policy and energy independence, and future outlook.

21–2 Early History

Since the beginning of time, humans have been grappling with the balancing of energy requirements with the environment. In trying to provide food and shelter for themselves and their families, they had to cope with energy demands. One of the first major impacts on the comfort index was the discovery and application of fire for cooking and heating. Depending on how well the fire was controlled, humans also impacted the environment by emitting particulate and acid gases to the atmosphere and leaving the residue on the ground. With so few people on earth and the fact that early humans were nomadic, the problem of searching for fuel for energy was readily solved. As the population grew and people began to live in organized settlements, it became apparent that the organization of energy needs of the masses was problematic, and was increasing with time.

In terms of early history, fossil fuels were the sole energy resource. As such, only fossil fuels—coal, oil, natural gas, shale, and tar sands—are discussed in this section. These fossil fuels stem from stores of carbon and hydrogen that built up in the plants and tiny animals that were located in shallow seas covering many parts of this planet 400 to 500 million years ago.

Coal

Since coal is derived from living things that flourished on Earth's surface millions of years ago, it is referred to as *fossil fuel*. Coal was formed from plants by chemical and geologic processes which occurred over millions of years. Layers of plant debris were deposited in wet or swampy regions under conditions that prevented exposure to air as the debris accumulated. Bacterial action, pressure, and temperature acted on the organic matter over time to form coal. The geochemical process that transformed plant debris to coal is defined as *coalification*. The first product of this process—peat—often contains partially decomposed stems, twigs, and barks and is not classified as coal. However, peat is progressively transformed to lignite, which eventually can become anthracite given the proper progression of geologic changes.

Various physical and chemical processes occurred during coalification. The heat and pressure to which the organic material was exposed caused chemical and structural changes; these changes included an increase in carbon content plus the loss of water, oxygen, and hydrogen. As one might expect, coal is very heterogeneous and can vary in chemical composition by location.

Oil

Although oil deposits have been formed by geologic processes similar to those that created coal seams, the origin of oil itself is more difficult to define. As mentioned above, many believe that it was probably derived from stores of carbon and hydrogen that built up in the plants and tiny animals which lived in the shallow seas covering many parts of Earth. As with coal, the remains of these creatures and plants eventually sank and mixed with mud on the sea bottom. As new layers of silt were washed into the seas by rivers and estuaries, the mixture was sealed off from the oxygen necessary for natural decomposition to take place, and released the stored energy back into the atmosphere in the form of heat. In addition, the pressure and heat caused by the buildup of new upper layers of material gradually turned the mud into a firm rock and, over millions of years, forced out the carbon and hydrogen. These components were again transformed by heat and pressure to become the fossil fuel known as *oil*. Oil, like coal, is a hydrocarbon fuel. The particular combination of hydrogen and carbon determines whether the fuel takes the solid form of coal or the liquid form of oil.

Oil has a higher ratio of hydrogen to carbon than does coal and is therefore lighter in weight. Interestingly, one ton of oil has about the same energy content as one and a half tons of coal.

Another name given to naturally occurring oil is *petroleum*. The word is Latin for *rock oil* and is an apt description. But the rock in which oil is deposited has a special characteristic—it is composed of millions of minute grains compressed against each other with void space in between. The liquid oil collected in this space over time.

Natural Gas

Natural gas has been chronicled for as long ago as 3000 years. The Chinese used bamboo pipes for the heating of pans of brine water to obtain salt. Both the Greeks (ancestors of this author) and Romans, as well as others, knew of natural gas before the birth of Christ. However, it was only used for ever-burning sacred lights or flames for religious purposes.

Before natural gas was commercialized, manufactured gas from coal was developed in Europe in the seventeenth century. However, it was not until the 1800s that gas was commercialized with manufactured gas from coal. The first recorded commercial use of natural gas was in Fredonia, New York in 1821.

The main application of natural gas was illumination in the early years. An important invention was made in 1855 by the German scientist R. V. Von Bunsen. The burner named for him is known to every student of chemical engineering and chemistry. This invention opened the way for natural gas to be used as a fuel for various industrial and other applications. The full commercialization of natural gas arrived with the development of a safe, economic method of piping; this occurred with the development of seamless steel pipes in the 1920s.

Shale Oil and Tar Sands

Oil deposits are not always buried deep down in Earth's crust. In some places oil has been formed below just a thin layer of oil and rock. This layer was too light to enable compression of the oil-bearing sands into the firm rock typical of the deep-underground oil reservoirs. However, there was sufficient heat at these shallow depths in Earth's crust for hydrogen and carbon from decomposing plant and marine life to combine over the aforementioned millions of years into crude oil similar to that found at much greater depths. Huge amounts of oil in this form are found along the Athabasca River in Alberta, Canada, where the deposits—known as oil sands—are near enough to the surface to be removed by a method similar to strip mining. This field is part of a much larger area of oil-bearing sands in Alberta that probably contain more oil than do all the oilfields of the Middle East. Interestingly, only 10 percent of this source was originally deemed suitable for surface extraction.

21–3 Energy Resources

The major energy resources may be classified into the following categories:

1. Natural gas
2. Liquid fuels (oil)
3. Coal
4. Shale
5. Tar sands
6. Solar

7. Nuclear (fission)
8. Hydroelectric
9. Wind
10. Geothermic
11. Hydrogen
12. Bioenergy
13. Waste

Extensive details on each of these energy resources are available on the Department of Energy (DOE) web site and the Internet as well as in the work of Skipka and Theodore.[1]

As with the previous section, this development will focus primarily on fossil fuels (although a complete listing and review of energy resources is presented later). Fossil fuels can be chemically bound in a complex manner, but they can be reduced to two elements: hydrogen and carbon. Both have relatively simple atoms with molecular weights of 2 and 12, respectively. Hydrogen exists at normal conditions as a diatomic gas, but is usually attached to another or other element(s). Hydrogen gas burns easily. Perhaps the most famous example of hydrogen combustion was the German dirigible, *Hindenburg*, which caught fire on landing in 1937 in Lakehurst, New Jersey. Carbon, the other element, is a solid at normal condition but can be difficult to ignite and burn in this state. An example of burning carbon is the conventional charcoal grill. Charcoal is mainly carbon with some ash. It is difficult to ignite without supporting fuel; when combustion is complete, the material remaining is *ash*.

Most fossil fuels are combinations of carbon and hydrogen ranging from a gas such as methane (CH_4), to a complex hydrocarbon of a mixture of carbon and hydrogen, often characterized as C_xH_y. Even waste fuels can be characterized by such a formula. A waste fuel such as municipal solid waste, known for its nonhomogeneous properties, can be approximated and represented chemically as cellulose, $C_xH_yO_z$. A challenge to the fuel technologist of the past was to describe and predict the physical and chemical attributes of these materials as a fuel so that one could utilize and convert their chemical energy to useful heat and power.

Heat and power are also defined as forms of energy and, in most cases, they result from the release of chemical energy from fuels. These fuels can take the form of solids, liquids, or gases. They release their stored energy as heat when they combine with oxygen from the air in a process referred to as *burning* or *combustion*. In order to quantify the ability of fuels to create energy, there must be a means to measure it. One form of energy has been quantified is *work*—the movement of a mass by a force over a specified distance. One method for quantifying the work of engines is to compare them to the approximate rate at which a horse could move weight; thus, the term horsepower is employed as a common way to rate engines. Another basic form of energy is *heat*. Units of equivalence between heat and work often are named after people who were able to devise a means of measuring energy. The surnames of James Joule and James Watt are assigned to units of energy. Some units have peculiar names, such as the *British thermal unit* (Btu). The SI-metric system counterpart is a *quad*, which is equivalent to 10^{15} Btu. Additional details on work and heat are available in Chap. 8.

21–4 Energy Quantity and Availability

An evaluation methodology was established by the author[2] for a comparative analysis of energy resources. Its purpose was to provide an answer to the question: "Can a procedure be developed that can realistically and practically quantify the overall advantages and disadvantages of the various energy resource options?" A list of 12 categories and parameters that affect the answer to the question for each energy category was prepared by the author. These are listed below; this analysis was applied to different sectors of the world, including the United States:[1]

1. Resource quantity
2. Resource availability
3. Energy quality (energy analysis)
4. Economic considerations
5. Conversion requirements
6. Transportation requirements
7. Delivery requirements
8. Operation and maintenance
9. Regulatory issues
10. Environmental concerns
11. Consumer experience
12. Public acceptance

Two of these categories are (1) resource quantity and (2) resource availability. Information on these topics is presented below with reference to several of the energy categories listed earlier.

Coal

It has been estimated that there are nearly one billion tons (nonmetric tons, i.e., 2000 lb each) of proven coal reserves worldwide. This level of availability could last over a century at the current rates of use. Coal is located in almost every country, with recoverable reserves in nearly 75 countries. The largest reserves are in the United States, Russia, China, and India. After centuries of mineral exploration, the location, size, and characteristics of most countries' coal resources are well known today. What tends to vary even more than the assessed level of the resource, i.e., the potentially accessible coal in the ground, is the level classified as proven recoverable reserves. *Proven recoverable reserves* is the tonnage of coal that has been verified by drilling and other means, and is economically and technically extractable. In recent years there has been a drop in the reserves versus production (R/P) ratio, which has prompted questions over whether the industry has reached a *coal peak*—the point in time at which the maximum global coal production rate is reached and after which the rate enters an *irreversible* decline. However, the recent decline in the R/P ratio can be attributed to the lack of incentives to justify increasing reserve numbers.

Where is all the coal in the United States? The two largest producing coal field regions are the Appalachian region, including Pennsylvania, West Virginia, Ohio, western Maryland, eastern Kentucky, Virginia, Tennessee, and Alabama, and the central states region, including Illinois, Indiana, western Kentucky, Iowa, Missouri, Kansas, Oklahoma, and Arkansas. However, two-thirds of the reserves lie in the Great Plains, the Rocky Mountains, and the western states. These coals are mostly subbituminous and lignite, which have low sulfur content. Therefore, some of these fields have been developed to meet the increasing demands of electric utilities. The low-sulfur coal permits more economical conformance to environmental regulation, including acid rain legislation.

The coal reserves of the United States constitute a vast energy resource. Based on earlier data from the Energy Information Administration (EIA), total coal resources, known and estimated, are about 4×10^{12} tons [metric tons (t), i.e., 1000 kg each]. Reserves that are likely to be mined range between 240 to 300×10^9 t.

Oil

Every estimate of reserves represents and is based on current knowledge. It is possible to increase reserves by new discoveries and enhanced recovery methods. Since oil is a nonrenewable resource, the end of the age of oil may be reached sooner rather than later.

Estimates of the total world resource of crude oil are approximately 2000 billion barrels (bbl). It is estimated that the Middle East has nearly half of this total. Russia and China are estimated to have approximately one-third of this total.

Information on the availability and distribution of oil in the United States is documented in the literature. The top *oil-producing* states are

1. Texas
2. Alaska
3. California
4. North Dakota
5. Louisiana

There are also rich deposits of oil on the outer continental shelf (OCS), especially off the Pacific coasts of California and Alaska and in the Gulf of Mexico. Numerous sources have been identified that contain significant oil (and natural gas) reserves. It is estimated that 30% of undiscovered U.S. oil (and natural gas) reserves are contained in the OCS. Offshore accounts for approximately 33% of U.S. oil production. As noted above, wells are located in the Gulf of Mexico, with additional wells off the coast of California. There is also significant oil in the Beaufort Sea off Alaska.

Natural Gas

Forecasters had been predicting only a 50-year supply of natural gas during the 40-year period of 1950 to 1990. Approximately 20 trillion ft^3 of natural gas was annually consumed during 1980 to 2000. Since then more is being consumed this century, and the predicted U.S. availability continues to be measured. The reason for additional availability has resulted from improved exploration and drilling techniques. There appears to sufficient natural gas available for the foreseeable future.

The twenty-first century will most probably be a boom century for the natural-gas industry throughout the world. Production, consumption, and international trade in natural gas should progressively increase in proportion to its resources. Predictions of availability in the past have shown to be more a factor of not only supply and demand but also technological advancements. In addition, gas has been produced from locations previously believed unrecoverable. The depth deemed practical to drill in the Gulf of Mexico was 200 to 1000 ft in the 1970s and increased to over a mile in the more recent decades. Also, recent advances in determining the exact location of natural gas has aided the discovery process.

Oil Shale and Tar Sands

In addition to its vast coal resources, the United States has a considerable fraction of the world's oil shale deposits (approximately 73% according to an earlier 1974 World Energy Conference). Virtually all of the commercial development plants for processing oil shale in the United States have focused on an area in Colorado, Wyoming, and Utah, referred to as the *Green River Formation*. While the oil shale industry will have significant impacts on the Green River area, the exact impacts will depend on the technology used to obtain oil from the shale. There are two possible approaches: conventional surface retorting processes and *in situ* processes. Use of both of these processes in developing the Green River resource is expected to continue.

Huge amounts of oil in this form are also found along the Athabasca River in Alberta, Canada, where the tar sand deposits are near enough to the surface to be removed by a method similar to strip mining. This field is part of a much larger area of oil-bearing sands in Alberta that contains more oil than all the oilfields of the Middle East, but it has been estimated that only ten percent is suitable for surface extraction.

Oil sand resources in the United States are located mostly in eastern Utah, with approximately 30 billion bbl of known and potential deposits. Interestingly, the U.S. tar sands are *hydrocarbon-wet,* while the Canadian sands are *water-wet.* Research continues to develop commercially viable production technology.

It is estimated that 90% of U.S. tar sand deposits are located in Utah. This amount is dwarfed, however, by the massive Athabasca tar sand deposits in Canada, where commercial activity is beyond the development stage. As noted, both surface mining and in situ recovery are being used to obtain oil from tar sand. Thus, considerable efforts today are being made to further develop processes to exploit these considerable reserves. Note that it is difficult to estimate the percentage of these reserves that are economically recoverable while adhering to environmental regulations. In addition, the characteristics of the Utah deposits are significantly different from the Athabascan deposits, as discussed above.

At the international level, over 50% of the world's tar sands are located in North America. Nearly 25% of the total reserves are located in South America; one of the large deposits is located in Venezuela. However, the reader should note that most of the tar sand data are incomplete. Other countries known to possess tar sand reserves included Russia, Siberia, the Republic of the Congo, Madagascar, Tsimiroro, and Bemolanga. Large-scale exploration is in the early planning stages with some of the latter countries.

Solar Energy

The practicing engineer and scientist can obtain information about the U.S. solar resources from the *National Solar Radiation Data Base* (*NSRDB,* Vol. 1, 1992), which was earlier developed as part of a national resource assessment project conducted by the National Renewable Energy Laboratory for the U.S. Department of Energy (DOE). A series of solar resource maps of the United States was produced from this database.[3]

The average solar radiation flux hitting Earth's surface is approximately 630 W/m^2. This is an *average* figure; the actual radiation received at a specific location can be higher or lower since no radiation is received at night, and a reduced amount is received at times other than noon. The solar energy received also varies significantly with latitude as well as with time of year. The solar energy received at a given location in the United States during a clear day in December is about 50% of that received in July. The region consisting of the southern two-thirds of New Mexico and Arizona and the bordering desert regions of Nevada and California receive an average insolation of about 260 W/m^2. This is about 40% more than the insolation of New York or the New England states.[4]

The top "solar" states are California, Nevada, Colorado, and Arizona.

Nuclear Energy

At the turn of the century, 30% of the nuclear energy in the world was generated in the United States. France, Japan, Russia, Germany, and Korea contributed 15, 10, 5, 5, and 5%, respectively, of the total. Other contributing nations included the United Kingdom,

Canada, Sweden, and the Ukraine. Their nuclear energy *capacity* values were, on a percent basis, roughly the same.

Nuclear energy generation steadily increased during the middle to latter part of the twentieth century. However, the addition of newly installed nuclear units has not increased. This trend reversal can be primarily attributed to the nuclear "accidents" that occurred earlier.

China presently has 25 nuclear power plants under construction, with plans to increase this number. The U.S. licenses of almost half its reactors have been extended to 60 years,[5] and plans to build another half-dozen are under serious consideration.[6] However, recent nuclear disasters have resulted in the change of nuclear energy policies in some countries. For example, Germany plans to close all its reactors by 2022, and Italy is seriously considering banning nuclear power. As a result, estimates of additional nuclear generating capacity for the next 25 years century have been reduced.[6]

France produces the highest percentage of its electrical energy from nuclear reactors—80% as of 2006. Nuclear energy provides 30% of the electricity in the European Union. Nuclear energy policy differs among European Union countries, and some, such as Austria, Estonia, and Ireland, have no active nuclear power plants.

The United States produces the most nuclear energy, with nuclear power providing nearly 20% of the electricity that it consumes.[6] The present nuclear energy consumption in the United States is estimated to be 8.55Q. Energy production is at 8.35Q.

The largest percent of electricity generated by nuclear energy by individual states is Vermont, 72.3%; New Jersey, 55.1%; Connecticut, 53.4%; South Carolina, 52.0%; and Illinois, 48.7%.

Hydroelectric

Brazil, Canada, New Zealand, Norway, Paraguay, Austria, Switzerland, and Venezuela are the only countries in the world where the majority of the internal electric energy production is from hydroelectric power. Paraguay produces 100% of its electricity from hydroelectric dams, and exports 90% of its production to Brazil and to Argentina. Norway also produces nearly 99% of its electricity from hydroelectric sources.[7]

At the domestic level, hydropower represents only approximately 7% of the electricity generated in the United States. The five leading hydropower producing states are Washington, Oregon, California, New York, and Alabama. At present (2013) there are nearly 100,000 dams in the United States, but there are only 2500 dammed hydroelectric plants.

Wind Energy

Many of the largest operational onshore wind farms are located in the United States. As of 2010, the Roscoe Wind Farm was the largest onshore wind farm in the world at 781 MW, followed by the Horse Hollow Wind Energy Center at 735 MW.

At the end of 2010, worldwide capacity of wind generators was 197 gigawatts (gW). Wind power now has the capacity to generate 430 TWh annually, which constitutes about 2.5% of worldwide electricity usage.[8] Since the average annual growth in new installations is approximately 25%, one can expect the market to increase by $(1.25)^5$ or 3.8 fold by 2015 or $(1.25)^{10}$ or 14.5-fold by 2020.[9] Several countries have nearly achieved relatively high wind power usage, including Denmark, Portugal, Spain, Ireland, and Germany. Nearly 100 countries are exploring wind power at this time.

The top 10 countries in wind power capacity as of 2010 were:

1. China
2. United States

3. Germany
4. Spain
5. India
6. Italy
7. France
8. United Kingdom
9. Canada
10. Denmark

Geothermal Energy

The areas with the highest underground temperatures are in regions with active or geologically young volcanoes. These "hot spots" occur at plate boundaries or at places where the crust is thin enough to allow the heat to pass. The Pacific Rim, often called the "ring of fire" for its many volcanoes, has many hot spots, including those in Alaska, California, and Oregon. In addition, Nevada has hundreds of hot spots, covering much of the northern part of the state. A similar "ring" extends through southern Europe and across the middle of Asia connecting with some of the South Sea Islands.

There are three main categories of geothermal resources:

1. *Hydrothermal resources*—underground reservoirs of hot water and/or steam.
2. *Hot dry rock resources*—typically at about a 3-mile depth, with little or no fluid present.
3. *Geopressured resources*—hot saline fluids found at very high pressures in porous formations. These are believed to contain large amounts of dissolved natural gas, and the production of natural gas and heat may be equally important.

In 2008, there were geothermal power plants in 19 countries, generating 60,435 MW of thermal energy in 24 countries. An additional 22 countries have new geothermal electricity projects in development.

In 2005, the International Geothermal Association (IGA) reported that 10,715 MW of geothermal power in 24 countries are online, which was expected to generate approximately 70,000 gWh of electricity in 2010.[10] This represents a 20% increase in online capacity since 2005. IGA projects growth to 18,500 MW by 2015, due to the projects presently under consideration, often in areas previously assumed to have little of this resource.

Of the geothermal reservoirs in the United States, most are located in the western states and Hawaii. California generates the most electricity from geothermal energy. The Geysers dry steam reservoir in northern California is the largest known dry steam field in the world and has been producing electricity since 1960. The United States led the world in geothermal electricity production in 2010 with 3000 MW of installed capacity from nearly 80 power plants with the largest group of geothermal power plants in the world located at Geysers in California. Most of the geothermal resources located in the United States are in the following western states:

California
New Mexico
Arizona
Utah

Nevada
Washington
Oregon
Idaho

Eighty percent of the 3000 MW produced is in California, where more than 40 geothermal plants provide nearly 5 percent of the state's electricity.

The bulk of thermal energy stored on Earth is located within its core. Some of this energy is transmitted from the interior by means of conduction[11] through the planet. The average temperature difference driving force is approximately 10°C/km. It is this gradient that transports this energy from earth's core to its surface. Since Earth's surface is roughly 510×10^{12} m^2, the total heat flow amounts to 32×10^{12} W. Only 1 percent of this total can be attributed to heat transfer by convection.[4,6]

21–5 General Conservation Practices in Industry

There are numerous general energy conservation practices that can be instituted at plants; 10 of the simpler ones are

1. Lubricate fans.
2. Lubricate pumps.
3. Lubricate compressors.
4. Repair steam and compressed air leaks.
5. Insulate bare steam lines.
6. Inspect and repair steam traps.
7. Increase condensate return.
8. Minimize boiler blowdown.
9. Maintain and inspect temperature measuring devices.
10. Maintain and inspect pressure measuring devices.

Providing details on fans, pumps, compressors, and steam lines is beyond the scope of this chapter. Descriptive information [12,13] and calculation procedures are available in the literature.[14] (See also Chap. 9.)

Eight energy conservation practices applicable to specific chemical operation are also provided below:

1. Recover energy from hot gases and/or liquids.
2. Cover tanks of heated liquid to reduce heat loss.
3. Reduce reflux ratios in distillation columns.
4. Reuse hot wash water.
5. Add effects to existing evaporators.
6. Use liquefied gases as refrigerants.
7. Recompress vapors or low pressure steam.
8. Generate low pressure steam from flash operations.

Providing details on distillation columns, evaporators, and refrigerators is also beyond the scope of this chapter. Descriptive information[12,13] and calculation procedures[14,15] are available in the literature. (Additional details are available in Part II of this book.)

For the purposes of implementing an energy conservation strategy, process changes and/or design can be divided into four phases, each presenting different opportunities for implementing energy conservation measures:

1. Product conception
2. Laboratory research
3. Process development (pilot plant)
4. Mechanical (physical) design

Energy conservation training measures that can be taken in the chemical process and other industries include:

1. Implementing a sound operation, maintenance, and inspection (OMI) program (see also Chap. 20)
2. Implementing a pollution prevention program (see also Chap. 25)
3. Instituting a formal training program for all employees

Obviously, a multimedia approach that includes energy conservation considerations requires a total-systems approach (see also Chap. 25). Much of the environmental engineering work in future years will focus on this area, since it appears to be the most cost-effective way of solving many energy problems.

Energy efficiency is a cornerstone of USEPA's pollution prevention strategy. If less electricity is used to deliver an energy service (such as lighting), the power plant that produces the electricity burns less fuel and thus generates less pollution. Lighting is not typically a high priority for conservation for U.S. industries. Because lighting is often the responsibility of facility management, it is viewed as an overhead item. For this reason, most facilities are equipped with the lowest first-cost, rather than the lowest lifecycle-cost lighting systems, and profitable opportunities to upgrade the systems are ignored or passed over in favor of higher visibility projects. As a result, industries pay needless overhead every year, reducing their own competitiveness and that of the United States. Moreover, as noted, wasteful electricity use becomes a particularly senseless source of pollution. For example, lighting accounts for 20 to 25% of all electricity sold in the United States. Lighting for industry, stores, offices, and warehouses represents 80 to 90% of total lighting electricity use, so the use of energy-efficient lighting has a direct effect on conservation. Every kilowatt·hour of lighting electricity not used prevents energy waste.

21–6 General Domestic and Commercial Property Conservation Applications

Domestic applications (discussed to some extent in Chap. 25) involving energy conservation are divided into six topic areas:

1. Cooling
2. Heating
3. Hot water
4. Cooking
5. Lighting
6. New appliances

Specific details are provided in the literature.[16] Action by Congress and state legislatures, rulings, pronouncements by influential people, or wishing alone cannot solve the energy problem. Individual efforts by everyone can make things happen and can help to win the battle against wasting energy. Each individual is an important participant in that battle. One person working alone or cooperating with neighbors, with schools and colleges, with industry, with government, and with nonprofit organizations can make a difference. Here are specific domestic suggestions that one can act on to help reduce energy waste.

1. Purchase energy-efficient automobiles.
2. Purchase energy-efficient recreation vehicles.
3. Purchase efficient appliances.
4. Purchase energy-efficient toys.
5. A well-tuned internal combustion engine makes a car, boat, lawnmower, or tractor more efficient and safer for the individual and the environment.
6. Carpooling, biking, walking, and using mass transit results in less pollution and energy savings.
7. Use natural ventilation in the automobile whenever possible.
8. Use natural ventilation in the home whenever possible.
9. Avoid unnecessary travel and trips.
10. Do not waste food.
11. Do not overeat.
12. Make a conscious effort to operate on an energy-efficient basis.

The author of this text is a former professor of chemical engineering. It was reported that a university mandated that faculty turn off lights in any room not occupied for over 15 min, close drapes and blinds during the summer, and keep window coverings open during the day in the winter to take advantage of solar gain. Coupled with changes in the building space and hot water temperature, these measures helped to reduce the facility's energy cost by over 25%.

Perhaps the major conservation measures in the future will occur in the commercial property arena since practitioners have come to realize the enormous financial investment possibilities that exist today. As a result, interest in this area has increased at a near-exponential rate. The initial energy efficiency investments made in the commercial real estate (CRE) market have been associated with lower cost improvements having relatively short payback periods (less than 2 to 3 years) and involving low technology risk. As a result, the CRE industry now has the opportunity to move from this initial phase of low-cost, short-payback energy efficiency improvements to the multifaceted second phase of implementing deep energy retrofits where the capital need is much more intensive and the payback period often longer. Thus, these technology changes can range from minor changes that can be implemented quickly and at low cost to major changes involving replacement of process equipment or processes at a very high cost. The challenge associated with deeper, more capital intensive energy efficiency retrofit improvements is complicated when internal financing is limited or not available. Fortunately, this is changing and the market ready, commercially attractive financing mechanisms that have arisen to meet these needs are also available.[17] As noted above, energy conservation projects involving commercial properties are certain to increase in the future. This activity will be enhanced by:

1. Convincing CRE properties to operate in a more efficient manner
2. Providing incentives, such as underwriting loans, to implement attractive energy-efficient projects

21–7 Architecture and the Role of Urban Planning

As energy concerns present some of the most pressing issues to the world, both professional and academic architects have begun to address how planning and *built form* affect society. Although the term *built environment* has come to mean different things to different people, one may state in general terms that it is the result of human activities that impact society. It essentially includes everything that is constructed or built, such as all types of buildings, chemical plants, utilities, arenas, roads, railways, parks, farms, gardens, and bridges. Thus, the built environment includes everything that one can describe as a structure or "green" space. Generally the built environment is organized into six interrelated components:

1. Products
2. Interiors
3. Structures
4. Landscapes
5. Cities
6. Regions

While *architecture* may appear to be one of the many contributors to the current environmental state, in reality, energy consumption and pollution associated with the materials, the construction, and the use of building contributes to most of the major environmental problems. In fact, architectural planning, design, and building can significantly contribute to the destruction of the rain forest, the extinction of plant and animal species, the depletion of nonrenewable energy sources, the reduction of the ozone layer, the proliferation of chlorofluorocarbons (CFCs), exposure to carcinogens and other hazardous materials, and *potential* global warming problems. Where one chooses to build, which construction materials are selected, how a comfortable temperature is maintained, or what type of transportation is needed to reach it—each issue decided by both architect and user—significantly impacts these overall conditions. Sadly, despite these opportunities to shape a healthier future, an analysis of American planning and building has in the past represented an assault on the existing energy ecological conditions.

Most architects today are committed to build green. Most new buildings will incorporate a range of green elements, including radiant ceiling panels that heat and cool, thus saving energy and improving occupant comfort; a cogeneration plant that utilizes waste heat; a green roof that is irrigated exclusively with rainwater in order to mitigate the heat island effect; and materials that are rapidly renewable and regionally manufactured. Additionally, buildings are being designed to maximize weather patterns, daytime lighting, and air circulation. For example, a birds nest's design has been employed that is efficient, withstanding wind loads and wind shear while simultaneously enabling light and air to move through it. Throughout the building process, construction, and demolition, energy is conserved and waste is recycled. Measurement and verification plans are also being employed to track utility usage for sustainability purposes.

Urban planners, who were previously called *architects*, are employing designs that operate like a wall of morning glories—adjusting to sunlight throughout the day, both regulating light and gathering solar energy. In effect, the design can often create an energy surplus that can be employed elsewhere in the system and/or process.

21–8 U.S. Energy Policy and Energy Independence

Energy is central to all current and future human activities, and, as such, this nation is approaching a crisis stage. Historically, the primary impetus for the emergence of the United States as a global leader has been the ability to satisfy energy needs independently.

Without a strategy for determining the most beneficial path to achieving future independence, the future is resigned to be at the mercy of those that will control energy.

The current energy policy being pursued by the United States is disjointed, random, and fraught with vested interests. Achieving *energy independence* by pursuing current approaches will be costly, inefficient, and disruptive, especially as resources diminish. In addition, environmental impacts will become major concerns, and the consequences of continuing on the current course may be surprisingly difficult to correct.

Skipka[18] has suggested one approach that will develop an energy policy and achieve energy independence. He has proposed investigations and solutions that will be performed under a three-phase study program. The first phase will be devoted to finding sponsorship, organizing research and analytical teams, defining teamwork scopes and objectives, literature reviews, outreach, matrix formulation, plan refinements, industrial-commercial-governmental coordination, and other tasks. The second phase will involve finalization of teamwork scopes and objectives; setting schedules; finalizing matrix criteria; coordinating teams; initiating work projects; scopes, periodic reviews and realignments; cost-benefit and fatal-flaw analyses; draft findings, documentation, industrial-commercial-governmental coordination; and other tasks. The final phase will be implementation of the optimal energy strategy defined during the first two phases. The study group will be supplemented with industrial, academic, and business consultants. The project will attempt to maximize long-term social return per unit of energy consumed. As such, other alternatives may also have important implications for the economic system, social processes, and lifestyles. Topics (where applicable) such as resource quantity and availability, economics, energy quality, conservation requirements, transportation requirements, delivery requirements, operation and manufacturing, regulatory issues, political issues, environmental concerns, cost consequences, advantages and disadvantages, and public acceptance, will be studied. Another feature of the project will include analysis that provides a rated quantitative detailed review and practical evaluation of all viable energy options, categories (see earlier section) and corresponding weighting factors that are contained in this analysis,[2] These considerations presently define the energy issues and can provide a means of solving and managing energy problems that exit today while at the same time defining the optimal course for future generations.[1]

21–9 Future Outlook

Many have been told that one can predict the future from the past. Perhaps. For starters, what caused the extinction of all life (as it was then) some 65 million years ago? Was it global warming? Or just the opposite? Was it a meteor? Perhaps. Was it something else? Should humankind plan for another catastrophe in the near future? Hopefully, space travel and/or ocean exploration will provide some answers to guide humans in the years ahead.

Of a more immediate concern, how does one predict the future needs of humankind? This is not an easy task, but it has been estimated that there will be four billion more people on the planet Earth by the middle of this century. This brings up problems associated with four major areas: energy, nonenergy material resources, land, and water. Superimposed on these four concerns will be quantitative, but related, issues such as employment, food and housing, and transportation. But, there's more, including the ecosystem, infrastructure problems with most cities, air traffic, and the ever-potential economic downturns.[18]

Various technologies have enabled humans to not only expand their knowledge of the universe but, more importantly, transform this Planet. More recently, the Internet has revolutionized nearly all forms of communication. Nanotechnology[19,20] will probably bring about changes (once) unheard of and unthinkable activities in the near future. Also, there is no doubt that there is more around the corner. However, the key problem at this time is energy; it will definitely need to be addressed, and hopefully technology will continue to spare Earth.

With regard to energy, the hope may lie with fuel cells, solar energy and nuclear energy. Nuclear, a source of unlimited energy is already here, but ill-informed environmentalists have thwarted its application. Technology will have to develop improved solar energy systems and fuel cells (that the author feels will probably not succeed)—developments which are at least 25 years away.[20]

The unified approach discussed in Sec. 21–8 can provide the most efficient, cost-effective use of energy, and the approach allows for a free interchange of ideas and selection of the best plan and evaluation. The proposed energy program must be initialized, since staying on the current course is reckless, inefficient, wasteful, and harmful. In addition, the loss of energy independence, economic stability, and competitive spirit will surely contribute to the demise of society as it is known today. Energy is central to all interests and all current and future human endeavors; the problems associated with continuing along the current path demand a redirection of efforts in the immediate future.

Creative innovative individuals will be needed for technology to solve the energy problems of the future. The author believes that the ingenuity will be there. But other problems may loom on the horizon, for if technology does succeed, it may both further expand the population explosion and introduce new counter productive activities. The future, with respect to energy, appears to be in flux.

References

1. K. Skipka and L. Theodore, *Energy Resources Availability, Management and Environmental Impacts*, CRC Press/Taylor & Francis Group, Boca Raton, Fla., 2014.
2. L. Theodore, personal notes, East Williston, NY, 2010.
3. U.S. Department of Energy, *National Solar Radiation Data Base NSRDN* Addison-Vol. 1, Washington, D.C., 1992.
4. R. Dorf, *Energy, Resources and Policy Addiction*, Mass., Wesley, Reading, MA, 1978.
5. Secondary Energy Information, Washington, D.C., 2010.
6. *Wikipedia, The Free Encyclopedia*, Nov. 21, 2011.
7. "Hydroelectric Power" (http://www.waterencyclopedia.com/Getty/Hydroelectric-Power), *Water Encyclopedia* (date unknown).
8. "Binge and Purge" (http://www.economist.com/displotory.cfm?storyid=12970769), *The Economist*, 2009.
9. S. Shaefer and L. Theodore, *Probability and Statistics for Environmental Science*, CRC Press/Taylor & Francis Group, Boca Raton, Fla., 2007.
10. Geothermal Energy Association, Geothermal Energy, International Market Update, May, 2010.
11. L. Theodore, *Heat Transfer Applications for the Practicing Engineer*, Wiley, & Hoboken, N.J., 2011.
12. J. Santoleri, J. Reynolds and L. Theodore, *Introduction to Hazardous Waste Incineration*, 2nd ed., Wiley, Hoboken, N.J., 2002.
13. J. Reynolds, J. Jeris, and L. Theodore, *Handbook of Chemical and Environmental Engineering Calculations*, Wiley, Hoboken, N.J., 2002.
14. P. Abulencia and L. Theodore, *Fluid Flow for the Practicing Engineer*, Wiley, Hoboken, N.J., 2009.
15. L. Theodore and F. Ricci, *Mass Transfer Operations for the Practicing Engineer*, Wiley, Hoboken, N.J., 2011.
16. M. K. Theodore and L. Theodore, *Introduction to Environmental Management*," CRC Press/Taylor & Francis Group, Boca Raton, Fla., 2010.

17. L. Theodore, unpublished article, "The Case for Energy Conservation: Commercial Property, Industrial Sector and Domestic Level," East Williston, N.Y., 2013.
18. K. Skipka, personal notes, RTP Consultants, Carle Place, N.Y., 2012.
19. L. Theodore and R. Kunz, *Nanotechnology: Environmental Implications and Solutions*, Wiley, Hoboken, N.J., 2005.
20. L. Theodore, *Nanotechnology: Basic Calculations for Engineers and Scientists*, Wiley, Hoboken, N.J., 2006.

22 Water Management

Chapter Outline

22–1 Introduction

The number one global *environmental* problem is the lack of potable water. Perhaps it is or will soon become the chief global problem. At a minimum, it will achieve greater significance in the years ahead.[1] Nations go to war over oil (or other natural resources), but are there resources that can replace water? There really is no substitute—and therein lies the problem, particularly with many of the undeveloped nations.

The U.S. Environment Protection Agency (EPA), in partnership with state and local governments, is responsible for improving and maintaining water quality. These efforts are centered around one theme: maintaining the quality of drinking water. This is addressed by monitoring and treating drinking water prior to consumption and by minimizing the contamination of surface waters and protecting against contamination of groundwater needed for human consumption.

The most severe and acute public health effects from contaminated drinking water, such as cholera and typhoid, have been eliminated in the United States. However, some less acute and immediate hazards remain in the nation's tap water. These hazards are associated with a number of specific contaminants in drinking water. Contaminants of special concern to the EPA are lead, radionuclides, microbiological contaminants, and disinfection by-products. These are detailed below.

The primary source of lead in drinking water is corrosion of plumbing materials such as lead service lines and lead solders, in water distribution systems, and in houses and larger buildings. Virtually all public water systems serve households with lead solders of varying ages, and most faucets contain materials that can contribute some degree of lead to drinking water.

Radionuclides are radioactive isotopes that emit radiation as they decay. The most significant radionuclides in drinking water are radium, uranium, and radon, all of which occur naturally in nature. While radium and uranium enter the body by ingestion, radon is usually inhaled after being released in the air during showers, baths, and other activities, such as washing clothes or dishes. Radionuclides in drinking water occur primarily in those systems that use groundwater. Naturally occurring radionuclides are seldom found in surface waters (such as rivers, lakes, and streams).[1]

Water contains many microbes: bacteria, viruses, and protozoa. Although some organisms are harmless, others can cause disease. Contamination continues to be a national concern because contaminated drinking water systems can rapidly spread disease.

Disinfection by-products are produced during water treatment by the chemical reactions of disinfectants with naturally occurring or synthetic organic materials present in untreated water. Since these disinfectants are essential to safe drinking water, the EPA is presently considering ways to minimize the risks from by-products contamination.

Drinking water safety cannot be taken for granted. There are many chemical and physical threats to drinking certain water supplies. Chemical threats include contamination from improper chemical handling and disposal, animal wastes, pesticides, human waste, and naturally occurring substances. Drinking water that is not properly treated or disinfected, or that travels through an improperly maintained distribution system, can also become contaminated and subsequently pose a health risk. Physical risks include failing water supply infrastructure and threats posed by tampering or terrorist activity.[2] These topics receive treatment in this chapter. The following topics are also discussed: physical and chemical properties of water, the hydrologic cycle, water usage, regulatory status, acid rain, treatment processes, and future concerns.

NOTE Much of the material presented in this chapter has been adapted from the work of Carbonaro.[2]

22–2 Physicochemical Properties of Water

The physical properties of water are fairly well documented. *Water chemistry* deals with the fundamental chemical properties of water itself, the chemical properties of other constituents that dissolve in water, and the countless chemical reactions that take place in water. These properties are addressed in this section.

Physical Properties

A regular (i.e., nondeuterated) water (H_2O) molecule consists of three atoms: two hydrogen atoms, each of which are bonded to a central oxygen atom. Water can exist in three states: solid, liquid, and gas. Water is a liquid at room temperature and atmospheric pressure, but below 0°C (32°F) it freezes and turns into ice. Water is present in the gaseous state above 100°C (212°F) and 1.0 atmosphere pressure (absolute); this is the boiling point of water, at which water will evaporate.

Water has a number of unusual physical properties that are a consequence of its hydrogen bonding among neighboring water molecules. The hydrogen bond is a weak bond that is the result of the dipolar nature of water molecules. The hydrogen bonds between water molecules provide water with a relatively large heat capacity, heat of vaporization, and heat of fusion than that expected for a molecule of its size. Hydrogen bonds are also important biologically. Bonding between adjacent base pairs holds double-stranded DNA together. Many proteins also utilize hydrogen bonding to maintain their three-dimensional shape and assist in enzyme-substrate binding.

Hydrogen bonding also causes water to slightly expand on freezing, which explains why ice water is less dense than liquid water. As a result, water collecting in the cracks of rocks will expand on freezing, which is a technique for mechanically breaking rocks apart. This mechanical weathering breaks rocks into smaller fragments, increasing their surface area. This, in turn, increases the breakdown of the rock by surface chemical reactions, a process known as *chemical weathering*.

Chemical Properties

Water is called the *universal solvent* because it is capable of dissolving many (but not all) substances. This chemical property of water arises from aforementioned dipolar nature of water molecules. Water molecules effectively surround positively charged ions (cations) and negatively charged ions (anions) which serve to prevent them from precipitating as solids. This indicates that wherever water goes, either through the ground or moving as a river or through one's body, it carries with it various solutes such as dissolved minerals, nutrients, organics, and heavy metals.

Even pure water contains some amount of hydrogen ion (H^+) and hydroxide ion (OH^-) as a result of chemical reaction known as the *autoionization* of water. The concentration of these ions change as acids and bases are added to water; however, the product of their concentrations is always constant. The relative presence of H^+ and OH^- is measured by the pH level, which was defined in Chap. 2 (Sec. 2–10) as the negative logarithm of the hydrogen ion (H^+) concentration in moles per liter (mol/L). A decrease in pH indicates an increase in the H^+ concentration and a decrease in the OH^- concentration.

The pH of water is an important chemical property which controls the distribution of chemical species among various forms. For example, pH is buffered at precise values in animal cells to maintain functionality of specific enzymes and proteins. Likewise, in the environment, pH controls the distribution of chemicals among their various forms and also controls the rate at which many chemical reactions occur. It is therefore often necessary to precisely measure the pH of water in a process to understand its water chemistry.

Pure water has a neutral pH of exactly 7.0. Values of pH less than this are considered *acidic*, while pH values greater than this are termed *alkaline* or *basic*. The typical pH range of water is from 0 to 14.

Chemical Composition of Natural Waters

The composition of natural waters is often described according to its physical qualities, chemical constituents, and/or its biological inhabitants. The discussion that follows focuses primarily on chemical constituents. Water-sampling programs are available and used to obtain information on the chemical characteristics of potential and existing water sources and the performance of water and wastewater treatment plants.[3] Typically, prior knowledge of the type of chemical constituent (i.e., organic vs. inorganic, dissolved vs. suspended) is required to design and implement effective sampling and treatment programs. Preservatives are often added to prevent degradation of certain constituents, and treatment holding times have been recommended by EPA and other agencies so as to maintain proper quality control.[4] The main constituents of natural water may be categorized as follows:

1. Dissolved minerals
2. Dissolved gases
3. Heavy metals
4. Organic constituents
5. Nutrients

Details on these and associated chemical reactions are provided by Carbonaro.[2]

22–3 The Hydrologic Cycle

Water is the original renewable resource. Although the total amount of water on Earth's surface has remained fairly constant over time, individual water molecules carry with them what the author describes as a rich history. The water molecules contained in the fruit one ate yesterday may have fallen as rain last year in a distant place or could have been used decades, centuries, or even millennia ago by one's ancestors.

Water may be assumed to be always in motion, and the hydrologic cycle describes this movement from place to place. The vast majority (96.5%) of water on the surface of this planet is contained in the oceans. With respect to the cycle, solar energy heats the water at the ocean surface, and some of it evaporates to form water vapor. Air currents take the water vapor up into the atmosphere along with water transpired from plants and evaporated from the soil. The cooler temperatures in the atmosphere cause the vapor to condense into clouds. Clouds traverse the world until the moisture capacity of the cloud is exceeded and the water falls as precipitation. Most precipitation in warm climates is displaced into the oceans or onto land where the water flows over the ground as *surface runoff*. Runoff can enter rivers and streams which transport the water to the oceans, accumulate, and be stored as freshwater for long periods of time. In cold climates, precipitation often falls as snow and can accumulate as ice caps and glaciers that can store water for thousands of years,

Throughout this cycle, water acquires contaminants originating from both naturally occurring and anthropogenic sources. Depending on the type and amount of contaminant present, water present in river, lakes, and streams or beneath the ground may become unsafe for use.

22-4 Water Usage

Society uses a significant quantity of water. On average, people living in the United States use 110 gal of water per day. Most of this water is used in the bathroom for showers, which consumes anywhere from 1.5 to 8.0 gal per minute, and toilet flushing, which uses 1.0 to 3.0 gal per flush.[5] This equates to approximately 1.1 billion gal of water per day for New York City's population of 8 million people. Factoring in other withdrawals of water for irrigation, thermoelectric power, and industry, the United States is estimated to use 40 billion gal per day of freshwater.[6] For example, the City of Yonkers obtains its drinking water from the New York City water supply system, and unfiltered surface water. Most of this water originates from two protected watershed areas, the Catskills and Delaware, located west of the Hudson River in upstate New York. The New York City Department of Environmental Protection (NYC DEP) Bureau of Water Supply, Quality and Protection oversees the operation, maintenance, and protection of this upstate reservoir system consisting of 19 reservoirs and three controlled lakes. On average, over a billion gallons per day of water travels down through two NYC DEP owned and operated aqueduct (tunnel) systems, the Catskills and Delaware, to feed the Kensico Reservoir located in Westchester County. Under normal operations the waters are blended here before traveling further south to the NYC Hillview Reservoir located in Yonkers. Before the water arrives at the Hillview Reservoir it enters the system at several locations. In addition, water also enters the Yonkers's System from the Westchester County Water District 1 (WCWD#1) Kensico Line. From these points of entry, the water enters 374 millions of distribution piping to serve the 200,000 residents of the City of Yonkers through approximately 30,000 metered service connections.

Natural waters consist of surface waters and groundwaters. *Surface water* refers to the freshwater in rivers, streams, creeks, lakes, and reservoirs, and saline water is present in inland seas, estuaries, and the oceans. The source of freshwater is vitally important to everyday life. The main uses of surface water include drinking water and other public uses, irrigation uses, and use by the thermoelectric power industry to cool electricity generating equipment.

The United States relies heavily on its surface water supplies, accounting for 79% of all the water usage.[6] The remaining 21% of the U.S water usage is from *groundwater*. The term *aquifer* is employed to define underground soil or fractured rock through which groundwater can move. Groundwater extracted from aquifers provides drinking water for more than 90% of rural populations. Even some major cities rely solely on groundwater for all their needs. Withdrawals of groundwater are expected to rise as the population increases and available sites for surface reservoirs become more limited.

Artificial recharge is the practice of increasing the amount of water that enters a groundwater aquifer; this involves the direction of water to the land surface through canals, followed by injection of water into the subsurface through wells. This water can then be called on when needed by pumping it back to the surface,

Saline water is not directly potable because of its dissolved salt content; however, its use is increasing. In 2000, the United States used about 62 billion gal per day of saline water, which was about 15% of all water used.[6] Currently, the main use is for thermoelectric power plant cooling, although saline water can be desalinated for use as drinking water with numerous treatment processes that are available to lower the amount of salt. The process traditionally has not been cost-effective, but by 2020, desalinized water is predicted to become a major contributor to the water supply.[7] There are currently over 250 U.S. desalination plants in states mostly with dense populations and arid climates, such as California and Texas. There are over 12,000 desalination plants worldwide, mostly concentrated in the Middle East, where freshwater is in short supply.

22-5 Regulatory Status

The discussion in this section addresses only the Safe Drink Water Act. More comprehensive literature is available[8], and additional material is presented in Chap. 25.

The Safe Drinking Water Act (SDWA)

The first legislation enacted in the United States to protect the quality of drinking water was the Public Health Service Act (PHSA) of 1912. The PHSA brought together the various federal health authorities and programs at that time, such as the Public Health Service and the Marine Hospital Service, under one statute. The PHSA authorized scientific studies on the impact of water pollution and human health, and introduced the concept of water quality standards. True national drinking-water standards were not established, however, until 60 years later with the SDWA.

The SDWA, originally passed by Congress in 1974, authorized the EPA to set national health-based standards for drinking water to protect against both naturally occurring and anthropogenic contaminants that may be found in drinking water. Since its enactment, there have been over 10 major revisions and additions; the most substantial changes occurred in earlier amendments in 1986 and 1996.

The SDWA applies to every public water supply system (PWS) in the United States, where approximately 87% of all water used is drawn from PWSs.[3] There are currently more than 160,000 U.S. PWSs. PWSs include municipal water companies, homeowner associations, schools, businesses, campgrounds, and shopping malls. The EPA works with these PWS systems, along with state and city agencies, to ensure that these standards are met. Originally, the SDWA focused primarily on treatment as the means of providing safe drinking water. The 1996 amendments greatly enhanced the existing law, which now includes source water protection, protection of wells and collection systems, making certain that water is treated by qualified operators, funding for water system improvements, and making information available to the public on the quality of their local drinking water.

Drinking-water standards (DWS) are regulations that EPA has established to control the concentration of contaminants in the drinking-water supply. In most cases, EPA delegates responsibility for implementing drinking-water standards to states and tribes. Drinking-water standards apply to PWSs, which provide water for human consumption through at least 15 service connections, or regularly serve at least 25 individuals.

The SDWA 1996 Amendments also required the EPA to identify potential drinking-water problems, establish a prioritized list of chemicals of concern, and set standards where appropriate. Peer reviewed science and data are required to support an intensive technological evaluation that includes many factors such as the occurrence of certain chemicals in the environment; human exposure and risks of adverse health effects in the general population and sensitive subpopulations; analytical methods of detection; technical feasibility; and, impacts of any regulations on water systems, the economy, and public health.

After reviewing health effect studies, EPA sets a *maximum contaminant level goal* (MCLG), which is defined as the maximum level of a contaminant in drinking water at which no known or anticipated adverse effect on human health would occur, and which allows for an adequate margin of safety. MCLGs are not enforced, but instead are public health goals. Since MCLGs consider only public health issues and not the limits of detection and treatment technology, they are sometimes set at a level that some water systems cannot meet. When determining an MCLG, EPA considers the risk to sensitive subpopulations (infants, children, the elderly, and those with compromised immune systems) of experiencing a variety of adverse health effects.

For chemicals that can cause noncancerous adverse health effects (noncarcinogens), the MCLG is based on the reference dose. A *reference dose* (RfD) is an estimate of the amount of a chemical that a person can be exposed to on a daily basis that is not anticipated to cause adverse health effects over that person's lifetime.[9] In RfD calculations, an uncertainty factor is used to account for sensitive subgroups of the population. The RfD is multiplied by typical adult body weight (70 kg) and divided by daily water consumption (2 L/day) to provide a *drinking water equivalent level* (DWEL). The DWEL is multiplied by a percentage of the total daily exposure contributed by drinking water (usually 20%) to determine a numeric value of the MCLG (in mg/L). Additional details are available in Chap. 25. Details of these calculations are available in the literature.[9,10]

If there is evidence that a chemical may cause cancer (carcinogenic), it is usually assumed that there is no dose below which the chemical is considered safe (i.e., no threshold), and the MCLG is set to zero.[9] For microbial contaminants that may present public health risk, the MCLG is also set at zero because ingestion of one protozoa, virus, or bacterium may cause an adverse health effect. (See also Chaps. 25 and 26.)

National Primary Drinking Water Regulations (NPDWRs or primary standards) are the legally enforceable MCLs and treatment techniques that apply to public water supply. The contaminants are divided up into the following groupings, according to the type of contaminant: inorganic chemicals, organic chemicals, microorganisms, disinfectants, disinfection by-products, and radionuclides. A list of these contaminants and their respective standard is available in the literature.[2,8]

National Secondary Drinking Water Regulations (NSDWRs or secondary standards) are nonenforceable guidelines regulating contaminants that may cause cosmetic effects (such as skin or tooth discoloration) or aesthetic effects (such as taste, odor, or color) in drinking water. EPA recommends secondary standards for water systems but does not require systems to comply; state and local agencies may choose to adopt these as enforceable standards. The SDWA also includes a process where new contaminants are identified that may require regulation in the future with a primary standard. EPA is required to periodically release a *Contaminant Candidate List* (CCL), which is used to prioritize research and data collection efforts to help determine whether a specific contaminant should be regulated.

The Clean Water Act (CWA)

Along with the SDWA, the CWA has played an important role in ensuring and maintaining the safety of both the sources and quality of drinking water. Growing public awareness and concern for controlling water pollution led to the enactment of the Federal Water Pollution Control Act Amendments of 1972. As amended in 1977, this law became commonly known as the CWA. The CWA established the basic structure for regulating discharges of pollutants into the waters of the United States. It gave EPA the authority to implement pollution control programs such as setting wastewater standards. The CWA also continued earlier requirements to set water quality standards for all contaminants in surface waters. The CWA made it unlawful for any person or organization to discharge any pollutant from a point source into navigable waters unless a permit was obtained that dictated the terms of the release. It also funded the construction of wastewater treatment plants under the Construction Grants Program. Pollutants regulated under the CWA include biochemical oxygen demand (BOD), total suspended solids (TSS), fecal coliform, oil and grease, and pH (conventional pollutants), toxic chemicals (priority pollutants), and various contaminants not identified as either conventional or priority (nonconventional pollutants).

The CWA introduced a *permit system* for regulating *point sources* of pollution. A *point source* is defined as a single identifiable and localized source of a contaminant.

Point source pollution can usually be traced back to a single origin or source. Examples of point sources include industrial facilities (manufacturing, mining, oil and gas extraction, etc.) and municipal and some agricultural facilities (e.g., animal feedlots). Point sources are not allowed to be discharged into surface waters without a permit from the *National Pollutant Discharge Elimination System* (NPDES). This system is managed by EPA in partnership with the pertinent state environmental agencies. EPA has authorized 45 states to issue permits directly to the discharging facilities. The EPA regional office issues permits directly in the remaining water quality-based standards for states and territories.

The CWA employs two general types of standards: technology-based standards, and in the case of a small number of toxic compounds, health-based effluent standards. EPA develops technology-based standards for categories of dischargers based on the performance of pollution control technologies without regard to the conditions of a particular receiving water body. The technology-based standards become the minimum regulatory requirement in a permit.[11] If the water quality is still impaired after application of technology-based standards in a permit for a particular water body, the permit agency will add water quality-based standards to that permit. The additional limitations are more stringent than the technology-based ones and require the permit applicant to install additional controls.[12]

Water quality standards (WQSs) are risk-based (see also Chaps. 25 and 26) requirements that set site-specific allowable pollutant levels for individual water bodies such as rivers, estuaries, lakes, streams, and wetlands. WQS defines the water quality goals of a water body by designating the intended use(s) of the water (e.g., recreation, water supply, aquatic life, agriculture), by setting criteria necessary to protect the users, and by preventing degradation of water quality through antidegradation provisions. The criteria are numeric pollutant concentrations similar to an MCL for drinking water. States adopt water quality standards to protect public health or welfare, enhance the quality of water, and serve the purposes of the CWA.

A *total maximum daily load* (TMDL) is defined as the maximum amount of a pollutant that a water body can receive and still meet WQSs. It is the collective sum of the allowable loads of a single pollutant from all contributing point and nonpoint sources. The calculation includes a *margin of safety* to ensure that the water body can be used for the purposes the state has designated. Since October 1, 1995, nearly 35,000 TMDLs have been approved by EPA.

Section 303(d) of the CWA required states to identify water bodies that do not meet WQSs and are not supporting their designated uses. Each state must submit an updated list, titled the 303(d) List of Impaired Water Bodies, every even-numbered year. The 303(d) list also identifies the pollutant or stressor causing impairment, and establishes a timeframe for developing a control to address the impairment. Placement of a water body on the 303(d) list triggers development of a TMDL for each pollutant listed for that water body.

Non-point source (NPS) pollution, unlike pollution from direct discharges, arises from many diffuse sources. NPS pollution is caused by rainfall or snowmelt traveling over the ground surface and through the ground. As the runoff moves, it picks up and carries away natural and anthropogenic pollutants, depositing them into lakes, rivers, wetlands, coastal waters, and groundwater. Pollutants associated with NPS include fertilizers, herbicides, and insecticides from agricultural lands and residential areas, oil and grease, toxic chemicals from urban runoff and energy production, sediment from improperly managed construction sites, crop and forest lands, eroding stream banks, salt from irrigation practices and acidity-induced damage from abandoned mines, and bacteria and nutrients from livestock, pet wastes, and faulty septic systems. Many of the sources of NPS pollution were not subject to the permit program as part of the original

1972 CWA. The Water Quality Act (WQA) attempts to address the stormwater problem by requiring that industrial stormwater discharges and municipal storm sewer systems obtain NPDES permits.

22-6 Acid Rain

Acid rain with a pH (see Sec. 2–10) below 5.6 is formed when certain anthropogenic air pollutants travel into the atmosphere and react with moisture and sunlight to produce acidic compounds. Sulfur and nitrogen compounds released into the atmosphere from different industrial sources are believed to play the biggest role in the formation of acid rain. The natural processes that contribute to acid rain include lightning, ocean spray causing plant decay and bacterial activity in the soil, and volcanic eruptions. Anthropogenic sources include those utilities, industries, businesses, and homes that burn fossils fuels, plus motor vehicle emissions. Sulfuric acid is the type of acid most commonly formed in areas that burn coal for electricity, while nitric acid is more common in areas that have a high density of automobiles and other internal combustion engines. There are several ways that acid rain affects the environment:[8]

1. Contact with plants can harm plants by damaging outer leaf surfaces and by changing the root environment.
2. Contact with soil and water resources damages the environment. Because of the acid in the rain, fishkills in ponds, lakes, and oceans, as well as effects on aquatic organisms, are common occurrences. Acid rain can cause minerals in the soil to dissolve and be leached away. Many of these minerals are nutrients for both plants and animals.
3. Acid rain mobilizes trace metals, such as lead and mercury. When significant levels of these metals dissolve from surface soils, they may accumulate elsewhere, leading to poisoning.
4. Acid rain has been known to damage structures and automobiles because of accelerated corrosion rates.

The extent of damage caused by acid rain depends on the total acidity deposited in a particular area and the sensitivity of the area receiving it. Areas with acid-neutralizing compounds in the soil, for example, can experience years of acid deposition without problems. Similar soils are common throughout the midwestern United States. On the other hand, the thin soils of the mountainous northeastern regions have very little acid-buffering capacity, making them vulnerable to damage from acid rain.

The adverse effects of acid rain are seen most clearly in aquatic ecosystems. The most common impact appears to be on reproductive cycles. When exposed to acidic water, female fish, frogs, salamanders, etc., may fail to produce eggs or may produce eggs that fail to develop normally. Low pH levels also impair the health of fully developed organisms. Some scientists believe that acidic water can kill fish and amphibian reptiles by altering their metabolism, but there is little evidence that this is occurring now. It is known, however, that acid rain plays a role in what scientists call the *mobilization* of toxic metals. These metals remain inert in the soil until acid rain moves through the ground. The acidity of this precipitation is capable of dissolving and mobilizing metals such as aluminum, manganese, and mercury. Transported by acid rain, these toxic metals can then accumulate in lakes and streams, where they may threaten aquatic organisms. Despite intensive research into most aspects of acid rain, scientists still have many areas of uncertainty and disagreement.

Although long-term trends in acid deposition cannot be determined exactly, it is possible to draw certain conclusions about current patterns. A comparison of the pH of U.S rainfall with the states producing the greatest SO_2 and NO_x emissions clearly shows a solid link between acidic emissions and acidic deposition. Data collected by several different monitoring networks show that the areas receiving the most acid rainfall are downwind and east of those states with the highest SO_2 and NO_x emissions.

The environmental effects of acid rain are usually classified into four general groups:

1. Aquatic
2. Terrestrial
3. Materials
4. Human

Although there is evidence that acid rain can cause certain effects in each category, the extent of those effects is uncertain. The risks these effects may pose to public health and welfare are also unclear and difficult to quantify.

22–7 Treatment Processes

Municipal wastewater is composed of a mixture of dissolved, colloidal, and particulate organic and inorganic materials. However, municipal wastewater contains 99.9% water. The total amount of the substances accumulated in a body of wastewater is referred to as *mass loading*. The concentration of any individual component is constantly changing as a result of sedimentation, hydrolysis, and microbial transformation and degradation of organic compounds.

Wastewater characteristics are described in terms of water flow conditions and chemical quality. The characteristics depend largely on the types of water usage in the community, and industrial and commercial contributions. During wet weather, a significant quantity of infiltration or inflow may also enter a municipal collection system. This will significantly change the characteristics of wastewater.

The characteristics of industrial wastewater may be obtained from the plant flow records and laboratory data at the municipal wastewater treatment plant. The data describing the wastewater characteristics should include at least the minimum, average and maximum dry weather flows, peak wet-weather flows, sustained maximum flows, and chemical parameters such as biological oxygen demand (BOD), total suspended solids (TSS), total dissolved solids (TDS), pH, total nitrogen, phosphorus content, and toxic chemicals. It is important that reliable estimates of the wastewater characteristics be made, since the municipal wastewater treatment plant will be treating these characteristics.

Wastewater treatment plants utilize a number of individual or unit operations and processes to achieve the desired degree of treatment. Processes are grouped together to provide what is known as *primary, secondary,* and *tertiary* (or advanced) treatment. The term *primary treatment* refers to physical unit operations. *Secondary treatment* refers to chemical and biological unit processes. *Tertiary treatment* refers to combinations of all three (see later paragraph for details).

Treatment methods in which the application of physical processes predominates are known as *physical unit operations*. These were the first methods used for wastewater treatment: screening, mixing, flocculation, sedimentation, flotation, thickening, and filtration, which are typical processes. Each of these processes removes the initial solid or *total suspended solids* (TSSs) from the raw sewage entering the facility.

Treatment methods in which the removal or conversion of contaminants almost always occurs with the addition of chemicals or through other chemical reactions known as *chemical unit processes*. Precipitation, adsorption, and disinfection are the most common examples used in wastewater treatment. The first two of these processes will form a solid particle for easier removal; the second process eliminates the discharge of any bacteria.

Treatment methods in which the removal of contaminants is brought about by biological activity are known as *biological unit processes*. Biological treatment is used primarily to remove and convert the biodegradable organic substances, colloidal or dissolved in wastewater, into gases that can escape to the atmosphere. The well nourished organisms are sequentially removed by allowing them to settle in a quiescent pond. Biological treatment can also be utilized to remove the nutrients in wastewater. (See also Chap. 25.)

Sludge arises when solids in the raw sewage settle prior to treatment. It can also be generated from filtration, aeration treatment, and chemical-addition sedimentation enhancement processes. The characteristics are greatly dependent on the type of treatment to which they have been subjected. Sludge typically consists of 1 to 7% of solids, with the remainder as wastewater. There are two basic types of sludge: settable sludge and biochemical sludge. Both are reviewed in the next two paragraphs.

Settable sludge is removed during the primary sedimentation in the primary settle tanks. It is fairly easy to manage and can be readily thickened or reduced of its water content by gravity or can be rapidly dewatered by other means. A higher solids capture and better dry sludge cake are obtained with primary settable sludge. Primary sludge production can be estimated by computing the quantity of TSSs entering the primary sedimentation tanks assuming a typical 70% efficiency of removal;[13] it is normally within the range of 100 to 300 mg/L of wastewater.

Biochemical sludge is produced in the advanced or secondary stage of treatment such as the activated sludge process from the aeration tanks. These sludges are more difficult to thicken; therefore, a portion of the sludge is recycled back into the activated sludge tank (aeration tank) in order to maintain a good population of microorganisms.

Biological nutrient removal of the inorganic constituents in the wastewater have received considerable attention. Excessive nutrients of nitrogen and phosphorus discharged to the receiving water can lead to eutrophication, causing excessive growth of aquatic plants, and indirectly deplete oxygen sources from the aquatic life and fish, There are also other beneficial reasons for biological nutrient removal, including monetary saving through reduced aeration capacity and the reduced expense of chemical treatment.

Advanced wastewater treatment, known as *tertiary treatment*, is designed to remove those constituents that may not be adequately removed by secondary treatment. This includes removal of nitrogen, phosphorus, and heavy metals.

A final treatment, if required, of a municipal wastewater is usually *disinfection*. Currently there are controversial issues as to what type of disinfection techniques should be employed. Historically, chlorine was the choice of many facilities. The disinfection by-products (of chlorides and bromides) that are formed are often toxic to the water environment. Alternate techniques that are being employed are *ozonation* and *ultraviolet* processes.

As with municipal wastewater management, numerous technologies exist for treating industrial wastewater. These technologies range from simple *clarification* in a settling pond to a complex system of advanced technologies requiring sophisticated equipment and skilled operators. Finding the proper technology or combination of technologies to treat a particular wastewater or meet federal and local requirements and still be cost-effective can be a challenging task for a chemical engineer.

Treatment technologies can be divided into three broad categories:

1. Physical
2. Chemical
3. Biological

Many treatment processes combine two or all three categories to provide the most economical treatment. There are a multitude of treatment technologies for each of these categories. Although the technologies selected for discussion above are among the most widespread employed for industrial wastewater treatment, they represent only a fraction of the available technologies.

Finally, the *biochemical oxygen demand* (BOD) test was developed in an attempt to reflect the depletion of oxygen that would occur in a stream due to utilization by living organisms as they metabolize organic matter. BOD is often used as the sole basis for determining the efficiency of the treatment plant in stabilizing organic matter. Effluent ammonia-nitrogen poses an analytical problem in measuring BOD. At 20°C (68°F), the nitrifying bacteria in raw domestic wastewater usually are significant in number and normally will not grow sufficiently during the 5-day BOD test to exert a measurable oxygen demand. To obtain a true measure of the treatment plant performance in removing organic matter, the BOD test may require correction for nitrification.

The *chemical oxygen demand* (COD) analysis is more reproducible and less time-consuming. The COD test and BOD test can be correlated, but the correlation ultimately gives a *qualitative* value. The COD test measure the nonbiodegradable as well as the ultimate biodegradable organics. A change in the ratio of biodegradable to nonbiodegradable organics affects the correlation between the COD and BOD. Such a correlation is specific for a particular waste but may vary considerably between treatment plant influent and effluent. (See also Chap. 25.) Additional information is available in the literature. [2,13,14]

22–8 Future Concerns

Is water conservation the answer? Water conservation has become an important topic in today's environmentally conscious society. Numerous urban centers are experiencing water shortages and are building dams to help curb the ever-increasing demand for water. These dams often have an undeniable impact on the natural environment, killing numerous species of animals and plants. In any event, conservation of this resource is very much needed and can be accomplished through personal conservation efforts, as well as conscious building design.

Domestic conservation recommendations are listed below:

1. Take shorter showers and save 5 to 7 gal. Fill the bathtub only halfway and save 10 to 15 gal.
2. Do not run the tap unnecessarily, e.g., while shaving, brushing teeth, and washing dishes. Flowing faucets use 2 to 3 gal/min.
3. Repair all leaks in the plumbing system (check all toilets and faucets). A slow-dripping faucet can waste up to 20 gal/day, and a running toilet can waste up to 100 gal/day.
4. Use the water meter to detect hidden leaks. Turn off all taps and water-using appliances. Then check the meter after 15 min; if it moved, there is a leak.
5. Limit watering the lawn to early morning and late evening hours when cooler temperature will not cause quick evaporation.
6. Do not cut the lawn too short; longer grass saves water.
7. Install low-flow showerheads, faucets, and toilets.
8. Shut faucets off tightly.
9. Run the dishwater only when full. Automatic dishwashers use 15 gal for every cycle.

10. Store drinking water in the refrigerator rather than letting the tap run every time.
11. Connect a shutoff nozzle to hoses so that water flows only when needed. When finished, turn it off at the spigot to avoid leaks.
12. Do not hose down the driveway or sidewalk. Use a broom to clean leaves and debris.
13. Never put water down the drain when there may be another use for it such as watering a plant or garden.
14. Wash the car with a bucket and hose with a nozzle.
15. Wash clothing in full loads.

One method of reducing a building's water consumption is through the use of low-volume toilets. The standard toilet uses as much as 5 gal of water per flush, whereas some water-saving models use as little as 2 quarts (qt). This can lead to substantial water savings, especially in public and commercial buildings. Another major source of water consumption is the irrigation of landscaped areas. This consumption may be reduced through the careful selection of landscape materials.

The reality is that the United States does not have a water problem at this time. However, the world needs to prepare for an insufficient and potentially depleted water supply. Although conservation is here for the present, desalination, disinfection, and nanoparticle-related treatment appear to be the major growth areas for the future. In any event, water will achieve a greater significance in the coming years and probably impact society in ways not presently imagined.[1]

Society is learning more and more each year about the impact that various sources of contamination can have on water. In fact, the emphasis on which source or area to concentrate regulatory efforts has changed drastically since the late 1980s. Thus, there is a critical need to give water resource protection the high national priority that it deserves and to encourage federal, state, and local agencies to develop the required strategies and programs needed to carry out this effort.

The future outlook for wastewater treatment involves upgrades and retrofits of existing plants to provide increased capacity and better removal through tertiary treatment. Existing plants will need to be updated and retrofitted to handle increased demand from growing populations. New treatment techniques such as vortex separators, membrane bioreactors (MBRs), and ultrafiltration have and will become more commonplace to provide either better or more efficient removal of wastewater constituents.

Water recycling and reuse have proved to be an effective and successful strategy for creating a new and reliable water source. Non-potable-water reuse is a widely accepted practice that is also expected to increase in the future. In many parts of the world, however, the uses of recycled water are expanding to accommodate the need of both the environment and growing water supply demands. Advances in wastewater treatment processes and additional information on the potential effects of recycled water on human health indicate that planned indirect potable reuse may become more acceptable and common in the future.

Improving the security of the nation's drinking water and wastewater infrastructures today has also become a top priority. Significant actions are underway to assess and reduce vulnerabilities to potential terrorist attacks, to plan for and practice response to emergencies and incidents, and to develop new security technologies to detect and monitor contaminants and prevent security breaches. The Bioterrorism Act of 2002[8] requires all systems serving populations of more than 3300 persons to conduct assessments of their vulnerabilities to terrorist attack or other intentional acts, and to defend against adversarial actions that might substantially disrupt the ability of a system to provide a safe and reliable supply of drinking water. In addition to the vulnerability assessments, the Act required the PWS to certify and submit a copy of the vulnerability assessment to the EPA administrator, prepare or revise an emergency response plan based on the results of the vulnerability assessment, and certify that an emergency response plan[9] has been completed or updated within 6 months of completing the assessment.

References

1. L. Theodore, "On Water," *The Williston Times*, East Williston, N.Y., Sept. 15, 2006.
2. R. Carbonaro, "Water Chemistry" and "Safe Drinking Water," in M.K Theodore and L. Theodore, eds., Chaps. 16 and 17, CRC Press/Taylor & Francis Group, Boca Raton, Fla., 2010.
3. MWH, *Water Treatment: Principles and Design*, Wiley, Hoboken, N.J., 2005.
4. L. Clesceri, A. Greenberg, and A.D. Eaton, *Standard Methods for the Examination of Water and Wastewater*, 20th ed., American Public Health Association Publications, Washington, D.C.,1998.
5. P. Gleick, *Water in Crisis: A Guide to the World's Fresh Water Resources*, Oxford University Press, N.Y., 1993.
6. S. Hutson, N. Barber, J. Kenny, K. Linsey, D. Lumia, and M. Maupin, *Estimated Use of Water in the United States in 2000*, U.S. Department of the Interior, U.S. Geological Survey, Reston, Va., 2004.
7. U.S. Bureau of Reclamation and Sandia National Laboratories, *Desalination and Water Purification Technology Roadmap*, U.S. Bureau of Reclamation and Sandia National Laboratories, Denver, Co., 2003.
8. L. Stander and L.Theodore, *Environmental Regulatory Calculations Handbook*, CRC Press/Taylor & Francis Group, Boca Raton, Fla., 2008.
9. L. Theodore and R. Dupont, *Environmental Health and Hazard Risk Assessment: Principles and Calculations*, CRC Press/Taylor & Francis Group, Boca Raton, Fla., 2012.
10. J. Reynolds, J. Jeris, and L. Theodore, *Handbook of Chemical and Environmental Engineering Calculations*, Wiley, Hoboken, N.J., 2004.
11. USEPA National Pollutant Discharge Elimination System (NPDES) *Permit Writer's Manual*, USEPA Office of Water, Washington, D.C., 1996.
12. D. Liu and N. Liprak, *Environmental Engineers' Handbook*, CRC Press/Taylor & Francis Group, Boca Raton, Fla., 1997.
13. G. Burke, B. Singh, and L. Theodore, *Handbook of Environmental Management and Technology*, 2nd ed., Wiley, Hoboken, N.J., 2000.
14. J. Jeris, lecture notes (with permission), Manhattan College, Bronx, N.Y., 1992.

23 Nanotechnology

Chapter Outline

23–1 Introduction[1]

Nanotechnology—everyone is talking about it. The financial markets and some readers don't quite know what to make of it. It is concerned with the world of invisible miniscule particles that are dominated by forces of physics and chemistry that cannot be applied at the macro- or human-scale level. These particles, when combined, have come to be defined by some as *nanomaterials*, and these materials possess unusual properties not present in traditional and/or ordinary material.

The word *nanotechnology* is derived from the Greek word *nano* (for dwarf) and *technology*. *Nano*, typically employed as a prefix, is defined as one-billionth of a quantity or term and is represented mathematically as 1×10^{-9} or simply as 10^{-9}. *Technology* generally refers to "the system by which a society provides its members with those things needed or desired." The term *nanotechnology* has come to be defined as those systems or processes that provide goods and/or services that are obtained from matter at the nanometer level, i.e., from sizes at or below one billionth of a meter. The new technology thus allows the engineering of matter by systems and/or processes that deal with atoms. One of the major problems that remain is the development of nanomachines that can produce other nanomachines in a manner similar to what may routinely describe as mass production.

Interestingly, the laws of chemistry and physics work differently when particles reach the nanoscale, as the powers of hydrogen bonding, quantum energy, and van der Waals forces endow some nanomaterials with some very unusual properties. Carbon nanotubes, for instance, discovered in the sooty residue of vaporized carbon rods, defy the standard laws of physics. Stronger and more flexible than steel, yet measuring about 10,000 times smaller than the diameter of a human hair, these cylindrical sheets of carbon atoms are useful as coatings on computers and other electrical devices. *Nanoparticles*, another manifestation of nanotechnology, are known to foster stubborn reactions (increase reaction rat) because they have enormous surface area relative to their volume.[2]

As noted above, the classical laws of science are different at the nanoscale. Nanoparticles possess large surface areas and essentially no inner mass; i.e., their surface-to-mass ratio is extremely high. This new "science" is based on the knowledge that particles in the nanometer range, and nanostructures or nanomachines that are developed from these nanoparticles possess special properties, and, in conjunction with their unique behavior, can significantly impact physical, chemical, electrical, biological, mechanical, and other functional quantities. These new "characters" can be harnessed and exploited by chemical engineers to engineer "Industrial Revolution II" processes.

This chapter discusses the following topics: early history, fundamentals and basic principles, nanomaterials, production methods, current applications, environmental concerns, current environmental regulations, and future prospects.

23–2 Early History[3]

Nanoparticles arrived on scene immediately following the Big Bang some 14 billion years ago. However, it is not clear when humans first began to take advantage of nanosized materials. It is known that in the fourth-century AD Roman glassmakers were fabricating glasses containing nanosized metals. Michael Faraday published a paper in 1857 titled the "Philosophical Transactions of the Royal Society," which attempted to explain how metal particles affect the color of church windows. Gustav Mie was the first to provide an explanation of the dependence of glass color on the type and size of metal. A century later, Richard Feynman presented a lecture titled "There is Plenty of Room at the Bottom," where he speculated on the possibility and potential of nanosized materials. He proposed manipulating individual atoms to create new small structures having different properties. Groups at Bell Laboratories and IBM fabricated the first two-dimensional quantum cells in the early 1970s. They were made by thin-film (epitaxial) growth techniques that build a semiconductor layer one atom at a time; this work initiated the development of the zero-dimensional quantum dot, which is now one of the nanotechnologies in commercial applications. In 1996, a number of government agencies led by the National Science Foundation commissioned a study to assess the current worldwide status of trends, research, and development in nanoscience and nanotechnology. This NSF activity provided the necessary impetus for the future for this industry to expand and flourish.

23-3 Fundamentals and Basic Principles

Matter is anything that has mass and can be physically observed. All matter is composed of *atoms* and *molecules*; it consists of a finite number of elements, often represented as *building blocks*. *Atoms* are small particles that cannot be made smaller, while *molecules* are groups of atoms bound together, but possessing properties different from those of an atom.

The atom is composed of a small core defined as a *nucleus* that is surrounded by *electrons*. The nucleus is composed of two types of particles: *protons* and *neutrons*; however, there is significant space between the electrons and the nucleus. Protons and neutrons are themselves made up of even smaller particles, known as *quarks*. One generally depicts the changes in units of electron charge, so that the charge of an electron is written as −1 and that of a proton is written as 1+.

Matter has physical and chemical properties that are related to its size; the properties of most solids depend on the size range over which they are measured,[3] the size range can be macroscopic, microscopic, or molecular. One may view these sizes as finite, differential, and molecular, respectively. The object of this chapter is to discuss these characteristics at the molecular, or nanometer, level. In the macro- or large-scale range ordinarily studied in traditional fields of physics such as mechanics, electricity, magnetism, and optics, the sizes of the objects under study range from millimeters to kilometers, i.e., finite sizes. The properties that one associates with these materials are averaged properties, such as the density and thermal conductivity.

When familiar materials such as metals, metal oxides, ceramics and polymers, and novel forms of carbon such as carbon nanotubes and fullerenes (also called "buckyballs,") named after R. Buckminster Fuller are converted or produced into infinitesimally small particle sizes (and, in the case of carbon nanotubes and buckyballs, unique structural geometries, as well), the resulting particles have an order-of-magnitude *increase* in available surface area. It is this remarkable surface area of particles in the nanometer range that confers on them some unique material properties, especially when compared to macroscopic particles of the same material.[4] (There is an inverse relationship between particle size and surface area.)

One hallmark of nanotechnology is the desire to produce and use nanometer-sized particles of various materials in order to explore the remarkable characteristics and performance attributes that many materials exhibit at these infinitesimally small (particle) sizes.

23-4 Nanomaterials

Nanomaterials originated from entities that are now defined as *prime materials*. These prime materials essentially consist of (pure) elements and compounds. The elements and compounds that have been successfully produced and deployed as nanometer-sized particles include:

1. Metals such as iron, copper, gold, aluminum, nickel, and silver
2. Oxides of metals such as iron, titanium, zirconium, aluminum, and zinc

3. Silica sols, and fumed and colloidal silica
4. Clays such as talc, mica, smectite, asbestos, vermiculite, and montmorillonite
5. Carbon compounds, such as fullerenes, nanotubes, and carbon fibers

Each of these types of materials, along with the manufacturing methods used to render them into nanoscale particles, is discussed in the next section. Some of the information presented was adapted from the literature.[5]

Metals[6]

As noted earlier, macrosized metals are produced in the nanometer range; they exhibit properties not found in large particle sizes. This includes properties such as quantum effects, the ability to sinter at temperatures significantly below their standard melting points, increased catalytic activity due to higher surface area per mass, and higher (more rapid) chemical reaction rates.Additionally, when nanoscale particles of metals are incorporated into other larger structures, they often exhibit increased strength, hardness, and tensile strength when compared to structures formed from conventional *micrometer-sized* powders.

However, reducing some metals to nanoscale powder presents some problems; in particular, when reduced to sufficiently small particle sizes, many of the chemical properties of metals become more reactive and subject to oxidation—often explosively. Copper and silver powders are among the few metallic nanoparticles that are not explosive and thus can be handled in air. Many others must be stabilized with a passivation layer or handled in an inert, blanketed (neutral) environment. Moreover, in some applications, effort is required to minimize unwanted particle agglomeration.

Metals that have served as prime materials include the following:

1. Iron
2. Aluminum
3. Nickel
4. Silver
5. Gold
6. Copper

Mixed oxides include the following:

1. Iron oxides (Fe_2O_3 and Fe_3O_4)
2. Silicon dioxide (silica; SiO_2)
3. Titanium dioxide (titania; TiO_2)
4. Aluminum oxide (alumina; Al_2O_3)
5. Zirconium dioxide (zirconia; ZrO_2) and zinc oxide (ZnO)

Additional details on both metals and metal oxides are available in the literature.[6]

23-5 Production Methods

In general, there are six widely used methods for producing nanoscaled particles:

1. Plasma arc and flame hydrolysis methods (including flame ionization)
2. Chemical vapor deposition (CVD)
3. Electrodeposition techniques

4. Sol-gel synthesis
5. Mechanical crushing via ball milling
6. Use of naturally occurring nanomaterials

High-temperature processes such as plasma-based and flame hydrolysis methods involve the use of a high temperature plasma or a flame ionization reactor. As an electrical potential difference is imposed across two electrodes in a gas, the gas, electrodes, or other materials ionize and vaporize if necessary and then condense as nanoparticles, either as separate structures or as surface deposits. An inert gas or vacuum is used when volatilizing the electrodes. During flame ionization, a material is sprayed into a flame to produce ions. In general, high temperature flame processes for creating nanoparticles are divided into two classifications: gas-to-particle or droplet-to-particle methods. In *gas-to-particle* processes, individual molecules of the product material are made by chemically reacting precursor gases or rapidly cooling a superheated vapor. In droplet-to-particle processes, liquid atomization is used to suspend droplets of a solution or slurry in a gas at atmospheric pressure. The solvent is evaporated from the droplets, leaving behind solute crystals, which are then heated to change their morphology; spray drying, pyrolysis, electrospray, and freeze-drying equipment are typically used in the droplet-to-particle production process.

In *chemical vapor deposition* (CVD), a starting material is vaporized and then condensed on a surface, usually under vacuum conditions. The deposit may be the original material or a new and different species formed by chemical reaction.

Using *electrodeposition*, individual species are deposited from solution, with the objective of laying down a nanoscaled surface film in a precisely controlled manner.

Sol-gel processing is a wet-chemical method that allows high-purity, high-homogeneity nanoscale materials to be produced at lower temperatures compared to competing high temperature methods. A significant advantage that sol-gel science affords over more conventional materials processing routes is the mild conditions that the approach employs. Two main routes and chemical classes of precursors have been used for sol-gel processing:[6]

1. The *inorganic route* (colloidal route), which uses metal salts in aqueous solution (chloride, oxychlorides, nitrate) as raw materials.
2. The *metallorganic route* in organic solvents. This route typically employs metal alkoxides M(OR)Z as the starting materials, where M is Si, Ti, Zr, Al, Sn, or Ce; OR is an alkoxy group; and Z is the valence or the oxidation state of the metal. Metal alkoxides are preferred because of their commercial availability and the high lability of the M-OR bond.

In general, the sol-gel process consists of the following steps:

1. Sol formation
2. Gelling
3. Shape forming
4. Drying
5. Densification

Mechanical crushing via high energy *ball milling* involves size reduction or pulverization using a conventional ball mill. High energy ball milling is in use today, but its use is considered by some to be limited because of the potential for contamination problems. However, the process is simple and inexpensive.

Naturally occurring materials, such as zeolites, can be used as found or synthesized and modified by conventional chemistry. A *zeolite* is a caged molecular structure containing large voids that can admit molecules of a certain size and deny access to other larger molecules. They find application as catalysts as well as adsorbents and other uses.[2,4,8]

The ongoing challenge for industry is to continue to devise, perfect, and scale up viable production methodologies that can cost-effectively and reliably produce the desired nanoparticles with the desired particle size, particle size distribution, purity, and uniformity in terms of both composition and structure.

To summarize, there are six major methods of producing nanomaterials: (1) plasma arcing over flame hydrolysis, (2) chemical vapor deposition (CVD), (3) electrodeposition, (4) sol-gel synthesis, (5) ball milling, and (6) the use of naturally occurring nanoparticles.

23-6 Current Applications

Present-day applications of nanoparticles include chemical products, that include plastics, specialty metals, and powders; and computer chip and computer systems.[1] Specific examples of nanotechnology in actual commercial use today are:[4]

1. Semiconductor chips and other microelectronics applications
2. High-surface-to-volume catalysts, which promote chemical reactions more efficiently and selectively
3. Ceramics, lighter weight alloys, metal oxides, and other metallic compounds
4. Coatings, paints, plastics, fillers, and food packaging applications
5. Polymer-composite materials, including tires, with improved mechanical properties
6. Transparent composite materials, such as sunscreens containing nanosized titanium dioxide and zinc oxide particles
7. Use in fuel cells, battery electrodes, communications applications, photographic film developing, and gas sensors
8. Nanobarcoding
9. Tips for scanning probe microscopes
10. Purification of pharmaceuticals and enzymes

Other applications are certain to emerge in the future.

23-7 Environmental Concerns[9,10]

Any technology can have various and impacting effects on the environment and society. Nanotechnology is no exception, and the results will be determined by the extent to which the technical community manages this technology. This is an area that has, unfortunately, been seized upon by a variety of environmental groups.

An environmental implication of nanotechnology has been dubbed by many in this diminutive field as "potentially negative." The reason for this label is as simple as it is obvious; the technical community is dealing with a significant number of unforeseen effects that could have disturbingly disastrous impacts on society. Fortunately, it appears that the probability of such dire consequences actually occurring is near zero … but *not* zero. This finite, but differentially small probability, is one of the reasons why this section was included in this chapter; and it is the key topic that is addressed in the material to follow.

Air, water, and land (solid waste) concerns with emissions from nanotechnology operations in the future, as well as accompanying health and hazard risks, are a concern. All of these issues arose earlier with the Industrial Revolution; the development, testing, and deployment of the atomic bomb; the arrival of the Internet; Y2K (arrival of the new millennium); and other events, and all were successfully (relatively speaking) resolved by the engineers and scientists of their period.

To the author's knowledge, there are no documented nanoscale human health hazards. Statements in the literature refer to *potential* health problems. The author has also speculated on the need for future nanoregulations (see next section). His suggestions and potential options are provided,[11] while noting that the ratio of pollutant nanoparticles (from conventional sources such as power plants) to engineered nanoparticles being released into the environment may be as high as a trillion to one[12] (i.e., 10^{12}:1). If this is so, the environmental concerns for nanoparticles can almost certainly be dismissed.

23-8 Current Environmental Regulations

Many environmental concerns are addressed by existing health and safety legislation. Most countries, including the United States, require a health and safety assessment for any new chemical before it can be marketed. In addition, the European Union (EU) has introduced the world's most stringent labeling system. Prior experience with materials such as PCBs and asbestos, and a variety of unintended effects of drugs such as thalidomide indicates that both companies and governments have incentives to closely monitor potentially negative health and environmental effects.[13] (The significance of this concern provides the rationale for including this section in the chapter.) Detailed analysis of various U.S. and EU laws and regulations are available in the literature.[14–16]

Many environmental concerns are addressed through existing health and safety legislation. Most countries require a health and safety assessment for any new chemical before it can be marketed. For example, the EU adopted a regulation (EC 1907/2006) in 2006 on chemicals and their safe use and established the European Chemicals Agency in Helsinki. This agency manages the *Registration, Evaluation, Authorization and Restriction of Chemical* (REACH) substances systems, which is a database of information provided by manufacturers and importers on the properties of their chemical substances.

It should be noted that there are no nanoscale or nanoscale-related environmental regulations in the United States or the European Union at this time that require controls on process releases, production activities, or specific workspace safety measures. Completely new legislation and regulatory efforts may be necessary to protect the public and the environment from the potentially adverse effects of nanotechnology.

The principal U.S. agencies concerned with environmental risks are the U.S. Environmental Protection Agency (EPA) and the Occupational Safety and Health Administration (OSHA). EPA's mission is to protect human health and the environment. One of its major goals is to ensure that all Americans are protected from significant risks to human health and the environment where they live, learn, and work, i.e., to closely monitor and control any potentially toxic or biohazardous conditions that could pose an unreasonable risk to human health or the environment. The mission of OSHA is to ensure safe and healthful working conditions for working men and women by setting and enforcing standards and by providing training, outreach, education, and assistance. Both agencies are, therefore, directly concerned with the environmental implications of nanotechnology.

It is very difficult to predict what future regulations might come into play for nanomaterials. In the past, regulations have been both a moving target and confusing. What can be said is that there will be regulations, and there is a high probability that they will be contradictory and confusing. Past and current regulations provide a measure of what can be expected. Control of the production and use of nanomaterials is most likely to occur under the Clean Air Act (CAA) and the Toxic Substances Control Act (TSCA), as discussed below, both of which are concerned with environmental health impacts.

Under the CAA, no specific requirements or regulatory procedures currently exist for nanoparticles. The use and production of nanomaterials could be regulated under the following circumstances (neither of which is under consideration at this time):

1. An installation that manufactures or uses nanomaterials may become subject to requirements of State Implementation Plans, which were developed to ensure attainment and maintenance of National Ambient Air Quality Standards for criteria pollutants [including particulate matter with a diameter of <2.5 μm (PM2.5)]. While emissions of nanoparticles may now be specifically subject to various requirements, the processes involved with producing such substances may result in emissions of criteria pollutants that must be controlled.

2. An installation that is using or manufacturing nanomaterials may become subject to the requirements of Section 112 of the Clean Air Act should such materials become identified as hazardous air pollutants. The Clean Air Act provides a list of 189 substances that have been determined to be hazardous air pollutants. The CAA also prescribes procedures for adding and deleting substances from this list. If adverse health and environmental effects are encountered as a result of emission from the use or manufacture of nanomaterials, EPA will be compelled to list such substances as hazardous air pollutants and require emission controls.

Commercial applications of nanotechnology are more likely to be regulated under TSCA, which authorizes EPA to review and establish limits on the manufacture, processing, distribution, use, and/or disposal of new materials that pose an unreasonable risk of injury to human health or the environment. The term *chemical* is defined broadly under TSCA. Unless a particular nanomaterial qualifies for an exemption under the law, a prospective manufacturer with low-volume production, with low-level environmental releases along with low volume, or with plans for limited test marketing would be subject to the full evaluation procedures. This would include a submittal of a remanufacturing notice, along with toxicity and other data to EPA at least 90 days prior to commencing production of the substance, followed by required recordkeeping and reporting. Requirements will differ, depending on whether EPA determines that a particular application constitutes a "significant new use" or a "new chemical substance." EPA can impose limits on production, including an outright ban when it is deemed necessary for adequate protection against "an unreasonable risk of injury to health or the environment." EPA may revisit a chemical's status under TSCA and change the degree or type of regulation when new health or environmental data warrant.[17–19] If the experience with genetically engineered organisms is any indication, there will probably be a push for not only EPA but also OSHA to update regulations in the future to reflect changes, advances, and trends in nanotechnology.

For example, waste from a commercial scale nanotechnology facility would be regulated under RCRA, provided that it meets the criteria for an RCRA waste. RCRA requirements could be triggered by a listed manufacturing process or RCRA's specified hazardous waste characteristics. The type and extent of regulation would depend on how much hazardous waste is generated and whether the wastes generated are treated, stored, or disposed of on site.

23-9 Future Prospects

For nanotechnology's most ardent supporters, the scope of this emerging field seems to be limited only by the imaginations of those who would dream within these unprecedented dimensions. However, considerable technological and financial obstacles still need to be reconciled before nanotechnology's full promise can be realized.

Ranking high among the challenges is the ongoing need to develop and perfect reliable techniques to produce (and mass produce) nanoscaled particles that have not only the desirable particle size and particle size distributions but also a minimal number of structural defects and acceptable purity levels, since these latter attributes can drastically alter the anticipated behavior and properties of the nanoscaled particles. Experience to date indicates that scale-up issues associated with moving today's promising nanotechnology-related developments from laboratory and pilot scale demonstrations to full scale commercialization can be considerable.[6]

Most believe that the major impact of nanotechnology will be on war, crime, terrorism, and the massive companion industries, particularly security and law enforcement. The military has a significant interest in nanotechnology, including such areas as optical systems, nanorobotics, nanomachines, "smart" weapons, nanoelectronics, virtual reality, massive memory, specialty materials for armor, nanobased materials for stopping bullets, and bionanodevices to deter and destroy chemical and biological agents. Most of this activity is concerned with protection against attack and minimizing risk to military personnel, e.g., devices that may be able to repair defective airframes or the hulls of ships before any major problems develop. But make no mistake, the rush is on (as with development of the atomic bomb) to conquer this technology; the individual, organization, or country that successfully conquers this technology will almost certainly conquer the world. Society, as well as the technical community, must understand that the misuse of this new technology can lead to and cause catastrophic damage. Alternately, nanotechnology could be used to provide not only sophisticated sensor and surveillance systems to identify military threats but also weapons (or the equivalent) that will eliminate these threats.[20]

Regarding crime, the techniques of nanoscience will have much to offer forensic investigations, for biological analysis, and for materials and chemical studies as well. Portable instruments with sophisticated nanosensors will be able to perform accurate high level analyses at crime sites. These instruments should greatly improve conviction rates and the ability to locate real clues. Nanotechnology will also stop money laundering by imprinting every computer digit.[21]

Nanotechnology may open up new ways of making computer systems and message transfers secure using special hardware keys that are impervious to any form of hacking. Very few current computer protection systems are able to keep out really determined hackers. Nanoimprinting, which already exists, could be used to make "keys" or even special nano-based biosensors coded with a dynamic DNA sequence. Nanoscale imprinting is already used to make bank notes virtually impossible to forge by creating special holograms in the clear plastic. Forgery would be possible only if the master stamps were actually stolen, but then a new hologram could be made.[21]

In addition, nanoparticle-related developments are being actively pursued to improve fuel cells, batteries and solar devices, advanced data storage devices such as computer chips and hard drives, magnetic audio and videotapes, and sensors and other analytical devices. Meanwhile, nanotechnology-related developments are also being hotly pursued in other medical applications, such as the development of more effective drug delivery mechanisms and improved medical diagnostic devices, to name just a few.

The unbridled promise of nanotechnology-based solutions has motivated academic, industrial, and government researchers throughout the world to investigate nanoscaled materials and systems with the hope of realizing commercial scale production and implementation in the future. Today, the private sector companies that have become involved run the gamut from established, global leaders throughout the chemical process industries to countless small, entrepreneurial start-up companies, many of which have been spun off from targeted research and development efforts at universities. Beyond the efforts to produce and use nanometer-size particles of various materials, some nanotechnology-related scientists and engineers are pursuing far more ambitious—and some would say fantastic or futuristic-applications of this powerful new technological paradigm. For instance, the research community is working toward being able to design and manipulate nanoscaled objects, devices, and systems by the manipulation of individual atoms and molecules. Such forward looking researchers hope that by using atom-by-atom construction techniques, they will eventually be able to create not just substances with remarkable functionality but also tiny, bacterium-sized devices and machines (thus far dubbed "nanobots"), that could be programmed, for example, to repair clogged arteries, destroy cancer cells, and even reverse cellular damage due to aging.[22,23]

The future of nanotechnology in therefore not known. Scientists, engineers, and even manufacturers can only speculate on its implication or the magnitude of its impact on environmental health. What some might view as a learned prediction of what the future will bring, others might consider as science fiction. The same is true with regard to future legal and regulatory approaches to managing environmental health and hazard risks. As menioned previously, many environmental scientists, attorneys, and others have speculated and offered their perspective on future regulatory activities.[16,17] Others offer alternative viewpoints. The author has speculated on the need for future regulations for nanomaterials. His suggestions and potential options were discussion earlier.[20] As noted, the ratio of nanoparticles that are currently being emitted from conventional sources such as power plants to present-day engineered nanoparticles being released into the environment may be as high as a trillion to one (i.e., $1 \times 10^{12}:1$, or more simply $10^{12}:1$).[12] If this ratio is correct, the environmental concerns associated with today's nanoparticles should not be a concern.

References

1. L. Theodore, "Nanotechnology I," *The Williston Times*, East Williston, N.Y., April 16, 2004.
2. R. D'Aquino, personal communication, 2005.
3. C. Poole and F. Owens, *Introduction to Nanotechnology*, Wiley, Hoboken, N.J., 2003.
4. S. A. Shelley with G. Ondrey, "Nanotechnology—the Sky's the Limit," *Chem. Eng.*, N.Y., December 2002, pp. 23–27, 72.
5. A. Boxall, Q. Chaudhry, A. Jones, B. Jefferson, and C. D. Watts, *Current and Future Predicted Environmental Exposure to Engineered Nanoparticles*, Central Science Laboratory, Sand Hutton, U.K., 2008.
6. L. Theodore and R. Kunz, *Nanotechnology: Environmental Implications and Solutions*, Wiley, Hoboken, N.J., 2005.
7. L. Theodore, *Air Pollution Control Equipment Calculations*, Wiley, Hoboken, N.J., 2008.
8. L. Theodore, *Nanotechnology: Basic Calculations for Engineers and Scientists*, Wiley, Hoboken, N.J., 2006.
9. M. K. Theodore and L. Theodore, *Major Environmental Issues Facing the 21st Century* (originally published by Prentice-Hall), Theodore Tutorials, East Williston, N.Y., 1996.
10. G. Burke, B. Singh, and L. Theodore, *Handbook of Environmental Management and Technology*, 2nd ed., Wiley, Hoboken, N.Y., 2000.
11. L. Theodore, *Waste Management of Nanomaterials*, USEPA, Washington, D.C., 2006.
12. L. Theodore, personal notes, 2006.
13. National Center for Environmental Research, *Nanotechnology and the Environment: Applications and Implications*, STAR Progress Review Workshop, Office of Research and Development, National Center for Environmental Research, Washington, D.C., 2003.
14. D. Fiorino, *Voluntary Initiatives, Regulations, and Nanotechnology Oversight: Charting a Path*, Woodrow Wilson International Center for Scholars, Washington, D.C., Nov. 2010.
15. L. Breeggin, R. Falkner, N. Jasoers, J. Pendergrass, and R. Porter, *Securing the Profits of Nanotechnologies: Towards Transatlantic Regulatory Cooperation*, Chatham House, London, Sept. 2009.

16. USEPA, Laws and Regulations, (U.S. Environmental Protection Agency, Washington, D.C.), http://www.epa.gov'lawsregs/laws/index.html (accessed Nov. 23, 2010).
17. L. Bergeson, "Nanotechnology Trend Draws Attention of Federal Regulators," *Manufacturing Today*, March/April 2004.
18. L. Bergeson and B. Auerbach, "The Environmental Regulation Implications of Nanotechnology," *BNA Daily Environment Reporter*, pp. B-1–B-7, April 14, 2004.
19. USEPA, *Toxic Substances Control Act—Inventory Status of Carbon Nanotubes* (notice), U.S. Environmental Protection Agency, Washington, D.C., Oct. 31, 2008, pp. 64946–64947.
20. L. Theodore, personal notes, East Williston, N.Y., 2006.
21. L. Theodore, "Nanotechnology II," *The Williston Times*, East Williston N.Y., April 23, 2004.
22. G. H. Reynolds, "The Science of the Small," *Legal Affairs*, July 1, 2003.
23. K. E. Drexler, "Machine-Phase Nanotechnology," *Sci. Am.* Sept. 16, 2001.

24 Project Management

Chapter Outline

24–1 Introduction

The author[1] has estimated that at least 75% of chemical engineers are involved with project management activities, and that percentage will continue to increase in the future. The need for the traditional chemical engineer who designed heat exchangers, specified pumps, predicated the performance of multicomponent distillation columns, etc., has virtually disappeared. This is a fact that the profession has difficulty in accepting.[2,3] But, the reality is that chemical engineers in the future will require some understanding of project management.

Project management has come to mean different things to different people. The term *project* implies a time-constrained endeavor undertaken to create a unique product, service, result, or activity; it is of temporary duration, with a defined beginning and end. The term *management* refers to an activity concerned with bringing a group of people (generally more than two) together to accomplish a desired or set goal. Management can include a host of activities, some of which are discussed in subsequent sections. *Project management* may be viewed as the process of "managing" multiple related projects, or as a discipline involved with the planning, organizing, securing, and managing resources to achieve specified goals. The challenge of project management thus is to achieve all (where possible) of the project goals and objectives. Perhaps key to project management is selecting the most appropriate projects and then applying appropriate project management techniques and approaches.

Participants are referred to as the *project team*. In effect, it is the management team leading the project, and provides all the necessary services to the project. The project manager should be a professional in the field of project management. This individual is primarily responsible for planning, executing, and terminating the project, and for the project plan—the formal, approved document used to guide both *project execution* and *project control*. It documents planning assumptions and decisions; facilitates communication among team members; and, documents approved scope, cost, and schedule *baselines*.

Another important element in project management is *risk management*. Its objective is to reduce different risks related to various activities to an acceptable level. It may be related to numerous other types of risk arising because of the environment, technology, economics, people involved, organizations, politics, and so on. Another important element in project management is the risk management associated with the probability of specific eventualities.[1] The risk may be related to financial or environmental considerations to reduce different risks related to a preselected activity to a level accepted by society. These too may be related to numerous other types of risk arising because of the environment, technology, people, organizations, and politics.

In terms of history, project management has been practiced since early civilization. The Great Pyramid of Giza was completed in 2570 BC. Some records indicated that the work was managed; e.g., there were managers of each of the four faces of the pyramid. The first major construction of the Great Wall of China occurred in 208 BC. Prior to the twentieth century, civil engineering projects were generally managed by creative architects, engineers, and so-called master builders. It was in the 1950s that technical organizations began to apply fundamental management principles and techniques to engineering projects and activities. Some have claimed that the forefather of project management was Henry Gantt, who is famous for the development and application of the Gantt chart as a project management tool. Gantt's approach first appeared in 1910.

There is a general theory that the history of modern project management started around 1950. The *critical-path method* (CPM) was developed in the 1950s, and the *program evaluation and review technique* (PERT) method was developed in 1958.

Industry has expanded rapidly since then; a detailed description of these activities is available in the literature.[4,5]

There are a number of approaches to managing project activities, including agile, interactive, incremental, and phased approaches. Each requires careful consideration with respect to overall project objectives, timeline, and cost, as well as the roles and responsibilities of all participants and stakeholders. The traditional phase activity involves a "railroad" approach that consists of five developmental components of the project:

1. Initiation
2. Planning and design
3. Execution and construction
4. Monitoring-controlling system
5. Completion

Not all projects will involve every stage, as some projects can be terminated before they reach completion. Some projects may not follow a structured (railroad) planning and/or monitoring process. Some projects are interactive and will repeat steps 2, 3, and 4 multiple times. It should be noted that many industries use variations of this approach. For example, these stages may be supplemented with go/no-go decisions, at which point the project's continuation is defined and decided.

Another relatively simple project management approach could take the following form:

1. Develop an overall project schedule, work plan, and task outline(s).
2. Develop project budget estimates.
3. Solicit and evaluate equipment vendor proposals.
4. Compare capital and operating costs.
5. Prepare cost-benefit analyses and understand the sensitivity of these analyses to the factors involved.
6. Monitor and report project status with respect to both schedule and budget.
7. Prepare a (final) report.

There have been several attempts to develop international management standards. The International Organization for Standardization (ISO) develops ISO standards; ISO 9000 are a family of standards for quality management systems. This group was founded in 1947, and is responsible for standardizing everything from paper size to film speeds. [The American National Standards Institute, (ANSI) is the U.S. representative to ISO.] The standards are a voluntary series of guidelines designed to address management issues. They focus on management in the belief that effective management will lead to better performance.

The chapter consists of seven additional sections. The discussion below will address the following (some of which are noted above) project management topic areas:

1. Initiation
2. Planning and scheduling
3. Executing and implementing
4. Monitoring and controlling
5. Closing
6. Small-scale project management
7. Preparing reports

24-2 Initiation

The initiating process generally determines the nature and scope of the project. If this stage is not performed well, the project will unlikely be successful in meeting its objectives. Following the selection of the project manager and team, this stage should include a plan that addresses the following areas:

1. Project objectives
2. Analyzing the business needs and requirements in terms of measurable goals
3. Reviewing present operations
4. Financial analysis of the costs and benefits
5. Stakeholder analysis, including users, and support personnel for the project

Projects have beginnings, middles, and ends. The project schedule or cycle generally determines what should be accomplished and what options are available. Since projects arise out of need, the whole project management process begins when an issue or a task needs to be addressed or accomplished. For project selection, choices have to be made. Some projects are selected, while others are rejected. Decisions are made on the basis of available resources, the number of needs which must be addressed, the cost of fulfilling those needs, and the relative importance of satisfying one set of needs and ignoring others.

The key player during this initiating stage is the project manager. Since (in a general sense) project management consists in the planning, organizing, directing, and controlling company resources for a relatively short term objective that has been established to complete specific goals and objectives, the project manager is responsible for coordinating and integrating activities related to the project. The two most important roles of project managers are to select the team and complete the project on time, within budget, and according to any specification(s). His/her basic roles are

1. Technical supervision
2. Initial project planning
3. Project team organization
4. Project team direction
5. Controlling technical quality, schedule, and costs
6. Early communication with client(s) and vendor(s)

A strong project manager is the nucleus of the project and can ensure the success of the project. However, individual experts must be accordingly coordinated. Team managers are often asked to deal with specified constraints of time and economics, sometimes under great stress. The project manager needs to give technical guidance and management expertise, plus enthusiasm and support to the team. Interestingly, project managers seldom have the authority to pick their own teams, but it is the author's contention that they should be given this opportunity.

Assignment of the individuals mentioned above to the project presents the project manager with a unique challenge: to obtain a commitment to the project from the team members, motivate them to achieve the project goals in a timely and cost-effective manner, and influence them to identify with the team and its objective. To meet this challenge, the project manager needs to be skilled in persuasion, motivational techniques, leadership techniques, and the ability to influence others.[6] One means of increasing the probability of success for the project is to convince all team members individually that the project is an essential part of their job. In addition, the project manager needs to establish communication channels whereby information can be exchanged in a timely and accurate way.

24–3 Planning and Scheduling

This obviously is the most important stage of the project, and understandably receives the bulk of the treatment. Planning is conducted throughout the duration of the project. Project milestones are identified, and tasks are laid out. Many tools exist to assist the project manager in devising the formal project plan, including work breakdown structures, Gantt charts, network diagrams, resource allocation charts, and cumulative cost distributions. As the project is carried out, the plan may undergo considerable modification as it adapts to unanticipated circumstances. Project plans are *time variable*, allowing the project staff to manage change in an orderly and systematic fashion.

The primary objective of this stage is to plan time, cost, and resources adequately in order to estimate the work required and to effectively manage any potential risk(s) during project execution. As with the initiation step, a failure to adequately plan greatly reduces the project's chances of successfully accomplishing its goals.

Project planning generally includes the following 10 steps:[7]

1. Obtaining approval from management to initiate the project
2. Identifying deliverables and creating the work breakdown structure
3. Identifying the activities needed to complete deliverables
4. Networking the activities in a logical order
5. Estimating the resource requirements for the project, including time and cost
6. Developing the project schedule
7. Developing the project budget
8. Investigating potential risks
9. Identifying all roles and responsibilities
10. Conducting an initial informal, introductory meeting

Any work plan should address the following questions:

1. What is to be done?
2. Who will do it?
3. When is it to be done?
4. How much it will cost?

Time, money, and human and material resources all come into play at this time. Time (management) is handled through the use of schedules. Money is handled by budgets which indicate how project funds are to be allocated, while human and material resources are concerned with how best to allocate (potentially limited) resources on projects.

An integral part of project planning involves scheduling the project tasks in such a way that the project is carried out both logically and efficiently. The schedule serves as a master plan from which the client, as well as management, can have an up-to-date picture of the progress of the project. Schedules (as noted earlier) normally include lists of tasks and activities, dates when those tasks are to be performed, durations of those tasks, and other information related to the timing of project activities. That information can be displayed in several ways. The most popular form is the bar chart, originally developed by Gantt (and which carries his name). The *Gantt chart* is a type of bar chart that illustrates a project schedule. It illustrates the beginning and ending dates of the terminal elements and summary elements of a project. Terminal elements and summary elements constitute the work breakdown structure of the project. Figure 24–1 illustrates a type of Gantt chart. By reading the time data from the horizontal axis (abscissa), the project staff will know the planned start and finish dates for different tasks. The chart is

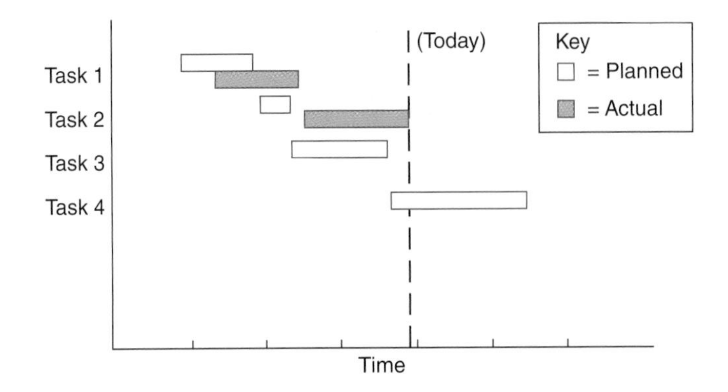

Figure 24–1
Gantt chart.

also useful for project control when actual start and finish times are added. It then allows for a visual comparison of plans with actual achievements, thereby enabling the determination of schedule variance. Figure 24–1 shows, for example, that the project is off schedule from the very beginning, when task 1 begins later than planned. Note that the actual duration of task 1 is equal to the planned duration, so the schedule slippage is entirely accounted for, despite the fact that it started late. With task 2, it is clear that not only did the task begin late, but it took longer to accomplish than planned. Schedule slippage here is caused by both the late start and sluggish performance that stretched out the task's planned duration.

Figure 24–2 illustrates a different approach to the employment of a Gantt chart. The basic data here are *identical* to those given in Fig. 24–1. However, they are presented in a different manner. With this approach, planned start dates are pictured as hollow upright triangles, and actual finish dates are solid downward-pointing triangles. Actual start dates are solid upright triangles, and actual finish dates are solid downward-pointing triangles. A comparison of the two charts indicates that they both give the same information. These Gantt charts are widely used for the planning and control of schedules on projects. Their popularity lies in their simplicity.[8]

One weakness of Gantt charts is that, while it portrays the start and finish dates for tasks, they do not show how schedule changes can have projectwide consequences. Two techniques were developed that allowed the project team to examine the consequences of changing task start and finish dates on the overall project schedule: the CPM and the PERT techniques mentioned in Sec. 24–1. Both approaches are based on similar flowcharts. In fact, there is very little distinction between the two techniques. As noted in the literature,[8] the first step in creating a CPM/PERT schedule is to create a *work breakdown structure* (WBS) for the project. The next step is to create a flowchart from the information contained in the WBS. What CPM/PERT networks do is incorporate scheduling information into a basic flowchart diagram. Tasks are laid out according to the sequence in which they should occur, and their interrelationships are indicated by lines.

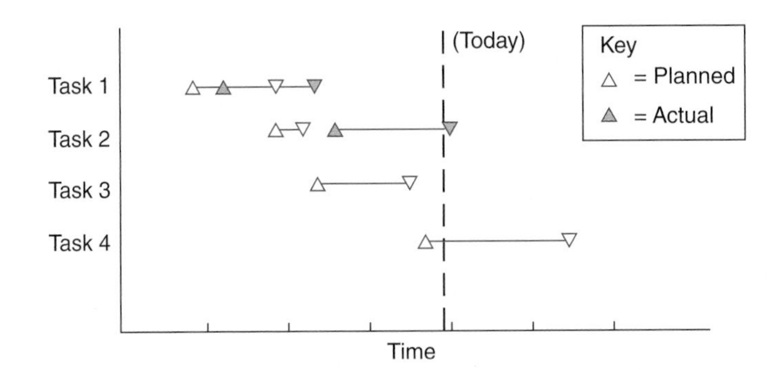

Figure 24–2
Gantt chart, alternate form.

An important concept necessary for an understanding of CPM/PERT networks is that associated with the critical path. The *critical path* in a schedule is the procedure that takes the longest time to complete, and for this reason, the critical path has no slack at all. In fact, if there is a schedule slippage along the critical path, the slippage will be reflected in the project as a whole. Thus, if a task on the critical path takes 3 days longer to complete than anticipated, the overall project schedule will slip by 3 days. Because activities off the critical path have some slack associated with them, they can tolerate some slippage in schedule. Davidson provides additional details.[8]

Another major responsibility of the project manager and team is to adhere to the project budget. The success or failure of the project will often be judged according to whether the project comes in under budget, on budget, or over budget. Exceeding the budget can have serious consequences for the organization. For example, if a project is funded through a contract, a cost overrun may lead to litigation, penalties, and financial losses. If the project is funded internally, an overrun may lead to a drain of organizational resources (see also Chap. 15).

The potential sources of financing for projects are varied and depend on a host of factors. In developing nations (see later section), additional funding sources, such as multinational agencies, play a significant role. Examples of these agencies are the World Bank, World Health Organization, and the Agency for International Development.

Economics, finance, various costs, and budgetary details receive treatment in Chap. 28.

24-4 Execution and Implementation

When a formal plan has been devised, the project is ready to be carried out. In a sense, implementation is at the heart of a project, since it entails doing the things that need to be done, as spelled out in the project plan.

Execution consists of the processes used to complete the work defined in the project plan to accomplish the project's objectives. The execution process involves coordinating people and resources, as well as integrating and performing the activities of the project management plan. The deliverables are produced as outputs from the process(es) performed as defined in the project management plan and other frameworks that might be applicable to the type of project at hand.

This stage also involves direct (or indirect) communication between all members of the team and with upper-level management. The procurement process must be clearly spelled out and adhere to *quality assurance–quality control* (QA/QC) needs to be implemented, and this involves both this and the next stage to be discussed.

The implementation phase of a project essentially consists of performing the actual work necessary to complete the project. The project manager performs a variety of functions to ensure the project is effectively implemented, considering the constraints of schedule and budget.

Successful project completion depends on a number of factors, including:

1. Updating the project plan, as necessary
2. Staying within the scope of work specified by the project plan
3. Obtaining authorization for changes
4. Providing deliverables in stages
5. Conducting project review meetings
6. Frequently checking the technical work being performed in order to provide guidelines and direction as needed
7. Keeping current project information
8. Reviewing the performance of the project staff and providing assistance if necessary.

Another technique used to assist in effective project implementation is to conduct regular meetings. Periodic review meetings are an integral component of project implementation and maintaining control over project execution. The general purposes for these project meetings are quality review, schedule review, cost review, and keeping the project team informed. Review meetings with appropriate corporate and functional managers and the customer to review the current status of the project are essential. It is important to focus on lessons learned rather than placing blame. Each delayed task should be evaluated for its individual impact on the project as a whole. The collective impact of all delayed (or slipped) activities should also be evaluated against the project schedule and budget.

Generally, when projects fail, one of the chief factors contributing to failure is poor performance by project personnel. Good communication skills, particularly during review meetings, can help the project manager avoid failures by identifying concerns and potential problems before they develop into actual problems.

24-5 Monitoring and Controlling

Monitoring and controlling includes those processes performed to observe project implementation to identify any potential problems at an early stage and in a timely manner so that corrective action can be taken. The main objective is for project performance to be observed and measured regularly to identify any variances from the project management plan. These control systems are required not only for cost, communication, time, changes, and procurement issues but also for risk, quality, and human resources. Thus, this stage may be viewed as an independent element in the overall plan. In multiphase projects, the monitoring/control process also provides feedback between project phases in order to implement corrective or preventive actions to bring the plan into compliance with the objective of the project management plan.

Project control is based on establishing an effective project plan with cost, time, and technical baselines. During the implementation phase, control is accomplished by measuring, evaluating, and reporting actual performance against these baselines. It is important to remember that project management is a team effort and must rely on the project team to successfully complete their project activities; this will help maintain control of the project.

The project manager may perform various functions to help maintain project control. These include the following:

1. Frequently review project team activities.
2. Provide the support the team needs to complete assigned activities.
3. Inform the project team of the status of the project, and what needs to be done next.
4. If necessary, prioritize tasks for the team members.
5. Communicate problems or changes to the customer, management, and the project team as soon as possible.

While the main objective of project control is good project planning, and monitoring, project execution and control is facilitated by using effective control methods such as:

1. Updating schedule and cost estimates
2. Conducting project team meetings
3. Using work authorizations

In developing an estimate of equipment capital and operating costs, it is vital to understand the accuracy of the estimation methods used and the degree of accuracy required at each stage of a project. Since it is not practical to determine the accuracy of each factor that is considered in preparing a cost estimate, an estimate of overall accuracy is typically made. This topic receives treatment in Chaps. 15 and 28. Some additional factors follow here.

When a cost problem is recognized, action must be taken early before the problem grows and while there are still funds assigned to the project. Normally, the action needed is to determine how to do the work for lower direct costs than anticipated. The following actions might be considered as means of reducing project direct costs:

1. Examine project staffing. Is there a more cost-effective way to get the job done?
2. Get the project staff to work more efficiently; more employees are willing to buckle down and do whatever is needed if they understand the problem and if the requests are reasonable.
3. Examine the technical approach to the scope of work. For example, are there steps that can be simplified?
4. Look at ways to better manage other direct costs such as travel or consultants; often savings can be achieved without incurring a noticeable impact on the project.
5. Prepare revised budget sheets reflecting changes made to ensure that sufficient action has been taken to solve any problems.

Experience has shown that extra time should be allocated in the bid submittal process to discussing the project with the vendors, answer their questions, and possibly follow up with other vendors. By taking the time to ask questions about a proposal, a better understanding of the abilities of both the equipment/hardware and the vendor can be obtained. It is likely that a vendor may raise an issue that had not been considered previously, or provide some new insight into the development of a project strategy. Of course, all vendors generally claim that their equipment is superior to that of all competitors, so vendor information must be weighed for accuracy. A vendor guarantee should be requested, e.g., catalyst life, along with the terms and conditions of that guarantee. Finally, direct discussions with vendors can help ascertain their level of expertise and willingness to help. One should also remember that the vendors' time is being supplied free of charge at this point, so information requests should be kept within reason. Despite one's best efforts to define the design parameters and method of bid submittal, the bids will invariably arrive in a confusing array of formats. To bring order out of this chaos, it may be desirable to establish a standard evaluation format.

Because unknown factors will inevitably arise during a project, and because of the inherent inaccuracies of the cost-estimating methods employed, additional costs should be allowed for as *contingencies.* For example, these unforeseen costs could include increases in equipment prices, construction labor shortages or stoppages, and damage due to weather. It is common to allow for *at least an* additional 3% to the cost of a project for these unknown and unforeseen obstacles.

Finally, over the course of any construction project, the work plan may change. Change is a normal and expected part of the process. Changes result because of design modifications, differing site conditions, material availability, contractor-requested changes, value engineering, and impacts from other parties. This is referred to as *change management.* Managing change during the course of project implementation involves using existing information to make decisions as changes occur in a project. The project plan can help the team anticipate the consequences of change and prepare a "what if" plan to handle those changes.[1] Changes are often small enough that project team members can respond to the fluctuations without much disturbance. However significant changes may require a more detailed analysis.

24-6 Closing (Project Termination)

Projects ultimately come to an end. Sometimes this end is abrupt and premature, such as when it is decided to kill a project before its scheduled termination. However, it is always hoped that the project will meet a more natural ending. In any case, when projects end, the project manager's responsibilities continue. There are various closeout duties to be performed. For example, if equipment was used, this equipment must be accounted for and possibly reassigned to new uses. On contracted projects, a determination must be made as to whether the project deliverables satisfy the contract. Final reports may have to be written, a topic discussed in the last section of this chapter.

The closing procedure includes the formal acceptance of the project and the ending thereof. Administrative activities include the archiving of the files and documenting any lessons learned. In fact, this final phase consists of two closures:

1. Finalize all activities across all of the process groups in order to formally close the project or a project phase.
2. Complete and settle each contract (including the resolution of any unresolved items) and terminate each contract applicable to the project (or project phase).

24-7 Small-Scale Project Management

Project management activities normally involve small-scale operations. One rarely thinks of the management and planning and implementation of a day, week, month, year, or decade (or even a lifetime) as a project management activity. For example, the author regularly plans the next day's activities the previous evening, usually immediately prior to sleeping. So it is with industry and business—small-scale projects are regularly (from a time perspective) the norm.

Small-scale project management is specifically applied for small-scale projects. These projects are characterized by factors such as short duration, a small team, and a small budget. There is also normally a balance between the time committed to delivering the project and the time committed to managing the project. Otherwise, these projects are not unique. The final product is the same as those of large-scale projects. They are unique in the time expended as well as quality of the results.

For example, USEPA differentiates between large-scale generators (of pollutants) and *small-quantity generators* (SQGs). The latter are defined as facilities generating between 100 and 1000 kg of hazardous waste per calendar month, while a *conditionally exempt small-quantity generator* generates by definition <100 kg of hazardous waste per calendar month. These sources represent a wide variety of industrial groups, and present unique challenges in terms of regulation and technical assistance. These generators generally have fewer than 5 to 10 employees and are managed and staffed by individuals with limited training in the identification and management of hazardous wastes or pollution prevention. (While these approximately 600,000 to 700,000 SQGs in the United States do not individually contribute significant quantities of materials to the waste stream, the nature of their waste materials and their aggregate are a significant part of the total hazardous-waste stream, generating as much as 1 million metric tons (t) of hazardous wastes annually.)[9]

As with the SQGs, small-scale projects are by far the most common form of project enacted by institutions and large-scale organizations who may utilize small projects in order to accomplish a range of small-order tasks. For those businesses, especially in the creative industries, the running of small projects may be an essential component of their core business. Individuals and groups regularly utilize small-scale projects as a means to deliver a range of outcomes.

The balance between *process* versus *output* is a key factor to consider when determining the effectiveness of applying project management methodologies to smaller projects. As noted above, other factors commonly associated with large-scale project management, such as time, cost, and quality, still apply, but to a lesser degree. In effect, these are generally viewed as leaner projects.

24-8 Reports

The most important project management report is the *final* report. These reports can take any of several forms, most of which were discussed in Chap. 15 and to a lesser degree in Chap. 4. As noted in Chap. 15, the most important part of this final report is the *abstract* or what some refer to as to *executive summary. This is the most important part of the report.* It should briefly summarize (without referring to the body of the report) the project's description and objectives, project activities, and the results, findings, and conclusions.

However, there are other earlier reports, and one should consider appending them to the final report. Those other reports generally fall into one of three categories: progress reports, contact reports, or project status reports.

Project progress reports provide one of the best ways to track the progress of a project; these reports are used to monitor the activities and compare them to schedule and budget. Periodic progress reports should contain the information on the status of project activities, schedule, budget, goals achieved, goals not achieved, goals due, important meetings, correspondence, release or delivery of deliverables, other reports, and (if applicable) equipment specifications and designs.

Contact reports are used to document project-related communication with the client. They are particularly useful in documenting any changes in project schedule, feedback on deliverables, and changes in project scope. Contact reports are normally sent to project team members and functional managers. They provide an excellent opportunity to keep everyone abreast of project progress. Contact reports are also used to confirm telephone conversations with clients. When sent to the client, they place in writing an understanding of the client's comments and planned actions. When used in this manner, contact reports minimize misunderstandings. However, it is neither necessary nor desirable to document everything in contact reports; to do so would undermine their value. Routine project communication can also be recorded on a contact or telephone log.

One of the many challenges facing a project manager is to accurately *assess project status*. The accuracy of the assessment will position both the project manager and the team to take the appropriate steps to complete the project successfully. Elements of these reports can include estimating the progress of each task, estimating project expenditures and budget status, and determining the present overall schedule and budget status.

References

1. L. Theodore, personal notes, East Williston, N.Y., 2006.
2. L. Theodore, "The Challenge of Change," *Chem. Eng. Prog.*, N.Y., Jan. 2007.
3. L. Theodore, letter to the editor (author response), "Changing the ChE Curriculum: How Much Is Too Much?" *Chem. Eng. Prog.*, N.Y., March 2007.
4. D. Lock, *Project Management,* 9th edition, Gower Publishing, London, 2007.
5. Y. Kwak, *A Brief History of Project Management: The Story of Managing Projects*, Greenwood Publishing Group, Westport, Conn., 2005.

6. J. Knitson and I. Bitz, *Project Management—How to Manage Successful Projects*, American Management Association, New York, N.Y., 1991.
7. L. Theodore, personal notes, East Williston, N.Y., 2012.
8. J. Davidson, *Managing Projects in Organizations*, Jossey-Bass Publishers, San Francisco, 1998.
9. L. Stander and L. Theodore, *Environmental Regulatory Calculations Handbook*, Wiley, Hoboken, N.J., 2008.

PART IV Accreditation Board for Engineering and Technology (ABET) Topics

Part Outline

25 Environmental Management

Chapter Outline

25–1 Introduction

Since the late 1960s there has had been an increased awareness of a wide range of environmental issues covering all sources: air, land, and water. More and more people are becoming aware of these environmental concerns, and it is important that chemical engineers, many of whom do not possess an understanding of environmental problems or have the proper information available when involved with environmental issues, develop capabilities in this area. All professional—not only chemical—engineers should have a basic understanding of the technical and scientific terms related to these issues as well as the regulations involved. Hopefully this chapter will serve the needs of chemical engineers, increasing their awareness of (and help solve) the environmental problems facing society.

Since the early 1970s there have been environmental tragedies as well as a heightened environmental awareness. The oil spills of the Exxon Valdez in 1989, the Gulf War of 1991, and the British Petroleum oil disaster showed how delicate our oceans and their ecosystems truly are. The disclosures of Love Canal in 1978 and Times Beach in 1979 made the entire nation aware of the dangers of hazardous chemical wastes. Discovery of the acquired immunodeficiency syndrome (AIDS) virus and the beach washups of 1985 brought the issue of medical waste disposal to the forefront of public consciousness. Nuclear accidents placed the spotlight on Three Mile Island, Chernobyl, and Fugasaki, and to this day, society is still seeing the effects of the last two events.

In addition to these events, the current human population on Earth is approximately 7 billion, and it almost certainly will increase in the future. The influence and effects of human activities on the environment have become increasingly evident at the local, state, national, and international levels, while issues such as environmental degradation, health, safety, green chemistry and engineering, and sustainability have become more pervasive and pressing. The net result is that there has been a more significant increase in both awareness and interest about the environment since the turn of this century.

The following topics are discussed in this chapter: environmental regulations; air, water, and solid waste management; classification, sources, and effects of pollutants; multimedia concerns; ISO 14000; energy conservation; the pollution prevention concept; green chemistry and green engineering; and sustainability.

NOTE Parts of the material presented in this chapter have been adapted from Theodore[1] and Burke et al.[2] Also, although the subject of communicating risk[1,2] is not addressed in this chapter, it does receive treatment in both references cited above. As is the case for most environmental issues of a volatile and personal nature, public perception of risk is unfortunately often based on fear rather than facts. The strategy to employ in communicating risk is therefore another important area of concern in environmental management.

25–2 Environmental Regulations

Environmental regulations are not simply a collection of laws (see also Chap. 18) on environmental topics. They are an organized system of statutes, regulations, and guidelines that minimize, prevent, and punish those responsible for the consequences of damage to the environment. This system requires each individual—whether an engineer, physicist, chemist, attorney, or consumer—to be familiar with its concepts and case-specific interpretations. Environmental regulations deal with the problems of human activities and the environment, and the uncertainties of law associated with them.

The National Environmental Policy Act (NEPA), enacted on January 1, 1970, is considered a political anomaly by some. NEPA was not based on specific legislation; instead, it is concerned in a general manner with environmental and quality-of-life issues. The Nixon Administration at that time became preoccupied with not only trying to pass more extensive environmental legislation but also implementing the laws. Nixon's White House Commission on Executive Reorganization proposed in the Reorganizational Plan 3 of 1970 that a single independent agency be established separate from the Council for Environmental Quality (CEQ). The plan was sent to Congress by President Nixon on July 9, 1970, and this new U.S. Environmental Protection Agency (EPA) began operation on December 2, 1970. EPA was officially born.

In many ways, EPA is the most far-reaching regulatory agency in the federal government because its authority is very broad. It is charged with protecting the nation's land, air, and water systems. Under a mandate of national environmental laws, it continues to strive to formulate and implement actions that lead to a compatible balance between human activities and the ability of natural systems to support and nurture life.[3]

EPA works with both state and local governments to develop and implement comprehensive environmental programs. Federal laws such as the Clean Air Act (CAA), the Safe Drinking Water Act (SDWA), the Resource Conservation and Recovery Act (RCRA), and the Comprehensive Environmental Response, Compensation, and Liability Act (CERCLA;[4] also called the *Superfund*) all mandate involvement by state and local governments in the details of implementation.

Along with an increasing awareness of the importance of the environment has come a greater understanding that the environment is not only a national issue but also a global issue. For example, the inhabitants of Earth share the same air (although there is minimal exchange between the northern and southern hemispheres), and some of the air over the United States might be over India in a few months, and vice versa. The harmful effects of air pollutants (agricultural damage, global climate change, acid deposition, etc.) know no political and geographic boundaries.

Continued industrial development will surely generate new challenges to human health and safety (see also Chap. 26) and to the well-being of Earth's environment as a whole. All development must be accompanied by research and evaluation of the potential environmental and human health consequences of each new process, product, and by-product. Addressing the associated problems requires a global perspective, and solving the problems will take a global commitment.

25-3 Air, Water, and Solid Waste Management

Air

Air management issues involve several different areas related to air pollutants and their control. Atmospheric dispersion of pollutants can be mathematically modeled (generally in equation form) to predict where pollutants emitted from a particular source, such as a utility stack, will settle to the ground and at what concentration. Pollution control equipment can be added to various sources to reduce the amount of pollutants before they are emitted into the air. Acid rain, the greenhouse effect, and global warming are all indicators of adverse effects to the air, land, and sea, which result from excessive amounts of pollutants being released into the air. (Note that some scientists—including the author—do not believe that either the greenhouse effect or global warming belong in this category.) One topic that few people are aware of is the issue of indoor air quality. Inadequate ventilation systems in homes and businesses directly affect the health of the building occupants. For example, the episode of Legionnaires' disease, which occurred in Philadelphia in the 1970s, was related to microorganisms that grew in the cooling water of the air conditioning system. Noise pollution, although not traditionally an air pollution issue, is included in this topic. The effects of noise pollution are seldom noticed until hearing is impaired. Although hearing impairment is a commonly known result of noise pollution, few people realize that stress is also a significant result of excessive noise exposure. The human body enacts its innate physiologic defensive mechanisms under conditions of loud noise, and the fight to control these physical instincts can cause significant stress on the individual.

Water

Pollutant dispersion in water systems and wastewater treatment is discussed in this subsection. Additional details are available in Chap. 23. Pollutants entering rivers, lakes, and oceans originate from a wide variety of sources, including stormwater runoff, industrial discharges, and accidental spills. It is important to understand how these substances disperse in their environment in order to determine how to control them. Studies on pollutants in municipal and industrial wastewater treatment systems, industrial use systems, drinking water supply, and other water systems are needed. Often, wastewater from industrial plants must be pretreated before it can be discharged into a municipal treatment system.

Solid Waste

Solid waste management issues address treatment and disposal methods for municipal, medical, and radioactive wastes. Programs to reduce and dispose of municipal and industrial waste include reuse, reduction, recycling, and composting, in addition to incineration and landfilling. Potentially infectious waste generated in medical facilities must be specially packaged, handled, stored, transported, treated, and disposed of to ensure the safety of both the waste handlers and the general public. Radioactive waste from any of a number of sources may have serious impacts on human health and the environment, and treatment and disposal requirements for radioactive substances must be strictly adhered to. Incineration has been a typical treatment method for hazardous waste for many years. The Superfund (CERCLA) was enacted in 1984 to identify and remedy uncontrolled hazardous waste sites. It also attempted to place the burden of cleanup on the generator rather than on the federal government. Asbestos, metals, and underground storage tanks either contain or are inherently hazardous materials that require special handling and disposal. Further, it is important to realize that both small and large generators of hazardous waste are regulated.

Pollution Prevention

Pollution prevention covers domestic and primarily industrial means of reducing pollution. This can be accomplished through (1) proper residential and commercial building design; (2) proper heating, cooling, and ventilation systems; (3) energy conservation; (4) reduction of water consumption; and (5) attempts to reuse or reduce materials before they become wastes. Domestic and industrial solutions to environmental problems can be addressed by considering ways to make homes and workplaces more energy efficient as well as ways to reduce the amount of wastes generated within them. (This topic is discussed in Sec. 25–8.)

25–4 Classification, Sources, and Effects of Pollutants

Not long ago the nation's natural resources were exploited indiscriminately. Waterways served as industrial pollution sinks; skies dispersed smoke from factories and power plants; and, the land proved to be a cheap and convenient place to dump industrial and urban wastes. However, society is now more aware of the environment and the need to protect it. The American people have been involved in a great social movement known broadly as *environmentalism*. Society has thus been concerned with the quality of the air that one breathes, the water that one drinks, and the land on which one lives and works. While economic growth and prosperity are still important goals, opinion polls indicate overwhelming public support for pollution controls and a pronounced willingness to pay for them. This section presents the reader with information on pollutants and categorizes their effects.

Pollutants are various noxious chemicals and refuse materials that impair the purity of the atmosphere, water, and soil. In the author's judgment the area most affected by pollutants is the atmosphere (or air). Air pollution occurs when wastes pollute the air. Artificially or synthetically created wastes are the main sources of air pollution. They can be in the form of gases or particulates which result from the burning of fuel to power motor vehicles and to heat buildings. More air pollution can be found in densely populated areas, e.g., urban areas. The air over largely populated cities often becomes so concentrated with pollutants that it harms not only human health but also plants, animals, and materials of construction.

Water pollution occurs when wastes are dumped into the water. This polluted water can spread typhoid fever and other diseases. In the United States, water supplies are disinfected to kill disease-causing germs. The disinfection process in some instances does not and often cannot remove all the chemicals and metals that may cause health problems in the distant future.

Wastes that are dumped into the soil are a form of land pollution that damages the thin layer of fertile soil that is essential for agriculture. In nature, cycles work to keep soil fertile. Wastes, including dead plants and wastes from animals, form a substance in the soil called *humus*. Bacteria then decay the humus and convert it into nitrates, phosphates, and other nutrients that can feed growing plants.

25-5 Multimedia Concerns

The current approach to environmental waste management requires some rethinking. A multimedia approach helps the integration of air, water, and land pollution controls and seeks solutions that do not violate the laws of nature as it integrates air, water, and land into a single concern. It seeks a solution to pollution that does not endanger society or the environment. The obvious advantage of a multimedia pollution control approach is its ability to manage the transfer of pollutants to prevent them from causing other pollution problems. Among the possible steps in the multimedia approach are understanding the cross-media nature of pollutants, modifying pollution control methods so as not to shift pollutants from one medium to another, applying available waste reduction technologies, and training environmental professionals in a total environmental concept. The challenge for the future environmental professional include:

1. Conservation of natural resources
2. Control of air, water, and land pollution
3. Regulation of toxics and disposal of hazardous wastes
4. Improvement in the quality of life

It is now increasingly clear that some treatment technologies, while solving one pollution problem, have created others. Most contaminants, particularly toxics, present problems in more than one medium. Since nature does not recognize neat jurisdictional compartments, these same contaminants are often transferred across media. Air pollution control devices and industrial wastewater treatment plants prevent waste from entering the air and water, but the toxic ash and sludge that these systems produce can also become hazardous waste problems themselves. For example, removing trace metals from a flue gas during the control phase usually transfers the products to a liquid or solid phase. Does this exchange an air quality problem for a liquid or solid waste management problem? The reader should ponder this question. Waste disposed on land or in deep wells can contaminate groundwater, and evaporation from ponds and lagoons can convert solid or liquid waste into air pollution problems.[5] Other examples include acid deposition, residue management, water reuse, and hazardous waste treatment and/or disposal.

Control of cross-media pollutants cycling in the environment is therefore an important step in the overall management of environmental quality. Pollutants that do not remain where they are released or deposited move from source to receptors by many routes, including air, water, and land. Unless information is available on how pollutants are transported, transformed, and accumulated after they enter the environment, they cannot be controlled realistically and effectively. A better understanding of the cross-media nature of pollutants and their major environmental processes—physical, chemical, and biological—is required.

Environmental quality and natural resources are under extreme duress in many industrialized nations and in virtually every developing nation as well. Environmental pollution is closely related to not only population density, energy, transportation demand, and land use patterns but also industrial and urban development. The main reason for environmental pollution is the increasing rate of waste generation in terms of quantity and toxicity that has exceeded society's ability to properly manage it. Another reason is that the environmental management approach has focused on the media-specific and the end-of-pipe strategies.

There is increasing reported evidence of the socioeconomic and environmental benefits realized from multimedia pollution prevention.[6] The prevention of environmental pollution in the 21st century will require not only enforcement of government regulations and controls but also changes in manufacturing processes and products that will ultimately impact lifestyles and behavior throughout society. Education will play a key role in the future in achieving the vital goal of multimedia pollution prevention.

25-6 ISO 14000

The International Organization for Standardization (ISO) is a private, nongovernmental, international standards body based in Geneva, Switzerland. Founded in 1947, ISO promotes international harmonization and the development of manufacturing, product, and communications standards. It is a nongovernmental organization; however, governments are allowed to participate in the development of standards, and many governments have chosen to adopt the ISO standards as their regulations. ISO also closely interacts with the United Nations.[7] ISO has promulgated over 16,000 internationally accepted standards for everything from paper sizes to film speeds. Over 130 countries participate in the ISO as either *Participating* or *Observer* members. The United States is a full-voting participating member and is officially represented in the ISO by the American National Standards Institute (ANSI).

Many people have noticed the apparent lack of consistency between the official title when used in full form; *International Organization for Standardization* and the short form, ISO. Should not the acronym be IOS? That would have been the case if it were an acronym. However, ISO is a word derived from the Greek word *isos,* meaning *equal.* From *equal* to *standard*, the line of thinking that led to the choice of *ISO* as a name of the organization is easy to follow. In addition, the acronym ISO is used around the world to denote the organization, thus avoiding the plethora of acronyms resulting from the translation of *International Organization for Standardization* into the different national languages of members, e.g., IOS in English or OIN in French. Whatever the country, the short form of the organization's name is now ISO.[8]

Over the years ISO has expanded the scope of their standards to incorporate areas such as the environment, energy conservation, service sectors, security, and managerial and organizational practice. There are currently more than 16,000 standards applying to three areas of sustainable development: economic, environmental, and social.[7] ISO's environmental mission is to promote the manufacturing of products in a manner that is effective, safe, and clean.[9] ISO hopes to achieve this goal through the dedication and participation of more countries.

ISO 14000 may be viewed as a generic environmental management standard. It can be applied to any organization and focuses on the processes and activities conducted by the company. It consists of standards and guidelines regarding environmentally managed systems (EMSs). The idea for it first evolved from the United Nations conference on Environment and Development (UNCED), held in Rio de Janeiro in 1992. The topic of sustainable development (see Sec. 25–10) was also discussed there, and ISO made a commitment to support this subject.[10]

ISO 14000 standards were first written in 1996 and have subsequently been amended and updated. Their intended purpose is to assist companies and organizations in minimizing their negative effect on the environment and in complying with any laws, regulations, or environmental requirements that have been imposed on them. ISO can also help to establish an organized approach to reducing any environmental impacts the company can control. Businesses that comply with these standards are eligible for certification. This certification is awarded by third-party organizations instead of ISO.[11]

25–7 Energy Conservation[12]

The reader is referred to Chap. 21 for additional details on this subject, some of which are repeated below for the reading audience.

Energy is the keystone of American life and prosperity as well as a vital component of environmental rehabilitation. The environment must be protected and the quality of life improved, but, at the same time, economic stability must also be maintained. Both of these objectives will be prime factors in determining domestic and foreign energy policies for years to come. Since energy consumption is a major contributor to environmental pollution, decisions regarding energy policy alternatives require comprehensive environmental analysis. Environmental impact data must be developed for all aspects of an energy system and/or energy conservation program and must not be limited to separate components.

Because energy has been relatively cheap and plentiful in the past, many energy-wasting practices have been allowed to develop and continue in all sectors of the economy. Industries have wasted energy by discharging hot process water instead of recovering the heat, and by wasting the energy discharged from power plant stacks.[13] Waste hydrocarbons have been discharged or combusted with little consideration for recovering their energy value.[14] There are many other examples, too numerous to mention. Elimination of these practices will, at least temporarily, partially reduce the rate of increase in energy demand. If conservation can reduce energy demand, it can reduce the associated pollution.

The most dramatic environmental improvements can be developed by *energy conservation* in the industrial sector of the economy. Industry accounts for approximately 40% of the energy consumed in the United States. Also, industry might be considered more dynamic, progressive, and strongly motivated by the economic incentives offered by conservation than the other energy-consuming sectors (residential, commercial, and transportation).

The environmental impacts of energy conservation and consumption are far reaching, affecting not only air, water, and land quality but also public health. Combustion of fossil fuels such as coal, oil, and natural gas is responsible for air pollution in urban areas; acid rain is damaging lakes and forests; and some of the nitrogen pollution is harming estuaries. Although data show that annual average ambient levels of all criteria air pollutants were down nationwide during 1977 to 1989 in the previous century, 96 major metropolitan areas still exceeded the national health-based standard for ozone, and 41 metropolitan areas exceeded the standard for carbon monoxide.[12]

Energy consumption also *appears* to be the primary anthropogenic contribution to global warming, referred by to some as the *greenhouse effect*. EPA has concluded that energy use—through the formation of carbon dioxide during combustion processes—has contributed approximately 50% to the global warming that has occurred since the late 1990s. Although the scientific community is not unanimous in regard to the causes of global warming, most individuals (not including this author) and groups have concluded that a "reasonable" chance of anthropogenic climatic change exists and have already begun to define the potential implications of such changes, many of which are catastrophic. In light of this situation, the Alliance to Save Energy has challenged Congress to pass meaningful legislation to promote and achieve energy efficiency.[12–14]

25–8 The Pollution Prevention Concept

The amount of waste generated in the United States has reached staggering proportions; according to EPA, nearly 300 million tons of solid waste alone are generated annually. Although both the Resource Conservation and Recovery Act (RCRA) and the Hazardous and Solid Waste Act (HSWA) encourage businesses to minimize the wastes they generate, the majority of current environmental protection efforts are centered around treatment and pollution cleanup.

The passage of the Pollution Prevention Act of 1990 has redirected industry's approach to environmental management; *pollution prevention* became the environmental option of the 1990s and the first decade of the twenty-first century. Whereas typical waste management strategies concentrate on "end of pipe" pollution control, pollution prevention attempts to handle waste at the source (i.e., source reduction). As waste handling and disposal costs increase, the application of pollution prevention measures is becoming more attractive than ever before. Industry continues to explore the advantages of multimedia waste reduction and developing agendas to improve environmental design while lowering production costs.

There are significant opportunities for both the individual and industry to prevent the generation of waste; indeed, pollution prevention is today primarily stimulated by economics, legislation, liability concerns, and the enhanced environmental benefit of managing waste at the source. The EPA Pollution Prevention Act of 1990 established pollution prevention as a national policy, declaring that "waste should be prevented or reduced at the source wherever feasible, while pollution that cannot be prevented should be recycled in an environmentally safe manner."[15] The EPA policy establishes the following hierarchy of waste management:

1. Source reduction
2. Recycling and reuse
3. Treatment
4. Ultimate disposal

These hierarchical categories are prioritized so as to promote the examination of each individual alternative prior to investigation of subsequent options (i.e., the most preferable alternative should be thoroughly evaluated before a less accepted option is considered). Practices that decrease, avoid, or eliminate the generation of waste are considered source reduction and can include the implementation of procedures as simple and economical as good housekeeping. Recycling is the use, reuse, or reclamation of waste and/or materials and may involve the incorporation of waste recovery techniques (distillation, filtration, etc.). Recycling can be performed at the facility (i.e., on-site), or away from the facility (i.e., off-site at an reclamation facility). Treatment involves the destruction or detoxification of wastes into nontoxic or less toxic materials by chemical, biological, or physical methods, or in a combination of these methods. Disposal has been included in the hierarchy because it is recognized that residual wastes will exist; the EPA's so-called ultimate disposal options in the past included landfilling, land farming, ocean dumping, and deep-well injection. However, the term *ultimate disposal* is a misnomer, but is included here because of its earlier adaptation by the EPA.

Table 25–1 provides a rough timetable demonstrating the national approach to waste management. Note how waste management has begun to shift from pollution *control* to pollution *prevention*.[16,17]

The author codeveloped a popular pollution prevention calendar for home or office use. Each of the two calendars contains 365 one-line suggestions (one for each day) dealing with waste reduction, energy conservation and health, safety, and accident prevention. Each topic receives 4 months of coverage. The calendar is available in either hard-copy or electronic format.[18]

The development of waste management practice in the United States has now moved toward securing a new pollution prevention ethic. The performance of pollution prevention assessments and their subsequent implementation will encourage increased research into methods that will further aid in the reduction of waste and pollution.

It appears evident that the majority of the present day obstacles to pollution prevention are based on either a lack of information or an anxiety associated with economic concerns. However, it has strengthened the exchange of information among businesses, and a better understanding of the unique benefits of pollution prevention will certainly be realized in the future.

Table 25–1

Waste Management Timetable

Time frame	Control
Prior to 1945	No control
1945–1960	Little control
1960–1970	Some control
1970–1975	Greater control (EPA founded)
1975–1980	More sophisticated control
1980–1985	Beginning of waste reduction management
1985–1990	Waste reduction management
1990–1995	Pollution Prevention Act
1995–2000	Pollution prevention activities
2000–2010	Sustainability and green energy
2010–2015	Industrial ecology
2015–	???

Sources: Refs. 16 and 17.

25–9 Green Chemistry and Green Engineering

Activities in the field of green engineering and green chemistry are increasing at a near-exponential rate. This section attempts to familiarize the reader with these topics by defining and presenting principles of each; future trends are also discussed. Before beginning this section it is important that the term *green* should not be considered a new method or type of chemistry or engineering. Rather, it should be incorporated into the way scientists and engineers design for categories that include the environment, manufacturability, disassembly, recycling, serviceability, and compliance. Today, the major green element is to search for technology to reduce and/or eliminate waste from operations and processes, with an important priority of not creating waste in the first place.

Green Chemistry

Green chemistry, also called *clean chemistry*, refers to that field of chemistry dealing with the synthesis, processing, and use of chemicals that reduce risks to humans and the environment.[19] It is defined as the invention, design, and application of chemical products and processes to reduce or to eliminate the use and generation of hazardous substances.[20]

A baker's dozen principles of green chemistry are provided below:[20]

1. *Prevention*—it is better to prevent waste than to treat or clean up waste after it has been generated.
2. *Atom economy*—synthetic methods should be designed to maximize the incorporation of all materials used in the process through to the final product.
3. *Less hazardous chemical syntheses*—whenever practicable, synthetic methods should be designed to use and generate substances that possess little or no toxicity to human health and the environment.
4. *Designing safer chemicals*—chemical methods should be designed to preserve efficacy of function while minimizing toxicity.
5. *Safer solvents and auxiliaries*—the use of auxiliary substances (e.g., solvents, separation agents) should be made unnecessary whenever possible and innocuous when being used.
6. *Design for energy efficiency*—energy requirements should be recognized for their environmental and economic impacts and be minimized. Synthetic methods should be conducted at ambient temperature and pressure whenever possible.
7. *Use of renewable feedstocks*—a raw material or feedstock should be renewable rather than depleting, wherever and whenever technically and economically practicable.
8. *Reduce derivatives*—unnecessary derivatization (blocking group, temporary modification of physico/chemical processes, etc.) should be avoided whenever possible because such steps require additional reagents and can generate waste.
9. *Catalysis*—catalytic reagents (that should be as selective, or discriminating, as possible) are superior to stoichiometric reagents.
10. *Biocatalysis*—enzymes and antibodies should be used to mediate reactions.
11. *Design for degradation*—chemical products should be designed in a way that at the end of their function they break down into innocuous degradation products and do not persist in the environment.

12. *Real-time analysis for pollution prevention*—analytical methods need to be further developed to allow for real-time, in-process monitoring and control prior to the formation of hazardous substances.
13. *Inherently safer chemistry for accident prevention*—substances and the form of a substance used in a chemical process should be chosen so as to minimize the potential for chemical accidents, including releases, explosions, and fires.

Green Engineering

Green engineering is similar to green chemistry in many respects, as witnessed by the underlying urgency of attention to the environment seen in both sets of principles.

A baker's dozen principles of green engineering are provided below.[21]

1. *Benign rather than hazardous*—designers must ensure that all material and energy inputs and outputs are as inherently nonhazardous as possible.
2. *Prevention instead of treatment by recycle or reuse*—it is better to prevent waste than to treat or clean up waste after it is generated.
3. *Design for separation*—separation and purification operations should be a component of the design framework.
4. *Maximize efficiency*—system components should be designed to maximize mass, energy, and temporal efficiency.
5. *Output-pulled versus input-pushed*—components, processes, and systems should be output-pulled rather than input-pushed through the use of energy and materials.
6. *Conserve complexity*—energy conservation must also consider entropy. Embedded entropy and complexity must be viewed as an investment when making design choices on recycle, reuse, or beneficial disposition.
7. *Durability rather than immortality*—targeted durability, not immortality, should be a design goal.
8. *Meet need, minimize excess*—design for unnecessary capacity or capability should not be considered a design flaw; this includes engineering "one size fits all" solutions.
9. *Minimize material diversity*—multicomponent products should be designed for material unification to promote disassembly and value retention (to minimize material diversity).
10. *Integrate material and energy flows*—design of processes and systems must include integration of interconnectivity with available energy and material flows.
11. *Design for afterlife*—performance metrics should include designing for performance in (commercial) afterlife.
12. *Renewable rather than depleting*—design should be based on renewable and readily available inputs throughout the life cycle.
13. *Engaging communities*—actively engage communities and stakeholders in development of engineering solutions.

A reasonable question to ask is: What is the difference between green engineering and green chemistry? From the definitions given above, one might conclude that green engineering is concerned with the design, commercialization, and use of all types of processes and products, whereas green chemistry covers only a very small subset of this—the development of chemical processes and products. Thus, green chemistry may be viewed as a subset of green engineering. It is, in fact, a very broad field, encompassing everything from improving energy efficiency in manufacturing processes to developing plastics from renewable resources.

Future Trends

Chemists and chemical engineers have the unique ability to affect the design of molecules, materials, products, processes, and systems at the earliest possible stages of their development. With much of the research occurring now in these two fields, chemists and chemical engineers must ask themselves the following questions:[22]

1. What will be the human health and the environmental impacts of the chemicals put into the marketplace and society?
2. How efficiently will the product manufacturing systems be?
3. What will tomorrow's innovations look like, and from what materials will they be created?

Three green problem areas stand out:[23]

1. Inventing technology to support the expanded availability and use of renewable energy.
2. Developing renewable feedstocks and products based on them.
3. Creating technology that does not produce pollution.

Some very important steps that must be taken in the near future must include: implementing greatly improved technologies for harnessing the fossil and nuclear fuels to ensure that their use, if continued, creates much lower environmental and social impact; developing and deploying the renewable energy sources on a much wider scale; and, effecting major improvements in the efficiency of energy conversion, distribution, and use.[24]

Green chemistry and green engineering are emerging twenty-first-century issues which come under the larger multifaceted spectrum of sustainable development. *Sustainable development* represents a change in consumption patterns toward environmentally more benign products, and a change in investment patterns toward augmenting environmental capital.[25] In this respect, sustainable development is feasible. It requires a shift in the balance of the way economic progress is pursued (see also next Sec. 25–10).

Environmental concerns must also be properly integrated into economic policy from the highest (macroeconomic) level to the most detailed (microeconomic) level. The environment must be viewed as a valuable, frequently essential input to human well-being. The fields of green chemistry and engineering is certain to solve problems that are of great significance to the future of humanity.

25–10 Sustainability

The term *sustainability* has many different meanings to different people. To *sustain* is defined by some as to "support without collapse." Discussion of how sustainability should be defined was initiated by the Bruntland Commission. This group was assigned a mission to create a "global agenda for change" by the General Assembly of the United Nations in 1984. They defined *sustainable* very broadly:[26] "Humanity has the ability to make development sustainable—to ensure that it meets the needs of the present without compromising the ability of future generations to meet their own needs."[27]

Sustainability involves simultaneous progress in four major areas: (1) human, (2) economic, (3) technological, and (4) environmental. Sustainability requires conservation of resources while minimizing depletion of nonrenewable resources, and using sustainable practices for managing renewable resources. However, there can be no product development or economic activity of any kind without available resources. Except for solar and nuclear energy, the supply of resources is finite. Efficient designs conserve resources while also reducing impacts caused by materials extraction and related activities. Depletion of nonrenewable resources and overuse of otherwise renewable resources limits their availability to future generations

Another principal element of sustainability is maintenance of the structure and function of the ecosystem. Because the health of human populations is linked to the health of the natural world, the issue of ecosystem health is a fundamental issue of concern to sustainable development. Thus, sustainability requires that the health of all diverse species as well as their interrelated ecological functions be maintained. As only one species in a complex web of ecological interactions, humans cannot separate their survivability from that of the total ecosystem.

Over the next 50 years (i.e., until 2063 or so), projections suggest that the world's population could increase by 50%. Global economic activity is expected to increase by 500%. Concurrently, global energy consumption and manufacturing activity may rise to approximately three times current levels. These trends could have serious social, economic, and environmental consequences unless a way can be found to use fewer resources more efficiently. The task ahead in this century is to help shape a sustainable future in a cost-effective manner, recognizing that economic and environmental considerations, supported by innovative science and technology, can work together and promote societal benefits. However, unless humans embrace sustainability, they will ultimately deplete Earth's resources and damage its environment to such an extent that conditions for human existence on the planet will be seriously compromised or even become impossible. This is a reality that society must come to grips with.

As stated above, sustainable development is feasible. Since *sustainable development* means a change in consumption patterns toward environmentally more benign products, and a change in investment patterns, it will require a shift in the balance of the way economic progress is pursued. Environmental concerns must be properly integrated into rearrangement policies and the environment must be viewed as an integral part of human well-being.

Summarizing, *sustainability* is a term that is used with greater frequency in the environmental community and with sometimes varied meanings. In short, it means the capacity to endure. A more specific definition of sustainability refers to the long-term maintenance of well-being, which has environmental, economic, and social dimensions, and includes the responsible management of resource use.

References

1. M. K. Theodore and L. Theodore, *Introduction to Environmental Management*, CRC Press/Taylor & Francis Group, Boca Raton, Fla., 2010.
2. G. Burke, B. Singh, and L. Theodore, *Handbook of Environmental Management and Technology*, Wiley, Hoboken, N.J., 1992.
3. *Webster's New World Dictionary*, 2nd Collegiate Edition, Prentice-Hall, Upper Saddle River, N.J., 1971.
4. L. Stander and L. Theodore, *Environmental Regulatory Calculations Handbook*, Wiley, Hoboken, N.J., 2008.
5. G. Burke, B. Singh, and L. Theodore, *Handbook of Environmental Management and Technology*, 2nd ed., Wiley, N.Y., 2000.
6. N. Schecter and G. Hunt, *Case Summaries of Waste Reduction by Industries in the Southeast*, Waste Reduction Resource Center, Raleigh, N.C., 1989
7. ISO in Brief, Aug. 2006. http://www.iso.org/iso/isoinbrief_2006-en.pdf
8. Introduction to ISO, www.iso.ch/infoe/intro
9. General Information on ISO, http://www.iso.org/iso/support/faqs/faqs_general_information_on_iso.htm
10. Summary of ISO 14000, Lighthouse Consulting, University of Rhode Island, 2003, www.crc.uri.edu/download/12_ISO_1400_summary_ok.pdf

11. ISO 14000, http://en.wikipedia.org/wiki/ISO_1400
12. K. Skipka and L. Theodore, *Energy Resources: Past, Present and Future Management*, CRC Press/Taylor & Francis Group, Boca Raton, Fla. (in press), 2014.
13. L. Theodore, F. Ricci and T. VanVliet, *Thermodynamics for the Practicing Engineer*, Wiley, Hoboken, N.J., 2009.
14. L. Theodore, *Heat Transfer Applications for the Practicing Engineer*, Wiley, Hoboken, N.J., 2011.
15. USEPA, *Pollution Prevention Fact Sheet* (author unknown), Washington, D.C., March 1991.
16. L. Theodore, personal notes, East Williston, N.Y., 2011.
17. D. Green and R. Perry, eds., *Perry's Chemical Engineers' Handbook*, 8th ed., McGraw-Hall, N.Y., 2008.
18. M. K. Theodore, *Pollution Prevention Calendar*, Theodore Tutorials, East Williston, N.Y., 2000 (and subsequent years).
19. P. Anastas and T. Williamson, in P. T. Anastas and T. C. Williamson, eds., *Green Chemistry: An Overview, Green Chemistry: Designing Chemistry for the Environment*, ACS Symposium Series 626, American Chemical Society, Washington, D. C., 1996, pp. 1–17.
20. P. Anastas and J. Warner, *Green Chemistry: Theory and Practice*, Oxford University Press, N.Y., 1998.
21. P. Anastas and J. Zimmmerman, "Design through the Twelve Principles of Green Engineering, Environ. Sci. Technol., **37**, 94A–101A, 2003.
22. http://portal.acs.org/portal/acs/corg/content?_nfpb=true&_pageLabel=PP_SUPERARTICLE&node_id=1415&use_sec=url_var=region1
23. http://www.chem.cmu.edu/groups/Collins/ethics/ethics06.html
24. G. Boyle, B. Everett, and J. Ramage, *Energy Systems and Sustainability*, Oxford University Press, Oxford, U.K., 2003.
25. D. Pearce, A. Mankandya and E. Barbier, *Blueprint for a Green Economy*, Earthscan, London, 1989.
26. P. Bishop, *Pollution Prevention*, Waveland Press, Prospect Heights, Ill., 2000.
27. United Nations, Report of the World Commission on Environment and Development, General Assembly Resolution 42/187, Dec. 11, 1987 (retrieved Oct. 31, 2007).

26 Health, Safety, and Accident Management

Chapter Outline

26–1 Introduction

This chapter not only discusses the dangers posed by hazardous substances but also examines the general subject of health, safety, and accident management. The laws and legislation passed to protect workers, the public, and the environment from the effects of these chemicals and accidents are also reviewed. The chapter also discusses the regulations, with particular emphasis on emergency planning. In effect, the chapter addresses topics that one would classify as health, safety, and accident prevention. The bulk of the material has been adapted from Theodore et al.[1] and Theodore and Theodore.[2]

Two general types of potential chemical health, safety, and accident exposures or concerns exist. These are classified as

1. *Chronic*. Continuous exposure occurs over long periods of time, generally several months to years. Concentrations of inhaled contaminants are usually relatively low, direct skin contact by immersion, by splash, or by contaminated air involving contact with substances exhibiting low dermal activity.
2. *Acute*. Exposures occur for relatively short periods of time, generally seconds or minutes to 1 or 2 days. Concentration of contaminants is usually high relative to their protection criteria. In addition to inhalation, airborne substances might directly contact the skin, or liquids and sludge may be splashed on the skin or into the eyes, leading to toxic effects.

In general, acute exposures to chemicals in air are more typical in transportation accidents, explosions, and fires, or releases at chemical manufacturing or storage facilities. High concentrations of contaminants in air rarely persist for long periods of time. Acute skin exposure may occur when workers come in close contact with the substances in order to control a release e.g., while patching a tank car, off-loading a corrosive material, uprighting a drum, or containing and treating a spilled material.

Chronic exposures, on the other hand, are usually associated with longer-term removal and remedial operations. Contaminated soil and debris from emergency operations may be involved in the round-the-clock discharge of pollutants to the atmosphere, soil and groundwater. They may be polluted, or temporary impoundment systems may contain diluted chemicals. Abandoned waste sites typically represent chronic exposure problems. As activities start at these sites, personnel engaged in certain operations such as sampling, handling containers, or bulking compatible liquids, face an increased risk of acute exposure. These exposures stem from splashes of liquids or from the release of vapors, gases, or particulates that might be generated.

In any specific incident, the hazardous and/or toxic properties of the materials may be the only problem that represents a potential risk. For example, if a tank car containing liquefied natural gas is involved in an accident, but remains intact, the risk from fire and explosion is low. In other incidents, the risks to response personnel are high, e.g., when toxic or flammable vapors are released from a ruptured tank truck. The continued concern for health and safety of response personnel requires that the risks, both real and potential at an accident be assessed, and appropriate measures instituted to reduce or eliminate the threat to response personnel.

From a health perspective, specific chemicals and chemical groups affect different parts of the body. One chemical, such as one with either a very low or a very high pH, may affect the skin, whereas another, such as carbon tetrachloride, might attack the liver. Some chemicals will affect more than one organ or system. When this occurs, the organ or system under attack is referred to as the *target organ*. The damage done to a target organ can differ in severity depending on chemical composition, length of exposure, and the concentration of the chemical.

Another health effect is involved when two different chemicals enter the body simultaneously; the result can be intensified or compounded. A *synergistic effect* results when one substance intensifies the damage done by the other. Synergism complicates almost any exposure due to a lack of toxological information. For just one chemical, it may typically take a toxological research facility approximately 2 years of studies to generate valid data. The data produced in that 2-year time frame applies only to the effect of that one chemical acting alone. With the addition of another chemical, the original chemical may have a totally different effect on the body. This fact results in a great many unknowns when dealing with toxic substances, and therefore increases risks due to a lack of dependable information.

The National Institute of Occupational Safety and Health (NIOSH) recommends standards for industrial exposure that the Occupational Safety and Health Administration (OSHA) uses in its regulations. The *NIOSH Pocket Guide to Chemical Hazards* contains a wealth of information of specific chemicals such as:

1. Chemical name, formula, and structure
2. Trade names and synonyms
3. Chemical and physical properties
4. Time-weighted average threshold limit values (TLVs)
5. Exposure limits
6. Lower explosive limit (LEL)
7. "Immediately dangerous to life and health" (IDLH) concentrations
8. Measurement methods
9. Personal protection and sanitation guidelines
10. Health problems information

The topics covered in this chapter include safety and accidents; emergency planning and response; regulations; introduction to environmental risk assessment; and, health and hazard risk assessment.

26-2 Safety and Accidents

The previous section discussed problems associated with exposure to chemicals. In the chemical industry, there is a high risk of accidents due to the nature of the processes and the materials used. Although precautions are taken to ensure that all processes run smoothly, there is always (unfortunately) room for error, and accidents will occur. This is especially true for highly technical and complicated operations, as well as processes operating under extreme conditions such as high temperatures and pressures. In general, accidents are caused by one or more of the following factors:

1. Equipment breakdown
2. Human error
3. Terrorism
4. Fire exposure and explosions
5. Control system failure(s)
6. Natural causes
7. Utilities and ancillary system outage
8. Faulty siting and plant layout

These causes are usually at the root of most industrial accidents. Although there is no way to guarantee that these problems will not arise, steps can be taken to minimize the number, as well as the severity, of incidents.

In an effort to reduce occupational accidents, measures should be taken in the following areas:[3]

1. *Training.* All personnel should be properly trained in the use of equipment and made to understand the consequences of misuse. In addition, operators should be rehearsed in the procedures to follow should something go wrong.
2. *Design.* Equipment should be used only for the purposes for which it was designed and intended. All equipment should be periodically checked for damage or errors in the design.
3. *Human Performance.* This should be closely monitored to ensure that proper procedures are followed. Also, working conditions should be such that the performance of workers is improved, thereby simultaneously reducing the chance of accidents. Periodic medical examinations should be provided to ensure that workers are in good health, and that the environment of the workplace is not causing undue physical stress. Finally, under certain conditions, it may be advisable to test for the use of alcohol or drugs—conditions that severely handicap judgment, and therefore render workers accident-prone.

Each day, on average, nearly 10,000 U.S. workers sustain disabling injuries on the job; 16 workers die from an injury suffered at work, and approximately 125 workers die from work-related diseases. The Liberty Mutual 2005 Workplace Safety Index estimated that employees spent $50.8 billion in 2003 on wage payments and medical care for workers not on the job.

26–3 Emergency Planning and Response

The extent of the need for emergency planning is significant, and continues to expand as new regulations on safety are introduced. Planning for an industrial emergency must begin before the plant is constructed. The new plant will have to pass all safety measures and OSHA standards. This is emphasized by Armenante, who states that "The first line of defense against industrial accidents begins at the design stage. It should be obvious that it is much easier to prevent an accident rather than to try to rectify the situation once an accident has occurred."[4]

Successful emergency planning begins with a thorough understanding of the event or potential disaster being planned for. The impacts on public health and the environment must also be estimated. Some of the types of emergencies that should be included in the plan are:[1–3]

1. Natural disasters such as earthquakes, tornadoes, hurricanes, and floods
2. Explosions and fires
3. Acts of terrorism
4. Hazardous chemical leaks
5. Power or utility failures
6. Radiation accidents
7. Transportation accidents

The affected area or emergency zone must be studied in depth in order to estimate the impact on the public or the environment. A hazardous gas leak, fire, or explosion may cause a toxic cloud to spread over a great distance as it did in Bhopal, India.[3] An estimate of the minimum affected area, and thus the area to be evacuated, should be based on an atmospheric dispersion model. Various models can be used. While the more complex models produce the most accurate results, simpler models are faster and usually still provide adequate data and information for planning purposes.[3]

The main objective for any plan should be to prepare a procedure to make maximum use of the combined resources of the community in order to accomplish the following:

1. Safeguard people during emergencies.
2. Safeguard people during an act of terrorism.
3. Minimize damage to property and the environment.
4. Initially contain and ultimately bring the incident under control.
5. Effect the rescue and treatment of casualties.
6. Provide authoritative information to the news media who will communicate the facts to the public.
7. Secure (the safe rehabilitation of) the affected area.

26–4 Regulations

Each company must develop a health-and-safety program for its workers. For example, OSHA has regulations governing employee health and safety at hazardous waste operations and during emergency responses to hazardous substance releases. These regulations (29 CFR 1910.120) contain general requirements for

1. Safety and health programs
2. Training and educational programs
3. Work practices along with personal protective equipment
4. Site characterization and analysis
5. Site control and evacuation
6. Engineering controls
7. Exposure monitoring and medical surveillance
8. Materials handling and decontamination
9. Emergency procedures
10. Illumination
11. Sanitation

The EPA Standard Operating Safety Guides supplement these regulations. However, OSHA's regulations must be used for specific legal requirements for an industry. Other OSHA regulations pertain to employees working with hazardous materials or at hazardous waste sites. These, as well as state and local regulations, must also be considered when developing worker health and safety programs.[5]

The OSHA Hazard Communication Standard was first promulgated on November 25, 1983, and can be found in 29 CFR Part 1910.120. The standard was developed to inform workers who are exposed to hazardous chemicals of the risk associated with specific chemicals. The purpose of the standard is to ensure that the hazards of all chemicals produced or imported are evaluated, and that information concerning chemical hazards is conveyed to *both* employers and employees.

Information on chemical hazards must be dispatched from the manufacturers to employers via the *material safety data sheets* (MSDSs) and container labels. These data must then be communicated to employees by means of comprehensive hazard communication programs, which include training programs, as well as the MSDSs and container labels.

The basic requirements of the Hazardous Communications Standard are as follows:[3]

1. There must be an MSDS on file for every hazardous chemical present or used in the workplace.
2. MSDSs must be readily available during each work shift, and all employees must be told how to obtain the information. If employees "travel" on the shift, the MSDSs may be kept in a central location at the primary job site, as long as the necessary information is immediately available in the event of an emergency.
3. It must be ensured that every container holding hazardous chemicals in the workplace is clearly and properly labeled and includes appropriate hazard warnings.
4. Labels of incoming hazardous chemical containers must not be removed or defaced in any way.
5. Prior to initial assignments, employees must be informed of the requirements of the standard operations in their work area where hazardous chemicals are present.
6. Employers must develop, implement, and maintain a written communication program for each workplace that describes how MSDSs, labeling, and employee information and training requirements will be met. This written program must also include a list of hazardous chemicals present in the workplace and the methods that will be used to inform employees of the hazards associated with performing nonroutine tasks.
7. Employers must train employees how to identify and protect themselves from chemical hazards in the work area, as well as how to obtain the employer's written hazard communication program and hazard information.

Companies with multiemployer workplaces must include (with the MSDS) methods that the employer will use for the contractors at the facility. These employers must also describe how they will inform the subcontractors' employees about precautions which must be followed and the specific labeling system used in the workplace.

The Superfund Amendments and Reauthorization Act (SARA) of 1986 renewed the national commitment to correcting problems arising from previous mismanagement of hazardous wastes. Title III of SARA, specifically known as the Emergency Planning and Community Right-to-Know Act, forever changed the concept of environmental management. Planning for emergencies became law. While SARA was similar in many respects to the original law, it also contained new approaches to the program's operation. The 1986 Superfund legislation accomplished the following:

1. Reauthorized the original program for 5 more years, dramatically increasing the cleanup fund from $1.6 to $8.5 billion.
2. Setting of specific goals and standards, stressing permanent solutions.
3. Expansion of state and local involvement in decision-making policies.
4. Provision for new enforcement authorities and responsibilities.
5. Strengthened the focus on human health issues caused by hazardous waste sites.

This law was more specific than the original statute with regard to such measures as remedies to be used at Superfund sites, public participation, and accomplishment of cleanup activities.

The Emergency Planning and Community Right-to-Know Act is undeniably the most important part of SARA in terms of public acceptance, participation, and support. Title III addresses the most important issues regarding community awareness and participation in the event of a chemical release. Title III establishes requirements for emergency planning, hazardous emissions reporting, emergency notification, and "community right-to-know." For instance, it is now law that companies release any data that a local or community planning committee needs in order to develop and implement its emergency plan.[6]

The objectives of Title III are to improve local chemical emergency response capabilities, primarily through improved emergency planning and notification, and to provide citizens and local governments with access to information about chemicals in their area. Title III has four major sections that aid in the development of contingency plans:

1. Emergency Planning (Secs. 301–303)
2. Emergency Notification (Sec. 304)
3. Community Right-to-Know Reporting Requirements (Secs. 311–312)
4. Toxic Chemicals Release Reporting—Emissions Inventory (Sec. 313)

Title III has also developed timetables for implementation of the Emergency Planning and Community Right-to-Know Act of 1986. Although the material discussed above was first published in 1986, much of this was still applicable at the time of the preparation of this chapter in 2012, and is available in the EPA literature.

The lessons learned from past accidents have impacted industry's approach to accident and emergency management. It is essential for industry to abide by stringent safety procedures. The more knowledgeable the personnel, from the management to the operators of a plant, and the more information that is available to them, the less likely that a serious incident will occur. The new regulations, and especially Title III of 1986, help ensure that safety practices are up to standard. However, these regulations should be viewed as a *minimum* standard. It should be up to companies, and specifically the plants, to ensure that every possible measure is taken to also ensure the safety and well-being of the community and the environment in the surrounding area. It is also up to the community itself, under Title III, to be aware of what occurs inside local industry, and to prepare for any problems that might arise.

The future promises to bring more attention to the topics discussed in the above paragraphs. In addition, it appears that there will be more research in the area of risk assessment, including fault-tree, event-tree, and cause-consequence analysis. Details on these topics are beyond the scope of this book, but are available in the literature.[3]

26-5 Introduction to Environmental Risk Assessment

Risk-based decision making and risk-based corrective action (RBCA) are decision-making processes for assessing and responding to a chemical release. The processes take into account effects on human health and the environment, inasmuch as chemical releases vary greatly in terms of complexity, physical and chemical characteristics, and in the risk that they may pose. RBCA was initially designed by the American Society for Testing and Materials (ASTM) to assess petroleum releases, but the process may be tailored for use with any chemical release. For example, in the 1980s, to satisfy the need to start corrective-action programs quickly, many regulatory agencies decided to uniformly apply, at underground storage tank (UST) cleanup sites, regulatory cleanup standards developed for other purposes. It became increasingly apparent that applying such standards without considering the extent of actual or potential human and environmental exposure was an incomplete means of providing adequate protection against the risks associated with UST releases. EPA now believes that risk-based corrective-action processes are tools that can facilitate efforts to clean up sites expeditiously, as necessary, while still ensuring the protection of human health and the environment.[7]

EPA and several state environmental agencies have developed similar decision-making tools. EPA refers to the process as the aforementioned "risk-based decision making." While the ASTM RBCA standard deals exclusively with human health risk, EPA advises that, in some cases, ecological goals must also be considered in establishing cleanup goals.

For the purposes of the next section, a few definitions of common terms will suffice. *Health risk* is the probability that persons or the environment will suffer adverse health consequences as a result of an exposure to a substance. The amount of risk is determined by a combination of the concentration the person or the environment is exposed to, the rate of intake or dose of the substance, and the toxicity of the substance. *Risk assessment* is the procedure used to attempt to quantify or estimate this risk. *Risk-based decision making* also distinguishes between the terms *point of exposure* and the *point of compliance*. The *point of exposure* is the point at which the environment or the individual comes into contact with the chemical release. An individual may be exposed by methods such as inhalation of vapors, as well as physical contact with the substance. The *point of compliance* is a point in between the point of release of the chemical (i.e., the source), and the point of exposure. The point of compliance is selected to provide a safety buffer for effected individuals and/or environments.

Understanding risk communication dynamics is also essential to successful risk communication efforts. Two-way communication with stakeholders (regulatory agencies, local residents, employees, etc.) prevents costly rework and permit delays, and provides information useful for prioritizing risk management efforts. As communities have become more interested and concerned about environmental issues, the role of the environmental manager has expanded to include communications with key audiences. The interest and concern is certain to expand in the future. In addition to addressing the technical aspects of environmental and health risk, efforts to address process, health, and lifestyle concerns has become more critical to the success of environmental projects and risk management.

26-6 Health Risk Assessment

As noted in Sec. 26–5, there are many definitions for the word *risk*. It is a combination of uncertainty and damage; a ratio of health problems to safeguards; a triple combination of event, probability, and consequences; or even a measure of economic loss or human injury in terms of both the incident likelihood and the magnitude of the loss or injury.[8] People face all kinds of risks every day, some voluntarily and others involuntarily. Therefore, risk plays a very important role in today's world. Studies on cancer caused a turning point in the world of risk because it opened the eyes of risk scientists-engineers, and health professionals to the world of risk assessments.

Since 1970, the field of risk assessment has received widespread attention within the engineering, scientific, and regulatory committees. It has also attracted the attention of the public. Properly conducted risk assessments have received fairly broad acceptance because they put into perspective the terms *toxic*, *health problems*, and *risk*. *Toxicity* is an inherent property of all substances. It states that all chemical and physical agents can produce adverse health effects at some dose or under specific exposure conditions. In contrast, exposure to a chemical that has the capacity to produce a particular type of adverse effect represents a health problem. As noted, risk is the probability or likelihood that an adverse outcome will occur in a person or a group that is exposed to a particular concentration or dose of the hazardous agent. Therefore, health risk is generally a function of exposure and dose. Consequently, *health risk assessment* is defined as the process or procedure used to estimate the likelihood that humans or ecological systems will be adversely affected by a chemical or physical agent under a specific set of conditions.[4]

The term *risk assessment* has been used to describe or predict the likelihood of an adverse response to a chemical or physical agent but also or of any unwanted event. This subject is treated in more detail in the next section. These include risks such as explosions or injuries in the workplace; natural catastrophes; injury or death due to various voluntary activities such as skiing, ski diving, flying, or bungee jumping; diseases; death due to natural causes; and many others.[9]

Risk assessment and *risk management* are two different processes, but they are closely related. Risk assessment and risk management provide a framework not only for setting regulatory priorities but also for making decisions that cut across different environmental areas. *Risk management* generally refers to a decision-making process that involves such considerations as risk assessment, technology feasibility, economic information about costs and benefits, regulatory requirements, and public concerns. Therefore, risk assessment supports risk management in that the choices on whether and to what extent to control future exposure to the suspected hazards may be determined.[10] Regarding both risk assessment and risk management, this section will address this subject primarily from a health perspective; Sec. 26–7 will examine risk assessment from a safety and accident perspective.

Health risk assessment provides an orderly, explicitly, and consistent way to deal with scientific issues in evaluating whether a health problem exists and what the magnitude of the problem might be. This evaluation typically involves large uncertainties because the available scientific data are limited, and the mechanisms associated with adverse health impacts or environmental damage are only imperfectly understood. When one examines risk, how does one decide how safe is safe, or how clean is clean? To begin with, the chemical engineer has to examine both sides of the risk equation, i.e., both the toxicity of a pollutant and the event of exposure. Information is required for both current and potential exposures, considering all possible exposure pathways. In addition to human health risks, one often needs to look at potential ecological or other environmental effects. In conducting a comprehensive risk assessment, one should remember that there are always uncertainties, and these assumptions must be included in the analysis.[10]

Several guidelines and handbooks have been written to help explain the approaches for performing health risk assessments. As discussed by a special National Academy of Sciences committee convened in 1983, most human or environmental health problems can be evaluated by dissecting the analysis into four parts: (1) health problem identification, (2) dose-response assessment, (3) exposure assessment, and (4) risk characterization (see also Fig. 26–1). Regarding *identification*, a health problem is defined as a toxic agent or a set of conditions that has the potential to cause adverse effects to human health or the environment. Identification involves an evaluation of various sources of information in order to identify the different problems. For some perceived health problem, the risk assessment might stop with the first step, health problem identification. If no adverse effect is identified or if an agency elects to not take regulatory action further analysis is eleminated.[9] *Dose-response* or *toxicity assessment* is required in an overall assessment; responses and effects can vary widely since all chemicals and contaminants vary in their capacity to cause adverse effects. This step frequently requires that assumptions be made regarding experimental data from animals and humans. *Exposure assessment* involves the determination of the magnitude, frequency, duration, and routes of exposure of human populations and ecosystems. Finally, in *risk characterization*, toxicology and exposure data/information are combined to obtain a qualitative or quantitative expression of risk.

Risk assessment involves the integration of the information and analysis associated with the four steps listed in the preceding paragraph to provide a complete characterization of the nature and magnitude of risk and the degree of confidence associated with this characterization. A critical component of the assessment is a qualitative or quantitative

Figure 26–1

Health risk assessment process.

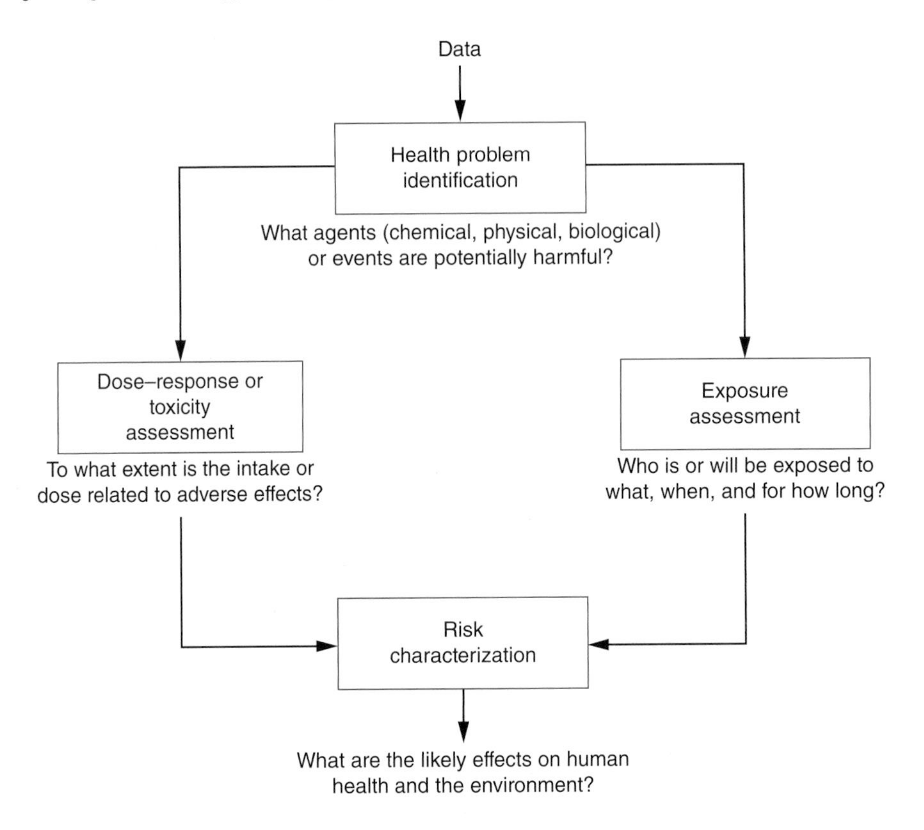

understanding of the *uncertainties* associated with each of the major steps. All the essential problems of toxicology are encompassed under this broad concept of risk assessment. It should treat uncertainty not by the application of arbitrary safety factors, but by stating them in quantitatively explicit terms so that they are not hidden from decision makers. Risk assessment, defined in this broad way, forces an assessor to confront all the scientific uncertainties and to set forth in explicit terms the means used in specific cases to deal with these uncertainties.[11]

26–7 Hazard Risk Assessment

Risk evaluation of accidents serves a dual purpose. It estimates the probability that an accident will occur and also assesses the severity of the consequences of an accident. Consequences may include damage to the surrounding environment, financial loss, or injury to life. This section is primarily concerned with the methods used to identify hazards and the causes and consequences of accidents. Issues dealing with health risks have been explored in the previous section. Risk assessment of accidents provides an effective way to help ensure that a mishap does not occur or reduces the likelihood of an accident. The result of a risk assessment can also allow concerned parties to take precautions to help prevent an accident before it happens.

The first thing a chemical engineer needs to understand is exactly what an accident is. An *accident* is defined as an unexpected event that has undesirable consequences.[12] The causes of accidents have to be identified in order to help prevent accidents from occurring. Any situation or characteristic of a system, plant, or process that has the potential to cause damage to life, property, or the environment is considered a hazard. A hazard can also be defined as any characteristic that has the potential to cause an accident. The severity of a hazard plays a large part in the potential amount of damage a hazard can cause if it occurs. Risk is the probability that human injury, damage to property, damage to the environment, or financial loss will occur.

An *acceptable risk* is a risk whose probability is unlikely to occur during the lifetime of the problem or plant or process. An acceptable risk can also be defined as an accident that has a high probability of occurring, but with negligible consequences. Risks can be ranked qualitatively in categories of high, medium, and low. Risk can also be ranked quantitatively as the annual number of fatalities per million affected individuals. This is normally denoted as a number times one millionth, i.e., 3×10^{-6}; this representation indicates that on average, three individuals will die every year for every million individuals. For every million individuals, a quantitative approach that has become popular in industry is the *fatal accident rate* (FAR) concept. This determines or estimates the number of fatalities over the lifetimes of 1000 workers. The *lifetime* of a worker is defined as 10^5 h, which is based on a 40-h work week for 50 years. A reasonable FAR for a chemical plant is 3.0, with 4.0 usually taken as a maximum. A FAR of 3.0 means that there are three deaths for every 1000 workers over a 50-year period.[13] Interestingly, the FAR for an individual at home is approximately 3.0.

There are several steps in evaluating the risk of an accident (see also Fig. 26–2). If the system in question is a chemical plant, the following guidelines apply:

1. A brief description of the equipment and chemicals used in the plant is needed.
2. Any hazard in the system has to be identified. Hazards that may occur in a chemical plant include:
   a. Corrosion
   b. Explosions
   c. Fires
   d. Rupture of a pressurized vessel
   e. Runaway reactions
   f. Slippage corrosion
   g. Unexpected leaks
3. The event or series of events that will initiate an accident must be identified. An event could be a failure to follow correct safety procedures, to improperly repair equipment, or the failure of a safety mechanism.
4. The probability that the accident will occur has to be determined. For example, if a nuclear power plant has a 10-year life, what is the probability that the temperature in a reactor will exceed the specified temperature range? The probability can be ranked qualitatively from low to high. A low probability means that it is unlikely for the event to occur in the life of the plant. A medium probability suggests that there is a possibility that the event will occur. A high probability means that the event will probably occur during the life of the plant.
5. The severity of the consequences of the accident must be determined.
6. If the probability of the accident and the severity of its consequences are low, then the risk is usually deemed acceptable and the plant should be allowed to operate. If the probability of occurrence is too high or the damage to the surroundings is too great, then the risk is usually unacceptable and the system needs to be modified to minimize these effects.

The heart of the hazard risk assessment algorithm provided is enclosed in the dashed-line box in Fig. 26–2. This algorithm allows for reevaluation of the process if the risk is deemed unacceptable (the process is repeated starting with step 1 or 2).

The reader should note that health assessment and hazard risk assessment plus accompanying calculations receives an extensive treatment by Theodore and Dupont.[13]

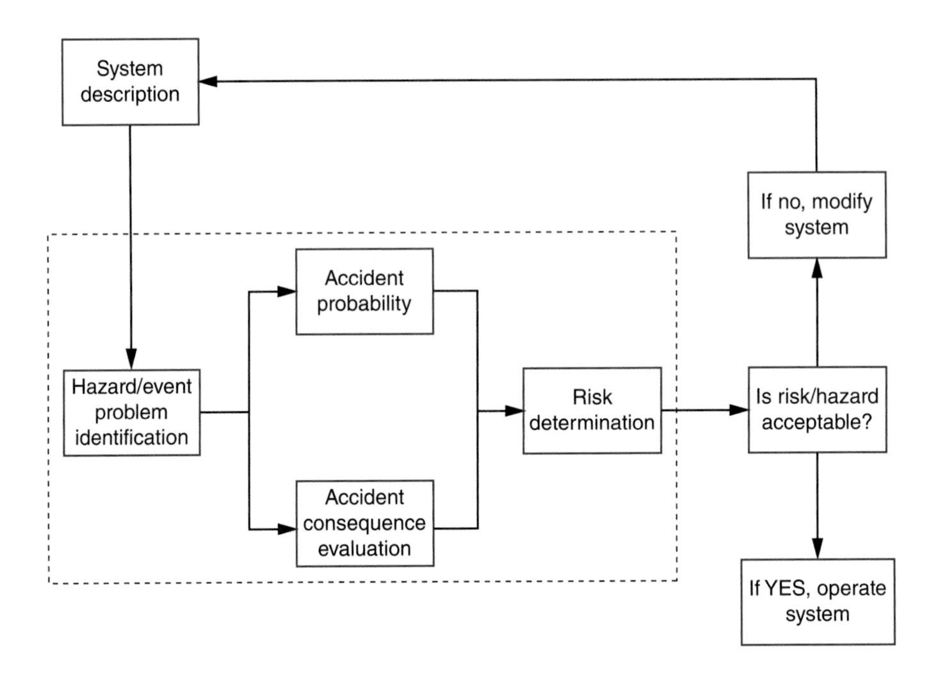

Figure 26–2

HZRA flowchart for a chemical plant.

References

1. L. Theodore, J. Reynolds, and K. Morris, *Accident and Emergency Management*, A Theodore Tutorial, Theodore Tutorials, East Williston, N.Y., 1994, originally published by USEPA/APTI, Research Triangle Park, N.C., 1996.
2. M. K. Theodore and L. Theodore, *Introduction to Environmental Management*, CRC Press/Taylor & Francis Group, Boca Raton, Fla., 2010.
3. A. M. Flynn and L. Theodore, *Accident and Emergency Management in the Chemical Process Industries*, CRC Press/Taylor & Francis Group, Boca Raton, Fla., 2004.
4. P. Armenante, *Contingency Planning for Industrial Emergencies*, Van Nostrand Reinhold, N.Y., 1991.
5. EPA Office of Emergency and Remedial Division, *Standard Operating Safety Guides*, Washington D.C., July 1988.
6. F. Lees, *Safety Cases within the Control of Industrial Major Accidents Hazards Regulations*, Butterworths, London, 1989.
7. USEPA, *Use of Risk-Based Decision Making*, OSWER Directive 9610.17, USEPA, Washington, D.C., March 1995.
8. AIChE, *Guidelines for Chemical Process Quantitative Risk Analysis*, Center for Chemical Process Safety of the American Institute of Chemical Engineers, New York, 1989.
9. D. Paustenbach, *The Risk Assessment of Environmental and Human Health Hazards: A Textbook of Case Studies*, Wiley, Hoboken, N.J., 1989.
10. G. Burke, B. Singh, and L. Theodore, *Handbook of Environmental Management and Technology*, 2nd ed., Wiley, N.Y., 2000.
11. J. Rodricks and R. Tardiff, *Assessment and Management of Chemical Risks*, American Chemical Society, Washington, D.C., 1984.
12. AIChE, *Guidelines for Hazard Evaluation Procedures*, Batelle Columbus Division for the Center for Chemical Process Safety of the American Institute of Chemical Engineers, N.Y.,1985.
13. L. Theodore and R. Dupont, *Environmental Health Risk and Hazard Risk Assessment: Principles and Calculations*, CRC Press/Taylor & Francis Group, Boca Raton, Fla., 2012.

27 Probability and Statistics

Chapter Outline

27–1 Introduction

The topic of this chapter is probability and statistics. Webster[1] defines *probability* as "the quality or state of being probable; likelihood; something probable; math; the number of times something will probably occur over the range of possible occurrences expressed as a ratio," and the term *statistics* as "facts or data of a numerical kind, assembled, classified, and tabulated so as to present significant information about a given subject; the science of assembling, classifying, tabulating, and analyzing such facts or data." There are obviously many other definitions.

The key area of interest to chemical engineers is, however, statistics. The problem often encountered is usually related to interpreting limited data and/or information. This can entail any one of several options:

1. Obtaining additional data or information
2. Deciding which data or information to use
3. Generating a mathematical model (generally an equation) to represent the data or information
4. Generating information about unknowns, a process often referred to as *inference*

The material persented in this chapter is divided into seven sections dealing with the following topics: probability definitions and interpretations, introduction to probability distributions, discrete probability distributions, continuous probability distributions, contemporary statistics, regression analysis, and analysis of variance.

NOTE No attempt was made in this chapter to introduce and apply packaged computer programs that are presently available; the emphasis is to provide the reader with an understanding of fundamental principles in order to learn how statistical methods can be used to obtain answers to question within one's subject matter specialty; becoming "computer literate" in this field was not the object. Also, the bulk of the material presented in this chapter was drawn from Theodore and Taylor.[2]

27–2 Probability Definitions and Interpretations

Probabilities are nonnegative numbers associated with the outcomes of so-called random experiments. A *random experiment* is an experiment whose outcome is uncertain. Examples include throwing a pair of dice, tossing a coin, counting the number of defectives in a sample from a lot of manufactured items, and observing the time to failure of a tube in a heat exchanger or a seal in a pump or a bus section in an electrostatic precipitator. The set of possible outcomes of a random experiment is called the *sample space* and is usually designated by S. Then $P(\text{A})$, the probability of an event A, is the sum of the probabilities assigned to the outcomes constituting the subset A of the space S. A *population* is a collection of *objects* having observable or measurable characteristics defined as *variates*, while a *sample* is a group of objects drawn from a population (usually random) that are equally likely to be drawn.

Consider, for example, tossing a coin twice. The sample space can be described as

$$\text{S} = \{\text{HH, HT, TH, TT}\}$$

If probability ¼ is assigned to each element of S and A is the event of at least one head (H), then

$$\text{A} = \{\text{HH, HT, TH}\}$$

The sum of the probabilities assigned to the elements of A is ¾. Therefore, $P(\text{A}) = 3/4$. The description of the sample space is not unique. The sample space S in the case of tossing a coin twice could be described in terms of the number of heads obtained. Then

$$S = \{0, 1, 2\}$$

Suppose that probabilities ¼, ½, and ¼ are assigned to the outcomes 0, 1, and 2, respectively. Then A, the event of at least one head, would have for its probability,

$$P(\text{A}) = P(1,2) = 3/4$$

How probabilities are assigned to the elements of the sample space generally depends on the desired interpretation of the probability of an event. Thus, P(A) can be interpreted as *theoretical relative frequency*, i.e., a number about which the relative frequency of event A tends to cluster as n, the number of times that the random experiment is performed, increases indefinitely; this is the objective interpretation of probability. Under this interpretation, to say that P(A) is ¾ in the aforementioned example means that if a coin is tossed n times, the proportion of times that one or more heads occurs clusters about ¾ as n increases indefinitely.

As another example, consider a single valve in a piping system that can stick in an open (O) or closed (C) position. The sample space can be described as follows:

$$S = \{O, C\}$$

Suppose that the valve sticks twice as often in the open position as it does in the closed position. Under the theoretical relative frequency interpretation, the probability assigned to element O in S would be 2/3, twice the probability assigned to the element C. If two such valves are observed, the sample space S can be described as

$$S = \{OO, OC, CO, CC\}$$

Assuming that the two valves operate independently, a reasonable assignment of probabilities to the element of S, as just listed, should be 4/9, 2/9, 2/9, and 1/9. If A is the event of at least one valve sticking in the closed position, then

$$A = \{OC, CO, CC\}$$

The sum of the probabilities assigned to the element of A is then 5/9. Therefore, $P(\text{A}) = 5/9$.

Probability P(A) can also be interpreted *subjectively* as a measure of degree of belief, on a scale from 0 to 1, that the event A occurs. This interpretation is frequently used in ordinary conversation. For example, if someone says, "The probability that I will go to a racetrack tonight is 90%," then 90% is a measure of the person's belief that he or she will go to a racetrack (a site regularly visited by the author). This interpretation is also used when, in the absence of concrete data needed to estimate an unknown probability on the basis of observed relative frequency, the personal opinion of an expert is sought. For example, a chemical engineer might be asked to estimate the probability that the seals in a newly designed pump will leak at high pressures. The estimate would be based on the expert's familiarity with the history of pumps of similar design.

27-3 Introduction to Probability Distributions

The probability distribution of a random variable concerns the distribution of probability over the range of the random variable. The distribution of probability, i.e., the values of random variables together with their associated probabilities, is specified by the *probability distribution function* (pdf). This section is devoted to providing general properties of the pdf in the case of discrete and continuous random variables, as well as an introduction to *cumulative distribution function* (cfd). Special pdf's finding extensive application in chemical engineering analysis are considered in the next two sections.

The pdf of a *discrete* random variable X is specified by $f(x)$, where $f(x)$ has the following essential properties:

1. $F(x) = P(X = x)$ = probability assigned to the outcome corresponding to the number x in the range of X (i.e., X is specifically designated value of x).
2. $F(x) \geq 0$.
3. $\Sigma f(x) = 1$.

Property 1 indicates that the pdf of a discrete random variable generates probability by substitution. Properties 2 and 3 restrict the values of $f(x)$ to nonnegative real numbers and numbers whose sum is 1, respectively.

The pdf of a *continuous* random variable X has the following properties:

$$\int_a^b f(x)dx = P(a < X < b) \tag{27–1}$$

$$F(x) \geq 0 \tag{27–2}$$

$$\int_{-\infty}^{\infty} f(x)dx = 1 \tag{27–3}$$

Equation (27–1) indicates that the pdf of a continuous random variable generates probability by integration of the pdf over the interval whose probability is required. When this interval contracts or reduces to a single value, the integral over the interval becomes zero. Therefore, the probability associated with any particular value of a continuous random variable is *zero*. Consequently, if X is continuous, then

$$\begin{aligned} P(a < X \leq b) &= P(a < X \leq b) \\ &= P(a \leq X < b) \\ &= P(a \leq X < b) \end{aligned} \tag{27–4}$$

Equation (27–2) restricts the values of $f(x)$ to nonnegative numbers. Equation (27–3) follows from the fact that

$$P(-\infty < X < \infty) = 1 \tag{27–5}$$

The expression $P(a < X < b)$ can be interpreted geometrically as the area under the pdf curve over the interval (a, b). Integration of the pdf over the interval yields the probability assigned to the interval. For example, the probability that the time in hours between successive failures of an aircraft air-conditioning system is greater than 6 but less than 10 is $P(6 < X < 10)$.

Another function used to describe the probability distribution of a random variable X is the *cumulative distribution function* (cdf) mentioned above. If $f(x)$ specifies the pdf of a random variable X, then $F(x)$ is used to specify the cdf. For both discrete and continuous random variables, the cdf of X is defined by

$$F(x) = P(X \geq x) \qquad (-\infty < x < \infty) \tag{27–6}$$

Note that the cdf is defined for all real numbers, not just the values assumed by the random variable. It is helpful to think of $F(x)$ as an accumulator of probability as x increases through all real numbers. In the case of a discrete random variable, the cdf is a step function increasing by finite jumps at the values of x in the range of X. In the case of a continuous random variable, the cdf is a continuous function.

The following properties of the cdf of a random variable X can be deduced directly from the definition of $F(x)$:

$$\begin{aligned} F(b) - F(a) &= P(a \le X \le b) \\ F(+\infty) &= 1 \\ F(-\infty) &= 0 \end{aligned} \tag{27–7}$$

Also note that $F(x)$ is a nondecreasing function of X. These properties apply to the cases of both discrete and continuous random variables.

Shaefer and Theodore[3] provided numerous illustrative examples of probability distribution.

27–4 Discrete Probability Distributions

There are numerous discrete probability distributions. However, this section examines the three distributions that chemical engineers are most likely to encounter in their careers:

1. The binomial distribution
2. The hypergeometric distribution
3. The Poisson distribution

Each receives treatment in the subsections to follow. (See also Shafer and Theodore.[3])

The Binomial Distribution

Consider n independent performances of a random experiment with mutually exclusive outcomes that can be classified as success or failure. The words *success* and *failure* are to be regarded as labels for two mutually exclusive categories of outcomes of the random experiment. In the development (i. e., text and equations) that follows, they do not necessarily have the ordinary connotation of success or failure.

Assume that p, the probability of success on any performance of the random experiment, is constant. Let Q be the probability of failure, so that

$$Q = 1 - p \tag{27–8}$$

The probability distribution of X, the number of successes in n performances of the random experiment, is the *binomial distribution* with a *probability distribution function* (pdf) specified by

$$f(x) = \frac{n!}{x!(n-x)!} p^x Q^{n-x} \qquad x = 0.1, \ldots, n \tag{27–9}$$

where $f(x)$ is the probability of x successes in n performances. One can show that the *expected value*[3] of the random variable X is np and its *variance*[3] is npQ.

The *binomial* distribution can be used to calculate the reliability of a *redundant system*. A redundant system consisting of n identical components is a system that fails only if more than r components fail. Familiar examples include single-usage equipment such as missile engines, short-life batteries, flashbulbs, and numerous electrical components which are required to operate for one time period and are not reused.

Once again, associate *success* with the failure of a component. Assume that the n components are independent with respect to failure, and that the reliability of each is $1-p$. Then the number of failures X has the binomial pdf in Eq. (27–9) and the reliability of the redundant system is

$$P(X \leq r) = \sum_{x=0}^{r} \frac{n!}{x!(n-x)!} p^x Q^{n-x} \tag{27–10}$$

The binomial distribution occurs in problems in which samples are drawn from a *large* population with specified success or failure probabilities and there is a desire to evaluate instances of obtaining a certain number of successes in the sample. It has applications in quality control, reliability studies, environmental management, consumer sampling, and many other cases.

Hypergeometric Distribution

The *hypergeometric distribution* is applicable to situations in which a random sample of r items is drawn without replacement from a set of n items. *Without replacement* means that an item is not returned to the set after it is drawn. Recall that the binomial distribution is frequently applicable in cases where the item is drawn *with replacement.*

Suppose that it is possible to classify each of the n items as a success or a failure. Again, the words *success* and *failure* do not have the usual connotation. They are merely labels for two mutually exclusive categories into which n items have been classified. Thus, each element of the population may be dichotomized as belonging to one of two disjointed classes.

In the development that follows, set a as the number of items in the category labeled success. Then $n - a$ will be the number of items in the category labeled failure. Let X denote the number of successes in a random sample of r items drawn without replacement from the set of n items. Then the random variable X has a hypergeometric distribution whose probability distribution function (pdf) is specified as follows:

$$f(x) = \frac{\{a!/[x!(a-x)!]\}\{(n-a!)/[(r-x)!(n-a-r+x)!]\}}{n!/[r!(n-r)!]} \qquad [x = 0,1,\ldots,\min(a,r)] \tag{27–11}$$

The term $f(x)$ is the probability of x successes in a random sample of n items drawn without replacement from a set of n items, where the terms a are classified as successes and n–1 as failures. The term min(a,r) represents the smaller of the two numbers of a and r, namely, $\min(a,r) = a$ if $a < r$ and $\min(a,r) = r$ if $r \leq a$.

The hypergeometric distribution is applicable in situations similar to those when the binomial distribution is used, except that samples are taken from a *small* population. Examples arise in sampling from small numbers of chemical samples as well as medical and environmental samples, and from manufacturing lots.

Poisson Distribution

The probability distribution function (pdf) of the *Poisson distribution* can be derived by taking the limit of the binomial pdf as $n \to \infty$, $P \to 0$, with $nP = \mu$ remaining constant. The Poisson pdf is then given by

$$f(x) = \frac{e^{-\mu}\mu^x}{x!} \qquad (x = 0,1,2,\ldots) \tag{27–12}$$

Here $f(x)$ is the probability of x occurrences of an event that occurs on the average μ times per unit of space or time. *Both* the mean and the variance of a random variable X having a Poisson distribution are μ.

The Poisson pdf can be used to estimate probabilities obtained from the binomial pdf given earlier for large n and small P. In general, good approximations will result when n exceeds 100 and nP is less than 10.

If λ is the failure rate (per unit of time) of each component of a system, then λt is the average number of failures rate (per unit of time) of each component of a system. The probability of x failures in a specified unit of time is obtained by substituting $\mu = \lambda t$ in Eq. (27–12) to obtain

$$f(x) = \frac{e^{-\lambda t}(\lambda t)^x}{x!} \qquad (x = 0,1,2,\ldots) \qquad (27\text{–}13)$$

In addition to the applications cited, the Poisson distribution can be used to obtain the reliability R of a standby redundancy system in which one unit is in the operating mode and n identical units are in standby mode. This relation often occurs in chemical engineering practice. Unlike parallel systems, the standby units are inactive; the reliability of the standby redundancy system may be calculated by employing the Poisson distribution under the following conditions:

1. All units have the same failure rate in the standby mode.
2. Unit failures are independent.
3. Standby units have zero failure rate in the standby mode.
4. There is perfect switchover to a standby when the operating unit fails.

The Poisson distribution is a distribution which arises in many other different situations. For instance, it provides probabilities of specified numbers of environmental sampling procedures, of a given number of defects per unit area of glass, textiles, sheet metals, piping, heat exchangers, papers, etc., and of various numbers of bacterial colonies per unit volume, etc.

27–5 Continuous Probability Distributions

There are numerous continuous probability distributions. However, this section examines four distributions that chemical engineers are most likely to encounter in their careers;

1. The exponential distribution
2. The Weibull distribution
3. The normal distribution
4. The lognormal distribution

The Exponential Distribution

The *exponential distribution* is an important distribution in that it represents the distribution of the time required for a single event from a Poisson possess to occur. In particular, in sampling from a Poisson distribution with parameter μ, the probability that no events occurs during $(0,t)$ is $e^{-\lambda t}$. Consequently, the probability that an event will occur during $(0,t)$ is

$$F(t) = 1 - e^{-\lambda t} \qquad (27\text{–}14)$$

This represents the cumulative distribution function (cdf) of t. One can therefore show that the probability distribution function (pdf) is

$$f(t) = e^{-\lambda t} \tag{27–15}$$

Note that the parameter $1/\lambda$ (sometimes denoted as μ) is the expected value. Normally, the reciprocal of this value is specified and represents the expected value[3] of $f(t)$. Because the exponential function appears in the expression for both the pdf and cdf, the distribution is justifiably called the *exponential distribution*.

Alternatively, the cumulative exponential distribution can be obtained from the pdf (with x replacing t):

$$F(x) = \int_0^x \lambda e^{-\lambda t} dx = 1 - e^{-\lambda t} \tag{27–16}$$

All that remains is a simple evaluation of the negative exponent in Eq. (27–16).

One often encounters a random variable's conditional failure density or hazard function $g(x)$ in statistical and reliability engineering applications. In particular, $g(x)dx$ is the probability that a "product" will fail during $(x, x + dx)$ under the condition that it had not failed before time x. Consequently

$$g(x) = \frac{f(x)}{1 - F(x)} \tag{27–17}$$

If probability density function $f(x)$ is exponential, with parameter λ, it follows from Eqs. (27–15) and (27–16) that

$$g(x) = \frac{\lambda e^{-\lambda t}}{1-(1-e^{-\lambda t})} = \frac{\lambda e^{-\lambda t}}{e^{-\lambda t}} \\ = \lambda \tag{27–18}$$

Equation (27–18) indicates that the failure probability is constant, irrespective of time. It implies that the probability of failure that a component whose time-to-failure distribution is exponential during the first hour of its life is the same as the probability that it will fail during an instant in the thousandth hour, presuming that it has survived up to that instant. It is for this reason that the parameter λ is usually referred to in life-test applications as the *failure rate*. This definition generally has meaning only with an exponential distribution.

The natural association with life testing and the fact that it is very tractable mathematically makes the exponential distribution attractive as representative of the life distribution of a complex system or several complex systems. In fact, the exponential distribution is as prominent in reliability analysis as the normal distribution is in other branches of statistics.

The Weibull Distribution

Unlike the failure rate of the exponential distribution, equipment frequently exhibits three stages:

1. A break-in stage with a declining failure rate
2. A useful-life stage characterized by a fairly constant failure rate
3. A wear-out period characterized by an increasing failure rate

Many industrial parts and components follow this path. A failure curve exhibiting these three phases (see Fig. 27–1) is called a "bathtub curve."

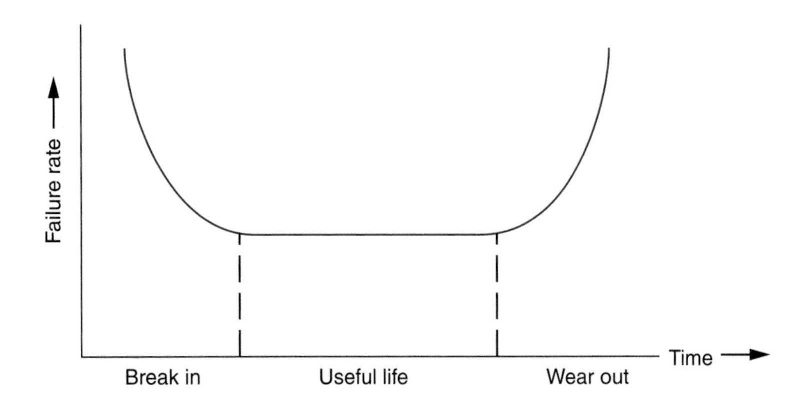

Figure 27–1

Bathtub curve (Weibull distribution).

Weibull introduced the distribution, which bears his name principally on empirical grounds, to represent certain life-test data. The Weibull distribution provides a mathematical model of all three stages of the bathtub curve. This is now discussed. An assumption about the failure rate $Z(t)$ that reflects all three stages of the bathtub stage is

$$Z(t) = \alpha\beta t^{\beta-1} \qquad (t > 0) \tag{27–19}$$

where α and β are constants. For $\beta < 1$, the failure rate $Z(t)$ decrease with time. For $\beta = 1$, the failure rate is constant and equal to α. For $\beta > 1$, the failure rate decreases with time. Translating the assumption about failure rate into a corresponding assumption about the pdf of T, time-to-failure, one obtains

$$f(t) = \alpha\beta t^{\beta-1}\exp\left(\int_0^t \alpha\beta t^{\beta-1}dt\right) = \alpha\beta t^{\beta-1}\exp(-\alpha t^{\beta}; t > 0; \alpha > 0, \beta > 0 \tag{27–20}$$

Equation (27–20) defines the pdf of the Weibull distribution. The exponential distribution discussed in the preceding subsection, whose pdf is given in Eq. (27–15), is a special case of the Weibull distribution with $\beta < 1$.

Various assumptions about failure rate and the probability distribution of time to failure that can be accommodated by the Weibull distribution make it especially attractive in describing failure-time distributions in industrial and chemical process plant applications, particularly as they apply to accidents.[4] Theodore is currently developing an improvement to the Weibull model.

The Normal Distribution

The initial presentation in this subsection on *normal distributions* will focus on failure rate, but can be simply applied to all other applications involving normal distributions. When time-to-failure, has a normal distribution, its probability distribution function (pdf) is given by

$$f(t) = \frac{1}{\sqrt{2\pi}\sigma}\exp\left[-\frac{1}{2}\left(\frac{t-\mu}{\sigma}\right)^2\right] \qquad (-\infty < t < \infty) \tag{27–21}$$

where μ is the mean value of T and σ is its standard deviation. The graph of $f(t)$ is the familiar bell-shaped curve (see also Fig. 27–2). The reliability function corresponding to the normal distributed failure time is given by

$$R(t) = \frac{1}{\sqrt{2\pi}\sigma}\exp\int_t^{\infty}\left[-\frac{1}{2}\left(\frac{t-\mu}{\sigma}\right)^2\right]dt \tag{27–22}$$

Figure 27–2

Areas under a standard normal curve.

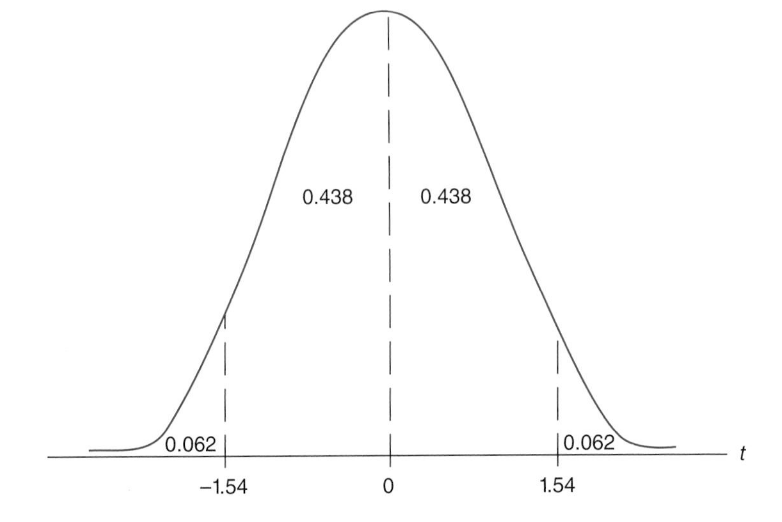

If T is normally distributed with mean μ and standard deviation σ, then the random variable $(T-\mu)/\sigma$ is normally distributed with mean 0 and standard deviation 1. The term $(T-\mu)/\sigma$ is called a *standard normal curve* that is represented by Z, not to be confused with the failure rate $Z(t)$. Tables 1–10 and 1–11 (in Chap. 1) are a tabulation of areas under a standard normal curve to the right of z_0. Probabilities about a standard normal variable Z can be determined from this table. For example, $P(Z > 1.54) = 0.062$ is obtained directly from the table as the area to the right of 1.54. As presented in Fig. 27–2, the symmetry of the standard normal curve about zero implies that the area to the right of zero is 0.5. If X is not normally distributed, then $\bar{X}$, the mean of a sample of n observations on X, is *approximately* normally distributed with mean μ and standard deviation $\sigma/\sqrt{n}$, provided the sample size n is large (>30). This result is based on an important theorem in probability called the *central limit theorem*.

Actual (experimental) data have shown many physical variables to be normally distributed. There are numerous examples of industrial physical measurements on living organisms, molecular velocities in an ideal gas, scores on an intelligence test, the average temperatures in a given locality, etc. Other variables, although not normally distributed per se, sometimes approximate a normal distribution after an appropriate transformation such as taking the logarithm (see next section) or square root of the original variable. The normal distribution also has the advantages of being tractable mathematically. Consequently, many of the techniques of static inference have been derived under the assumption of underlying normal variants.

The Lognormal Distribution

A nonnegative random variable X has *a lognormal distribution* whenever ln X, that is, the natural logarithm of X, has a normal distribution. The probability distribution function (pdf) of a random variable X having a lognormal distribution is specified by

$$f(x) = \begin{cases} \dfrac{1}{\sqrt{2\pi}\beta} x^{-1} \exp\left[-\dfrac{(\ln x - \alpha)^2}{2\beta^2}\right] & \text{for } x > 0 \\ f(x) = 0 \; ; elsewhere \end{cases} \tag{27–23}$$

The mean and variance of a random variable X having a lognormal distribution are given by

$$\mu = e^{\alpha+(\beta^2/2)} \tag{27–24}$$

$$\sigma^2 = e^{\alpha+(\beta^2/2)}(e^{\beta^2} - 1) \tag{27–25}$$

Probabilities concerning random variables having a lognormal distribution can be calculated from the previously employed tables of the normal distribution (see also Tables 1–10 and 1–11, Chap. 1). If X has a lognormal distribution with parameters α and β, then the ln X has a normal distribution with $\mu = \alpha$ and $\sigma = \beta$. Probabilities concerning X can therefore be converted into equivalent probabilities concerning $\ln(X)$. Estimates of the parameters α and β in the pdf of a random variable X having a lognormal distribution can be obtained from a sample of observations on X by making use of the fact that ln X is normally distributed with mean α and standard deviation β. Therefore, the mean and standard deviation of the natural logarithms of the sample observations of X furnish estimates of α and β.

The lognormal distribution has been employed as an appropriate model in a wide variety of situations including chemical engineering, biology, and economics. Additional applications include the distributions of personal incomes, inheritances, bank account deposits, and also the distribution of organism growth subject to many small impurities. Perhaps the primary application of the lognormal distribution has been to represent particle size distribution in gaseous emissions and chemical products from many industrial processes.[5]

27–6 Contemporary Statistics

This section examines seven topics that Shafer and Theodore[3] have classified as contemporary statistics. The following subject areas are reviewed:

1. Confidence intervals for means
2. Confidence intervals for proportions
3. Hypothesis testing
4. Hypothesis test for means and proportions
5. Chi-square χ^2 distribution
6. The F distribution
7. Nonparametric tests

Shafer and Theodore[3] provide numerous illustrative examples associated with these topics.

Confidence Intervals for Means

As described in a earlier section, the sample mean $\bar{X}$ constitutes a so-called point estimate of the mean μ of the population from which the same was selected at random. Instead of a *point* estimate, an *interval* estimate of μ may be required along with an indication of the confidence that can be associated with the interval estimate. Such an interval estimate is called a *confidence interval*, and the associated confidence is indicated by a *confidence coefficient*. The length of the confidence interval varies directly with the confidence coefficient for fixed values of the sample size n; the larger the value of n, the shorter the confidence interval. Thus, for fixed values of the confidence coefficient, the limits that contain a parameter with a probability of 95% (or some other stated percentage) are defined as 95% (or any other percentage) confidence limits for the parameter; the interval between the confidence limits is the aforementioned confidence interval.

For normal distributions, the confidence coefficient Z can be obtained from the standard normal table for various confidence limits or corresponding levels of significance for μ. This can be employed to obtain the probability of a value falling inside or outside the range $\mu \pm Z\sigma$. For example, when $Z = 2.58$, one can say that the level of significance is 1%. The statistical interpretation of this is as follows. If an observation deviates from the mean by at least $\pm 2.58\sigma$, the observation is significantly *different* from the body of data on which the describing normal distribution is based. Further, the probability that this statement is in error is 1%, i.e., the conclusion drawn from a rejected observation will be wrong 1% of the time.

One can also state that there is a 99% probability that an observation will fall *within* the range $\mu \pm 2.58$. This degree of confidence is referred to as the 99% *confidence level*. The limits, $\mu - 2.58\sigma$ and $\mu + 2.58\sigma$, are defined as the *confidence limits*, whereas the difference between the two values is defined as the *confidence interval*. Once again, this essentially states that the actual (or true) mean lies within the interval $\mu - 2.58\sigma$ and $\mu + 2.58\sigma$ with a 99% probability of being correct. The foregoing analysis can be extended to provide the difference of two population means, that is, $\overline{X}_2 - \overline{X}_1$. For this case, the confidence limits are

$$\overline{X}_2 - \overline{X}_1 \pm Z\sqrt{\frac{\sigma_2^2}{n_2} + \frac{\sigma_1^2}{n_1}} \tag{27–26}$$

This subsection also serves to introduce Student's t distribution (named after Student, pen name of statistician W. Gosset). It is common to use this distribution if the sample size is small. For a random sample of size n selected from a normal population, the term $(\overline{X} - \mu)/(s - \sqrt{n})$ has Student's distribution with $(n-1)$ degrees of freedom.[3] *Degree of freedom* is a label used for the parameter appearing in Student's distribution pdf. (See also Fig. 1–12 in Chap. 1.) Details are provided by Shaefer and Theodore.[3]

Confidence Intervals for Proportions

Consider a random variable X that has a binomial distribution. For large n, the random variable X/n is approximately normally distributed with mean p and standard deviation equal to the square root of pQ/n. If n is the size of a random sample from a population in which p is the proportion of elements classified as *success* because of the possession of a specified characteristic, this serves as the basis for constructing a confidence interval for P, the corresponding population proportion. The term n is *large* means that np and $n(1-p)$ are both greater than 5.

One can generate an inequality whose probability equals the desired confidence coefficient using the large sample distribution of the sample proportion. Note, once again, that X/n, the sample proportion, is approximately normally distributed with mean p and standard deviation equal to the square root of pQ/n. Therefore, $(X/n-p)/(pQ/n)^{1/2}$ is *approximately* a standard normal variable and

$$P\left(-Z < \frac{(X/n) - p}{\sqrt{(pQ)/n}} < Z\right) = \text{confidence interval; fractional basis} \tag{27–27}$$

where Z is obtained from the table of the standard normal distribution as the value such that the area over the interval from $-Z$ to Z is the stated fractional basis. This inequality can then be converted into a statement about P:

$$\frac{X}{n} - Z\sqrt{\frac{pQ}{n}} < P < \frac{X}{n} + Z\sqrt{\frac{pQ}{n}} \tag{27–28}$$

The endpoints of the statement may then be evaluated after replacing p and Q in the endpoints by the observed value of X/n and $1 - X/n$, respectively.

The large sample distribution of the sample proportion X/n also provides the basis for determining the sample size n for estimating the population proportion p with maximum allowable error E and specified confidence.

Hypothesis Testing

In *hypothesis testing*, one makes a statement (hypothesis) about an unknown population parameter and then uses statistical methods to determine (test) whether the observed sample data support that statement. Note that a statistical hypothesis is an assumption made about some parameter, i.e., about a statistical measure of a population. One may also define *hypothesis* as a statement about one or more parameters. Synonyms for hypothesis could include *assumption* or *guess*. A hypothesis can be tested to verify its credibility.

An important notion to understand is the relationship of the null hypothesis to the *alternative hypothesis* (or hypotheses). In all applications, the *null hypothesis* and alternatives are written in terms of population parameters. For example, if μ represents the population mean (e. g., temperature), one could write

Null hypothesis is H_0: $\mu = 100$

Alternative hypothesis is H_1: $\mu \neq 100$

Since statistics are only estimates of a "true" population parameter based on a sample of observations, it is reasonable to expect that the value of the estimate will deviate from the value of the true population parameter. One uses a method called *hypothesis testing* to decide whether the value of the statistic is consistent with a hypothesized value from the population parameter. When an estimate is made, a hypothesis is tested concerning an assumed population parameter.

In hypothesis testing, the statistician and/or chemical engineer uses various techniques to decide whether to accept or reject hypotheses. If, for example, someone assumed a population mean of 70 (e. g., a thermometer reading), and a sample from that population were selected that had a mean of 80, one might want to perform a test or calculation to see if the population mean of 70 could still be reasonably assumed. The individual would then employ a statistical technique to either accept or reject the hypothesis that the population mean equals 70.

In the above example, a sample mean equal to 80 is being used to test a hypothesized population mean of 70. The notation employed is as follows:

$$\bar{X} = 80 \text{ (sample mean)}$$

and the hypothesis is

$$\mu = 70 \text{ (population mean)}$$

As indicated earlier, each null hypothesis is accompanied by an alternative hypothesis. The hypothesis being tested in this example can be written as follows:

$$\mu = 70$$

Hypothesis Test for Means and Proportions

As described in the previous subsection, testing a statistical hypothesis concerning a population mean involves the setting of a rule for deciding, on the basis of a random sample from the population, whether to reject the hypothesis being tested; this is the so-called aforementioned null hypothesis. The test is formulated in terms of a test statistic, i.e., some function of the sample observations. In this case of a large sample ($n > 30$), testing a hypothesis concerning the population mean μ involves use of the following test statistic:

$$Z = \frac{\overline{X} - \mu_0}{\sigma/\sqrt{n}} \quad (27\text{–}29)$$

where $\overline{X}$ = sample mean
μ_0 = value of μ specified by null hypothesis H_0
σ = population standard deviation
n = sample size

When σ is unknown, the sample standard deviations may be employed to estimate it. This test statistic is approximately distributed as a standard normal variable when the sample originates from a normal population whose standard deviation σ is known.

The values of the test statistic for which the null hypothesis H_0 is rejected constitute the *critical region*. The critical region depends on the alternative hypothesis H_1 and the tolerated region for three alternative hypotheses (see Table 27–1). The term Z_α is the value which a standard normal variable exceeds with the probability α, and $Z_{a/2}$ is defined similarly. Typical values for the tolerated probability of a type I error[3] are 0.05 and 0.01, with 0.05 the more common selection. These are summarized in Table 27–1.

Chi-Square (χ^2) Distribution

A random variable X is said to have a *chi-square distribution* with v degrees of freedom if its probability distribution function (pdf) is specified by

$$f(\chi^2) = \frac{1}{2^{v/2}\Gamma(v/2)}(\chi^2)^{(v/2)-1} e^{-X^2/2} \qquad (\chi^2 > 0; \Gamma = \text{gamma function}) \quad (27\text{–}30a)$$

or

$$f(\chi) = \frac{1}{2^{v/2}\Gamma(v/2)}\chi^{(v/2)-1} e^{-\chi/2} \qquad (\chi > 0; \Gamma = \text{gamma function}) \quad (27\text{–}30b)$$

Equations (27–30) may also be written as

$$f(\chi^2) = A(\chi^2)^{(v/2)-1} e^{-\chi^2/2} \quad (27\text{–}31)$$

Here, $v = n-1$ is once again the number of degrees of freedom and A is a constant depending on a value of v such that the total area under the curve is unity. For random samples from a normal population with variance σ^2, the statistic

$$\chi^2 = \frac{(n-1)s^2}{\sigma^2} \quad (27\text{–}32)$$

where s^2 is the sample variance and n, the sample size, has a chi-square distribution with $n-1$ degrees of freedom. The variance may be calculated using procedures outlined earlier.

Table 27–1

Normal Distributions: Critical Regions

Alternative hypothesis	Critical region
$\mu < \mu_0$ (one tail)	$Z < -Z_a$
$\mu > \mu_0$ (one tail)	$Z > Z_a$
$\mu \neq \mu_0$ (two tail)	$Z < Z_{a/2}$ or $Z > Z_{a/2}$

A test of the hypothesis H_0: $\sigma^2 = \sigma_0^2$ utilizes the test statistic

$$\frac{(n-1)s^2}{\sigma_0^2} \tag{27–33}$$

The critical region is determined by the alternative hypothesis H_1 and α, the tolerated probability of a type I error.[3]

As with the normal and t distributions, one can define 95%, 99%, or other confidence limits and intervals for χ^2 by using a table of the χ^2 distribution.[3] In this manner, one can estimate within specified limits of confidence the population standard deviation σ in terms of a sample standard deviation s.

The F Distribution

As noted several times earlier, *variance* measures variability. Comparison of variability, therefore, involves comparison of variance. Suppose that σ_1^2 and σ_2^2 represent the unknown variances of two independent normal populations. The null hypothesis H_0: $\sigma_1^2 = \sigma_2^2$ asserts that the two populations have the same variance and are therefore characterized by the same variability. When the null hypothesis is true, the test statistic s_1^2/s_2^2 has an *F distribution* with parameters n_1-1 and n_2-1 called *degree of freedom*. The term s_1^2 is the sample variance of a random sample of n_1 observations from the normal population having variance σ_1^2, while s_2^2 is the sample variance of a random sample of n_2 observations from a normal population having variance σ_2^2. When s_1^2/s_2^2 is used as a test statistic, the critical region depends on the alternative hypothesis H_1 and α, the tolerated probability of a type I error.

The term F_α is the value that a random variable having an F distribution with n_1-1 and n_2-1 degrees of freedom exceeds with probability α. The terms $F_{\alpha/2}$ and $F_{1-\alpha/2}$ are defined similarly. Also note that

$$F_{\alpha;v_1v_2} = \frac{1}{F_{1-\alpha;v_1v_2}} \tag{27–34}$$

where v_1 and v_2 are again the degrees of freedom. Tabulated values of the F distribution are available in the literature.[3] In addition, the ratio s_1^2/s_2^2 must necessarily be greater than unity: $s_1^2/s_2^2 > 1.0$.

Nonparametric Tests

Classical tests of statistical hypotheses are often based on the assumption that the populations sampled are normal. In addition, these tests are frequently confined to statements about a finite number of unknown parameters on which the specification of the probability distribution function (pdf) of the random variable under consideration depends. Efforts to eliminate the necessity of the restrictive assumptions about the population sampled have resulted in statistical methods called *nonparametric* methods, directing attention to the fact that these methods are not limited to inferences about population parameters. These methods are also referred to as *distribution free*, emphasizing their applicability in cases in which little is known about the functional form of the pdf of the random variable observed. Nonparametric tests of statistical hypotheses are tests whose validity generally requires only the assumption of continuity of the cumulative distribution function (cdf) of the random variable involved.

The *sign test* for paired comparison provides a good illustration of the nonparametric approach. In the case of paired comparisons, the data consist of pairs of observations on a random variable. Examples are observations on the air pollution chemically measured by two different instruments, and observations on the weights of subjects before and after a weight reduction program. Interest centers on whether the paired observations "differ slightly."

A t test using Student's distribution assumes the difference of paired observations to be independently and normally distributed with a common variance; the hypothesis test is one in which the mean of the normal population of differences is equal to zero. In the sign test no assumption is required concerning the form of the distribution of the differences; the hypothesis tested is that each of the differences has a probability distribution with the median equal to zero and, therefore, the probability of a plus-sign difference as opposed to a minus-sign difference is equal to 0.5. Interestingly, the probability distribution need not be the same for all differences. The test statistic becomes the number of plus-sign differences. When the null hypothesis is true, the test statistic has a binomial distribution with P equal to 0.5 and n equal to the number of pairs of observations. (Details of this test are provided by Shafer and Theodore.[3])

The reader should note that the sign test makes use only of the sign of the difference—not the magnitude. This inefficiency is compensated for, however, by the aforementioned freedom from restrictive assumptions concerning the probability distribution of the sample observations, by the adaptability of many nonparametric tests to situations in which the observations are not susceptible to quantitative measurement, and by usually increased computational simplicity, especially in the case of small samples.

The reader should also note that when ties occur in ranking, the mean value of the ranking numbers is applied to each of the tied observations. For example, if two observations are tied for 10th rank, they are both ranked as 10.5 and the next observation is ranked number 12. If three observations in another application are tied for 11th rank, they are all ranked 12, and the next observation after the ties is rank 14.

A large group of nonparametric tests employ the method of replacing observations in a combined sample by their ranks after these observations have been arranged in ascending order of magnitude. Rank 1 is assigned to the smallest (numerical) observation, rank 2 to the next, and so on.

27–7 Regression Analysis

It is no secret that many statistical calculations are now preformed with spreadsheets or packaged programs. This statement is particularly true for *regression analysis*, a topic of interest to all chemical engineers. The result of this approach has been to reduce or eliminate one's fundamental understanding of this subject. This section attempts to correct this shortcoming.

Chemical engineers and scientists in numerous disciplines often encounter applications that require the need to develop a mathematical relationship between data for two or more variables. For example, if Y (a dependent variable) is a function of or is dependent on X (an independent variable), that is

$$Y = f(X) \tag{27–35}$$

one may need to express this (X, Y) data in equation form. This process is referred to as *regression analysis*, and the regression method most often employed is the method of *least squares*.

An important step in this procedure (which is often omitted) is to prepare a plot of Y versus X. The result, referred to as a *scatter diagram* (or *scatterplot*), could take on any form. Three such plots are provided in Fig. 27–3. The first plot (a) suggests a linear relationship between X and Y:

$$Y = a_0 + a_1X \tag{27–36}$$

Figure 27–3

Scatter diagrams: (*a*) linear relationship; (*b*) parabolic relationship; (*c*) dual-linear relationship.

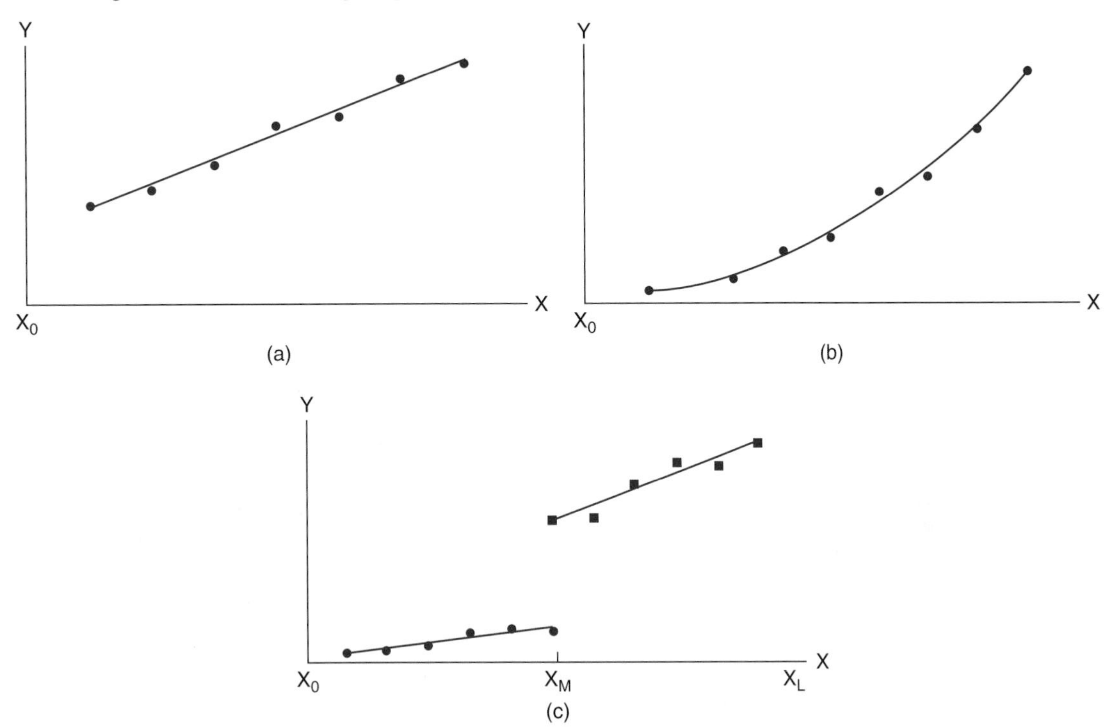

The second graph (*b*) appears to be best represented by a second-order (or parabolic) relationship:

$$Y = a_0 + a_1 X + a_2 X^2 \qquad (27\text{–}37)$$

The third plot (*c*) suggests a linear model that applies over two different ranges; thus, it could represent the data

$$Y = a_0 + a_1 X \qquad (X_0 < X < X_M) \qquad (27\text{–}38)$$

and

$$Y = a_0' + a_1' X \qquad (X_M < X < X_L) \qquad (27\text{–}39)$$

This multiequation model finds application in representing adsorption equalibria, multiparticle size distributions,[6] and quantum energy relationships. In any event, a scatter diagram and individual judgment can suggest an appropriate model at an early stage in the analysis.

Some of the models often employed by chemical engineers are as follows:

$$Y = a_0 + a_1 X \qquad \text{(linear)} \qquad (27\text{–}40)$$

$$Y = a_0 + a_1 X + a_2 X^2 \qquad \text{(parabolic)} \qquad (27\text{–}41)$$

$$Y = a_0 + a_1 X + a_2 X^2 + a_3 X^3 \qquad \text{(cubic)} \qquad (27\text{–}42)$$

$$Y = a_0 + a_1 X + a_2 X^2 + a_3 X^3 + a_4 X^4 + \ldots \qquad \text{(higher order)} \qquad (27\text{–}43)$$

Procedures for evaluating the regression coefficients a_0, a_1, a_2, and so on, are available in the literature. The technique provides numerical values for the regression coefficients a_i such that the sum of the squares of the difference (error) between the actual Y and the Y_e for all data points predicted by the equation or model is minimized.

The correlation coefficient provides information on how well the model, or line of regression, fits the data; it is denoted by r. The procedure for calculating r and its properties is available in the literature.[3] The correlation coefficient provides information only on how well the model fits the data. It is emphasized that r provides *no* information on how good the model is or, to reword this, whether this is the correct or best model for describing the functional relationship of the data. This topic receives treatment in the next and last section.

27-8 Analysis of Variance

Analysis of variance is a special statistical technique that chemical engineers may apply during their careers. It features the splitting of a measure of total variation of data into components measuring variation attributable to one or more factors or combinations of factors. The simplest application of analysis of variance involves data classified in categories (levels) of one factor.

Suppose, for example, that k different pharmaceutical drugs for patients are to be compared with respect to blood pressure elevation. Let X_{ij} denote the blood pressure elevation of the jth patient in a random sample of n_i patients receiving the ith drug; assume $i = 1, \ldots, k$ and $j = 1, \ldots, m_i$. Let n, $\overline{X}_i$, and $\overline{X}$ denote the total sample size, the mean of the observations in the ith sample, and the mean of the observations in all k samples, respectively. Then

$$n = \sum_{i=1}^{k} m_i = km \tag{27–44}$$

$$\overline{X}_i = \frac{\sum_{j=1}^{m} X_{ij}}{m_i} \tag{27–45}$$

$$\overline{X} = \frac{\sum_{i=1}^{k}\sum_{j=1}^{m} X_{ij}}{n} = \frac{\sum_{i=1}^{k} m_i \overline{X}_i}{n} \tag{27–46}$$

The term $\sum_{i=1}^{k}\sum_{j=1}^{m}(X_{ij} - \overline{X}_i)^2$, a measure of the total variation of the observations, can be algebraically split into components (derivation not provided here) as follows:

$$\begin{aligned}\sum_{i=1}^{k}\sum_{j=1}^{m}(X_{ij} - \overline{X}_\iota)^2 &= \sum_{i=1}^{k}\sum_{j=1}^{m}[(X_{ij} - \overline{X}_\iota) + (X_{ij} - \overline{X}_\iota)^2] \\ &= \sum_{i=1}^{k}\sum_{j=1}^{m}(X_{ij} - \overline{X}_\iota)^2 + \sum_{i=1}^{k} m_i (X_{ij} - \overline{X}_\iota)^2\end{aligned} \tag{27–47}$$

Or, in abbreviated form

$$\text{TSS} = \text{RSS} = \text{GSS} \tag{27–48}$$

where TSS = total sum of squares
RSS = within-group sum of squares
GSS = among-group sum of squares

Thus, the measure of the total variation above has been expressed in terms of two components, the first measuring variation *within* the k samples, and the second measuring the variation *among* the k samples.

In the terminology of analysis of variance (ANOVA), the within-group sum of squares is also called the *residual sum of squares* (RSS) or the *error sum of squares.* For example, the ANOVA statistic for testing the null hypothesis of no significant differences among food supplements with respect to milk yield features comparison of the among-group sum of squares (GSS) versus the within-group sum of squares (RSS). If the mean GSS, that is, the GSS divided by $k–1$ (the degrees of freedom), is large relative to the mean RSS, that is, GSS divided by $n–k$, there is evidence for contradiction of the null hypothesis regarding the food supplements.

The among group of squares (GSS) and the residual sum of squares (RSS) may also be expressed as

$$\text{GSS} = \sum_{i=1}^{k} \frac{T_{i.}^2}{m_i} - \frac{T_{..}^2}{n} \qquad (27\text{–}49)$$

$$\text{RSS} = \sum_{j=1}^{m_i} \sum_{i=1}^{k} (X_{ij})^2 - \sum_{i=1}^{k} \frac{T_{i.}^2}{m_i} \qquad (27\text{–}50)$$

where $T_{i.}$ = total of j values in group
$j = 1, 2, \ldots, m_i$
$T_{..}$ = grand total

The total sum of squares (TSS) is given by the sum of the results of Eqs. (27–49) and (27–50). A summary of the analysis is provided in Table 27–2.

The precise form of the ANOVA test statistic (ANVT) is

$$\text{ANVT} = \frac{n-k}{k-1} \frac{\sum_{i=1}^{k} m_i \left(\bar{X}_i - \bar{X}\right)^2}{\sum_{i=1}^{k} \sum_{j=1}^{m} (X_{ij} - \bar{X}_l)^2} \qquad (27\text{–}51)$$

According to the null hypothesis, this statistic has an F distribution with $k – 1$ and $n – k$ degrees of freedom. A random variable having such a distribution is indicated by $F_{k-1,n-k}$. When the observed value of the test statistic exceeds $F_{k-1,n-k,\alpha}$, where $P(F_{k-1,n-k} > F_{k-1,n-k,\alpha}) = \alpha$, the tolerated probability of a type I error, the null hypothesis is rejected.[3]

This topic receives extensive treatment by Shaefer and Theodore.[3] Interested readers should definitely review other literature resources related to statistics.

Table 27–2
ANOVA Summary

Group(i), row(j)	$i = 1$	$i = 2$	$i = i$	$i = k$
$j = 1$	X_{11}	X_{12}	$X_1 i$	$X_1 k$
$j = 2$	X_{21}	X_{22}	$X_2 i$	$X_2 k$
$j = j$	Xj_1	Xj_2	Xji	Xjk
$j = m$	Xm_1	Xm_2	Xmi	Xmk
	T_1	T_2	T_3	T_4
	M_1	M_2	M_3	M_4

Note: $T_1 = X_{11} + X_{21} + \cdots + Xj_1 + \cdots + Xm_1$; $T_2 = X_{12} + X_{22} + \cdots + Xj_2 + \cdots + Xm_2$; $T_3 = X_{13} + X_{23} + \cdots + Xj_3 + \cdots + Xm_3$; $T.. = X_1. + X_2. + \cdots + Xj. + \cdots + Xm.$; $M_1 = T_1/m$; $M_2 = T_2/m$; $Mi = Ti/m$; $Mk = Tk/m$; and $n = mk$.

References

1. *Webster's New World Dictionary*, 2nd Collegiate Edition, Prentice-Hall, Upper Saddle River, N.J., 1971.
2. L. Theodore and F. Taylor, *Probability and Statistics*, A Theodore Tutorial, Theodore Tutorials, East Williston, N.Y., originally published by USEPA/APTI, Research Triangle Park, N.C., 1993.
3. S. Shaefer and L. Theodore, *Probability amd Statistics Applications for Environmental Science*, CRC Press/Taylor & Francis Group, Boca Raton, Fla., 2007.
4. L. Theodore and R. Dupont, *Environmental Health and Hazard Risk Assesment: Principles and Calculations*, CRC Press/Taylor & Francis Group, Boca Raton, Fla., 2012.
5. L. Theodore, *Air Pollution Control Equipment Calculations*, Wiley, Hoboken, N.J., 2008.

28 Economics and Finance

Chapter Outline

28–1 Introduction

An understanding of the economics and finances involved in chemical engineering is important in making decisions at both the engineering and management levels. Every chemical engineer should be able to execute an economic evaluation of a proposed project. If the project is not profitable, it should obviously not be pursued, and the earlier such a project can be identified, the fewer are the resources that will be wasted.

Before the cost of a unit or process or facility can be evaluated, the factors contributing to the cost must be recognized. There are two major contributing factors: capital costs and operating costs; these are discussed in the next two sections. Once the total cost of the unit or process has been estimated, the chemical engineer must determine whether the project will be profitable. This involves converting all cost contributions to an annualized basis, a method that is discussed in the following sections. If more than one project proposal is under study, this method provides a basis for comparing alternate proposals and for choosing the best proposal. Project optimization is the subject of a later section, where a brief description of a perturbation analysis is presented.

Detailed cost estimates are beyond the scope of this text. Such procedures are capable of producing accuracies in the neighborhood of ±5 percent; however, such estimates generally require many months of engineering work. This chapter is designed to give the reader a basis for a *preliminary cost analysis* only, with an expected accuracy of approximately ±20 percent. (See also Chaps. 15 and 24.)

The following topics are discussed in this chapter: capital and operating costs, project evaluation, perturbation studies in optimization, cost-benefit analysis, and principles of accounting.

NOTE The material presented in this chapter was adapted primarily from Shen et al.,[1] Theodore and Neuser,[2] Theodore and Moy,[3] and Santoleri et al.[4] The reader should also note that much of material has appeared elsewhere in the book.

28–2 Capital Costs

Equipment cost is a function of many variables, one of the most significant of which is capacity. Other important variables vary with the cost of equipment or process. Preliminary estimates are often based on simple cost-capacity relationships that are valid when the other variables are confined to narrow ranges of values; these relationships can be represented by an approximate linear (on log-log coordinates) cost equipment relationship of the form.[5]

$$C = \alpha Q^{\beta} \tag{28–1}$$

where C = cost
Q = some measure of equipment or process capacity
α, β = empirical "constants" that depend mainly on equipment type

It should be emphasized that this procedure is suitable for rough estimation only; actual estimates from vendors are preferable. Only major pieces of equipment are included in this analysis; smaller peripheral equipment such as pumps and compressors may not be included. Similar methods for estimating costs are available in the literature.[5] If greater accuracy is needed, however, actual quotes from vendors should be used.

Again, the equipment cost estimation model just described is useful for a very preliminary estimation. If more accurate values are needed and old price data are available, the use of an indexing method may be preferable, although a bit more time consuming. The method consists of adjusting the earlier cost data to present values using factors that correct for inflation.

A number of such indices are available; some of the most commonly used are

1. Chemical Engineering Fabricated Equipment Cost Index (FECI),[6] past values of which are listed in Table 28–1
2. Chemical Engineering Plant Cost Index[7] listed in Table 28–2
3. Marshall and Swift (M&S) Equipment Cost Index,[8] listed in Table 28–3

Other indices for construction, labor, buildings, engineering, and so on are also available in the literature.[6–8] Generally, it is not wise to use past cost data older than 5 to 10 years as in Tables 28–1 to 28–3, even when using the cost indices. Within that time span, the technologies used in developing indices may have changed drastically. Using the indices could cause the estimates to greatly exceed the actual costs. Such an error might lead to the choice of an alternative proposal other than the least costly one. Note that the three indices listed in Tables 28–1 to 28–3 range over the years only to the end of the twentieth century. The base years for Tables 28–1 and 28–2 are 1957–1959 with an index of 100. One may compare the fabricated equipment index to costs available from the time frame between 1993 and 1999. Another comparison would be the plant cost index from 1989 to 1999. The Marshall and Swift index uses 1926 as a base year with a cost index of 100. Comparing the ratios for any 2 years, as calculated from Tables 28–2 and 28–3, one notes that the change in index varies from 1 to 3%. Since these methods are used for preliminary cost estimates with an expected accuracy of approximately ±20%, use of either index will not significantly impact the final estimate.

Table 28–1

Fabricated Equipment Cost Index

Year	Index
1999	434.1
1998	435.6
1997	430.4
1996	425.5
1995	425.4
1994	401.6
1993	391.2
1957–1959	100

Source: Ref. 6.

Table 28–2

Plant Cost Index

Year	Index
1999	390.6
1998	389.5
1997	385.5
1996	381.7
1995	381.1
1994	368.1
1993	359.2
1992	358.2
1991	361.3
1990	357.6
1989	355.4
1957–1959	100

Source: Ref. 7.

Table 28–3

Marshall & Swift Equipment Cost Index

1999	1068.3
1998	1061.9
1997	1056.8
1996	1039.1
1995	1027.5
1994	993.4
1993	964.2
1992	943.1
1991	930.6
1990	915.1
1926	100

Source: Ref. 8.

The usual technique for determining the *capital costs* (i.e., total capital costs, which include equipment design, purchase, and installation) for a facility is based on the *factored method* of establishing direct and indirect installation costs as a function of the known equipment costs. This is basically a *modified Lang method*, in which cost factors are applied to known equipment costs. [9,10] This is a three-step method.

1. Obtain directly from vendors (or, if less accuracy is acceptable, from one of the estimation techniques discussed above) the purchase prices of the primary and auxiliary equipment. The total base prices, designated by X, which should include instrumentation, control, taxes, and freight costs, serves as the basis for estimating the direct and indirect installation costs. The installation costs are obtained by multiplying X by the cost factors, which can be adjusted to more closely model the proposed calculation by using adjustment factors that may be available in order to into account for the complexity and sensitivity of the system.[9,11]

2. Estimate the *direct installation cost* by summing all the cost factors involved in the direct installation costs, which can include piping, insulation, foundation, and supports. The sum of these factors is designated as the *direct installation cost factor* (DCF). The direct installation costs are then the product of the DCF and X.

3. Once the direct and indirect installation costs (ICF) have been calculated, the total *capital cost* (TCC) may be evaluated as

$$\text{TCC} = X + (\text{DCF})(X) + (\text{ICF})(X) \tag{28–2}$$

This cost is then converted to *annualized* capital costs with the use of the *capital recovery factor* (CRF), which is described in a later section. The *annualized capital cost* (ACC) is the product of the CRF and TCC and represents the total installed equipment cost distributed over the lifetime of the facility.

28–3 Operating Costs

Operating costs can vary from site to site, plant to plant, and equipment to equipment, since these costs, in part, reflect local conditions, e.g., staffing practices, labor, and utility costs. Operating costs, like capital costs, may be separated into two categories: direct and indirect costs. *Direct costs* cover material and labor and are directly involved in operating the facility. These can include labor, materials, maintenance, labor, maintenance supplies, replacement parts, waste (e.g., residues) disposal fees, utilities, and laboratory costs. *Indirect costs* are operating costs associated with but not directly involved in operating the unit or facility in question; costs such as overhead (e.g., building-land leasing and office supplies), administrative fees, local property taxes, and insurance fees fall into this category.[12, 13]

The major direct operating costs are usually associated with labor and materials. *Materials* costs usually involve the cost of chemicals needed for the operation of the system(s). *Labor* costs differ greatly, depending on whether the facility is located on- or off-site and the degree of controls and/or instrumentation. Typically, there are three working shifts per day plus the standby shift used for rotation on a weekly basis with one supervisor per shift. On the other hand, it may be controlled by a single operator for only one-third or one-half of each shift; usually only an operator, supervisor, and site manager are necessary to run a facility. Salary costs vary from state to state and depend significantly on the location of the facility. The cost of *utilities* generally consists of costs for electricity, water, fuel, compressed air, and steam. The annual costs are estimates. Annual maintenance costs can be estimated as a percentage of the capital equipment costs. The life expectancies can be found in the literature. Laboratory costs (if applicable) depend on the number of samples tested and the extent of these tests; these costs may be estimated as 10 to 20% of the operating labor costs, if applicable.

The indirect operating costs consist of overhead, local property tax, insurance, and administration, less any credits. The overhead can include payroll, fringe benefits, health costs, social security, unemployment insurance, and other compensation that is indirectly paid to the plant personnel. This cost can be estimated as 70 to 90% of the operating labor, supervision, and maintenance costs.[14,15] Local property costs can be estimated as 2% of the TCC.

The *total operating cost* is the sum of the direct and indirect operating costs less any credits that may be recovered (e.g., the value of recovered by-products such as steam). Unlike capital costs, operating costs are almost always calculated on an annual basis.

28-4 Project Evaluation

Although this section primarily deals with a plant or process, it may be applied to equipment or other economic issues. In comparing alternate processes or different options of a particular process from an economic perspective, the total capital cost should be converted to an annual basis by distributing it over the projected lifetime of the facility. The sum of both the *annualized capital costs* (ACCs) and the annual operating costs (AOCs) is known as the *total annualized cost* (TAC) of the facility. The economic merit of the proposed facility, process, or scheme can be examined once the total annual cost is available. Alternate facilities or options may also be compared. A small flaw in this procedure is the assumption that the operating costs remain constant throughout the lifetime of the facility. However, since the analysis is geared to comparing different alternatives, the changes with time should be somewhat uniform among the various alternatives, resulting in little loss of accuracy.

The conversion of the total capital cost to an annualized basis involves an economic parameter known as the *capital recovery factor* (CRF), an approach routinely employed by the author in the past. These factors can be found in any standard economics text[16–18] or can be calculated directly from

$$\text{CRF} = \frac{i(1+i)^n}{(1+i)^{n-1} - 1} \tag{28–3}$$

where n = projected lifetime of the system, years
i = annual interest rate (expressed as a fraction)

The CRF calculated from Eq. (28–3) is a positive, *fractional* number. The ACC is computed by multiplying the TCC by the CRF. The annualized capital cost reflects the cost associated with recovering the initial capital outlay over the depreciable life of the system.

Investment and operating costs can be accounted for in other ways, such as the popular *present-worth* analysis. However, the capital recovery method is preferred because of its simplicity and versatility. This is especially true when comparing systems having different depreciable lives. There are usually other considerations in such decisions besides the economics, but if all the other factors are equal, the proposal with the lowest total annualized cost should be the most attractive.

If an industrial system is under consideration for construction, the total annualized cost should be sufficient to determine whether the proposal is economically attractive in comparison to other proposals. If, however, a commercial process is being considered difficulty the profitability of the proposed operation becomes an additional factor. One difficulty in this analysis is estimating the revenue generated from the facility because both technology and costs can change from year to year. Another factor affecting the revenue generated is the availability and raw materials to be handled by the facility. If, the revenue that will be generated from the facility can be reasonably estimated, a rate of return can be calculated. This method of analysis is known as the *discounted cash flow method using an end-of-year convention*, where the cash flows are assumed to be generated at the end of the year, rather than throughout the year (the latter obviously is the real case). An expanded explanation of this method can be found in any engineering economics text.[6] The data required for this analysis are the TCC, the annual after-tax cash flow (A), and the working capital (WC).

Usually, an after-tax *rate of return* on the initial investment of at least 30% is desirable. The method used to arrive at a rate of return will be discussed briefly. An *annual after-tax cash flow* can be computed as the annual revenues (R) less the annual operating cost (AOC) and minus the income taxes (IT). Income taxes are often estimated at 50% (this number may vary depending on the current tax laws) of taxable income (TI):

$$\text{IT} = 0.5\ (\text{TI}) \tag{28–4}$$

The taxable income can be obtained by subtracting the AOC and the depreciation of the plant (D) from the revenues generated (R):

$$\text{TI} = R - \text{AOC} - D \tag{28–5}$$

For simplicity, straight-line depreciation is assumed; in other words, the plant will depreciate uniformly over the life of the plant. For a 10-year lifetime, the facility will depreciate by 10% each year, i.e.,

$$D = 0.1\ (\text{TCC}) \tag{28–6}$$

The annual after-tax cash flow (A) is then

$$A = R - \text{AOC} - \text{IT} \tag{28–7}$$

This procedure involves a trial-and error solution. There are both positive and negative cash flows. The positive cash flows consist of A and the recoverable working capital in year 10. Both should be discounted *backward* to time = 0, the year when the facility begins operation. The negative cash flows consist of the TCC and the initial WC. In actuality, the TCC can be assumed to be spent evenly, for example, over a (2-year) construction period. Therefore, one-half of this flow is adjusted *forward* from after the first construction year (time = −1 year) to the year when the facility begins operating ($t = 0$). The other half, plus the WC, is assumed to be expended at time = 0. Forward adjustment of the 50% TCC is accomplished by multiplying by an economic parameter known as the *single-payment compound amount factor F/P*, given by

$$\frac{F}{P} = (1+i)m \tag{28–8}$$

where i = rate of return (fraction)
m = number of years (in this case, 1 year)

For the *positive* cash flows, the annual after-tax cash flow (A) is discounted backward by using a parameter known as the *uniform series present worth factor* (P/A). This factor is dependent on both interest rate (rate of return) and lifetime of the facility and is defined by

$$\frac{P}{A} = \frac{(1+i)^n - 1}{i(1+i)^n} \tag{28–9}$$

where n = lifetime of facility (in this case, 10 years)

The reader should note that the P/A is essentially the inverse of the CRF (capital recovery factor). The recoverable working capital at year 10 is discounted backward by multiplying WC by the *single present worth factor* (P/F), which is given by

$$\frac{P}{F} = \frac{1}{(1+i)^n} \tag{28–10}$$

The positive and negative cash flows can now be equated, and the value of i, the rate of return, may be determined by a trial and error calculation.

28–5 Perturbation Studies in Optimization

Once a particular process scheme (or project) has been selected, it is common practice to optimize the process from a capital cost and operation-and-maintenance standpoint. There are many optimization procedures available, most of them too detailed for meaningful application to most studies. These sophisticated optimization techniques, some of which are routinely used in the design of conventional chemical and petrochemical plants, invariably involve computer calculations. Use of these techniques in many chemical engineering applications is not warranted, however.

One simple optimization procedure that is recommended is the perturbation study. This involves a systematic change (or perturbation) of variables, one by one, in an attempt to locate the optimum design from a cost-and-operation perspective. To be practical, this often means that the chemical engineer must limit the number of unknowns by assigning constant values to those process variables that are known beforehand to play an insignificant role. Reasonable guesses and simple or shortcut mathematical methods can further simplify the procedure. Much information can be gathered from this type of study since it usually identifies those variables that significantly impact on the overall performance of the process and also helps identify the major contributors to the total annualized cost.

28–6 Cost-Benefit Analysis

An ability to estimate the cost of a project or system is important for many reasons. From a business standpoint, it is essential to have sufficient knowledge of the cost so that adequate funds can be secured to carry out the project. Where several options exist, it is often useful to determine the costs of each candidate system to determine which approaches are affordable and which are most cost-effective. It may also become apparent that, rather than spending funds on a particular project or process, a more cost-effective approach may involve changing the nature of the process.

Cost-benefit analysis (CBA) is a tool which provides a technique for economic evaluation of policy and/or projects. It has been used primarily as an instrument for evaluating programs and projects. The technique provides quantitative information to analysts on the costs and benefits of various choices which they might make.

The basic premise of CBA is rather simple and easy to accept: benefits and costs of a program should be weighted prior to deciding on a particular choice. However, performing CBA is a demanding technique which requires a great deal of information; quantification of the potential benefits and costs of a program is not an insignificant exercise. Information must first be identified and structured so as to allow comparison in common measures of value and time. The value of benefits and costs are generally measured in monetary terms, while the time value of money must be considered. Establishing monetary values by time allows a cost or a benefit for one year to be compared with that cost or benefit at some other time.

A common difficulty in measuring both costs and benefits of a program is the fact that the analysis is forward looking and thus requires an estimate of what a particular strategy will cost, which is more difficult than quantifying what a program actually does cost. All costs must be considered for each alternative program. Measuring benefits can also be a difficult process. Many benefits can also be difficult to process. Benefits that are generated from a project are also often difficult, if not impossible, to identify and quantify.

Consider the benefits which may arise from a pollution control program. How does one quantify the benefits derived from improved health, improved visibility, or reduction in damage to plants, animals, and properties? Chemical engineers sometimes look to the "willingness to pay" in trying to place a value on intangible or incommensurable benefits. Obviously, this approach is quite subjective and prone to error and uncertainty. However, to ignore intangible benefits of a project or program is to bias the results; the fact that benefits are intangible does not mean that they are unimportant.[18]

Cost-benefit analysis provides an economic tool to assist chemical engineer decision makers in evaluating alternative program or policy decisions for their economic feasibility. It establishes a systematic approach for decision making which attempts to comprehensively address all consequences of deciding on a particular alternative. However, there are many limitations to the method, which should be recognized. The estimation of costs and benefits is sometimes difficult. All benefits and costs must be identified individually, with appropriate monetary values assigned and time periods determined. Establishing a monetary value for intangible or incommensurate benefits is, as noted above, uncertain at best.

A *sensitivity analysis* will involve adjusting inputs into the model to evaluate the effects on the ultimate outcome. For example, a range of discount rates may be applied to the analysis to evaluate the sensitivity of the model to a change in the discount rate. Programs that are more sensitive to changes may be scrutinized further before taking action point. CBA may be internally or externally driven. A facility or plant faced with major expenditures for updating equipment would naturally look for the solution that provides the greatest benefit for the least expense. While this would typically be the lowest-cost solution, other factors may lead to consideration of higher-cost options.

There is much more to selecting a system or equipment or unit than simply picking one with the lowest purchase cost. Although the initial costs must certainly be considered relative to the project budget, long-term annual costs are typically used in comparing the costs for the purpose of selecting the most suitable strategy. In addition to the initial purchase and installation costs, those long-term annual costs are greatly influenced by equipment life expectancy, utility and environmental (if applicable) costs, and maintenance and labor costs.

28-7 Principles of Accounting

Accounting is the science of recording business transactions in a systematic manner. Financial statements are both the basis for and the result of management decisions; practicing chemical engineers are rarely involved. Such statements can tell a manager or a chemical engineer a great deal about a company, provided that one can interpret the information correctly.

Since a fair allocation of costs requires considerable technical knowledge of operations in the chemical process industry, a close liaison between the senior process engineer (usually a chemical engineer) and the accountants in a company is desirable. Indeed, the success of a company depends on a combination of financial, technical, and managerial skills.

Accounting has also been defined as the language of business. The different departments of management use it to communicate within a broad context of financial and cost terms. Chemical engineers who do not bother to learn the language of accountancy deny themselves the most important means available for communicating with top management. They may be regarded by upper management as lacking business knowhow. Some chemical engineers have only themselves to blame for their low status within the company hierarchy since they seem determined to hide themselves from business realities behind the screen of their specialized technical expertise. However, an increasing number of chemical engineers are becoming involved in decisions that are related to business.

Chemical engineers involved in feasibility studies and detailed process evaluations are dependent on financial information from the aforementioned company accountants, especially information regarding the way the company intends to allocate its overhead costs. It is vital that the chemical engineer correctly interpret such information or data and be able, if necessary, to make the accountant understand the effect of the chosen method of allocation.

The method of allocating overheads can seriously affect the assigned costs of a project and hence the apparent cash flow for that project. Since these cash flows are often used to assess profitability by such methods as present net worth (PNW) (see earlier section), unfair allocation of overhead costs can result in a wrong choice between alternative projects.

In addition to understanding the principles of accounting and obtaining a working knowledge of its practical techniques, chemical engineers should be aware of possible inaccuracies of accounting information in the same way that they allow for errors in any technical information data.

At first acquaintance, the language of accounting appears illogical to most engineers. Although the accountant normally expresses information in tabular form, the basis of all calculation can be simply expressed by

$$\text{Capital} = \text{assets} - \text{liabilities} \tag{28–11}$$

or

$$\text{Assets} = \text{capital} + \text{liabilities} \tag{28–12}$$

Capital, often referred to as *net worth*, is the money value of the business since assets are the money values of things that the business owns while liabilities are the money value of the things the business owes.

Most chemical engineers have great difficulty in thinking of capital (also known as *ownership*) as a liability. This may be alleviated if it is realized that a business is a legal entity in its own right, owing money to the individuals who own it. This realization is absolutely essential when considering large companies with stockholders, and is used for consistency even for sole ownerships and partnerships. If a person, say, Mary K. Theodore (MKT) puts up $10,000 capital to start a tutorial business, then that business has a liability to repay $10,000 to that person (MKT).

It is even more difficult to consider profit as being a liability. *Profit* is the increase in money available for distribution to the owners, and effectively represents the interest or revenue obtained on the capital. If the profit is not distributed, it represents an increase in capital by the normal concept of compound interest. Thus, if the business makes a profit of $5000, the liability is increased to $15,000. With this concept in mind, Eq. (28–12) can be expanded to

$$\text{Assets} = \text{capital} + \text{liabilities} + \text{profit} \tag{28–13}$$

where capital is once again considered as the cash investment in the business and is distinguished from the resultant profit in the same way that principal and interest are separated.

Profit (as defined above) is the difference between the total cash revenue from sales and the total of all costs and other expenses incurred in making those sales. With this definition, Eq. (28–13) can be further expanded to

$$\text{Assets} + \text{expenses} = \text{capital} + \text{liabilities} + \text{profit} + \text{revenue from sales} \tag{28–14}$$

Some chemical engineers have the greatest difficulty in regarding an expense as being equivalent to an asset, as is implied by Eq. (28–14). However, consider MKT's earnings. During the period in which she made a profit of $5000, her total expenses, excluding her earnings, were $8000. If she assessed the worth of her labor to the business at $12,000, then the revenue required from sales would be $25,000. Thus, effectively, MKT made a personal income of $17,000 in the year but she apportioned it to the business as $12,000 expense for her labor and $5000 return on her capital. In larger businesses, some people will receive salaries but not hold stock and therefore, receive no profits, and there may be stockholders who receive profits but no salaries. Thus, the difference between expenses and profits is very practical.

The period covered by the published accounts of a company is usually *one year*, but the details from which these accounts are compiled are often entered *daily* in a *journal*. The journal is a chronological listing of every transaction of the business, with details of the corresponding or associated income or expenditure. For the smallest businesses, this may provide sufficient documentation but, in most cases, the unsystematic nature of the journal can lead to computational errors. Therefore, the usual practice is to keep accounts that list the transactions related to a specific topic such as "Purchase of raw materials account." This account would list the cost of each purchase of raw materials together with the date of purchase, as extracted from the journal.

The traditional work of accountants is becoming increasingly concerned with future planning. Modern accountants can roughly be divided into two branches: financial accountancy and management accountancy.

Financial accountancy is concerned with *stewardship*. This involves the preparation of balance sheets and income statements that represent the interest of stockholders and bond holders and are consistent with the existing legal requirements. Taxation is an important element of financial accounting.

Management accounting is concerned with decision making and control. This is the branch of accountancy closest to the interest of most (process) chemical engineers.

Management accounting is concerned with standard costing, budgetary control, and investment decisions.

Accounting statements only present facts that can be expressed in financial terms. They do not indicate whether a company is developing new products that will ensure a sound business future. A company may have impressive current financial statements and yet may be heading for bankruptcy in the near future if there is no provision for the introduction of sufficient new products or services.

References

1. T. Shen, Y. Choi, and L. Theodore, EPA Manual *Hazardous Waste Incineration*, (USEPA Manual) USEPA, Research Triangle Park, N.C., 1985.
2. L. Theodore and K. Neuser, *Engineering Economics and Finance*, A Theodore Tutorial, Theodore Tutorials, East Willinston, N.Y., originally published by USEPA/APTI, Research Triangle Park, N.C., 1996.
3. L. Theodore and E. Moy, *Hazardous Waste Incineration* , USEPA, Research Triangle Park, N.C., 1992.
4. J. Santoleri, J. Reynolds, and L. Theodore, *Introduction to Hazardous Waste Incineration*, 2nd ed., Wiley, Hoboken, N.Y., 2009.
5. R. McCormick and R. DeRosier, *Capital and O&M Cost Relationships for Hazardous Waste Incineration*, Acurex Corporation, USEPA, Cincinnati, report date unknown
6. Economic Indicators, "Chemical Engineering Plant Cost Index," *Chem. Eng.* **99**, 178, July 1992; **101**, 206, May 1994; **102**, 180, April 1995; **102**, 162, May 1995; **103**, 168, May 1996; **104**, 229, May 1997; **105**, 189, June 1998; **107**, 164, N.Y., May 2000.
7. Economic Indicators, "Chemical Engineering Plant Cost Index," *Chem. Eng.* **103**, 172, Nov. 1996; **107**, 164, N.Y., May 2000.
8. Economic Indicators, "Marshall and Swift Equipment Cost Index," *Chem. Eng.*, **103**, 172, Nov. 164, N.Y., May 1996.
9. R. Neveril, *Capital and Operating Costs of Selected Air Pollution Control Systems*, Card, Inc., Niles, Ill., US EPA Report 450/5-80-002, Dec. 1978.
10. W. Vatavuk and R. Neveril, "Factors for Estimating Capital and Operating Costs," *Chem. Eng.*, 157–162, N.Y., Nov. 3, 1980.
11. G. Vogel and E. Martin, "Hazardous Waste Incineration, Part 1—Equipment Sizes and Integrated-Facility Costs," *Chem. Eng.*, 143–146, N.Y., Sept. 5, 1983.
12. G. Vogel and E. Martin, "Hazardous Waste Incineration, Part 2—Estimating Costs of Equipment and Accessories," *Chem. Eng.*, 75–78, N.Y., Oct. 17, 1983.
13. G. Vogel and E. Martin, "Hazardous Waste Incineration, Part 3—Estimating Capital Costs of Facility Components," *Chem. Eng.*, 87–90, N.Y., Nov. 28, 1983.
14. G. Ulrich, *A Guide to Chemical Engineering Process Design and Economics*, Wiley, Hoboken, N.J., 1984.
15. G. Vogel and E. Martin, "Hazardous Waste Incineration, Part 4—Estimating Operating Costs," *Chem. Eng.*, 97–100, N.Y., Jan. 9, 1984.
16. E. DeGarmo, J. Canada, and W. Sullivan, *Engineering Economy*, 6th ed., Macmillan, N.Y., 1979.
17. C. Hodgman, S. Selby, and R. Weast, eds., *CRC Standard Mathematical Tables*, 12th ed., Chemical Rubber Company, Cleveland, 1961 (presently CRC Press, Boca Raton, Fla.).
18. T. Tielenberg, *Environmental and Natural Resource Economics*, Scott Foreman, Glenview, Ill., 1984.
19. F. Holland, F. Watson, and J. Wilkinson, "Financing Principles of Accounting," *Chem. Eng.*, N.Y., July 8, 1974.

29 Ethics

Chapter Outline

29-1 Introduction

The following statement was adapted from Theodore[1] and obviously represents the author's opinion on what has become a controversial issue. "Unfortunately, ethics has definitely come to mean different things to different people." *Ethics* is a philosophical discipline that draws on human reason and analysis to assess moral choices; Webster talks of "conforming to moral standards" where "moral" relates to decisions that have an impact on the lives of other people.

There is currently a renaissance or grass-roots movement in academia to make students aware of the FOLD (fabricating data, omitting information, lying and acting in a deceitful manner) principle.[1] It is happening because many colleges and/universities are now including ethics training in their curriculum. The Accreditation Board of Engineering and Technology (ABET), which accredits engineering schools, now required ethics training to be incorporated in the curricula. In fact, all engineering programs need to address ethical issues in order to receive accreditation. For example, Manhattan College has been formally sensitizing students since 1992 to possible ethical dilemmas which they might encounter. This is also happening because many have come to realize that ethical and moral decadence will ultimately destroy this country.

The chapter contains sections concerned with the present state, moral issues, engineering ethics, and environmental justice.

NOTE A significant portion of the sections to follow have been drawn from the work of Wilcox and Theodore.[2]

29-2 The Present State

Since the topic of this chapter is ethics, there is a need for information on definitions. Here are four that might help the reader:

1. *Ethics* is defined as that branch of philosophy dealing with the rules of right conduct.
2. *Environmental ethics* deal with the moral issues of conduct with respect to the environment.
3. *Ethical theory* is an attempt to answer certain questions about standards of conduct or ethics. It also attempts to provide a framework for decisions regarding what moral principles are correct and how one should treat each other, the environment, other species, and so on.
4. *Morality* is concerned with reasons for the desirability of certain kinds of actions and the undesirability of others. To say that an act is right is not to express a mere feeling or bias, but instead to assert that the best moral reasons support doing it. *Moral reason* is a reason that requires individuals to respect other people, and to care for others' good as well as one's own. In addition, moral reasons are such that they set limits to the legitimate pursuit of self-interest. They can be used to evaluate, praise, and criticize laws.

With respect to the above, MBAs and CPAs of industry have spawned financial abuses and excesses. These individuals have turned what were previously moral and ethics questions into legal technicalities. Industry executives are more likely to ask what one can get away with legally rather than to be concerned with what is right, honest, and/or fair.

The above abuses and excesses have spread like a cancer into society, particularly with lawyers and individuals in government. Lawyers have become adept at creating controversy, smokescreens, and/or obstruction of the truth, and using the complexity of the laws for their own aggrandizement. Elected officials, on the other hand, have used their entrusted position of power to further and maintain their careers… rather than to represent their constituency.[1]

Consider if a chemical engineer, acting as an expert witness in a court of law, conjures up an idea that creates reasonable doubt even though it is highly unlikely. This type of conduct is clearly defined as fraud in any engineering code of ethics. Given the above, the reader should consider a lawyer's conduct in and out of court.[1]

Well, what's the answer? Is there an answer? The reality may be that the above-referenced corruption is just too widespread and amorphous—like humidity in southern Florida in the summer. It seems that the only thing that can ultimately turn things around on this issue is for each individual to take a stand for basic values and virtues, and return to the likes of integrity, responsibility, and selflessness. If a grass-roots movement is to succeed, it will ultimately rest on each individual's ability and willingness to discard the practices of so many lawyers and elected officials. As Albert Schweitzer put it, "Man must cease attributing his problems to his environment, and learn again to exercise his will—his personal responsibility in the realm of faith and morals."

29-3 Moral Issues

It is generally accepted that any historical ethic can be found to focus on one of four different underlying moral concepts:[4]

1. *Utilitarianism* focuses on good consequences for all.
2. *Duties ethics* focus on one's duties.
3. *Rights ethics* focus on human rights.
4. *Virtue ethics* focus on virtuous behavior.

Note that *duties ethics* and *rights ethics* are often combined as *deontological ethics.*)

Utilitarians hold that the most basic reason why actions are morally right is that they lead to the greater good for the greatest number. "Good and bad consequences are the only relevant considerations, and, hence all moral principles reduce to one 'we ought to maximize utility.'"[5]

Duties ethicists concentrate on an action itself rather than the consequences of that action. To these ethicists, there are certain principles of duty such as "do not deceive" and "protect innocent life" that should be fulfilled, even if the most good does not result. The list and hierarchy of duties differ from culture to culture, religion to religion. For Judeo-Christians, the Ten Commandments provide an ordered list of duties imposed by their religion.[5]

Often considered to be linked with duties ethics, *rights ethics* also assesses the act itself rather than its consequences. Rights ethicists emphasize the rights of the people affected by an act rather than the duty of the person(s) performing the act. For example, because a person has a right to life, murder is morally wrong. Rights ethicists propose that duties actually stem from a corresponding right. Since each person has a "right" to life, it is everyone's "duty" not to kill. It is because of this link and their common emphasis on the actions themselves that rights ethics and duties ethics are often grouped under the common heading: *deontological ethics*.[6]

The display of virtuous behavior is the central principle governing *virtue ethics*. An action would be wrong if it expressed or developed vices—for example, negative character traits. Virtue ethicists, therefore, focus on becoming a morally good person.

To display the different ways that these moral theories view the same situation, one can explore their approach to the following scenario that Martin and Schinzinger present:[5]

"On a midnight shift, a botched solution of sodium cyanide, a reactant in organic synthesis, is temporarily stored in drums for reprocessing. Two weeks later, the day shift foreperson cannot find the drums. The plant manager discovers that the batch has been illegally dumped into the sanitary sewer. He severely disciplines the night shift foreperson. On making discreet inquiries he determines that no apparent harm has resulted from the dumping. Should the plant manager inform government authorities, as is required by law in this kind of situation?"

If a representative of each of the four different ethics theories just mentioned were presented with this dilemma, their decision-making process would focus on different principles.

The *utilitarian* would assess the consequences of his options. If he told the government, his company might suffer immediately under any fines administered and later (perhaps more seriously) from exposure of the incident by the media. If he chose not to inform authorities, he risks heavier fines (and perhaps even worse press coverage) in the event that someone discovers the cover-up. As a consequences, the utilitarian is only employing consideration in his decision-making process.

The *duties ethicist* would weigh her duties, and her decision would probably be more clear cut than her utilitarian counterpart. She is obliged foremost by her duty to obey the law and must inform the government.

The *rights ethicist's* mind frame would lead to the same course of action as the duties ethicist—not necessarily because he has a duty to obey the law, but because the people in his community have the right to informed consent. Even though the plant manager's inquiries informed him that no harm resulted from the spill, he knows that the public around the plant has the right to be informed of how the plant is operating.

Vices and virtues would be weighed by the *virtue ethicist* plant manager. The course of her thought process would be determined by her own subjective definition of what things are virtuous, what things would make her a morally good person. Most likely, she would consider both honesty and obeying the law virtuous, and withholding information from the government and public as virtueless and would, therefore, inform the authorities.

Numerous case studies of this nature are provided by Wilcox and Theodore.[2]

29-4 Engineering Ethics[3,7]

The ethical behavior of chemical engineers as well as other professionals is more important today than at any time in the history of the profession. Chemical engineers' ability to direct and control the technologies they master has never been stronger. In the wrong hands, the scientific advances and technologies of today's chemical engineer could become the worst form of corruption, manipulation, and exploitation. Chemical engineers, however, are bound by a code of ethics that carry certain obligations associated with the profession. A baker's dozen of these obligations follow:

1. Support one's professional society.
2. Guard privileged information and data.
3. Accept responsibility for one's actions.
4. Employ proper use of authority.
5. Maintain one's expertise in a state-of-the-art world.
6. Build and maintain public confidence.
7. Avoid improper gifts and/or gift exchange(s).
8. Avoid conflict of interest.
9. Apply equal opportunity employment.
10. Maintain honesty in dealing with employers, clients, and government.
11. Practice conservation of resources, pollution prevention, and sustainability.

12. Practice energy conservation.
13. Practice health, safety and accident prevention and management.

Many codes of ethics have appeared in the literature. The preamble for one of these codes is provided below:[5]

"Engineers in general, in the pursuit of their profession, affect the quality of life, for all people in our society. Therefore, an Engineer, in humility and with the need for divine guidance, shall participate in none but honest enterprises. When needed, skill and knowledge shall be given without reservation for the public good. In the performance of duty and in fidelity to the profession, Engineers shall give utmost."[5]

Regarding environmental ethics, Taback[8] defined ethics as, "the difference between what you have the right to do and the right thing to do." More recently, he has added that the environmental engineer-scientists should "recognize situations encountered in professional practice with conflicting interests that test one's ability to take the 'right' action. Then, take each situation to a trusted colleague to determine the best course of action consistent with the above precepts and that would have the least adverse impact on all stakeholders."

The Air & Waste Management Association (A&WMA), primarily through the efforts of Taback, developed The Code of Ethics in 1996. Details are provided in the next paragraphs.

"In the pursuit of their profession, environmental (many of whom are chemical engineers) professionals must use their skill and knowledge to enhance human health and welfare and environmental quality for all. Environmental professionals must conduct themselves in an honorable and ethical manner so as to merit confidence and respect, as well as to maintain the dignity of the profession. This code is to guide the environmental professional in the balanced discharge of his or her responsibilities to society, employers, clients, coworkers, subordinates, professional colleagues, and themselves.

As an environmental professional, I shall regard my responsibility to society as paramount and shall endeavor to:

1. Direct my professional skills toward conscientiously chosen ends I deem to be of positive value to humanity and the environment; decline to use those skills for purposes I consider to conflict with my moral values.
2. Inform myself and others, as appropriate, of the public health and environmental consequences, direct and indirect, immediate and remote, of projects in which I am involved, consistent with both standards of practice in industry and government, as well as laws and regulations that currently exist.
3. Comply with all applicable statutes, regulations, and standards.
4. Hold paramount the health, safety, and welfare of the public, speaking out against abuses of the public interest that I may encounter in my professional activities, as deemed appropriate per professional standards and existing laws and regulations.
5. Inform the public about technological developments, the alternatives they make feasible, and possible associated problems wherever known.
6. Keep my professional skills up-to-date and endeavor to be aware of current events, as well as environmental and societal issues pertinent to my work.
7. Exercise honesty, objectivity, and diligence in the performance of all my professional duties and responsibilities.
8. Accurately describe my qualifications for proposed projects or assignments.
9. Act as a faithful agent or trustee in business or professional matters, provided such actions conform to other parts of this code.
10. Keep information on the business affairs or technical processes of an employer or client in confidence while employed and later, as required by contract or applicable

laws, until such information is properly released, and provided such confidentiality conforms to legal requirements and other parts of this code.

11. Avoid conflicts of interest and disclose those known that cannot be avoided.
12. Seek, accept, and offer honest professional criticism, properly credit others for their contributions, and never claim credit for work I have not done.
13. Treat coworkers, colleagues, and associates with respect and respect their privacy.
14. Encourage the professional growth of colleagues, coworkers, and subordinates.
15. Report, publish, and disseminate information freely, subject to legal and reasonable proprietary or privacy restraints, provided such restraints conform to other parts of this code and do not unduly impact public health, safety, and welfare.
16. Promote health and safety in all work situations.
17. Encourage and support adherence to this code, never giving directions that could cause others to compromise their professional responsibilities.

It is the duty of every member to adhere to a code of conduct as may be adopted by the board of directors. Such code shall include the association's code of ethics. The Code of Conduct shall be published periodically by the Association and shall be provided to new members." Additional details are available in the literature.[9]

Although the environmental movement has grown and matured since the early 1970s, its development is far from stagnant. To the contrary, change in individual behavior, corporate policy, and governmental regulations are occurring at a dizzying pace.

Because of the federal sentencing guidelines, the defense industry initiative, as well as a move from compliance to a value-based approach in the marketplace, corporations have inaugurated companywide ethics programs, hotlines, and senior line positions responsible for ethics training and development. The sentencing guidelines allow for mitigation of penalties if a company has taken the initiative in developing ethics training programs and codes of conduct.

Regarding education, the ABET 2000 accreditation guidelines mandate programs have to clearly demonstrate that students are exposed to an ethics education; they also have to perform outcome assessments. Despite indicators that reveal the value of an ethics education, few large universities require an ethics course. Ideally, a student would take an ethics course and would also be exposed to ethics training in several other courses each year.

29–5 Environmental Justice

The environmental policy of USEPA has historically had two main points of focus: (1) defining an acceptable level of pollution and (2) creating the legal rules to reduce pollution to a specified level. Understandably, it seems that the program has been most concerned with economic costs and efficiency.[9] This appears to some to lack considerations of equity, both distributional and economic. While EPA's two main points of focus are important considerations, relying on such criteria in the formation of an environmental protection policy can potentially neglect to account for the potential inequalities of capitalism and its effects throughout the policy process.

The history of environmental policy making illustrates to some the incompatibility of equity and efficiency. Economic pressures of environmental regulation have motivated corporations to seek new ways to reduce costs. Industries have attempted to maximize profits by "externalizing" the environmental costs.[11] It has been suggested that this redistribution of costs is more regressive in its effects than the general sales tax.[12] To date, major corporate polluters sometimes have more to gain financially by continuing pollution practices than in obeying regulations. In some instances, the result of increased environmental costs has paradoxically caused negative impacts on environmental regulations.

As long as corporations feel unaffected by such environmental degradation, they have little incentive other than altruism to end these debilitating practices.

Generally, poverty and poor living conditions go hand in hand. As a result, less affluent citizens of the United States are more affected by pollution and environmental hazards. Evidence of the effects and concentration of environmental pollution in low-income communities has fueled a grass-roots environmental movement since the early 1980s. While this movement "can be regarded as a socioeconomic upheaval, intended to improve living conditions for the impoverished, it has unfortunately acquired serious racial and cultural undertones." The movement calls for grassroots, multiracial, and multicultural activism to redress the distributional inequalities that have resulted from past policy and to prevent the same inequalities from occurring in future policy. The movement advocates that minorities use historically nonexercised political and legal power to force the EPA to address concerns and to further oppose policies that further impoverish the poor.

The environmental justice movement is committed to political empowerment as a way to challenge inequities and injustices. *Empowerment* is a term that has been used to denote the inclusive involvement and education of community members by equipping them with skills for self-representation and defense. Organized activism at the grass-roots level could circumvent the power structures that may have underrepresented particular communities in the first place. Community activists want to participate in the decision making that affects their communities. Further, they have argued for increased pay for community members who engage in environmentally hazardous labor, for better working conditions in factories, and for the requirement of more job safety precautions. Activists contend that industries must contribute to community development if they are to detract from the community in other ways. For example, activists suggest industrial investment in community projects and educational systems.[13] The use of legal power is also encouraged. The most profound way of ensuring that law-breaking corporations are reprimanded is to utilize the American justice system to assist in settling legal precedents.

29-6 The Pros and Cons of Environmental Justice

Environmental protection policy has attempted to reduce environmental risks overall; however, and as noted above, in the process of protecting the environment, risks have been redistributed and concentrated in particular segments of society. Although federal regulations to protect the environment are not explicitly discriminatory, some argue that environmental protection policies have not been sensitive to distributional inequalities. Others insist that they have not adequately addressed specific minority environmental concerns. Like many programs of reform and activism, environmental justice was started with principally good intentions. However, ground rules need to be set before any meaningful discussion regarding environmental justice can be presented. One problem is that environmental justice has come to mean different things to different people, at different times at different locations, and for different situations. There appears to be no clear-cut decision regarding this term, but the EPA defines it as "the fair treatment of people of all races, cultures, and incomes with respect to development, implementation, and enforcement of environmental laws and policies, and their meaningful involvement in the decision-making process of government." According to this definition, there appear to be three major components of environmental justice:

1. Environmental racism
2. Environmental equity
3. Environmental health

Details of each are available in the literature.[9] It should be noted that the author does not support current EPA policies related to environmental justice.

References

1. L. Theodore, "On Ethics I," *The Williston Times*, East Williston N.Y., Nov. 28, 2003.
2. J. Wilcox and L. Theodore, *Environmental and Engineering Ethics: A Case Study Approach*, Wiley, New York, 1998.
3. L. Theodore, "On Ethics II," *The Williston Times*, East Williston, N.Y., March 26, 2004.
4. M. K. Theodore and L. Theodore, *50 Major Issues Facing the 21st Century*, Theodore Tutorials (originally published by Prentice-Hall), East Williston, N.Y., 1995.
5. M. Martin and R. Schinzinger, *Ethics in Engineering*, McGraw Hill, New York, N.Y., 1989.
6. I. Barbour *Ethics in an Age of Technology*, Harper, San Francisco, 1993.
7. H. Taback, *AWMA's Code of Ethics*, EM, pp. 42–43, Pittsburgh, Sept., 2007.
8. www.awma.org/about/index/html
9. M. K. Theodore and L. Theodore, *Introduction to Environmental Management*, CRC Press/Taylor & Francis Group, Boca Raton, Fla., 2010.
10. R. Lazarus, "Pursuing Environmental Justice: The Distributional Effects of Environmental Protection," *Northwestern University Law Review*, Chicago, 1987.
11. L. Cole, "Empowerment as the Key to Environmental Protection: The Need for Environmental Poverty Law," *Ecolo. Law Quart.*, 1992.
12. N. Dorfman and A. Snow, "Who Will Pay for Pollution Control? The Distribution by Income of the Burden of the National Environmental Protection Program," *National Tax J.* 1972–1980.

30 Open-Ended Problems

Chapter Outline

30–1 Introduction

In terms of introduction, the cliché of the creative individual has unfortunately been aptly described throughout history—the einsteinian wild hair, being locked in a room for days at a time, mumbling to one's self, eating sporadically, being lost in a fog of conflicting thoughts, not paying attention to one's hygiene, working diligently until those times when the "light goes on" moment of discovery, etc. This chapter will provide (among other things) specific suggestions on how to develop and improve one's critical thinking abilities.

Engineering is one of the noblest of professions, and the author is extremely proud to be part of it. He is fortunate to have served as a chemical engineering educator during his career. A good part of his effort in recent years was directed to improving critical thinking skills of students. Check any engineering school Web site and locate its mission statement. Many of these will carry the phrase "fosters creativity and innovation" among its students. But do they really? The author hopes so. But then again, how does one teach it?

As a chemical engineering educator, one is required to teach traditional basic scientific and technical principles in courses such as thermodynamics, heat transfer, reaction kinetics, and environmental management. But along with the lectures, one should include an emphasis on creativity, problem solving, and failure(s). These three terms are definitely interrelated. Finding solutions to problems is a creative activity. Failure comes into play since there are often solutions with high uncertainty and many or no correct answers.

The remainder of this chapter addresses a host of topics involved with open-ended problems and approaches. The following issues are addressed: the author's approach, earlier experiences, developing students' power of critical thinking, creativity and brainstorming, and inquisitive minds. The chapter concludes with a section on applications.

30–2 The Author's Approach

Here is what the author stressed to his students in terms of developing problem-solving skills and other creative thinking:

1. Carefully define the problem.
2. Gather all pertinent data and information.
3. Generate an answer or solution.
4. Evaluate as many other alternatives as possible, including "what if" scenarios.
5. Reflect on the above over time.

Step 4 is important in terms of critical thinking; words and actions such as *magnify* or *expand*, *nullify* or *reduce*, *rearrange*, *replace*, *combine*, *reverse*, and *eliminate*, often need to be applied in an attempt to be creative or achieve a successful solution to a detailed, difficult, and/or complicated problem. With problem solving, one finds that good judgment and sound decisions often come from experience. Unfortunately, experience usually comes from poor judgment and unwise decisions. In addition, failure is an integral part of being creative; it can teach individuals that early mistakes can lead later to simple solutions and/or innovative answers.

Many now believe that creative thinking should be part of every student's education. Here are some ways that have proven to nudge the creative process along:

1. Break out of the one-and-only answer rut.
2. Use creative thinking techniques and games.
3. Foster creativity with assignments and projects.
4. Be careful not to punish creativity.

These suggested activities will ultimately help develop a critical thinker who:

1. Raises vital questions/problems, formulating them clearly and precisely.
2. Gathers and assesses relevant information, using abstract ideas to interpret it effectively.
3. Comes to well-reasoned conclusions and solutions, testing them against relevant criteria and standards.
4. Thinks open-mindedly within alternative systems of thought, recognizing, and assessing, as need be, his/her assumptions, implications, and practical consequences.
5. Communicates effectively with others in figuring out solutions to complex problems.

30–3 Earlier Experience[2,3]

The educational literature provides frequent references to individuals, particularly engineers and those in other technical fields, who have different learning styles; and, in order to successfully draw on these different styles, various of approaches can be employed. One such approach for educators involves the use of *open-ended* problems.

The term open-ended has come to mean different things to different people. It basically describes an approach to the solution of a problem and/or situation for which there is rarely a unique solution. Three literature sources[4–6] provide sample problems that can be used when this educational tool is employed.

The author of this book has applied this somewhat unique approach and has included numerous open-ended problems in several chemical engineering course offerings at Manhattan College. Student comments for a general engineering graduate course titled "Accident and Emergency Management" were recently tabulated (see also Chap. 26). Student responses to the question "What aspects of this course were most beneficial to you?" are listed below:

1. "The open-ended questions gave engineers a creative license. We don't come across many of these opportunities."
2. "Open-ended questions allowed for candid discussions and viewpoints that the class may not have been otherwise exposed to."
3. "The open-ended questions gave us an opportunity to apply what we were learning in class with subjects we have already learned and gave us a better understanding of the course."
4. "Much of the knowledge that was learned in this course is applicable to everyday situations and our professional lives."
5. "Open-ended problems made me sit down and research the problem to come up with ways to solve them."
6. "I thought the open-ended problems were inventive and made me think about problems in a better way."
7. "I felt that the open-ended problems were challenging. I, like most engineers, am more comfortable with quantitative problems than qualitative."

In effect, the approach requires teachers to ask questions, to not always accept things at face value, and to select a methodology that provides the most effective and efficient solution. Those who conquer this topic have probably taken the first step toward someday residing in an executive suite.

30–4 Developing Students' Power of Critical Thinking[7]

It has often been noted that chemical engineers are living in the middle of an information revolution. For more than a decade, that revolution has had an effect on teaching and learning. Teachers are hard-pressed to keep up with the advances in their fields. Often their attempts to keep the students informed are limited by the difficulty of making new material available.

The basic need of both teacher and student is to have useful information readily accessible. Then comes the problem of how to use this information properly. The objectives of both teaching and studying such information are (1) to ensure comprehension of the material and integrate it with the basic tenets of the field it represents, and (2) use comprehension of the material as a vehicle for *critical thinking*, *reasoning*, and *effective argument*.

Information is valueless unless it is put to use; otherwise it becomes mere data. To use information most effectively, it should be taken as an instrument for analytical development. The process of this utilization works on a number of incremental levels. Information can be absorbed, comprehended, discussed, argued in reasoned fashion, written about, and integrated with similar and contrasting information.

The development of critical and analytical thinking is the key to the understanding and use of information. It is what allows the student to discuss and argue points of opinion and points of fact. It is the basis for the student's formation and development of independent ideas. Once formed, these ideas can be written about and integrated with both similar and contrasting information.

30–5 Creativity and Brainstorming

Chemical engineers bring mathematics and other sciences to bear on practical problems and applications, molding materials, and harnessing technology for human benefit. *Creativity* is often a key component in this synthesis; it is the spark, motivating efforts to devise solutions to novel problems, design new products, and improve existing practices. In the competitive marketplace, it is a crucial asset in the bid to win the race to build better machines, decrease product delivery times, and anticipate the needs of future generations.[8]

One of the keys to the success of a chemical engineer or a scientist is to generate fresh approaches, processes, and products; in short, they need to be creative. Gibney[8] has detailed how some schools and institutions are attempting to use certain methods that essentially share the same objective: open students' minds to their own creative potential.

Gibney[8] provides information on "the art of problem definition" developed by Rensselaer Polytechnic Institute. To stress critical thinking, they teach a seven-step methodology for creative problem development:

1. Define the problem.
2. State objectives.
3. Establish functions.
4. Develop specifications.
5. Generate multiple alternatives.
6. Evaluate alternatives.
7. Build.

In addition, Gibney[8] identified the phases of the creative process set forth by psychologists. They essentially break the process down into five basic stages:

1. Immersion
2. Incubation

3. Insight
4. Evaluation
5. Elaboration

Psychologists have ultimately described the creative process as *recursive*. At any one of these stages, one can double back, revise ideas, or gain new knowledge that reshapes one's understanding. For this reason, being creative requires patience, discipline, and hard work.

Delia Femina[9] outlined five secrets regarding the creative process:

1. Creativity is ageless.
2. You don't have to be Einstein.
3. Creativity is not an 8-hour job.
4. Failure is the mother of all creativity.
5. Dead men don't create.

Panitz[10] has demonstrated how *brainstorming strategies* can help engineering students generate an outpouring of ideas. Brainstorming guidelines include:

1. Carefully define the problem up front.
2. Allow individuals to consider the problem before the group tackles it.
3. Create a comfortable environment.
4. Record all suggestions.
5. Appoint a group member to serve as a facilitator.
6. Keep brainstorming groups small.

A checklist for change was also provided, as detailed below:

1. Adapt.
2. Modify.
3. Magnify.
4. Minify.
5. Find other uses.
6. Substitute.
7. Rearrange.
8. Reverse.
9. Combine.

30-6 Inquiring Minds

In an exceptional and well-written article[11] on inquiring minds, Lih commented on inquiring minds by saying "you can't transfer knowledge without them." His thoughts (which have been edited) on the inquiring or questioning process follow:

1. Inquiry is an attitude—a very important one when it comes to learning. It has a great deal to do with curiosity, dissatisfaction with the status quo, a desire to dig deeper, and having doubts about what one has been told.
2. Questioning often leads to believing—there is a saying that has been attributed to Confucius: "Tell me, I forget. Show me, I remember. Involve me, I understand." It might also be fair to add: "Answer me, I believe."

3. Effective inquiry requires determination to get to the bottom of things.
4. Effective inquiry requires wisdom and judgment. This is especially true for a long-range intellectual pursuit that is at the forefront of knowledge.
5. Inquiry is the key to successful lifelong learning. If one masters the art of questioning, independent learning is a breeze.
6. Questioning is good for the questionee as well. It can help clarify issues, uncover holes in an argument, correct factual and/or conceptual errors, and eventually lead to a more thoughtful outcome.
7. Teachers and leaders should model the importance of inquiry. The teacher or leader must allow and encourage questions and demonstrate a personal thirst for knowledge.

Ultimately, the degree to which one succeeds (or fails) is often based in part on one's state of mind or attitude. As President Lincoln once said: "Most people are about as happy as they make their minds to be." William James once wrote: "The greatest discovery of my generation is that human beings can alter their lives by altering their attitude of mind." So, no matter what one does, it is in the hands of that individual to make it a meaningful, pleasurable, and a positive experience. This experience will almost definitely bring success.

30–7 Applications

Several open-ended illustrative examples are available in the literature[1,2] and in class notes from Theodore.[3] Three such applications follow.

- Comment on the idea that the kinetic energy of a moving gas stream normally discharged from a plant can be recovered as part of an energy conservation measure. On the surface, this appears to be a project worthy of consideration. However, the energy content of a gas stream, even at high extremely high velocities, possesses an insignificant (relatively speaking) amount of energy. One notes that if 10,000 acfm at (60°F, 1 atm) of a gas stream (that may be considered air) is discharged from a stack at 50 ft/s, its kinetic energy (KE) is as follows:

$$\mathrm{KE} = \frac{1}{2}mv^2 = \frac{10{,}000}{379}\frac{(29)}{32.2}\frac{(50)^2}{(60)}$$
$$= 1200\ (\mathrm{ft \cdot lb_f/s}) = 2.0\ \mathrm{Btu/s}$$

  This would therefore *not* appear to be a cost-effective option for recovering energy.

- Consider the following plant dilemma. A cooling-water pump is no longer capable of delivering the required flow rate to a nuclear (highly exothermic) reactor. Rather than purchase a new pump, you have been asked to list and/or describe what steps can be taken to resolve the problem. The obvious option is to replace the pump. Since that is not a viable option, nine other potential options can be to:

1. Carefully check the pump, including the clogging of screens and/or intakes, impellers, etc.
2. Increase pipe size(s), i.e., the diameter.
3. Employ smoother pipe.
4. Decrease pipe length.

5. Eliminate unimportant valves, expansion and contraction joints, and other components, in order to reduce the pressure drop.
6. Decrease the viscosity of the water by increasing its temperature. (Note that the viscosity of a gas increases with increasing temperature.)
7. Use a different cooling medium altogether or add an *additive* to the cooling water to decrease the viscosity and/or increase the heat capacity of the cooling medium.
8. Create an endothermic side reaction in the system.
9. Use synthetic lubricants to reduce the friction of the flowing medium.

- Finally, during the heat of the space race in the 1960s, the U.S. National Aeronautics and Space Administration (NASA) decided that it required a ballpoint pen to write in the zero-gravity confines of its space capsules. NASA reasoned that since the gravity is zero, a pressure force is needed to force the ink to flow out. A simple device was developed to resolve the problem; a sketch is provided in Fig. 30–1. When the screw moves, ink will be transferred to the edge of the ballpoint. After considerable research and development, the "astronaut pen" was developed by NASA at a cost of $1 million. The pen worked and also enjoyed some modest success as a novelty item back here on Earth. The Soviet Union, faced with the same problem, used a pencil.

The author leaves the subject of open-ended problems with the following thoughts:

1. Creativity usually dies a quick death in rooms that house large conference tables.
2. Do not let anyone talk you out of pursuing (or following up) on what you believe to be a great idea.
3. Do not delay acting on a good idea. Chances are someone else has also thought of it. Success often comes to the one who acts first.
4. Never ask a lawyer or an accountant for business advice. They are trained to find problems, not solutions.
5. In the long run, shoddy thinking is costly, in terms of both in money and quality of life.
6. Keep a note pad and pencil on your night table. Million-dollar ideas can arrive at 3:00 A.M. Have a note pad and pencil handy on weekends, holidays, and vacations.

Figure 30–1

Zero-gravity pen.

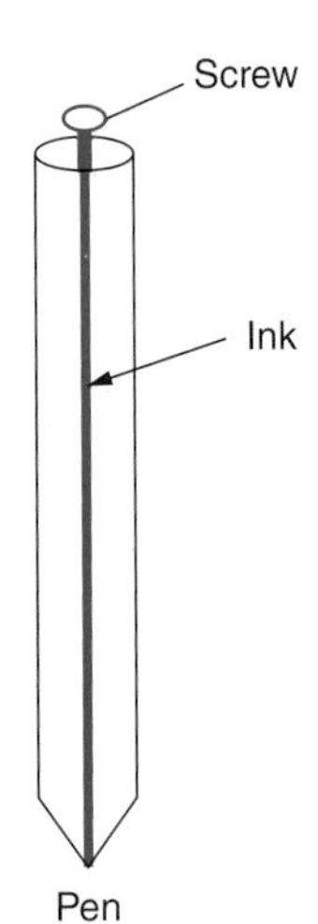

In addition, the author recently prepared a manuscript concerned with the open-ended problem approach.[13]

References

1. L. Theodore, "On Creative Thinking II," *Discovery*, East Williston, N.Y., Aug. 13, 2004.
2. L. Theodore, personal notes, East Williston, N.Y., 1995.
3. J. P. Abulencia and L. Theodore, *Fluid Flow for the Practicing Chemical Engineer*, Wiley, Hoboken, N.J., 2009.
4. A. Flynn and L. Theodore, "An Air Pollution Control Equipment Design Course for Chemical and Environmental Engineering Students Using an Open-Ended Problem Approach," paper presented at ASEE Meeting, Rowan University, Glassboro, N.J., 2001.
5. A. Flynn, J. Reynolds, and L. Theodore, "Courses for Chemical and Environmental Engineering Students Using an Open-Ended Problem Approach," paper presented at Air and Water Management Association (AWMA) Meeting, San Diego, 2003.
6. L. Theodore, class notes, East Williston, N.Y., 1999–2003.
7. Manhattan College Center for Teaching, "Developing Students' Power of Critical Thinking," Bronx, N.Y., Jan. 1989.
8. K. Gibney, "Awakening Creativity," American Society for Engineering Education (ASEE) Promo, Washington D.C., March 1988.
9. J. Delia Femina, "Jerry's Rules," *Modern Maturity,* March–April 2000.
10. B. Panitz, "Brain Storms," ASEE Promo, Washington D.C., March 1998.
11. M. Lih, "Inquiring Minds," ASEE Promo, Washington D.C., Dec. 1998.
12. *New York Times*, Nov. 11, 1993.
13. L. Theodore, *Open-Ended Problems: The Future Chemical Engineering Education Approach*, in preparation, East Williston, N.Y., 2013.

Index

D